Lecture Notes in Physics

Springer-Verlag
Berlin Heidelberg GmbH

The Editorial Policy for Proceedings

The series Lecture Notes in Physics reports new developments in physical research and teaching – quickly, informally, and at a high level. The proceedings to be considered for publication in this series should be limited to only a few areas of research, and these should be closely related to each other. The contributions should be of a high standard and should avoid lengthy redraftings of papers already published or about to be published elsewhere. As a whole, the proceedings should aim for a balanced presentation of the theme of the conference including a description of the techniques used and enough motivation for a broad readership. It should not be assumed that the published proceedings must reflect the conference in its entirety. (A listing or abstracts of papers presented at the meeting but not included in the proceedings could be added as an appendix.)
When applying for publication in the series Lecture Notes in Physics the volume's editor(s) should submit sufficient material to enable the series editors and their referees to make a fairly accurate evaluation (e.g. a complete list of speakers and titles of papers to be presented and abstracts). If, based on this information, the proceedings are (tentatively) accepted, the volume's editor(s), whose name(s) will appear on the title pages, should select the papers suitable for publication and have them refereed (as for a journal) when appropriate. As a rule discussions will not be accepted. The series editors and Springer-Verlag will normally not interfere with the detailed editing except in fairly obvious cases or on technical matters.
Final acceptance is expressed by the series editor in charge, in consultation with Springer-Verlag only after receiving the complete manuscript. It might help to send a copy of the authors' manuscripts in advance to the editor in charge to discuss possible revisions with him. As a general rule, the series editor will confirm his tentative acceptance if the final manuscript corresponds to the original concept discussed, if the quality of the contribution meets the requirements of the series, and if the final size of the manuscript does not greatly exceed the number of pages originally agreed upon. The manuscript should be forwarded to Springer-Verlag shortly after the meeting. In cases of extreme delay (more than six months after the conference) the series editors will check once more the timeliness of the papers. Therefore, the volume's editor(s) should establish strict deadlines, or collect the articles during the conference and have them revised on the spot. If a delay is unavoidable, one should encourage the authors to update their contributions if appropriate. The editors of proceedings are strongly advised to inform contributors about these points at an early stage.
The final manuscript should contain a table of contents and an informative introduction accessible also to readers not particularly familiar with the topic of the conference. The contributions should be in English. The volume's editor(s) should check the contributions for the correct use of language. At Springer-Verlag only the prefaces will be checked by a copy-editor for language and style. Grave linguistic or technical shortcomings may lead to the rejection of contributions by the series editors. A conference report should not exceed a total of 500 pages. Keeping the size within this bound should be achieved by a stricter selection of articles and not by imposing an upper limit to the length of the individual papers. Editors receive jointly 30 complimentary copies of their book. They are entitled to purchase further copies of their book at a reduced rate. As a rule no reprints of individual contributions can be supplied. No royalty is paid on Lecture Notes in Physics volumes. Commitment to publish is made by letter of interest rather than by signing a formal contract. Springer-Verlag secures the copyright for each volume.

The Production Process

The books are hardbound, and the publisher will select quality paper appropriate to the needs of the author(s). Publication time is about ten weeks. More than twenty years of experience guarantee authors the best possible service. To reach the goal of rapid publication at a low price the technique of photographic reproduction from a camera-ready manuscript was chosen. This process shifts the main responsibility for the technical quality considerably from the publisher to the authors. We therefore urge all authors and editors of proceedings to observe very carefully the essentials for the preparation of camera-ready manuscripts, which we will supply on request. This applies especially to the quality of figures and halftones submitted for publication. In addition, it might be useful to look at some of the volumes already published. As a special service, we offer free of charge LaTeX and TeX macro packages to format the text according to Springer-Verlag's quality requirements. We strongly recommend that you make use of this offer, since the result will be a book of considerably improved technical quality. To avoid mistakes and time-consuming correspondence during the production period the conference editors should request special instructions from the publisher well before the beginning of the conference. Manuscripts not meeting the technical standard of the series will have to be returned for improvement.

For further information please contact Springer-Verlag, Physics Editorial Department II, Tiergartenstrasse 17, D-69121 Heidelberg, Germany

Jesús Navarro Artur Polls (Eds.)

Microscopic Quantum Many-Body Theories and Their Applications

Proceedings of a European Summer School
Held at Valencia, Spain, 8–19 September 1997

Springer

Editors

Jesús Navarro
Facultat de Física
IFIC (CSIC-Universitat de València)
Avda. Dr. Moliner, 50
E-46100 Burjassot (València), Spain

Artur Polls
Facultat de Física
Dept. d'ECM
Diagonal, 647
E-08028 Barcelona, Spain

Cataloging-in-Publication Data applied for.

Die Deutsche Bibliothek - CIP-Einheitsaufnahme

Microscopic quantum many body theories and their applications : proccedings of a European summer school, held at Valencia, Spain, 8 - 19 September 1997 / Jesus Navarro ; Artur Polls (ed.). - Berlin ; Heidelberg ; New York ; Barcelona ; Budapest ; Hong Kong ; London ; Milan ; Paris ; Santa Clara ; Singapore ; Tokyo : Springer, 1998
(Lecture notes in physics ; Vol. 510)

ISSN 0075-8450
ISBN 978-3-662-14171-7 ISBN 978-3-540-69787-9 (eBook)
DOI 10.1007/978-3-540-69787-9

Originally published by Springer-Verlag Berlin Heidelberg New York in 1998
Softcover reprint of the hardcover 1st edition 1998

Typesetting: Camera-ready by the authors
Cover design: *design & production* GmbH, Heidelberg
SPIN: 10644181 55/3144-543210 - Printed on acid-free paper

Preface

This book contains the lectures delivered at the *European Summer School on Microscopic Many-Body Theories and their Applications*, held in Valencia (Spain) 8-19 September 1997.

The idea of this summer school was born in the European network *Microscopic Quantum Many-Body Theories: Applications to Traditional and Novel forms of Matter*, as a response to the rapid evolution of the different formulations of quantum many-body theories (QMBT) over the last decade and a serious lack of reference books on the most recent developments. The members of this network wanted to offer, to both their own postgraduate students and other young scientists, the opportunity to acquire training in the main subjects of QMBT, from an interdisciplinary perspective.

Indeed QMBT have nowadays become one of the most fundamental and exciting areas of modern physics, being at the same time a necessary tool for all physicists. Simply stated, the aims are to understand and predict the emergent properties of macroscopic matter that have their origins in the underlying interactions between the elementary constituents. By its very nature, the field is interdisciplinary and it spans a huge range of intelectual endeavour, from the development of fundamental theoretical techniques to practical applications.

Obviously, it is not possible to cover all the subjects in a single school, and so some choices have been made. The selected main topics were: *Coupled Cluster Method*, *Correlated Basis Function Theory* and *Montecarlo Methods*. The lectures on these topics were given by Professors R.F. Bishop, S. Fantoni, R. Guardiola and E. Krotscheck. More specific applications addressed to update the recent progress of QMBT in different topics were discussed in additional seminars by Professors J. Boronat, A. Fabrocini, U. Kaldor, D. Mukherjee and S. Pieper. Moreover, Professor J.P. Toennies presented a complete and exciting view of the recent and future experiments on helium clusters. However, his interesting talk has not been included here because of the theoretical character of this book.

The school was hosted by the Universidad Internacional Menéndez Pelayo (UIMP) in València. A creative and stimulating atmosphere naturally emerged among the sixty participants during the two weeks of the school, and we are sure that this positive experience will seed future collaborations.

The realization of the school was possible thanks to the finantial support provided by the Commision for the European Communities (under the Training and Mobility of Researchers programme), the Universidad Internacional Menéndez Pelayo in València, the Universitat de València, the Institució Valenciana d'Estudis i Investigació and the Instituto de Física Corpuscular.

The success of the school was mostly due to the efforts of many people who gave us their advice and time. We would like to thank the speakers for their efforts in preparing pedagogic and complete lectures. We are particularly indebted to Rafael Guardiola for his constant support during the preparation of the school and his invaluable help in the preparation of this book. Finally we wish to acknowledge Iñaki Moliner and Mariela Portesi who, besides attending the school, did a tremendous job by solving the daily problems of the other attendees and organizing the social events.

We hope that our effort will be fruitful not only for the participants of the school but also for the rest of the scientific community.

Burjassot, March 1998 J. Navarro and A. Polls

Contents

List of Participants

Khaled ABU QASEM kq@thp.uni-koeln.de
Institut für Theoretische Physik, Universität zu Köln, Zülpicher Str. 77, 50937 Köln, Germany

Luis ALVAREZ alvarez@condor.ific.uv.es
Departamento de Física Teórica, Facultad de Física, Dr. Moliner 50, 46.100 Burjassot, Spain

Alessandro AMICI amici@mpipks-dresden.mpg.de
Max-Plank-Institute für Physick Komplexer Systeme, Nöthnizer Str. 38, 01187 Dresden, Germany

Dilip ANGOM angom@iiap.ernet.in
Indian Institute of Astrophysics, Koramangala, 560034 Bangalore, India

Vesa APAJA vap@hopper.oulu.fi
Department of Physical Sciences, Theoretical Physics, University of Oulu, Linnanmaa, 90570 Oulu, Finland

Pilar BERNARDOS bernardp@besaya.unican.es
Departamento de Física Moderna, Facultad de Ciencias, Avda. de los Castros s/n, 39005 Santander, Spain

Raymond F. BISHOP rfbishop@fs1.ma.umist.ac.uk
Department of Physics, UMIST, PO Box 88, Manchester, M60 1DQ, UK

Jordi BORONAT boronat@fen.upc.es
Departament de Física i Enginyeria Nuclear, Universitat Politècnica de Catalunya, Campus Nord B4-B5, 08034 Barcelona, Spain

Michael BRÄUER brauer@al.bundy.uni-jena.de
IOMC FSU Jena, Humboldtstrasse 10, 07743 Jena, Germany

Mercedes CALBI calbi@df.uba.ar
Departamento de Física, Facultad de Ciencias Exactas y Naturales, Pabellón 1 - Ciudad Universitaria, 1428 Buenos Aires, Argentina

José CARO jose.caro@physik.tu-muenchen.de
Technische Universität München, 85747 Munich, Germany

Josep CHIRALT quifis@bohr.quifis.uv.es
Departamento de Química Física, Facultad de Química, Dr. Moliner 50 46.100 Burjassot, Spain

Christian COMMENDA chris@chioma.tphys.uni-linz.ac.at
Institut for Theoretical Physics, Johannes Kepler Universität, 4040 Linz, Austria

Paraskevi DEMETRIOU vivian@pv.infn.it
Istituto Nazionale di Fisica Nucleare, Sezione di Pavia, Via Basi 6, 27100 Pavia, Italy

Yves DEWULF yves@inwpent1.rug.ac.be
Department of Subatomic and Radiation Physics, Proeftuinstraat 86, 9000 Gent, Belgium

Susan DITTMAR sdm@thp.uni-koeln.de
Institut für Theoretische Physik, Universität zu Köln, Zülpicher Str. 77, 50937 Köln, Germany

Helder DOMINGOS helder@fteor6.fis.uc.pt
Departamento de Fisica, Universidade de Coimbra, 3000 Coimbra, Portugal

Christoph ELLERT Christoph.Ellert@nrc.ca
NRC/SIMS, 100 Sussex Dr. Rm 2087, K1A 0R6 Ottawa, Canada

Adelchi FABROCINI fabro@sp04.pi.infn.it
Dipartamento di Fisica, Università di Pisa, 56.100 Pisa, Italy

Stefano FANTONI fantoni@snoopy.ilas.sissa.it
SISSA, Via Beirut 2, 34014 Trieste, Italy

Damian FARNELL farnell@linux09.kri.thp.uni-koeln.de
Institut für Theoretische Physik, Universität zu Köln, Zülpicher Str. 77, 50937 Köln, Germany

Harald FORBERT forbert@phys.tamu.edu
Physics Department, Texas A&M University, 77843-4242 College Station, USA

Marcella GRASSO grasso@ct.infn.it
Dipartamento di Fisica, Corso Italia 57, Catania, Italy

Victor GRAU victor@montcau.upc.es
Departament de Fisica i Enginyeria Nuclear, EUETIT, Colom 1, 08222 Terrassa, Spain

Slava GREBENEV sgreben1@msfd1.dnet.gwdg.de
Max-Plank Institut für Strömunsforschung, Bunsenstrasse 10, 37073 Göttingen, Germany

Rafael GUARDIOLA Rafael.Guardiola@uv.es
Departamento de Física Atómica y Nuclear, Facultad de Física, Dr. Moliner 50, 46.100 Burjassot, Spain

Jani HALINEN jha@balrog.oulu.fi
Department of Physical Sciences, Theoretical Physics, University of Oulu, Linnanmaa, 90570 Oulu, Finland

Michael HEIM michael.heim@uni-tuebingen.de
Institut für Theoretische Physik, Auf der Morgenstelle 14, 72076 Tübingen, Germany

Eugen HOFER hofer@gwilli.tphys.uni-linz.ac.at
Institut for Theoretical Physics, Johannes Kepler Universität, 4040 Linz, Austria

Natalie JACHOWICZ natalie@inwfaxp4.rug.ac.be
Institute for Nuclear Physics, Proeftuinstraat 86, B-9000 Gent, Belgium

Uzi KALDOR kaldor@chemib2.tau.ac.il
School of Chemistry, University of Tel Aviv, 69978 Tel Aviv, Israel

Eckhard KROTSCHECK kro@grizzly.tphys.uni-linz.ac.at
Institut für Theoretische Physik, Johannes Kepler Universität, 4040 Linz, Austria

Norbert LIGTERINK n.ligterink@umist.ac.uk
Physics Department, UMIST, PO Box 88, Sackville Street, Manchester, M60 1QD, UK

Thomas LINDENAU tl@thp.uni-koeln.de
Institut für Theoretische Physik, Universität zu Köln, Zülpicher Str. 77, 50937 Köln, Germany

Theodore LIOLIOS theoliol@ccf.auth.gr
Department of Theoretical Nuclear Physics, University of Thessaloniki, 54006 Thessaloniki, Greece

Eugenio MARCO emarco@titan.ific.uv.es
Departamento de Física Teórica, Facultad de Física, Dr. Moliner 50, 46.100 Burjassot, Spain

Boris MARIC maric@ph-cip.uni-koeln.de
Institut für Theoretische Physik, Universität zu Köln, Zülpicher Str. 77, 50937 Köln, Germany

Holger MERLITZ holgi@pel4.mpi-hd.mpg.de
MPI für Kernphysik, Saupfersceckweg 1, 69117 Heidelberg, Germany

Iñaki MOLINER moliner@kate.fisato.uv.es
Departamento de Física Atómica y Nuclear, Facultad de Física, Dr. Moliner 50, 46.100 Burjassot, Spain

Lysiane MORNAS mornas@nanhp7.in2p3.fr
Departamento de Física Teórica, Facultad de Física, Dr. Moliner 50, 46.100 Burjassot, Spain

Debashis MUKHERJEE pcdm@iacs.ernet.in
Department of Physical Chemistry, Indian Association for the Cultivation of Science, 700 032 Calcutta, India

Stephan NATSCHLÄGER stn@garfield.tphys.uni-linz.ac.at
Institut for Theoretical Physics, Johannes Kepler Universität, 4040 Linz, Austria

Jesús NAVARRO Jesus.Navarro@uv.es
IFIC, Facultad de Física, Avda. Dr. Moliner 50, 46.100 Burjassot, Spain

Panagiota PAPAKONSTANTINOU ppapakon@atlas.uoa.gr
Physics Department, Nuclear and Particle Physics Section, University of Athens, Panepistimouploi, Ilissia, 15771 Athens, Greece

Assumpta PARREÑO assum@dirac.ecm.ub.es
Departament d'Estructura i Constituents de la Matèria, Facultat de Física, Diagonal 647, 08028 Barcelona, Spain
Guillermo PERIS peris@vents.uji.es
Departament de Ciències Experimentals, Ctra. Borriol s/n, Box 224, 12080 Castelló, Spain
Marianthi PETRAKI mpetraki@atlas.uoa.gr
Physics Department, Nuclear and Particle Physics section, University of Athens, Panepistimouploi, Ilissia, 15771 Athens, Greece
Steven PIEPER pieper@theory.phy.anl.gov
ANL, Physics Division, Bldg 203, Argonne, IL-60439, USA
Josep PITARCH pitarchj@bohr.quifis.uv.es
Departamento de Química Física, Facultad de Química, Avda. Dr. Moliner 50, 46.100 Burjassot, Spain
Artur POLLS artur@ecm.ub.es
Departament d'Estructura i Constituents de la Matèria, Facultat de Física, Diagonal 647, 08028 Barcelona, Spain
Mariela PORTESI portesi@kate.fisato.uv.es
IFIC, Facultad de Física, Avda. Dr. Moliner 50, 46.100 Burjassot, Spain
Alexis PREVOST prevost@ipno.in2p3.fr
Division de Physique Théorique, Institut de Physique Nucléaire, 91406 Orsay (CEDEX), France
Antonio PUENTE dfsapf8@ps.uib.es
Departament de Física, Universitat de les Illes Balears, 07071 Palma de Mallorca, Spain
Andreu RIGO dfsarg8@ps.uib.es
Departament de Física, Universitat de les Illes Balears, 07071 Palma de Mallorca, Spain
Eduardo RODRIGUES erodrig@ulb.ac.be
Departamento de Fisica, Universidade de Coimbra, 3000-Coimbra, Portugal
Stefan ROMBOUTS Stefan.Rombouts@rug.ac.be
Subatomaire en Stralingfysica, Proeftuinstraat 86, 9000 Gent, Belgium
Josi ROSENFELD mcbpgjr@afs.mcc.ac.uk
UMIST, 24 Everett Rd., Withingon, Manchester, M14 6BN , UK
Antonio SARSA asarsa@goliat.ugr.es
Departamento de Física Moderna, Facultad de Ciencias, Universidad de Granada, 18071 Granada, Spain
Erdmann SCHILLER erdmann.schiller@uni-tuebingen.de
Institut für Theoretische Physik, Auf der Morgenstelle 14, 72076 Tübingen, Germany
Karl SCHÖRKHUBER kars@idefix.tphys.uni-linz.ac.at
Institut for Theoretical Physics, Johannes Kepler universität, 4040 Linz, Austria

Hussain SHAFQAT shafqat.hussain@fys.uio.no
University of Oslo, Lindren, Oslo, Norway

Gil SILVA gil@fteor6.fis.uc.pt
Departamento de Fisica, Universidade de Coimbra, 3000 Coimbra, Portugal

Rita SINHA rita-ph@jmi.ernet.in
Department of Physics, Jamia Millia Islamia, Jamia Nagar, 110025 New Delhi, India

Ubaldo TAMBINI tambini@fe.infn.it
Physics Department, Ferrara University, 44100 Ferrara, Italy

J. Peter TOENNIES
Max-Plank Institut für Strömunsforschung, Bunsenstrasse 10, 37073 Göttingen, Germany

Isaac VIDANYA isaac@ecm.ub.es
Departament d'Estructura i Constituents de la Matèria, Facultat de Física, Diagonal 647, 08028 Barcelona, Spain

Michel WAROQUIER michel@inwfax2.rug.ac.be
Department of Subatomic and Radiation Physics, Proeftuinstraat 86, 9000 Gent, Belgium

Robert ZILLICH zil@akira.tphys.uni-linz.ac.at
Institut for Theoretical Physics, Johannes Kepler Universität, 4040 Linz, Austria

The Coupled Cluster Method

Raymond F. Bishop

Department of Physics, UMIST,
(University of Manchester Institute of Science and Technology),
Manchester M60 1QD, United Kingdom

Abstract. The coupled cluster method (CCM) is nowadays widely recognised as providing one of the most powerful, most universally applicable, and numerically most accurate at attainable levels of implementation, of all available *ab initio* methods of microscopic quantum many-body theory. The number of successful applications of the method to a wide range of physical and chemical systems is impressively large. In almost all such cases the numerical results are either the best or among the best available. A typical example is the electron gas, where the CCM results for the correlation energy agree over the entire metallic density range to within less than one millihartree per electron (or $< 1\%$) with the essentially exact Green's function Monte Carlo results.

What has since become known as the normal (NCCM) version of the method was invented some forty years ago to calculate the ground-state energies of closed-shell atomic nuclei. Extensions of the CCM have since been developed to calculate excited states, energies of open-shell systems, density matrices and hence other properties, sum rules, and the sub-sum-rules that follow from embedding linear response theory within the NCCM. Further extensions deal with the general dynamics of quantum many-body systems, and with their mixed states appropriate, for example, to their behaviour at nonzero temperatures. More recently, a so-called extended (ECCM) version of the method has been introduced. It has the same ability as the NCCM to describe accurately the local properties of quantum many-body systems, but it also has the potential to describe such global phenomena as phase transitions, spontaneous symmetry breaking, states of topological excitation, and nonequilibrium behaviour.

The role of the CCM within modern quantum many-body theory is first surveyed, by a comparison with, and discussion of, the alternative microscopic formulations. We then discuss the method and each of its individual components in considerable detail. Our overall aim is to stress the broad applicability of the method. To that end we introduce and exploit a very general theoretical framework in which to formulate the key ideas and to develop the theory. We end with a brief review of the applications of the method to date.

1 Introduction

1.1 Many-Body Systems

Many-body physics or quantum many-body theory (QMBT) is not the rather specialised subfield of physics that it is sometimes believed to be. On the contrary, we are nearly always faced with many-particle situations at the fundamental level. Thus, we are especially interested in formulations of QMBT at the fully microscopic or *ab initio* level, which are powerful enough to treat, both in principle and in practice, the full range of many-body and field-theoretic systems. Since these comprise nearly all of the physical world around us, according to our present degree of understanding, the topic of QMBT occupies a key position in modern theoretical physics.

Examples of quantum many-body systems abound. Thus, it is clear that in fields like molecular, solid-state, and nuclear physics most of the fundamental objects of discourse are interacting many-body systems. But even in elementary particle physics one is usually dealing with more than one particle. For example, at some level of reality a nucleon is made up of three quarks interacting via gluons and surrounded by a cloud of mesons, which are themselves composed of quark-antiquark pairs. Even more fundamentally, even the "physical vacuum" of any quantum field theory is endowed with an enormously complex infinite many-body structure due to the virtual excitation of the particles allowed by the theory under discussion. Table 1 gives an incomplete but illustrative set of examples of quantum many-body systems. It is a primary aim of modern QMBT to derive and explore methodologies which are broad enough to be able to be applied to all such systems, and powerful enough to describe their key emergent macroscopic properties from a knowledge only of the underlying microscopic interactions. The coupled cluster method (CCM) under discussion here now plays a central role in modern QMBT since, as we shall explore in detail, it has the potential to fulfill these key criteria amply.

1.2 Many-Body Methods

From our discussion above it should be clear that although techniques to deal with interacting many-body systems are likely to have separately arisen from specific physical contexts, they may also be of more general applicability and interest. The CCM and its birth some 40 years ago within nuclear physics [1,2] is just one example. Nevertheless, if we limit ourselves to comparisons with other fundamental and "universal" tools, the number of available methods is surprisingly small. Before we enumerate them, however, we emphasise that there exist also many other macroscopic or phenomenological or approximate or empirical descriptions and techniques, which have proven to be successful in one or other context. They are, nevertheless, excluded from our present considerations because, by their very nature, they are "specialist" tools which are tailor-made for a particular system or application, and because our emphasis here is on techniques that have a more universal appeal.

The main microscopic techniques fall essentially into seven main classes:

- **Quantum Monte Carlo (QMC) Methods:** These come in several varieties, e.g., Green's function Monte Carlo, diffusion Monte Carlo, path-integral Monte Carlo. Each aims to solve the N-body Schrödinger equation by stochastic simulation, typically for $N \lesssim 1000$, with extrapolation

Table 1. Examples of quantum many-body systems.

System	Particles	Interaction	Infinite Limit
Molecular clusters	e.g., CO_2 molecules	van der Waals type (i.e., repulsive core + attractive tail)	Gases; fluids; molecular crystals
Rare gas atom clusters	e.g., He or Ar atoms	Interatomic potentials (e.g., van der Waals type)	Rare gas; liquid; crystal
Metallic atom clusters	e.g., Na or Ag atoms	Interatomic potentials	Metallic crystal
Atoms and molecules	Electrons and nuclei	Mainly Coulombic (+ relativistic/QED effects)	Infinite Thomas–Fermi atom
Quantum spin clusters	Spins on a spatial lattice	e.g., Heisenberg interaction	Magnetic phases (e.g., antiferromagnets)
Electron clusters	Electrons (and holes) on a regular spatial lattice	e.g., Hubbard model, t–J model	Strongly interacting electron-lattice systems (e.g., high-T_c superconductors)
Nuclei ($A = 2Z$)	(i) α-particles	2-body phenomenological potentials (repulsive core + attractive tail); also 3-body potentials needed for accurate fits to experimental data	Alpha matter
	(ii) Nucleons		"Standard" nuclear matter
	(iii) Baryons and mesons	Field-theoretic interaction (sometimes replaced by boson-exchange potentials)	Nuclear matter as "baryon-meson soup"
	(iv) Quarks and gluons	(Hints from) QCD	Quark-gluon plasma
Elementary particles (models from quantum field theory)			
(i) Leptons and exchange bosons		Electroweak interaction (Weinberg–Salam theory)	Already there! (– even for the physical vacuum!)
(ii) Quarks and gluons		(Hints from) QCD	

to the thermodynamic limit, $N \to \infty$, using finite-size scaling or other techniques. The QMC methods typically require either that the many-body wave function is everywhere non-negative (as for bosonic ground states) or that its nodal surface structure is known or well approximated. The latter requirement leads to the infamous "fermion minus sign problem", which still exists as a serious limitation on the practical usefulness of QMC techniques.

- **Time-Independent Perturbation Theory (TIPT):** This method basically provides an expansion in powers of the interaction coupling constant. It finds a diagrammatic representation in terms of so-called Goldstone diagrams. Its main drawback is that one needs in numerical implementations to neglect all terms which are, somehow, deemed to be unimportant or which, more usually, simply cannot be handled in practice.
- **Green's Function (GF) or Propagator Methods:** These methods are basically a reformulation of time-dependent perturbation theory and, as such, can be given a diagrammatic representation in terms of so-called Feynman diagrams. They suffer similar drawbacks to those alluded to above in the case of TIPT. GF methods are formulated as time-dependent (i.e., dynamic) equations for matrix elements (or propagators) which describe the propagation of clusters of particles in the many-body medium.
- **Variational Methods:** The aim of all variational calculations is to compute the energy expectation value, $\langle H \rangle$, with a trial correlated many-body wave function, typical forms of which are the Jastrow or Jastrow–Feenberg varieties. Since $\langle H \rangle$ cannot usually be calculated in closed form, various techniques are used. Some typical ones are Monte Carlo evaluation of the implicit multi-dimensional integrals, various cluster expansions, and various partially re-summed cluster sub-series (e.g., the hypernetted chain approximation).
- **Correlated Basis Function (CBF) Method:** Variational calculations suffer from the main drawback that even if $\langle H \rangle$ is calculated exactly for a given trial wave function, one has only an upper bound on the exact energy and an approximation to the exact many-body wave function. CBF techniques aim to remove this restriction by providing a scheme in which to improve systematically on a trial wave function (typically of Jastrow or similar form) by introducing a *complete* set of correlated basis functions. CBF techniques typically enable the short-range correlations which are important for many physical applications to be efficiently incorporated from the outset.
- **Configuration-Interaction Method (CIM)** (or generalised many-body shell model): This method typically aims to diagonalise the Hamiltonian in a finite subspace of the full many-body Hilbert (or Fock) space. Despite its underlying simplicity, it suffers from the severe drawback for

many-body applications of not satisfying the important requirements of size-extensivity and size-consistency.

- **Coupled Cluster Method (CCM):** By utilising an exponentially correlated form of the wave function, the CCM cures the size-extensivity problem inherent in the previous CIM. The CCM has the nice feature that if its basic equations are iterated one regains TIPT and the associated Goldstone diagrams. Nevertheless, one need not, and usually does not, solve the equations this way, thereby circumventing the main drawback of TIPT methods. The characteristic exponential form of the CCM wave functions correctly counts independent clusters excitations from a suitably chosen model or reference state, which itself can take a variety of forms depending on the system and on its particular phase under study.

The QMC, variational, and CBF methods are all extensively discussed by other contributors to this volume. It is nevertheless worthwhile to spend a little while in discussing the salient features of all of the methods (apart from QMC techniques), in order to put them properly in context before turning to a detailed discussion of the CCM.

Both the TIPT and GF methods are perturbative in origin. In almost all cases of interest various infinite partial resummations are performed out of necessity, either because the series diverges badly or is otherwise ill-defined from the outset. Well-known examples include the ladder-diagram summation (for hard-core or other strongly repulsive short-range potentials) which leads to the Bethe–Brueckner–Goldstone G-matrix expansion or the hole-line expansion (and see, e.g., Refs. [3–5]); and the bubble- or ring-diagram summation of the random phase approximation (RPA) [6,7] (for long-range potentials of the Coulomb type). For accurate numerical calculations of most realistic many-body systems it is usually found to be necessary to incorporate at least the self-consistent sum of all ring and ladder diagrams.

Even in such rearranged or partially resummed examples, one is still ultimately forced in practice to neglect infinite classes of terms that are presumed to be unimportant or that cannot easily be incorporated into a higher approximation. Furthermore, it is usually difficult in realistic applications to justify the retention of certain terms at the expense of neglecting others. The history of microscopic nuclear theory, for example, is littered with incorrect or misleading calculations which amply illustrate the danger of the blind inclusion of extra diagrams, as has been forcefully pointed out many times by Kümmel and his collaborators [8,9]. In quantum chemistry, the TIPT techniques provide the usual means of doing many-body perturbation theory (MBPT), or what has increasingly become known in chemistry as Møller–Plesset perturbation theory. In chemical applications it is usual to include all terms through some given finite order in the electron interactions. Full fourth-order, MBPT(4), calculations are now commonplace, for example [10].

Various propagator or GF techniques are also still widely used in quantum chemistry (and see, e.g., Refs. [11,12]), where they are particularly used for molecular spectroscopic properties, as well as in other areas. Nevertheless, on the wider front, GF methods suffer from many of the same drawbacks as the TIPT techniques discussed above. Thus, with one important potential caveat, neither TIPT nor GF methods are nowadays generally considered within quantum many-body theory to be sufficiently accurate or sufficiently versatile to be candidates for the position of a universal, high-precision method. The only real exception to this viewpoint is provided by the so-called *parquet diagram,* or *planar theory,* approach within the broader GF scheme.

The ideas behind parquet theory were introduced [13] at around the same time that Coester first introduced the CCM. The specific context for their introduction was certain problems in particle physics. Similar equations were later used to study X-ray absorption and emission in metals [14]. However, it was not until 1979 that the potential of parquet theory for the study of strongly interacting condensed Bose systems was noted by Ripka [15]. He called attention to the fact that the hypernetted chain (HNC) equations of the Jastrow variational approach shared several very desirable features with the parquet equations. At the time the HNC approximation to the variational approach, which is discussed in detail in the contribution to this volume by Fantoni and Fabrocini, was meeting with considerable success in dealing with such extended bosonic systems as liquid ^{4}He. Ripka noted that just as the HNC equations of variational theory treated both the long- and short-range correlations simultaneously and consistently, so too would the parquet equations of perturbation theory, if only they could be solved.

This astute observation of Ripka provided the stimulus for the 1982 study of Jackson, Lande, and Smith [16] which examined the parquet equations for bosons in great detail. They also showed how these otherwise rather formidable equations could be rendered tractable, and hence solved numerically, by the introduction of simple localising approximations.

The basic concepts of parquet theory are both simple and appealing. The method first focusses on the effective two-body interaction, which it then expresses in terms of a large and physically interesting class of Feynman diagrams. These so-called parquet diagrams are a particular self-consistent sum of ring, ladder, and vertex correction terms for the two-body Green's function. Most important physically is the fact that the two-body (i.e., particle–particle and hole–hole) ladder diagrams and the particle–hole ring diagrams are iterated together *in a maximal fashion.* Although the full two-body parquet equations are highly complex, their local counterparts were applied to such model bosonic systems as liquid ^{4}He using the Lennard–Jones potential, and neutron matter treated as a Bose system interacting via the Reid soft-core $^1S_0$ interaction. Not surprisingly in view of Ripka's comments, good agreement was found with the corresponding optimised Jastrow HNC results.

At the formal level it was later demonstrated [17] that a similar, but nonetheless distinct, local form of the full two-body parquet equations is *identical* to optimised Jastrow HNC theory. This result was particularly important since it provided a bridge between the otherwise very disjoint perturbative approaches of TIPT and GF theory on the one hand, and variational approaches on the other. Later formal developments have included a possible extension to fermionic systems [18,19], the inclusion of three-body terms [20], and parquet perturbation theory [21] for bosons, as an expansion in the difference between the exact and approximate propagators, in order to improve systematically upon the local parquet equations. Interesting connections between parquet theory and Baym–Kadanoff theory [22] have also been made [23]. This latter approach is noteworthy in that it uses an initial approximation for the two-body vertex to construct from it a *conserving* vertex, namely one which conserves particle number, momentum, and energy. Despite what is now a fairly large corpus of *formal* developments, the parquet method has not yet been widely applied and tested. Nevertheless, it clearly deserves to be, and it must be considered as a potential candidate for the position of a universal tool in quantum many-body theory. We have stressed it here for just these reasons.

We turn next to the use of variational methods in quantum many-body theory. Perhaps the simplest of these approaches is based on a trial wave function of the (Bijl–Dingle–)Jastrow type [24]. The early calculations of this sort relied on various cluster expansions of the ensuing approximate matrix elements [24–26]. Later on it was realised that the variational approaches may themselves be formulated diagrammatically [15,27,28]. This feature has been of considerable help in constructing such powerful approximations as the Percus–Yevick and HNC summations (which have their origins in the classical theory of liquids) and their variants, both for bosons and fermions [15,29–31]. Various articles and reviews concerning the use of variational methods in quantum many-body theory exist (and see, e.g., Refs. [15,32–35]). The 1979 article by Clark [32] is particularly recommended as what is perhaps still the standard review of the variational theory of extended nuclear matter. Similar reviews (and see, e.g., Ref. [36]) also exist for applications to few-body systems. Finally, Fantoni and other contributors to this volume describe the variational techniques and their applications in some depth, from both a modern perspective and a pedagogical point of view.

The variational approaches sketched above suffer from two fundamental flaws. In the first place the various partial summations of the graphs considered by such approximation schemes as HNC give methods which lose one of the most attractive features of variational formulations, namely that they yield estimates for the energy which are upper bounds to the exact ground-state energy. Secondly, even a complete summation of graphs (or a variational Monte Carlo evaluation of the corresponding expectation values) for a given trial wave function, say of the Jastrow type, gives only the exact variational result and not the true ground state. This second deficiency may be removed in principle by including more general state-dependent correlations and higher-order correlation functions of the Feenberg type [37]. Alternatively, and more generally, one may extend the Jastrow wave function to a complete set of correlated basis functions (CBF). The CBF method was introduced nearly forty years ago by Feenberg and his collaborators [29,38], and has since been developed largely by Clark and his co-workers [32,39–45]. A very brief and qualitative survey of the method, which contains many more citations of the original CBF literature, has also been given by the present author [46]. Despite the fact that the method and its applications are also extensively discussed in the present volume by Fantoni, Fabrocini, and Krotscheck, it is useful for us also to review the key features of the method for purposes of later comparison with the CCM.

The central ingredient of the CBF method is the direct incorporation of the most important interparticle correlations which are believed to characterise the system under consideration, into the approximate wave functions on which the microscopic description is based. At its simplest level the method considers only a single configuration and hence reduces to ordinary variational theory. As we have seen above, the latter further simplifies to Jastrow theory if the simplest reasonable choice of correlation operator is made.

At its most general level, the CBF method constructs for the N-body system a multiconfigurational correlated basis $\{|\Psi_I\rangle\}$ of normalised but generally nonorthogonal state vectors,

$$|\Psi_I\rangle = \frac{F|\Phi_I\rangle}{\langle\Phi_I|F^\dagger F|\Phi_I\rangle^{\frac{1}{2}}} \ , \tag{1}$$

in terms of a correlation operator F applied to a complete orthonormal basis $\{|\Phi_I\rangle\}$ of model states. The latter usually carry the correct quantum statistics and any essential symmetries of the system. They would hence provide an adequate lowest-order description of the system if it were not strongly interacting. For an application to molecules, for example, the states $\{|\Phi_I\rangle\}$ would probably be chosen as a set of Slater determinants of some appropriate single-electron orbitals.

The correlation operator F is thus symmetric in the complete sets of single-particle quantum numbers used to denote the N particles. Most importantly, F also possesses the cluster decomposition property, namely that upon separating one subgroup of particles (say 1,2,...,n) far from the rest $(n+1, n+2, ..., N)$ in real space, the operator $F(1, 2, ..., N)$ decomposes into a product,

$$F(1, ..., N) \to F^{(n)}(1, ..., n)F^{(N-n)}(n+1, ..., N) \ . \tag{2}$$

It is this property which allows the natural definition of correlation operators $F^{(n)}(1, ..., n)$ for n-body subsystems with $1 \le n \le N$, in terms of a given

N-body operator $F \equiv F^{(N)}$. It is also the key ingredient in allowing the derivation of linked cluster expansions for physical quantities.

The simplest choice for F is the state-independent Jastrow form, $F \to F_J$, specified in its usual coordinate-space form as follows,

$$F_J = \prod_{1 \le i < j \le N} f(r_{ij}) \ , \tag{3}$$

where r_{ij} is the relative coordinate for particles i and j. The more general Feenberg form, $F \to F_F$, is given in terms of higher-order correlation functions as,

$$F_F = \prod_{i<j} f_2(r_{ij}) \prod_{i<j<k} f_3(r_{ij}, r_{ik}, r_{jk}) \cdots \ . \tag{4}$$

Whatever choice for F is adopted, the CBF method devolves onto the computation of the matrix elements H_{IJ} and N_{IJ} of the Hamiltonian H and the unit operator, respectively,

$$H_{IJ} \equiv \langle \Phi_I | F^\dagger H F | \Phi_J \rangle \ ; \quad N_{IJ} \equiv \langle \Phi_I | F^\dagger F | \Phi_J \rangle \ . \tag{5}$$

One assumes that any exact stationary energy eigenstate $|\Xi\rangle, H|\Xi\rangle = E|\Xi\rangle$, may now be (approximately) expanded in the multiconfigurational basis,

$$|\Xi\rangle = \sum_J c_J |\Psi_J\rangle \ . \tag{6}$$

The Schrödinger equation is then decomposed as usual into the coupled set of linear generalised eigenvalue equations,

$$\sum_J (H_{IJ} - E N_{IJ}) c_J = 0 \ , \tag{7}$$

which have a nontrivial solution for the coefficients $\{c_J\}$ if and only if E satisfies the secular equation,

$$\det(H_{IJ} - E N_{IJ}) = 0 \ . \tag{8}$$

Clearly, the accurate numerical evaluation of the matrix elements $\{H_{IJ}\}$ and $\{N_{IJ}\}$ is far from trivial. In this context various cluster-expansion techniques have been developed [26,47], which themselves also provided a basis for much of the later diagrammatic analysis of both diagonal elements (expectation values) [27–31] and off-diagonal elements [40]. This latter work led to the definition of the CBF effective interaction, and to many illuminating connections with conventional diagrammatic MBPT. Exploration of these connections has allowed techniques from standard TIPT as developed for weakly-interacting systems to be taken over into their CBF counterparts for application to strongly-interacting systems (and see, e.g., Ref. [44]).

Finally, however the matrix elements $\{H_{IJ}\}$ and $\{N_{IJ}\}$ are derived, the remaining generalised eigenvalue problem of (7) and (8) must then be attacked in a systematic, approximate, or heuristic manner. Various such methods have been developed within the general CBF framework. These include nonorthogonal perturbation theory [39]; straightforward CIM-type diagonalisation in the space spanned by some subset of the full correlated basis [39,48]; and a correlated version of the usual RPA [6], which now performs a semiclassical treatment of small-amplitude oscillations about an equilibrium correlated configuration [41,42,44,45]. There have even been attempts to formulate a generalised version of coupled cluster theory in the CBF basis [49].

We turn finally to the CIM [50] and the CCM, where the latter is our main concern. For present overview purposes we merely remind the reader that the CIM is basically a straightforward diagonalisation of the Hamiltonian in some multiconfigurational subspace of the full many-body Hilbert space. One of its main advantages is that it is also easily and equivalently formulated variationally, in a parametrisation that preserves the manifest Hermitian adjoint relationship between corresponding bra and ket wave functions. In this way we see that each of the set of approximate energy eigenvalues so obtained provides an upper bound to the corresponding member of the similar set of exact lowest eigenvalues. Furthermore, by the well-known interlacing (or Hylleraas–Undheim) theorem [51], an increase in the size of the multiconfigurational CIM subspace is guaranteed to lead to improved estimates, i.e., to ones which are lower (or, in the worst case, unchanged) in energy.

This advantage of the CIM (which is *not* shared by the CCM) is, however, offset by the disadvantage that the method is not size-extensive [52]. Thus, in diagrammatic language, the CIM contains terms which are not linked (connected).[1] For extended systems this deficiency is fatal. In order to guarantee the size-extensitivity of the energy of the many-body system, it is clearly necessary that the effective Hamiltonian is separable over the various possible subsystems which arise in the corresponding dissociation limits where they are asymptotically separated to large relative distances so that the interactions between them become vanishingly small. This will certainly be the case if the wave operator is multiplicatively separable. The importance of exactly maintaining the separability of the wave function *at any subsequent level of approximation* has been stressed by Primas [53], although its importance for energy calculations was perhaps first realised by Brueckner [54] when using nondegenerate Rayleigh–Schrödinger MBPT to calculate the ground-state energy of infinite nuclear matter. An obvious method of ensuring that the wave operator is multiplicatively separable is to write it as the exponential of the so-called cluster operator, which in turn is additively separable and hence

[1] In the present article we use the terms "linked" and "connected" as synonyms; the interested reader should beware, however, that in the context of open-shell versions of the CCM it is usual to distinguish between the two terms.

representable by a sum of connected diagrams. This is the central feature of the original CCM, as described more fully in Sect. 3.

From our discussion to date it should be apparent that the only existing formulations of quantum many-body theory which satisfy *all* of the desirable criteria of being:

- fully microscopic;
- widely applicable to a broad spectrum of both finite and extended systems of interest in physics (and quantum chemistry);
- equally applicable to systems in a spatial continuum or on a discrete (regular) spatial lattice;
- capable of systematic improvement via increasingly higher-order implementations of some well-defined scheme of hierarchical approximations;
- very accurate in practice, i.e., better than or comparable to the highest practicable precisions attainable by any alternative technique, however specialised to the system under discussion; and
- very widely and deeply tested,

and which can hence claim to be universal, high-precision tools, are the CBF and CCM formalisms. Parquet (or planar) theory may also potentially fall into this category, but it has not yet been sufficiently widely tested to be able to judge properly. Naturally, QMC methods also yield excellent results which, when the methods can be implemented, can often be viewed as exact benchmarks, as other contributions to this volume make clear. Nevertheless, it is still often convenient to view QMC techniques as providing a third paradigm in physics, sitting between experiment and microscopic (analytic or semi-analytic) theory.

We note that the CBF method, and the variational techniques which it subsumes and extends, have been widely applied to systems as diverse as finite nuclei [41,45,48]; nuclear matter [32,55–57]; neutron stars [58]; bulk liquid ^{4}He [55,59,60]; unpolarised [55,61] and polarised [44] bulk liquid ^{3}He; the electron gas [61]; various species of electron-spin-aligned bulk atomic deuterium [43]; the lattice Hubbard model of strongly interacting electrons [62]; electronic correlations in atoms [63]; Ising spins in a transverse magnetic field [64]; and $U(1)_3$ lattice gauge field theory in the pure gauge sector [65]. Other contributors to this volume discuss these and other applications in much more detail.

The CCM has been at least as widely and at least as successfully (and accurately) applied in both physics and chemistry as the CBF method, as we shall outline below in Sect. 4.

1.3 Relationships Between Methods

Before turning our attention to the CCM we end the present overview of many-body techniques by remarking that the fact that there exists only a small number of fundamental *ab initio* methods is not necessarily something to be regretted. Indeed, it is widely believed by quantum many-body theorists that what is much more important nowadays than the development of further techniques is the exploration of the interconnections between existing ones. This is clearly a reflection on the power and universality of the available methods. In this regard we note that there have already been several important confrontations and proposed marriages between existing methods, and we conclude by mentioning some of them.

In the first place, one of the earliest confrontations was provided by the so-called "crisis in nuclear matter theory" which occupied a central place in quantum many-body theory in the 1970s. Two fundamental issues were involved. Thus, there was a disagreement between the lowest-order Brueckner theory calculations of TIPT and various variational calculations, when both were performed with the same internucleon potential. Furthermore, neither method gave good saturation properties when the best available potentials were used. That crisis has now been essentially resolved (and see, e.g., Ref. [37]) to the mutual advantage of both methods. In particular, the later development of Brueckner theory that was necessary for the resolution has greatly added to our understanding of many-body theory and many-body systems. Another very beneficial outcome of this early confrontation was that the power of the variational and the CBF techniques was thereafter much more widely appreciated.

Secondly, we have already noted the important formal role played by parquet theory [17] in demonstrating the exact equivalence between the optimised hypernetted chain (HNC) approximation to the variational Jastrow theory and a particular (approximate) localised form of itself. This result provided a real insight into the relationships between perturbative expansions on the one hand and variational cluster expansions on the other. It also provided a bridge between the two methods, as well as a common language in which to interpret both their features and the results obtained from them.

Thirdly, the relationship of the CCM to TIPT as providing at given levels of truncation (e.g., in the SUBn scheme discussed in detail in Sect. 3) very sophisticated and intricate resummations of various physically important and infinite classes of Goldstone diagrams is by now well known to the practitioners, and will be fully explored in Sect. 3. In particular, the very natural grouping together within the CCM of vast classes of Goldstone diagrams has shown graphically how cautious one must be when using perturbation theory in order "not to split small quantities into large pieces" [66].

A fourth result concerns the relationship of the CIM to the CCM. In particular, the CIM together with the two main modern complete formulations of the CCM which we describe more fully in Sect. 3, and which are known respectively as the normal (NCCM) and extended (ECCM) versions [67], have been shown [68,69] to form a natural closed hierarchy of three increasingly sophisticated parametrisations in which the underlying amplitudes have increasing degrees of connectedness. In the same spirit, as we discuss in Sect. 3, each of these three *independent cluster methods* has also been completely reformulated [67–69] via a variational principle, thereby providing further bridges between perturbative and variational techniques.

Fifthly, and at the level of possible marriages between various of the methods, we have already mentioned the "correlated coupled cluster theory" [49], which was motivated by a desire to combine the best features of both the CCM and CBF approaches. It is unfortunate that, to the best of our knowledge, this proposed marriage has not yet really been put to the test of a real application.

More recently, however, two other such extensions of the CCM have been discussed. Thus, sixthly, in a recent series of papers [70–74] the CCM (and the CIM) have been the subjects of intense investigation in order to incorporate exactly the translational invariance property, which is vital for the accurate treatment of such light nuclei as the alpha-particle. As a consequence, the emphasis has been to formulate the CCM and the underlying correlations directly in coordinate space, rather than in the more usual multiconfigurational Fock-space representation discussed at length in Sects. 2 and 3 below. In this way contact has been made both with more traditional generalised (many-body) nuclear shell-model calculations of the CIM type, with the result that the number of independent configurations can be dramatically reduced [70,71], and also with variational approaches [72,73]. This work now holds out the possibility of combining some of the best elements of the CIM, CCM, and variational approaches. We also note in passing that similar functional forms of the CCM parametrisation of many-body wave functions, rather than the more standard Fock-space or operatorial forms, have also been used recently in the context of lattice field theory. To date, specific applications of the CCM have been made to both discrete and continuous lattice gauge field theories. The former includes the Z(2) case [75], while the latter include both the Abelian U(1) case [75–80] and the non-Abelian SU(2) case [77,81]. Very recent applications of the CCM have also been made to a latticised O(4) non-linear sigma model of chiral meson field theory [82], wherein the operatorial and functional forms of the CCM have been compared in considerable depth.

Finally, and as a seventh example, recent work [83,84] has been connected with the incorporation of Jastrow-type correlations into the higher-order cluster terms that are otherwise neglected entirely in the standard (e.g., SUBn) CCM truncation schemes discussed in Sect. 3. Indeed, much current effort is being expended in the general direction of inventing new approximation schemes for existing methods. One promising way to do this is by attempting to fuse the best features of different methods.

All of the above various partial mergers hold considerable promise for the future of quantum many-body theory. The interested reader is referred to Ref. [84] for a more detailed, yet still pedagogical, confrontation between the different microscopic techniques discussed above. This paper discusses in some depth the various ways that two-body correlations in quantum many-body systems are described within different microscopic theories. A particular application of each method is made to the exactly integrable (by the Bethe ansatz [85] method) Lieb model [86,87] of many bosons interacting in one dimension via repulsive pairwise delta-function potentials.

In the remainder of this article we now focus our attention on the CCM. However, before describing the method in Sect. 3, we first discuss the general question of the choice of a suitable reference state (or "generalised vacuum" state) for a given quantum many-body system, with respect to which the dynamic (multiparticle) correlations may be quantitatively described.

2 The Construction of Many-Body Reference States

For many purposes it is extremely convenient to distinguish some *model space*, $\mathcal{M}$, which is a subspace of the full many-body Hilbert or Fock space, $\mathcal{G}$, in which the quantum many-body or quantum field-theoretical problem under discussion is defined. Of particular interest is the case where the subspace is spanned by a suitably chosen set of $(D+1)$ normalised *reference states* $\{|\Phi_i\rangle;\ i=0,1,2,\cdots,D\}$, with $\langle\Phi_i|\Phi_i\rangle = 1$. We define two operators, P and Q, as follows,

$$P \equiv \sum_{i=0}^{D} |\Phi_i\rangle\langle\Phi_i| \ ; \quad Q \equiv \mathbb{1} - P \ , \tag{9}$$

where $\mathbb{1}$ is the identity operator in $\mathcal{G}$. If the reference states are orthonormalised,

$$\langle\Phi_i|\Phi_j\rangle = \delta_{ij} \ , \tag{10}$$

the operators P and Q are simply projection operators into and out of the model space, $\mathcal{M}$,

$$P^2 = P \ ; \quad Q^2 = Q \ ; \quad PQ = QP = 0 \ , \tag{11}$$

and we can decompose an arbitrary state $|\Psi\rangle \in \mathcal{G}$ as,

$$|\Psi\rangle = P|\Psi\rangle + Q|\Psi\rangle \ . \tag{12}$$

Although the above projection simply represents an expansion of a wave function in some (suitably chosen and suitably truncated) complete orthonormal set, if, in some sense to be defined below, the states $\{|\Phi_i\rangle;\ i=0,1,\cdots,D\}$ are "zeroth-order" or "starting" or "reference" vectors, then it can be very useful to make this explicit. Furthermore, and more generally, in order to describe many-body correlations quantitatively, we *always* need some appropriate state with respect to which the correlations are defined. Henceforth, we shall call the states $\{|\Phi_i\rangle;\ i=0,1,\cdots,D\}$ the *reference states* or *model states*, and the $(D+1)$-dimensional space $\mathcal{M}$ spanned by them the *reference space* or *model space.* Later, we shall see that the case $D = 0$ is appropriate to the cases of nondegenerate perturbation theory and to the so-called *single-reference version of the CCM*, while the case $D > 0$ is more appropriate to the cases of degenerate perturbation theory and to the so-called *multi-reference version of the CCM.*

For the moment, we restrict ourselves to the case $D = 0$ of a single model state $|\Phi_0\rangle$. For many purposes, and as we shall see specifically in Sect. 3 for the CCM, it is very useful or even vital for the model state $|\Phi_0\rangle$ to be a *generalised vacuum state,* with respect to which all of the states in $\mathcal{G}$ can be expressed in terms of many-body, creation correlation operators acting on $|\Phi_0\rangle$. Implicit in the above is the enormous simplification that arises if the algebra of all operators in $\mathcal{G}$ and its adjoint space $\mathcal{G}^*$ is spanned by the *two Abelian subalgebras* of (many-body) *multiconfigurational creation operators* $\{C_I^\dagger\}$ and their Hermitian-adjoint counterparts, namely the *multiconfigurational destruction operators* $\{C_I\}$. Both sets of operators are defined with respect to the given model state $|\Phi_0\rangle$. We note that implicit in the compact notation is the fact that the index I is in general a *set-index,* comprising a set of single-particle labels (in some suitable single-particle basis) which completely characterises a given many-body *configuration* in this basis. More specifically, the single-particle labels in the set-index I usually comprise only those needed to describe the states which differ from those occupied in the model state $|\Phi_0\rangle$. For this reason it is convenient to introduce the notation,

$$C_0^\dagger \equiv \mathbb{1} \equiv C_0 \quad . \tag{13}$$

In summary, we require from our sets $\{|\Phi_0\rangle; C_I^\dagger\}$ the following properties,

$$[C_I^\dagger, C_J^\dagger] = 0 = [C_I, C_J] \quad , \tag{14}$$

$$C_I|\Phi_0\rangle = 0 = \langle\Phi_0|C_I^\dagger \quad ; \quad \forall I \neq 0 \quad . \tag{15}$$

We further require that the two subalgebras and the state $|\Phi_0\rangle$ are *cyclic* in the following sense,

$$|\Psi\rangle = \sum_I \psi_I C_I^\dagger|\Phi_0\rangle \quad ; \quad \forall\, |\Psi\rangle \in \mathcal{G} \quad , \tag{16a}$$

$$\langle\tilde{\Psi}| = \sum_I \tilde{\psi}_I \langle\Phi_0|C_I \quad ; \quad \forall\, \langle\tilde{\Psi}| \in \mathcal{G}^* \quad , \tag{16b}$$

in terms of some sets of c-number expansion coefficients $\{\psi_I\}$ and $\{\tilde{\psi}_I\}$. The configuration-label space $\mathcal{I} \equiv \{I\}$ must thus be complete (for a given generalised vacuum or model state $|\Phi_0\rangle$) with respect to the many-body configurations. In general, the operators $\{C_I^\dagger\}$ will be products (or sums of products) of suitable single-particle operators. Although it is not vital, it is also convenient to impose the *orthonormality condition,*

$$\langle\Phi_0|C_I C_J^\dagger|\Phi_0\rangle = \delta_{IJ} \quad , \tag{17}$$

on the operators, where δ_{IJ} is a Kronecker delta symbol implying equality between the sets of single-particle (s.p.) labels I and J, i.e., equality under at least one permutation. Henceforth we assume that (17) holds, and hence we have the completeness relation in $\mathcal{G}$,

$$\sum_I C_I^\dagger|\Phi_0\rangle\langle\Phi_0|C_I = \mathbb{1} = |\Phi_0\rangle\langle\Phi_0| + \sum_I{}' C_I^\dagger|\Phi_0\rangle\langle\Phi_0|C_I \ , \tag{18}$$

where, here and henceforth, a prime on a sum over configuration labels I excludes the term $I = 0$.

We note that it is not, *a priori*, obvious that sets $\{|\Phi_0\rangle; C_I^\dagger\}$ exist with the above properties (14)–(18). We show below, by specific examples to various broad classes of important many-body systems, that not only can such sets generally be found rather easily, but that they are often non-unique in the sense that several sets can be found for use with the same space $\mathcal{G}$.

2.1 Field-Theoretic (Number-Nonconserving) Bose System

An obvious choice for $|\Phi_0\rangle$ in this case is the bare vacuum state, $|0\rangle$; and the multiconfigurational creation operators, $C_I^\dagger$, then become suitably normalised products of single-boson creation operators, $b_\alpha^\dagger$, in some particular single-particle basis,

$$|\alpha\rangle \equiv b_\alpha^\dagger|0\rangle \ . \tag{19}$$

The single-boson creation and destruction operators, $b_\alpha^\dagger$ and b_α, respectively, obey the usual bosonic commutation relations,

$$[b_\alpha, b_\beta] = 0 = [b_\alpha^\dagger, b_\beta^\dagger] \ ; \quad [b_\alpha, b_\beta^\dagger] = \delta_{\alpha\beta}\mathbb{1} \ . \tag{20}$$

In the symmetric boson Fock space $\mathcal{B}$ we have the fundamental completeness relation,

$$\mathbb{1} = |0\rangle\langle 0| + \sum_{n=1}^{\infty} \frac{1}{n!} \sum_{\{\alpha_i\}} b_{\alpha_1}^\dagger \cdots b_{\alpha_n}^\dagger |0\rangle\langle 0| b_{\alpha_n} \cdots b_{\alpha_1} \ , \tag{21a}$$

as a specific example of the relation (18). However, in (21*a*) any s.p. index in the products of operators may be repeated any number of times. Hence, it is often more convenient to re-label the s.p. indices as an *ordered* set $\{\bar{\alpha}_i\}$, e.g., $\bar{\alpha}_1 < \bar{\alpha}_2 < \bar{\alpha}_3 \cdots$, such that all members of this set are distinct. The completeness relation (21*a*) may then be conveniently rewritten in the equivalent form,

$$\mathbb{1} = |0\rangle\langle 0| + \sum_{n=1}^{\infty} \sum_{\{\bar{\alpha}_i\}} \sum_{\{m_i\}} \frac{1}{m_1! \cdots m_k!}$$
$$\times \left(b_{\bar{\alpha}_1}^\dagger\right)^{m_1} \cdots \left(b_{\bar{\alpha}_k}^\dagger\right)^{m_k} |0\rangle\langle 0| \left(b_{\bar{\alpha}_k}\right)^{m_k} \cdots \left(b_{\bar{\alpha}_1}\right)^{m_1} \ , \tag{21b}$$

where $\sum_{i=1}^{k} m_i = n$. Thus, in this example, the set-index I becomes the set of occupation numbers,

$$I \to (m^I_{\bar{\alpha}_1}, m^I_{\bar{\alpha}_2}, \cdots) \ ; \quad m^I_{\bar{\alpha}_i} = 0, 1, 2, \cdots \ , \tag{22}$$

and the generic multiconfigurational creation operators $C_I^\dagger$ become

$$C_I^\dagger \to \prod_i \left[(m^I_{\bar{\alpha}_i}!)^{-\frac{1}{2}} \left(b^\dagger_{\bar{\alpha}_i}\right)^{m^I_{\bar{\alpha}_i}} \right] \ , \tag{23}$$

where the superscript I in the occupation numbers, $m^I_{\bar{\alpha}_i}$, reminds us of the particular configuration to which they refer. In the coordinate-space basis, $|\mathbf{r}\rangle = b^\dagger(\mathbf{r})|0\rangle$, for example, we have

$$I \to (\mathbf{r}_1, \mathbf{r}_2, \cdots, \mathbf{r}_n) \ ; \quad n = 0, 1, 2, \cdots \ , \tag{24}$$

and

$$C_I^\dagger \longrightarrow (n!)^{-\frac{1}{2}} \prod_{i=1}^{n} b^\dagger(\mathbf{r}_i) \ , \tag{25}$$

in terms of the creation field operator, $b^\dagger(\mathbf{r})$.

We note that for the special case of *single-mode bosonic field theory*, in which only one mode exists, with a corresponding creation operator $b^\dagger$, we have

$$I \longrightarrow n \ ; \quad n = 0, 1, 2, \cdots \ , \tag{26}$$

and

$$C_I^\dagger \longrightarrow (n!)^{-\frac{1}{2}} (b^\dagger)^n \ . \tag{27}$$

Such single-mode bosonic field theories are useful nontrivial examples of real quantum field theories in $(d+1)$ dimensions, where d is the number of spatial dimensions, for the limiting case $d = 0$. Clearly, such $(0+1)$-dimensional theories also map precisely onto one-body quantum mechanics through the usual mapping of the single-mode creation and destruction operators, $b^\dagger$ and b, onto the usual position and momentum operators, $\hat{x}$ and $\hat{p}$, respectively,

$$\hat{x} = 2^{-\frac{1}{2}}(b^\dagger + b) \ ; \quad \hat{p} = 2^{-\frac{1}{2}}\mathrm{i}(b^\dagger - b) \ . \tag{28}$$

2.2 Number-Conserving Bose System

We now consider a system of N bosons in a box of volume Ω, appropriate to a system such as liquid ^{4}He. We shall usually be interested in such systems in the *thermodynamic limit*, where $N \to \infty$ and $\Omega \to \infty$, such that the density, $\rho \equiv N/\Omega$, remains constant. In such cases an obvious candidate for the model state $|\Phi_0\rangle$ in the fixed-N symmetric boson Hilbert space $\mathcal{B}_N$ is the *Bose condensate* $|B\rangle$, in which all N particles condense into the lowest-energy s.p. state, $b_0^\dagger|0\rangle$,

$$|\Phi_0\rangle \longrightarrow |B\rangle \equiv (N!)^{-\frac{1}{2}} (b_0^\dagger)^N |0\rangle \ . \tag{29}$$

If we now specifically single out the condensate index 0 from the complete set of s.p. labels $\{\alpha\}$ in Sect. 2.1, as follows,

$$\alpha = \begin{cases} 0 & ;\ \text{condensate state} \\ k \neq 0\,; & \text{non-condensate states} \end{cases} , \tag{30}$$

we find that the set-index I becomes

$$I \rightarrow \left(m^I_{\bar{k}_1}, m^I_{\bar{k}_2}, \cdots\right) \quad ; \quad m^I_{\bar{k}_i} = 0, 1, 2, \cdots \quad , \quad \forall i \ , \tag{31}$$

in the notation of Sect. 2.1, and the multiconfigurational creation operators $C^\dagger_I$ can be written as follows,

$$C^\dagger_I \longrightarrow \left[\frac{(N-n)!}{N!}\right]^{\frac{1}{2}} \prod_i \left[\left(m^I_{\bar{k}_i}!\right)^{-\frac{1}{2}} \left(b^\dagger_{\bar{k}_i}\right)^{m^I_{\bar{k}_i}} (b_0)^n\right] \ , \tag{32}$$

where $\sum_i m^I_{\bar{k}_i} = n$, and $n = 0, 1, 2, \cdots$. Alternatively, in the coordinate-space basis, the set-index I is as given by (24), and the generic multiconfigurational creation operators $C^\dagger_I$ become

$$C^\dagger_I \longrightarrow \left[\frac{(N-n)!}{N!}\right]^{\frac{1}{2}} \prod_{i=1}^{n} b^\dagger(\mathbf{r}_i)(b_0)^n \ . \tag{33}$$

2.3 Many-Fermion System

We now consider a comparable system of N fermions, such as liquid ^{3}He, finite nuclei, nuclear matter, the electron gas, etc. In this case an obvious (but not unique) candidate for the model state $|\Phi_0\rangle$ in the fixed-N antisymmetric Hilbert space $\mathcal{F}_N$ is the *filled Fermi sea*, $|F\rangle$. This is an $(N \times N)$ Slater determinant of s.p. states formed from some complete s.p. basis $\{|\alpha_i\rangle\}$, usually from the N states of lowest energy. The single-fermion creation and destruction operators, $a^\dagger_\alpha$ and a_α, respectively, obey the usual anticommutation relations,

$$\{a_\alpha, a_\beta\} = 0 = \{a^\dagger_\alpha, a^\dagger_\beta\} \quad ; \quad \{a_\alpha, a^\dagger_\beta\} = \delta_{\alpha\beta}\mathbb{1} \ , \tag{34}$$

and the s.p. states are obtained from the vacuum state $|0\rangle$ as,

$$|\alpha\rangle \equiv a^\dagger_\alpha|0\rangle \ . \tag{35}$$

In terms of the s.p. creation operators we may write the antisymmetric state $|F\rangle$ in the second-quantised form,

$$|\Phi_0\rangle \longrightarrow |F\rangle = \prod_{i=1}^{N} a^\dagger_{\nu_i}|0\rangle \ , \tag{36}$$

where we have adopted the standard notation,

$$\alpha_i = \begin{cases} \nu_i \; ; \; i = 1, 2, \cdots, N \; ; \; \text{if } |\alpha_i\rangle \in |F\rangle \\ \rho_i \; ; \; i = 1, 2, \cdots \quad ; \; \text{if } |\alpha_i\rangle \notin |F\rangle \end{cases} , \tag{37}$$

which is the fermionic analogue of the bosonic relation (30). Thus, we subdivide the s.p. states $|\alpha_i\rangle$ into two classes with repect to the model state $|F\rangle$, viz., those states $\{|\nu_i\rangle;\ i = 1, 2, \cdots, N\}$ from which $|F\rangle$ is constructed, and the remaining states $\{|\rho_i\rangle;\ i = 1, 2, \cdots\}$. For reasons in connection with creating new states from $|F\rangle$, the states $|\nu_i\rangle$ which are "occupied" in $|F\rangle$ are called hole states, and the states $|\rho_i\rangle$ which are "unoccupied" in $|F\rangle$ are called particle states. By analogy with the notation introduced in Sect. 2.1, the sets of particle-state indices $\{\rho_i\}$ and hole-state indices $\{\nu_i\}$ may be written as ordered sets, $\{\bar{\rho}_i;\ \bar{\rho}_1 < \bar{\rho}_2 < \cdots\}$ and $\{\bar{\nu}_i;\ \bar{\nu}_1 < \bar{\nu}_2 < \cdots < \bar{\nu}_N\}$ respectively.

In $\mathcal{F}_N$ we may now write the following equivalent completeness relations,

$$\begin{aligned} \mathbb{1} &= \frac{1}{N!} \sum_{\alpha_1, \cdots, \alpha_N} a^\dagger_{\alpha_1} \cdots a^\dagger_{\alpha_N} |0\rangle\langle 0| a_{\alpha_N} \cdots a_{\alpha_1} \\ &= \sum_{\bar{\alpha}_1 < \bar{\alpha}_2 < \cdots < \bar{\alpha}_N} a^\dagger_{\bar{\alpha}_1} \cdots a^\dagger_{\bar{\alpha}_N} |0\rangle\langle 0| a_{\bar{\alpha}_N} \cdots a_{\bar{\alpha}_1} \\ &= \sum_{m=0}^{N} \sum_{\bar{\rho}_1 < \bar{\rho}_2 < \cdots < \bar{\rho}_m} \sum_{\bar{\nu}_1 < \bar{\nu}_2 < \cdots < \bar{\nu}_m} a^\dagger_{\bar{\rho}_1} \cdots a^\dagger_{\bar{\rho}_m} a_{\bar{\nu}_1} \cdots a_{\bar{\nu}_m} |F\rangle\langle F| \\ &\qquad \times a^\dagger_{\bar{\nu}_m} \cdots a^\dagger_{\bar{\nu}_1} a_{\bar{\rho}_m} \cdots a_{\bar{\rho}_1} \\ &= \sum_{m=0}^{N} \frac{1}{(m!)^2} \sum_{\rho_1, \cdots, \rho_m} \sum_{\nu_1, \cdots, \nu_m} a^\dagger_{\rho_1} \cdots a^\dagger_{\rho_m} a_{\nu_1} \cdots a_{\nu_m} |F\rangle\langle F| \\ &\qquad \times a^\dagger_{\nu_m} \cdots a^\dagger_{\nu_1} a_{\rho_m} \cdots a_{\rho_1} \; . \end{aligned} \tag{38}$$

Hence, we see that with respect to the model state $|F\rangle$ the multiconfigurational creation operators $C_I^\dagger$ are products of particle–hole pair creation operators, $a^\dagger_{\rho_i} a_{\nu_i}$. More specifically the set-index I becomes

$$I \longrightarrow (n^I_{\bar{\nu}_1}, \cdots, n^I_{\bar{\nu}_N}; n^I_{\bar{\rho}_1}, n^I_{\bar{\rho}_2}, \cdots) \; ; \quad n^I_{\bar{\rho}_i}, n^I_{\bar{\nu}_j} = 0, 1 \; , \; \forall i, j \; , \tag{39}$$

and the multiconfigurational creation operators $C_I^\dagger$ become

$$C_I^\dagger \longrightarrow \prod_i \left(a^\dagger_{\bar{\rho}_i}\right)^{n^I_{\bar{\rho}_i}} \prod_{j=1}^{N} \left(a_{\bar{\nu}_j}\right)^{n^I_{\bar{\nu}_j}} \; ;$$

$$\sum_i n^I_{\bar{\rho}_i} = \sum_{j=1}^{N} n^I_{\bar{\nu}_j} = n \; ; \quad n = 0, 1, \cdots, N \; . \tag{40}$$

In the thermodynamic limit ($N \to \infty, \Omega \to \infty, \rho \equiv N/\Omega$ fixed), if the system remains homogeneous (and this is an assumption), the resulting translational invariance implies that the s.p. states are (best chosen as) plane waves,

$$|\alpha\rangle \longrightarrow |k\rangle \equiv |\mathbf{k}, i\rangle \ , \tag{41}$$

where $i = 1, \cdots, \nu$ labels the internal quantum numbers (e.g., spin, isospin) and ν is the associated degeneracy. In this case,

$$\begin{aligned} |\rho\rangle &\longrightarrow |\mathbf{k}, i\rangle \ ; \quad |\mathbf{k}| > k_F \\ |\nu\rangle &\longrightarrow |\mathbf{k}, i\rangle \ ; \quad |\mathbf{k}| < k_F \ , \end{aligned} \tag{42}$$

in terms of the *Fermi wavenumber*, k_F. Using the particle number operator, $\hat{N}$,

$$\hat{N} = \sum_{\mathbf{k},i} a^{\dagger}_{\mathbf{k}i} a_{\mathbf{k}i} \ , \tag{43}$$

and (42), we easily prove that

$$\frac{N}{\Omega} \equiv \rho = \frac{\nu k_F^3}{6\pi^2} \ . \tag{44}$$

It is worth noting that there also exist alternative choices to $|F\rangle$ for the model state $|\Phi_0\rangle$ for fermion systems. A common alternative is just the standard BCS state, $|S\rangle$, introduced by Bardeen, Cooper, and Schrieffer [88] in the context of superconductivity. The BCS state $|S\rangle$ is simply another Slater determinant analagous to $|F\rangle$, i.e., formed from quasiparticle states comparable to those in (35), but which are created by quasiparticle creation operators which are linear combinations of the previous particle and hole states. The corresponding Bogoliubov transformation is a canonical one, so that the quasiparticle creation and destruction operators obey the same fermion anticommutation relations (34).

2.4 Single Spin in an Irreducible Multiplet, J

It is important to realise that we can also handle, in the same broad general framework, systems of particles other than just bosons or fermions. Quantum-mechanical spins provide a good example. The simplest system of this type is just a single spin in an irreducible multiplet, J. In this case the Hilbert space $\mathcal{H}_J$ is now finite-dimensional and is spanned by the set of $(2J + 1)$ states, $\{|J, M_J\rangle;\ M_J = -J, -J+1, \cdots, J\}$. The fundamental underlying algebra, SU(2), is defined by the usual commutation relations between the operators, $J^{\pm} \equiv J^x \pm \mathrm{i}J^y$, and J^z,

$$[J^z, J^{\pm}] = \pm J^{\pm} \ ; \quad [J^+, J^-] = 2J^z \ . \tag{45}$$

We note that models of this category arise in various guises. One example is the well-known model many-body Hamiltonian of Lipkin, Meshkov, and Glick [89] (the LMG model), which was originally invented to simulate the collective monopole vibrations in spherical atomic nuclei. The original model comprises N identical fermions distributed between two energy levels, each of which is N-fold degenerate, and interacting via a simple pairing interaction. The interest in the LMG model and its variants, centres on the fact that they exhibit a phase or shape transition at some critical value of the coupling strength. In nuclear physics, the transition from spherical to deformed shape which occurs in the regime of the rare-earth nuclei is believed to be due to a collective quadrupole excitation of this type, and the LMG model was first introduced as its simpler monopole analogue.

A second reason for studying such anharmonic spin models is that they provide a finite-dimensional Hilbert space analogue of the corresponding anharmonic oscillator models which have infinite-dimensional Hilbert spaces, and whose mathematical properties can thereby be obscured. Thus, it is easy to verify that if we define a new set of operators,

$$\mathcal{J}_0 \equiv J^z + J \ ; \quad \mathcal{J}_\pm \equiv (2J)^{-\frac{1}{2}} J^\pm \ , \tag{46}$$

the commutation relations for large values of J between the operators $\mathcal{J}_+, \mathcal{J}_-$, and $\mathcal{J}_0$ are the same as those between bosonic operators $b^\dagger, b$, and $b^\dagger b$, respectively, except for terms of order $O(\langle b^\dagger b\rangle/J)$. Hence, for example, in the limit that $J \to \infty$, the anharmonic spin Hamiltonian, H_J, defined as

$$H_J = \frac{1}{2} + J + J^z + \frac{\lambda}{16J^2}(J^+ + J^-)^4 \ , \tag{47}$$

should, making use of (28), smoothly approach the anharmonic oscillator Hamiltonian,

$$\begin{aligned} H &= b^\dagger b + \frac{1}{2} + \frac{\lambda}{4}(b^\dagger + b)^4 \\ &= \frac{1}{2}\hat{p}^2 + \frac{1}{2}\hat{x}^2 + \lambda\hat{x}^4 \ , \end{aligned} \tag{48}$$

in the well-defined sense that any energy eigenvalue $E_n^{(J)}$ with fixed value of n should converge to the corresponding anharmonic oscillator energy eigenvalue E_n. Aalto *et al.* [90] have performed an ECCM analysis of both Hamiltonians (47) and (48).

We now consider the various possible choices of model state $|\Phi_0\rangle$ from the $(2J+1)$ independent states spanning $\mathcal{H}_J$.

Lowest State, $|\Phi_o\rangle \ \to \ |J,-J\rangle$: From the basic commutation relations (45) one may easily prove that

$$J^+|J,-J+m\rangle = [(m+1)(2J-m)]^{\frac{1}{2}}|J,-J+m+1\rangle \ ; \quad m = 0,1,\cdots,2J \ . \tag{49}$$

Equation (49) gives immediately that, in relation to the model state $|\Phi_0\rangle \to |J,-J\rangle$, the generic operators $C_I^\dagger$ take the following form,

$$C_I^\dagger \longrightarrow \left[\frac{(2J-m)!}{(2J)!m!}\right]^{\frac{1}{2}} (J^+)^m \ , \tag{50}$$

where the label I takes the values,

$$I \longrightarrow m \ ; \quad m = 0, 1, \cdots, 2J \ . \tag{51}$$

Highest State, $|\Phi_o\rangle \to |J, J\rangle$: In this case the role of the basic creation operator is now taken by $J^- \equiv (J^+)^\dagger$, and otherwise everything now goes exactly as in the previous case, but with $J^+ \to J^-$.

Intermediate State, $|\Phi_o\rangle \to |J, M\rangle$, with $M \neq \pm J$: In this case, although we can clearly construct all remaining states in $\mathcal{H}_J$ by letting either J^+ or J^- act on $|\Phi_0\rangle$ a sufficient number of times, we cannot now construct two Abelian subalgebras from their necessarily simultaneous use to span $\mathcal{H}_J$, since $[J^+, J^-] \neq 0$. Nevertheless, it is important to realise that this does *not* imply that such intermediate states cannot be chosen as model states. Rather, we need to redefine our creation operators, $C_I^\dagger$, accordingly. As a specific example of how this may be done more generally, let us consider the case $J = 1$, and attempt to use $|\Phi_0\rangle = |1, 0\rangle$ as a model state. If we now use the column vector notation for the angular momentum eigenstates,

$$|1,1\rangle \equiv \begin{pmatrix} 1 \\ 0 \\ 0 \end{pmatrix} \ ; \quad |1,0\rangle \equiv \begin{pmatrix} 0 \\ 1 \\ 0 \end{pmatrix} \ ; \quad |1,-1\rangle \equiv \begin{pmatrix} 0 \\ 0 \\ 1 \end{pmatrix} \ , \tag{52}$$

we may define two new operators as follows,

$$U^+ \equiv \begin{pmatrix} 0\,1\,0 \\ 0\,0\,0 \\ 0\,0\,0 \end{pmatrix} \ ; \quad D^+ \equiv \begin{pmatrix} 0\,0\,0 \\ 0\,0\,0 \\ 0\,1\,0 \end{pmatrix} \ , \tag{53}$$

whose mode of action on the model state is

$$U^+|1,0\rangle = |1,1\rangle \ ; \quad D^+|1,0\rangle = |1,-1\rangle \ . \tag{54}$$

However, whereas their mode of action on $|1, 0\rangle$ is thus identical to those of J^+ and J^-, respectively, the new operators now *do* commute, $[U^+, D^+] = 0$. Thus, we may use $|\Phi_0\rangle = |1, 0\rangle$ as a model state in terms of the set of mutually commuting creation operators,

$$\{C_I^\dagger\} \longrightarrow \{\mathbb{1}, U^+, D^+\} \ . \tag{55}$$

2.5 Quantum Spin Lattices, e.g., Antiferromagnets

In the case of several similar spins, we may simply take the tensor products of the operators introduced in Sect. 2.4 for our multiconfigurational creation operators, $C_I^\dagger$. Some examples are given below.

Spin-Half Particles on a Bipartite Lattice: A bipartite lattice is one on which there are two completely equivalent sublattices A and B such that all nearest neighbours of sites on sublattice A lie on sublattice B, and vice versa (e.g., a square lattice in two spatial dimensions, and a simple cubic lattice in three dimensions). On such lattices the *classical* ground state of many Hamiltonians is often the ideal Néel antiferromagnetic state in which all spins on sublattice A align along one direction (say, the negative z-direction) and all spins on sublattice B align along the opposite (i.e., positive z) direction. A simple example is provided by the Heisenberg Hamiltonian,

$$H = J \sum_{\langle i,j \rangle} \mathbf{s}_i \cdot \mathbf{s}_j \quad , \tag{56}$$

where the sum on $\langle i,j \rangle$ denotes a sum over all nearest-neighbour pairs, and $\mathbf{s}_i$ is the spin on site i, in the case where the exchange constant, $J > 0$. In the quantum-mechanical case the interest in such models is often concerned with whether the quantum fluctuations (or quantum correlations) will or will not completely destroy the perfect Néel long-range order of the classical antiferromagnet.

In such cases, an appropriate choice of model state $|\Phi_0\rangle$ for spin-half particles is the Néel state $|N\rangle$, which is simply the tensor product,

$$|N\rangle \equiv \bigotimes_{i \in A} |\downarrow\rangle_i \bigotimes_{j \in B} |\uparrow\rangle_j \quad ; \tag{57}$$

of all "down" states, $|\downarrow\rangle \equiv |\frac{1}{2}, -\frac{1}{2}\rangle$, on sites i of sublattice A with all "up" states, $|\uparrow\rangle \equiv |\frac{1}{2}, \frac{1}{2}\rangle$, on sites j of sublattice B. The configuration label (or set-index) I may now be taken as the set of sites,

$$I \longrightarrow (i_1, \cdots, i_m; j_1, \cdots, j_n) \quad ; \qquad m, n = 0, 1, \cdots , \tag{58}$$

on which the spins are reversed with respect to the Néel state $|N\rangle$. The mutually commuting set of multiconfigurational creation operators with respect to $|N\rangle$ may thus be taken as,

$$C_I^\dagger \longrightarrow s_{i_1}^+ \cdots s_{i_m}^+ s_{j_1}^- \cdots s_{j_n}^- \quad . \tag{59}$$

We note again that the choice $|N\rangle$ for the model state $|\Phi_0\rangle$ is not unique. In cases where the ground state of the quantum Hamiltonian is expected to be, for example, dimerised or trimerised, one may construct model states $|\Phi_0\rangle$ built from products of suitable (noninteracting) dimer or trimer states. The interested reader is referred to the literature [91,92] for further details of such states, which are particularly pertinent for a valence-bond description of quantum spin lattices.

Spin-One Particles on a Bipartite Lattice: In the case of spin-one particles, the Néel state (57) is still a possible choice of model state $|\Phi_0\rangle$, where $|\downarrow\rangle \equiv |1,-1\rangle$. However, the multiconfigurational creation operators (59) now include the possibility that a single-spin raising operator on a site $i \in A$ or a single-spin lowering operator on a site $j \in B$ may appear either once or twice on the same site, in order to include all three possible spin projections on each site. Similarly, the configuration label (or set-index) I given by (58) must be extended to allow any given site index to appear up to two times in the set.

By extension of the discussion in Sect. 2.4 we may also consider an alternative planar model state $|\Phi_0\rangle \to |P\rangle$, where $|P\rangle$ is the state in which the spin on every site is in the state $|0\rangle \equiv |1,0\rangle$ with zero projection along the quantisation z-axis. In this case a suitable set of mutually commuting multiconfigurational creation operators $C_I^\dagger$ is given by the tensor product of operators U_k^+ and D_k^+ given by (53) over a given subset of sites $\{k\}$. The set-index label I now comprises the subset $\{k\}$, together with a two-valued flag at each member of the subset which indicates whether the operator U_k^+ or D_k^+ acts at each site. Once again, many other more complicated model states, $|\Phi_0\rangle$, can also be devised.

3 Basic Structure of the CCM

After the discussion in Sect. 2 we are now in a position to explain the basic ingredients and the formal structure of the CCM at a very general level, without reference to specific many-body systems.

3.1 Exponential Structure of Many-Body Wave Functions

Exponential structures arise frequently in physics, and often for similar underlying fundamental reasons. For example, in statistical mechanics, the Ursell–Mayer cluster expansion for the thermodynamic partition function (from which all thermodynamic properties of a bulk system derive) has an underlying *exponential form.* This arises naturally from the basic compounding, *with the correct statistical multiplicities,* of all contributions to the partition function from the *independent* "atomic clusters" of a given size.

We shall now show that a comparable exponential form also underpins the natural parametrisation (i.e., the "correct" form) for a quantum-mechanical many-body wave function. Historically, this was probably first realised by Goldstone [93] and Hubbard [94] in the context of time-independent perturbation theory (TIPT), where it also takes the related form of the *Goldstone linked cluster theorem.* Similar ideas apply equally well to the well-known representation of eigenstates in quantum field theory provided by the Gell-Mann and Low theorem [95], although this seems not to have been properly or fully realised at that time.

For pedagogical purposes let us consider a "closed-shell" N-fermion system, i.e., one which is reasonably approximated in zeroth order by a s.p. shell-model Slater determinant $|\Phi_0\rangle = |F\rangle$ of the sort discussed in Sect. 2.3, and which can be expressed in the second-quantised form (36). The choice of s.p. states from which to construct this filled Fermi sea is, at this stage, completely arbitrary. For example, one may choose harmonic oscillator states, self-consistent Hartree-Fock states, or any other basis. In this context an important theorem of Thouless [96] asserts that the most general determinantal wave function $|\Phi_0'\rangle$ not orthogonal to a given Slater determinant $|\Phi_0\rangle$ has the form

$$|\Phi_0'\rangle = \mathrm{e}^{S_1}|\Phi_0\rangle \ , \tag{60}$$

for some suitable choice of the one-body operator S_1 which acts on $|\Phi_0\rangle$ to produce a one-particle/one-hole ($1p$–$1h$) excitation. Thus, in the notation of Sect. 2.3, we may write

$$S_1 = \sum_{\rho}\sum_{\nu}\langle\rho|S_1|\nu\rangle\, a_\rho^\dagger a_\nu \ . \tag{61}$$

The effect of allowing single particles to be independently elevated above the Fermi sea (which, as we shall see below, is what is encapsulated in the exponential operation in the relation (60)) is, thus, equivalent to changing the s.p. orbitals that comprise the Slater determinant.

Consider now the problem of obtaining the exact ground-state wave function, $|\Psi_0\rangle$, and energy, E_0, of the interacting system of fermions described by the many-body Hamiltonian, H,

$$H|\Psi_0\rangle = E_0|\Psi_0\rangle \ . \tag{62}$$

The *dynamical correlations* induced by the interactions contained in H will modify the model state $|\Phi_0\rangle$. We *assume* from the outset that $|\Psi_0\rangle$ has a nonzero overlap with $|\Phi_0\rangle$, and we shall henceforth work in the *intermediate normalisation* scheme wherein

$$\langle\Phi_0|\Psi_0\rangle = \langle\Phi_0|\Phi_0\rangle \equiv 1 \ . \tag{63}$$

Now, the simplest correlation, other than a single $1p$–$1h$ excitation discussed above in the context of the Thouless theorem, is a single two-particle/two-hole ($2p$–$2h$) excitation of the form $S_2|\Phi_0\rangle$, where

$$S_2 = \frac{1}{(2!)^2}\sum_{\rho_1,\rho_2}\sum_{\nu_1,\nu_2}\langle\rho_1\rho_2|S_2|\nu_1\nu_2\rangle_A\, a_{\rho_1}^\dagger a_{\rho_2}^\dagger a_{\nu_2} a_{\nu_1} \ , \tag{64}$$

and the suffix A on the matrix element $\langle\rho_1\rho_2|S_2|\nu_1\nu_2\rangle_A$ denotes an antisymmetrised element (in the notation where $|\nu_1\nu_2\rangle_A \equiv |\nu_1\nu_2\rangle - |\nu_2\nu_1\rangle$).

Now, it may also happen that two *independent* pairs of particles also excite out of $|\Phi_0\rangle$ due to dynamic correlations. Clearly, this process is described by applying the operator S_2 twice, but with the proviso that we must include a statistical weighting factor (or multiplicity) of $\frac{1}{2}$ to avoid double counting. The resulting contribution to the correlated wave function is $\frac{1}{2}S_2^2|\Phi_0\rangle$. This process of independent excitation of pairs out of the Fermi sea may be continued to obtain a contribution $(1/m!)S_2^m|\Phi_0\rangle$ from the excitation of m independent $2p$–$2h$ pairs. Hence, the total amplitude for the excitation of an arbitrary number (including zero) of independent $2p$–$2h$ excitations is

$$\sum_{m=0}^{\infty} \frac{1}{m!} S_2^m |\Phi_0\rangle = e^{S_2}|\Phi_0\rangle \ , \tag{65}$$

and we note, very importantly, that the exponential structure arises very simply and fundamentally from a simple *counting of independent events*. This exponential structure lies at the very heart of the CCM.

Next we consider processes involving the simultaneous excitation of three fermions from the Fermi sea. These may be described by a contribution $S_3|\Phi_0\rangle$ to the exact wave function. Similarly to the case of pairs, there will be a contribution $(1/n!)S_3^n|\Phi_0\rangle$ from the simultaneous excitation of n independent triplets. If we then take into account the possibility of the independent simultaneous excitation of m pairs and n triplets from the Fermi sea, the corresponding contribution to $|\Psi_0\rangle$ is $(1/m!n!)S_2^m S_3^n|\Phi_0\rangle$. It is important that we do not need to worry about the order of the operators S_2 and S_3 in this product because they describe independent processes and hence commute, as one can also verify from their specific second-quantised forms. Summing over all possible values of m and n then leads to the contribution $\exp(S_2+S_3)|\Phi_0\rangle$ to $|\Psi_0\rangle$. Proceeding in this way with the independent excitation of clusters of up to N particles we arrive at the CCM representation,

$$|\Psi_0\rangle = e^S|\Phi_0\rangle \ ; \quad S = \sum_{m=1}^{N} S_m \ , \tag{66a}$$

$$S_m = \frac{1}{(m!)^2} \sum_{\rho_1\cdots\rho_m} \sum_{\nu_1\cdots\nu_m} \langle \rho_1 \cdots \rho_m | S_m | \nu_1 \cdots \nu_m \rangle_A \, a^\dagger_{\rho_1} \cdots a^\dagger_{\rho_m} a_{\nu_m} \cdots a_{\nu_1} \ . \tag{66b}$$

3.2 The Configuration-Interaction Method (CIM)

For the same N-fermion case considered above, a more naive and (seemingly) simpler parametrisation of $|\Psi_0\rangle$ is that used in the CIM,

$$|\Psi_0\rangle = (1+F)|\Phi_0\rangle \ ; \quad F = \sum_{m=1}^{\infty} F_m \ , \tag{67}$$

$$F_m = \frac{1}{(m!)^2} \sum_{\rho_1\cdots\rho_m} \sum_{\nu_1\cdots\nu_m} \langle \rho_1 \cdots \rho_m | F_m | \nu_1 \cdots \nu_m \rangle_A \, a^\dagger_{\rho_1} \cdots a^\dagger_{\rho_m} a_{\nu_m} \cdots a_{\nu_1} \ . \tag{68}$$

Again, F_n excites n particles and n holes from the filled Fermi sea $|\Phi_0\rangle \equiv |F\rangle$, but in this case F_n generally contains *unlinked* (or *disconnected*) pieces which arise from the products of noninteracting subclusters. A comparison of the first few terms in the expansions (65) and (67) yields, for example,

$$F_1 = S_1 \; ; \; F_2 = S_2 + \frac{1}{2}S_1^2 \; ; \; F_3 = S_3 + S_2 S_1 + \frac{1}{6}S_1^3 \; ;$$

$$F_4 = S_4 + S_3 S_1 + \frac{1}{2}S_2^2 + \frac{1}{2}S_2 S_1^2 + \frac{1}{24}S_1^4 \; ; \; \text{etc.} \; , \tag{69}$$

with corresponding inverses,

$$S_1 = F_1 \; ; \; S_2 = F_2 - \frac{1}{2}F_1^2 \; ; \; S_3 = F_3 - F_2 F_1 + \frac{1}{3}F_1^3 \; ;$$

$$S_4 = F_4 - F_3 F_1 - \frac{1}{2}F_2^2 + F_2 F_1^2 - \frac{1}{4}F_1^4 \; ; \; \text{etc.} \tag{70}$$

More generally, by putting $F_n \to \lambda^n F_n$ and $S_n \to \lambda^n S_n$, and by equating powers of λ in the two expansions (65) and (67), we find

$$F_n = \sum_{l=1}^{n} \sum_{m_1} \cdots \sum_{m_l} \frac{1}{m_1! \cdots m_l!} S_1^{m_1} \cdots S_l^{m_l} \; ; \quad \sum_{i=1}^{l} i m_i = n \; , \tag{71}$$

where the restriction on the sums over $m_1, \cdots, m_l$ is such that one sums over all sets of non-negative integers $\{m_i \geq 0\}$ such that $\sum_{i=1}^{l} i m_i = n$. Since the number of ways of arranging k identical objects into $l \leq k$ classes is $k!/(m_1! \cdots m_l!)$, where m_i objects are in class i, and $\sum_i m_i = k$, we easily derive the alternative relation,

$$F_n = \sum_{k=1}^{n} \frac{1}{k!} \sum_{n_1} \cdots \sum_{n_k} S_{n_1} \cdots S_{n_k} \; ; \quad \sum_{i=1}^{k} n_i = n \; , \tag{72}$$

where the restriction on the sums over $n_1, \cdots, n_k$ is such that one sums over all sets of positive integers $\{n_i > 0\}$ such that $\sum_{i=1}^{k} n_i = n$. The corresponding inverse relations are also readily found by using $S = \ln(1+F)$. The analogues to (71) and (72) are thus,

$$S_n = \sum_{l=1}^{n} \sum_{m_1} \cdots \sum_{m_l} (-1)^{k-1} \frac{(k-1)!}{m_1! \cdots m_l!} F_1^{m_1} \cdots F_l^{m_l} \; ;$$

$$\sum_{i=1}^{l} i m_i = n \; , \quad \sum_{i=1}^{l} m_i = k \; , \tag{73}$$

and

$$S_n = \sum_{k=1}^{n} \frac{(-1)^{k-1}}{k} \sum_{n_1} \cdots \sum_{n_k} F_{n_1} \cdots F_{n_k} \; ; \quad \sum_{i=1}^{k} n_i = n \; . \tag{74}$$

We note that both the CCM and CIM parametrisations are exact and equivalent so long as no approximations (i.e., no truncations over the complete set of multiparticle configurations) are made. However, once we approximate the equivalence is lost. We note that whereas, by construction, all of the matrix elements of the operators S_n are linked, the matrix elements of the operators F_n mix linked with unlinked terms (e.g., F_4 mixes the linked $4p$–$4h$ term S_4 with unlinked terms like $\frac{1}{2}S_2^2$ which corresponds to two unlinked $2p$–$2h$ pairs). Although the unlinked terms ultimately cancel out in any exact theory, this is not true in the CIM when truncations are made. By contrast, the cancellation of unlinked terms is automatic in the CCM, whatever approximations are made to the *cluster correlation operator, S*.

In practice the CIM just amounts to truncating the number of possible many-body configurations excited out of $|\Phi_0\rangle$ to some number $\mathcal{N}$, and then diagonalising the Hamiltonian as an $\mathcal{N} \times \mathcal{N}$ matrix in this truncated multiconfigurational basis. The hope is then that as $\mathcal{N}$ is increased one obtains convergence. As we have explained in Sect. 1.2 the interlacing theorem [51] guarantees this (uniform) convergence in principle. However, for large numbers of particles ($N \to \infty$) the matrix becomes *ill-conditioned*, and the CIM becomes fatally flawed due to the inexact cancellation of all unlinked terms. What happens is that for such extensive variables as the energy, E_0, the linked terms lead properly to contributions which scale (properly) linearly in the particle number, N. By contrast terms containing m unlinked pieces scale as N^m, and the (in principle) exact cancellation of contributions from terms with $m \geq 2$ rapidly becomes numerically unstable with increasing N, unless the exact cancellation is incorporated from the outset, as in the CCM. These ideas are expressed more quantitatively later, but are perhaps first best related to the so-called *"size-extensivity problem"* of the CIM [52], already alluded to in Sect. 1.2.

3.3 Size-Extensivity

Consider a system of N particles, whose ground-state wave function $|\Psi_0\rangle$ is parametrised by operators $\{F_n\}$ in the CIM and $\{S_n\}$ in the CCM. Suppose we now separate the system spatially into two subsystems A and B, with N_A particles in A and N_B particles in B. We assume that as the spatial separation increases, $r_{AB} \to \infty$, the two subsystems cease to interact with each other, so that the Hamiltonian becomes,

$$H \underset{r_{AB}\to\infty}{\longrightarrow} H^A + H^B \quad , \qquad \text{with } [H^A, H^B] = 0 \ . \tag{75}$$

Hence, in this limit, the energy becomes additively separable,

$$E_0 \underset{r_{AB}\to\infty}{\longrightarrow} E_0^A + E_0^B \quad , \tag{76}$$

while the wave function becomes multiplicatively separable,

$$|\Psi_0\rangle \underset{r_{AB}\to\infty}{\longrightarrow} |\Psi_0^A\rangle \otimes |\Psi_0^B\rangle \ . \tag{77}$$

A system is said to be *size-extensive* when it obeys the above separability properties (76) and (77).

We can now see immediately that the exponential parametrisation of the wave function given by (65) in the CCM guarantees the separability relation (77), no matter how the cluster correlation operator S is truncated. For example, a common truncation scheme in both the CCM and CIM is the so-called SUBn approximation scheme, wherein the m-body partitions of the operator S in (65) or of the operator F in (67) are truncated so that one retains only the components with $m \leq n$, and sets the higher partitions with $m > n$ to zero. Thus, in the CCM SUBn scheme we have,

$$\begin{aligned}|\Psi_0\rangle &\approx \exp(S_1 + \cdots + S_n)|\Phi_0\rangle \\ &\underset{r_{AB}\to\infty}{\longrightarrow} \exp(S_1^{(A)} + S_1^{(B)} + \cdots + S_n^{(A)} + S_n^{(B)})|\Phi_0\rangle \\ &\equiv \exp(S_1^{(A)} + \cdots + S_n^{(A)})|\Phi_0^A\rangle \otimes \exp(S_1^{(B)} + \cdots + S_n^{(B)})|\Phi_0^B\rangle \ . \end{aligned} \tag{78}$$

By contrast, in the CIM SUBn scheme, we have

$$\begin{aligned}|\Psi_0\rangle &\approx (1 + F_1 + \cdots + F_n)|\Phi_0\rangle \\ &\underset{r_{AB}\to\infty}{\not\longrightarrow} (1 + F_1^{(A)} + \cdots + F_n^{(A)})|\Phi_0^A\rangle \otimes (1 + F_1^{(B)} + \cdots + F_n^{(B)})|\Phi_0^B\rangle \ . \end{aligned} \tag{79}$$

In this case the separability of the wave function is *not* preserved since the separated parametrisation in the second line of (79) would require excitations of up to $2n$ particle/hole pairs which are not contained in a SUBn scheme.

3.4 General Cumulant (or Connected Moment) Expansions

The exponential parametrisation (65) of the CCM is intimately connected with similar exponential structures underpinning general *cumulant expansions.* Thus, in probability theory one considers the moment generating function, $\phi(\lambda)$, defined as

$$\phi(\lambda) \equiv \langle \exp(\lambda x)\rangle = \sum_{n=0}^{\infty} \frac{\lambda^n}{n!}\mu_n \ ,$$

$$\mu_n \equiv \langle x^n\rangle \ , \quad n = 1, 2, \cdots \ ; \quad \mu_0 \equiv 1 \ . \tag{80}$$

The function $\phi(\lambda)$ thereby generates all of the positive integral *moments*, $\mu_n = \langle x^n\rangle$, of an *arbitrary random variable*, x, according to some *probability distribution* implied by the angular brackets. This average, $\langle\cdots\rangle$, can either be a "classical", e.g., thermal, average or a quantal expectation value, e.g., $x \to H$, and $\langle\cdots\rangle \to \langle\Phi_0|\cdots|\Phi_0\rangle$ gives the energy moments $\langle\Phi_0|H^n|\Phi_0\rangle$ in the state $|\Phi_0\rangle$. In practice, the state $|\Phi_0\rangle$ is an arbitrary trial state, which is required only to have a nonzero overlap with the exact ground state $|\Psi_0\rangle$, and which (thus usually) shares its principle symmetries.

In principle, the moments contain all of the information about the probability distribution. For example, the energy moments, $\mu_n \equiv \langle\Phi_0|H^n|\Phi_0\rangle, n \geq 1$, contain all of the dynamical information about the system, and one might expect to be able to construct all other quantities pertaining to the system from them. However, there are two important points to note about the moments, $\mu_n \equiv \langle x^n\rangle$, namely:

– they contain essentially redundant information, and
– the information is encapsulated in a very inefficient and obscure way.

These problems lead in practice to the fact that many naive inversion (or inversion-type) methods based on moments are badly ill-conditioned. This manifests itself in the problem that as one attempts to gain more information by using more moments, numerical rounding errors rapidly grow and frustrate these attempts, no matter what level of numerical precision is adopted.

A standard way to improve this situation is to use, instead of the moments themselves, the so-called *cumulants* (or, synonymously, the *connected moments* or *semi-invariants*) defined by [97],

$$\ln\langle\exp(\lambda x)\rangle \equiv \sum_{n=1}^{\infty} \frac{\lambda^n}{n!}\nu_n \ , \tag{81}$$

where $\nu_n \equiv \langle x^n\rangle_c$ are the connected moments or cumulants. Such connected moments have many simple properties. Especially important are:

– that they scale linearly with the size of the system (i.e., $\nu_n \propto N$, $\forall n$, as $N \to \infty$) or, equivalently,
– that the combined cumulants for two *independent* subsystems are the corresponding sums of the two separate cumulants.

This latter relation is trivial to prove. Thus, for two random variables x and y we have, in an obvious notation,

$$\ln\langle\exp[\lambda(x+y)]\rangle \equiv \sum_{n=1}^{\infty} \frac{\lambda^n}{n!}\nu_n^{(x+y)} \ . \tag{82}$$

However, if x and y are independent we have

$$\langle\exp[\lambda(x+y)]\rangle = \langle\exp(\lambda x)\rangle\langle\exp(\lambda y)\rangle \ , \tag{83}$$

and hence,

$$\nu_n^{(x+y)} = \nu_n^{(x)} + \nu_n^{(y)}, \ \ \forall n \ . \tag{84}$$

Similarly, for k *independent* variables $\{x_i;\ i = 1, \cdots, k\}$, we have

$$\nu_n^{(x_1+\cdots+x_k)} = \sum_{i=1}^{k} \nu_n^{(x_i)}, \ \ \forall n \ , \tag{85}$$

and all cross-terms vanish.

By direct comparison of the coefficients of equal powers of λ in both series (80) and (81), one readily checks that the lowest n cumulants determine all moments up to order n, and vice versa. The explicit relationships between the first few are as follows,

$$\mu_1 = \nu_1 \;\; ; \;\; \mu_2 = \nu_2 + \nu_1^2 \;\; ; \;\; \mu_3 = \nu_3 + 3\nu_2\nu_1 + \nu_1^3 \;\; ;$$
$$\mu_4 = \nu_4 + 4\nu_3\nu_1 + 3\nu_2^2 + 6\nu_2\nu_1^2 + \nu_1^4 \;\; ; \;\; \text{etc.} \;\; , \tag{86}$$

with corresponding inverses,

$$\nu_1 = \mu_1 \;\; ; \;\; \nu_2 = \mu_2 - \mu_1^2 \;\; ; \;\; \nu_3 = \mu_3 - 3\mu_2\mu_1 + 2\mu_1^3 \;\; ;$$
$$\nu_4 = \mu_4 - 4\mu_3\mu_1 - 3\mu_2^2 + 12\mu_2\mu_1^2 - 6\mu_1^4 \;\; ; \;\; \text{etc.} \;\; , \tag{87a}$$

or, equivalently,

$$\langle x\rangle_{\rm c} = \langle x\rangle \;\; ; \;\; \langle x^2\rangle_{\rm c} = \langle x^2\rangle - \langle x\rangle^2 \;\; ; \;\; \langle x^3\rangle_{\rm c} = \langle x^3\rangle - 3\langle x^2\rangle\langle x\rangle + 2\langle x\rangle^3 \;\; ;$$
$$\langle x^4\rangle_{\rm c} = \langle x^4\rangle - 4\langle x^3\rangle\langle x\rangle - 3\langle x^2\rangle^2 + 12\langle x^2\rangle\langle x\rangle^2 - 6\langle x\rangle^4 \;\; ; \;\; \text{etc.} \tag{87b}$$

Clearly, by comparison of (67) and (65) with (80) and (81), respectively, we see that there is a complete analogy with the CIM and CCM representations,

$$\mu_n \longleftrightarrow n!F_n \;\; ; \quad \nu_n \longleftrightarrow n!S_n \;\; . \tag{88}$$

By comparison with the results derived in Sect. 3.2, we may easily prove the explicit general relationships,

$$\begin{aligned}\mu_n &= \sum_{k=1}^{n} \frac{1}{k!} \sum_{n_1} \cdots \sum_{n_k} \frac{n!}{n_1! \cdots n_k!} \nu_{n_1} \cdots \nu_{n_k} \;\; ; \quad \sum_{i=1}^{k} n_i = n \\ &= n! \sum_{l=1}^{n} \sum_{m_1} \cdots \sum_{m_l} \prod_{i=1}^{l} \left[\frac{1}{m_i!} \left(\frac{\nu_i}{i!} \right)^{m_i} \right] \;\; ; \quad \sum_{i=1}^{l} i m_i = n \;\; ,\end{aligned} \tag{89}$$

and the corresponding inverses,

$$\begin{aligned}\nu_n &= \sum_{k=1}^{n} \frac{(-1)^{k-1}}{k} \sum_{n_1} \cdots \sum_{n_k} \frac{n!}{n_1! \cdots n_k!} \mu_{n_1} \cdots \mu_{n_k} \;\; ; \quad \sum_{i=1}^{k} n_i = n \\ &= n! \sum_{l=1}^{n} \sum_{m_1} \cdots \sum_{m_l} (-1)^{k-1}(k-1)! \prod_{i=1}^{l} \left[\frac{1}{m_i!} \left(\frac{\mu_i}{i!} \right)^{m_i} \right] \;\; ; \\ &\qquad \sum_{i=1}^{l} i m_i = n \;\; , \quad \sum_{i=1}^{l} m_i = k \;\; ,\end{aligned} \tag{90}$$

and where the summation variables are integers such that $n_i > 0$ and $m_i \geq 0$.

It is also not difficult to prove the recurrence relation,

$$\mu_{n+1} = \sum_{m=0}^{n} \frac{n!}{m!(n-m)!} \mu_{n-m} \nu_{m+1} \; ; \quad \mu_0 = 1 \; , \tag{91}$$

and its counterpart relating the CIM to the CCM correlation operators,

$$F_n = \sum_{m=1}^{n} \frac{m}{n} F_{n-m} S_m \; ; \quad F_0 = \mathbb{1} \; . \tag{92}$$

Relations (91) and (92) can be useful for numerical computations.

Such cumulants (or, equivalently the linked cluster correlation operators $\{S_n\}$ and their analogues) appear in many places in quantum many-body theory. They lie, for example, at the heart of the Goldstone linked cluster theorem [93], which we prove very easily below from within the CCM framework. More generally, they always lie at the heart of any classical or quantum-mechanical (diagrammatic or other) expansion which deals with the (physical) linked or connected quantities. In this very real sense the CCM parametrisation of ground-state wave functions is the quantal (i.e., operatorial) analogue of the classical (i.e., c-number) Ursell–Mayer [98] expansion for the partition function of classical statistical mechanics.

3.5 Formal Elements of the Single-Reference Normal CCM

By making use of the ideas and powerful notation introduced in Sect. 2, we are now in a position to present the key ingredients of the normal CCM (NCCM) formalism at a very general level. We restrict ourselves in the first instance to the single-reference (or "closed shell") version of the method, and to its zero-temperature formulation in terms of pure states.

Ket Ground State: In terms of a general model state, $|\Phi_0\rangle$, which plays the role of a cyclic vector with respect to a complete set of mutually commuting multiconfigurational creation operators, $C_I^\dagger$, as discussed in detail in Sect. 2, we now easily generalise the CCM parametrisation (66a,b) of a many-fermion ground state to the generic case. Thus, the exact many-body ket ground-state wave function (or, more generally, any exact ket eigenstate $|\Psi_0\rangle$ not orthogonal to $|\Phi_0\rangle$) is written as

$$|\Psi_0\rangle = \mathrm{e}^S |\Phi_0\rangle \; ; \quad S = {\sum_I}' s_I C_I^\dagger \; , \tag{93}$$

where the prime on the sum excludes the term $I = 0$ (corresponding to the case $C_0^\dagger = \mathbb{1}$), in order to preserve the intermediate normalisation scheme of (63). The operator S is a linked-cluster operator, by construction, and the quantities $\{s_I\}$ are a complete set of c-number cluster amplitudes. In general, $|\Phi_0\rangle$ will contain the important symmetries of the phase of the system under consideration, including the correct particle statistics, as appropriate.

Ground-State Energy: Insertion of the CCM parametrisation (93) into the ground-state Schrödinger equation (62) leads to the typical CCM form,

$$(\mathrm{e}^{-S}H\mathrm{e}^{S} - E_0)|\Phi_0\rangle = 0 \ , \tag{94}$$

in which the Hamiltonian appears in the similarity-transformed form which is a hallmark of the method. By taking the inner products of (94) with $|\Phi_0\rangle$ itself and with the remainder of the complete set of states $\{C_I^\dagger|\Phi_0\rangle;\ I \neq 0\}$ we are thus led respectively to an equation for $E_0 = E_0(s_I)$,

$$E_0 = \langle\Phi_0|\mathrm{e}^{-S}H\mathrm{e}^{S}|\Phi_0\rangle = \langle\Phi_0|H\mathrm{e}^{S}|\Phi_0\rangle \ , \tag{95}$$

and a set of formally exact, microscopic, coupled nonlinear equations for the cluster correlation coefficients $\{s_I\}$,

$$\langle\Phi_0|C_I\mathrm{e}^{-S}H\mathrm{e}^{S}|\Phi_0\rangle = 0 \ ; \ \forall I \neq 0 \ , \tag{96}$$

in which there appear no macroscopic quantities like the energy E_0. Just as for the specific case of a many-fermion system discussed in Sect. 3.3, so in the general case the energy eigenvalue, E_0, is guaranteed to be an extensive quantity, no matter how the formally exact sum over configurations $\{I\}$ is truncated.

We may now easily prove the important *Goldstone linked cluster theorem* [93], namely that no unlinked terms appear in the above equations (95) and (96). The proof relies on the elementary nested commutator expansion,

$$\mathrm{e}^{-S}H\mathrm{e}^{S} = H + [H,S] + \frac{1}{2!}[[H,S],S] + \cdots \ , \tag{97}$$

and the fact that all of the components of S commute with each other, by construction, so that each element of S in the parametrisation (93) is linked directly to the Hamiltonian. The reason for this is that the only non-vanishing contributions to each of the commutators in the series (97) arise from terms where one of the single-particle operators in the second-quantised form for the Hamiltonian H meets its Hermitian adjoint in one of the partitions of any one of the cluster operators S. Such non-vanishing components of the commutators form the *elementary links*. Thus, we see that the only links that can arise are between H and an operator S; there are no links between different S operators, and hence every operator S in every nested commutator must be linked to the Hamiltonian. Thus, we have proven that $\mathrm{e}^{-S}H\mathrm{e}^{S}$ is a *fully linked operator.* Unlinked or disconnected pieces simply cannot arise, even when approximations to S are made. This linkedness feature is often emphasised by writing

$$\mathrm{e}^{-S}H\mathrm{e}^{S} \equiv \{H\mathrm{e}^{S}\}_{\mathcal{L}} \ , \ \{HS^{n+1}\}_{\mathcal{L}} \equiv [\{HS^{n}\}_{\mathcal{L}},S] \ ; \ n = 0,1,\cdots \ . \tag{98}$$

Furthermore, we note that if H contains only up to n-body operators (i.e., it comprises sums of products of no more than $2n$ single-particle operators), the series (97) will exactly terminate with the term involving $2n$ cluster operators, S. For example, in the common case where $H = T + V$ comprises kinetic energy and pairwise potential terms only, the series (97) terminates at terms of fourth order in S. We note that the fact that the basic exact CCM equations are thus of finite order in the coefficients $\{s_I\}$ ensures that no further truncation of the CCM equations is needed after the operator S has itself been approximated. By contrast, were the energy E_0 to be calculated, as in a standard variational theory treatment, as an expectation value,

$$E_0 = \frac{\langle \Psi_0 | H | \Psi_0 \rangle}{\langle \Psi_0 | \Psi_0 \rangle} = \frac{\langle \Phi_0 | e^{S^\dagger} H e^S | \Phi_0 \rangle}{\langle \Phi_0 | e^{S^\dagger} e^S | \Phi_0 \rangle} , \tag{99}$$

the resulting expression is generally of infinite order in the correlation operator S and its adjoint $S^\dagger$, however S is truncated. Furthermore, although the cancellation of unlinked terms between the numerator and denominator of (99) may be proven [99,100], it is by no means transparent.

Bra Ground State: Up to this point we have parametrised only the ket ground state, and we have made use of the Schrödinger equation itself to extract the energy, E_0, without the need to consider the bra state. However, in order to calculate other ground-state properties we cannot avoid the use of the bra state to form expectation values. It is precisely at this point that the CCM formalism now divides into the normal (NCCM) and extended (ECCM) schemes [67]. The crucial feature of both schemes is that corresponding ket and bra states are parametrised *independently* in a way that does *not* explicitly preserve their Hermitian-adjoint relationship to one another, and which as a consequence actually breaks it at a general (e.g., SUBn) level of truncation. The underlying reasons for this have been given by Arponen and his co-workers [67,68,101], and are explained below.

We introduce the notation $\langle \tilde{\Psi}_0 |$ for the bra ground state,

$$\langle \tilde{\Psi}_0 | \equiv \frac{(|\Psi_0\rangle)^\dagger}{\langle \Psi_0 | \Psi_0 \rangle} \equiv \frac{\langle \Psi_0 |}{\langle \Psi_0 | \Psi_0 \rangle} , \tag{100}$$

and its corresponding NCCM parametrisation,

$$\langle \tilde{\Psi}_0 | = \langle \Phi_0 | \tilde{S} e^{-S} \quad ; \quad \tilde{S} = 1 + \sum_I{}' \tilde{s}_I C_I \ . \tag{101}$$

Equation (101) preserves the explicit normalisation,

$$\langle \tilde{\Psi}_0 | \Psi_0 \rangle = \langle \Phi_0 | \Psi_0 \rangle = \langle \Phi_0 | \Phi_0 \rangle = 1 \ . \tag{102}$$

Upon taking expectation values between $\langle \tilde{\Psi}_0 |$ and $|\Psi_0\rangle$ it also maintains intact the fundamental CCM similarity transformation, which, as we have seen, itself encodes the linked-cluster property. Although the operator $\tilde{S}$ formally satisfies the condition,

$$\langle \Phi_0 | \tilde{S} = \frac{\langle \Phi_0 | e^{S^\dagger} e^S}{\langle \Phi_0 | e^{S^\dagger} e^S | \Phi_0 \rangle} , \tag{103}$$

we maintain $\{s_I, \tilde{s}_I\}$ as a complete set of *independent* variables. If the multiconfigurational label set $\{I\}$ is truncated, exact Hermiticity will generally be violated. The full set of amplitudes $\{s_I, \tilde{s}_I\}$ provides a *complete* NCCM parametrisation of the ground state. For example, an arbitrary operator, $A \in \mathcal{G}$ has a ground-state expectation value,

$$\bar{A} \equiv \langle A \rangle \equiv \langle \tilde{\Psi}_0 | A | \Psi_0 \rangle$$

$$= \langle \Phi_0 | \tilde{S} e^{-S} A e^S | \Phi_0 \rangle = \bar{A}(s_I, \tilde{s}_I) \ . \tag{104}$$

We note that $\bar{A}$ is again fully linked (and hence scales properly with particle number, N), even though the operator $\tilde{S}$ itself contains unlinked pieces.

The bra-state coefficients $\{\tilde{s}_I\}$ are obtained via the ground-state Schrödinger equation written in the form,

$$\langle \tilde{\Psi}_0 | H = E_0 \langle \tilde{\Psi}_0 | \implies \langle \Phi_0 | \tilde{S} (e^{-S} H e^S - E_0) = 0 \ . \tag{105}$$

Projection onto the states $C_I^\dagger | \Phi_0 \rangle$ yields the equations,

$$\langle \Phi_0 | \tilde{S} (e^{-S} H e^S - E_0) C_I^\dagger | \Phi_0 \rangle = 0 \ ; \ \ \forall I \neq 0 \ . \tag{106a}$$

Alternatively, we can eliminate E_0 from (106a) by making use of the ket-state equation (94) projected with the state $\langle \Phi_0 | \tilde{S} C_I^\dagger$, to yield the equations,

$$\langle \Phi_0 | \tilde{S} e^{-S} [H, C_I^\dagger] e^S | \Phi_0 \rangle = 0 \ ; \ \ \forall I \neq 0 \ , \tag{106b}$$

where we have made use of (14) and (94) so that $[S, C_I^\dagger] = 0, \forall I$. Equations (106$a,b$) are equivalent linear sets of equations for the NCCM coefficients $\{\tilde{s}_I\}$, and use the $\{s_I\}$ coefficients as input. Formally, we may also solve (105) or, equivalently, (106a) or (106b), for the quantity $\langle \Phi_0 | \tilde{S}$. It is not difficult to prove the relation,

$$\langle \Phi_0 | \tilde{S} = \langle \Phi_0 | + \langle \Phi_0 | e^{-S} H e^S Q (E_0 - Q e^{-S} H e^S Q)^{-1} Q \ , \tag{107}$$

where, as in (9), the operator $Q \equiv \mathbb{1} - |\Phi_0\rangle\langle\Phi_0|$ is the projector out of our model space $\mathcal{M}$, which is now spanned by the single state $|\Phi_0\rangle$.

Hellmann–Feynman Theorem: We are now in a position to show that the NCCM bra-state parametrisation has the very important property that it is *derivable from the Hellmann–Feynman theorem* [102], in a sense which we will make clear below. The Hellmann–Feynman theorem, which is extremely simple to prove, states that if we perturb our Hamiltonian, $H \to H' = H + \lambda A$, where λ is an (infinitesimally) small quantity, such that the ground-state energy changes as $E_0 \to E_0' = E_0 + \lambda \mathrm{d}E_0/\mathrm{d}\lambda + O(\lambda^2)$, then

$$\mathrm{d}E_0/\mathrm{d}\lambda \equiv \bar{A} = \langle \Psi_0 | \mathrm{d}H/\mathrm{d}\lambda | \Psi_0 \rangle \ . \tag{108}$$

Now, instead of using the expectation value functional $\bar{H}$ to calculate E_0, let us use the functional

$$\mathcal{I}_H[S] \equiv \langle \Phi_0 | \mathrm{e}^{-S} H \mathrm{e}^{S} | \Phi_0 \rangle \ . \tag{109}$$

Equations (95) and (96) show that $\mathcal{I}_H[S] = E_0$ when S is calculated from the CCM equations (96). Let us now demand that the expectation value, $\bar{A}$, of the perturbing operator is calculated from $\mathcal{I}_H[S]$ by using the Hellmann–Feynman theorem. By making use of the relations $[S, C_I^\dagger] = 0$, and $\langle \Phi_0 | C_I^\dagger = 0$, we readily prove that

$$\begin{aligned} \bar{A} &= \frac{\mathrm{d}}{\mathrm{d}\lambda} \langle \Phi_0 | \mathrm{e}^{-S} (H + \lambda A) \mathrm{e}^{S} | \Phi_0 \rangle \\ &= \langle \Phi_0 | \mathrm{e}^{-S} A \mathrm{e}^{S} | \Phi_0 \rangle + \sum_I{}' \langle \Phi_0 | \mathrm{e}^{-S} H \mathrm{e}^{S} C_I^\dagger | \Phi_0 \rangle \frac{\delta s_I}{\delta \lambda} \ . \end{aligned} \tag{110}$$

Next, we calculate δs_I from the perturbed CCM equations (96) in which $H \to H + \lambda A$, $S \to S + \delta S$. By retaining only $O(\lambda, \delta S)$ terms, and using (96) to eliminate the unperturbed term, we readily find

$$\sum_J{}' \langle \Phi_0 | C_I (E_0 - \mathrm{e}^{-S} H \mathrm{e}^{S}) C_J^\dagger | \Phi_0 \rangle \frac{\delta s_J}{\delta \lambda} = \langle \Phi_0 | C_I \mathrm{e}^{-S} A \mathrm{e}^{S} | \Phi_0 \rangle \ ; \quad \forall I \neq 0 \ , \tag{111a}$$

where we have also made use of (95). By making use of the closure relation (18), we readily find that (111a) may be written in the equivalent form,

$$\sum_J{}' C_J^\dagger | \Phi_0 \rangle \frac{\delta s_J}{\delta \lambda} = Q (E_0 - Q \mathrm{e}^{-S} H \mathrm{e}^{S} Q)^{-1} Q \mathrm{e}^{-S} A \mathrm{e}^{S} | \Phi_0 \rangle \ . \tag{111b}$$

The use of (111b) in (110) then shows that $\bar{A}$ takes precisely the previously given form, $\bar{A} = \langle \Phi_0 | \tilde{S} \mathrm{e}^{-S} A \mathrm{e}^{S} | \Phi_0 \rangle$ with the operator $\tilde{S}$ as specified in (107).

Now, it is straightforward to show that each of equations (105) and (107) implies the other. Thus, from the above derivations we have shown that the use of the Hellmann–Feynman theorem with respect to the use of the functional $\mathcal{I}_H[S]$ to calculate E_0, and the use of the CCM equations (96) to calculate the linked-cluster correlation coefficients $\{s_I\}$, implies the NCCM bra-state parametrisation (101). Conversely, we note that no other bra-state parametrisation is compatible with the Hellmann–Feynman theorem in this way. In particular, using Hermiticity to relate $\langle\tilde{\Psi}_0|$ to $|\Psi_0\rangle$ will violate the theorem at anything less than an exact calculation. A different method of calculating ground-state expectation values $\bar{A}$ within the CCM by Kümmel [103] also violates the theorem.

Thouless [96] has pointed out that the Hellmann–Feynman theorem immediately implies that an expectation value $\bar{A}$ of an arbitrary operator A may be calculated diagrammatically from the same set of Goldstone diagrams as for the energy, $\bar{H}$, but in which each interaction potential is replaced in turn, one at a time, by the operator A. We shall see below that not only is the CCM parametrisation $\{S, \tilde{S}\}$ derivable from the Hellmann–Feynman theorem in the above sense, but all truncations are also consistent with it. Thus, if the CCM energy is calculated from $\bar{H}$ as defined in (104), then if the coefficients $\{s_I, \tilde{s}_I\}$ are truncated by retaining only some subset of them, and the resulting expression for $\bar{H}$ is expressed as the sum of some definite subset of Goldstone diagrams, then the average value $\bar{A}$ of an arbitrary operator, calculated by (104), corresponds to precisely the same subset of diagrams when each interaction is replaced in turn by the operator A.

We note that Monkhorst [104] was the first to formulate a CCM prescription for ground-state average values, $\bar{A}$, which is compatible with the Hellmann–Feynman theorem, by employing techniques of linear and higher-order response theory. It is interesting to note, however, that he never explicitly introduced a functional form for $\bar{A}$. Once supplemented [105] by the basis set effects that are needed for the practical evaluation of energy derivatives within quantum chemistry, the scheme of Monkhorst was successfully applied [106] to the prediction of vibrational spectra and to the location of transition states for decomposition reactions.

Static Variational Principle: When Arponen introduced the NCCM [67], it was basically a generalisation of the method of Monkhorst [104]. Importantly, he showed that the NCCM equations (96) and (106) for the coefficients $\{s_I, \tilde{s}_I\}$ can also be derived by requiring that the expectation value $\bar{H}$, defined as in (104), be stationary (i.e., $\delta\bar{H} = 0$) with respect to all variations in each of the *independent* variables $\{s_I, \tilde{s}_I\}$. We thus see very simply that the requirements $\delta\bar{H}/\delta\tilde{s}_I = 0$ and $\delta\bar{H}/\delta s_I = 0$ immediately yield the former equations (96) and (106*b*), respectively. We also note that at the stationary point, determined by the solutions to these NCCM equations, $\bar{H} \to \bar{H}|_{\text{stat}} = E_0$, where E_0 is given by (95), since at this stationary point equations (96) are fulfilled. The reader should beware, however, that although E_0 is thus derived from a variational (or, better, a bivariational) principle, the estimate obtained for it using a truncated set of amplitudes $\{s_I, \tilde{s}_I\}$ does *not* yield an upper bound, due to the loss of the Hermitian-adjoint relationship between the ground bra and ket states.

We may now also return to the Hellmann–Feynman theorem. If we perturb the Hamiltonian, $H \to H' = H + \lambda A$, as above, so that the new stationary point of $\bar{H}'$ is with NCCM parameters $\{s_I + \delta s_I, \tilde{s}_I + \delta \tilde{s}_I\}$, we readily see that at the perturbed stationary point,

$$\bar{H}'|_{\text{stat}} \approx \bar{H}|^0_{\text{stat}} + \sum_I{}' \left(\frac{\delta \bar{H}}{\delta s_I} \delta s_I + \frac{\delta \bar{H}}{\delta \tilde{s}_I} \delta \tilde{s}_I \right) \Bigg|^0_{\text{stat}} + \lambda \bar{A}|^0_{\text{stat}} \ , \tag{112}$$

correct to first-order changes, and where $\bar{A}|^0_{\text{stat}}$ now denotes the value of $\bar{A}$ at the unperturbed stationary point. Hence, since $\delta \bar{H}/\delta s_I = 0 = \delta \bar{H}/\delta \tilde{s}_I$ at the unperturbed stationary point, we immediately have that

$$\frac{\mathrm{d}}{\mathrm{d}\lambda} \langle H + \lambda A \rangle|_{\text{stat}} = \frac{\mathrm{d}E_0}{\mathrm{d}\lambda} = \bar{A}|^0_{\text{stat}} \ , \tag{113}$$

which is just the Hellmann–Feynman theorem. We see that the NCCM is thus fully consistent with (and derivable from) the Hellmann–Feynman theorem at any level of approximation by truncation to a subset of configurations $\{I\}$.

Near the stationary point of $\bar{H}$ we have, correct to terms of second order,

$$\bar{H} \approx E_0 + \sum_{I,J}{}' \left(\delta \tilde{s}_I E_{IJ} \delta s_J + \frac{1}{2} \delta s_I F_{IJ} \delta s_J \right) \ , \tag{114}$$

where

$$\begin{aligned} E_{IJ} &\equiv \delta^2 \bar{H} / \delta \tilde{s}_I \delta s_J |_{\text{stat}} \equiv (E^T)_{JI} \ , \\ F_{IJ} &\equiv \delta^2 \bar{H} / \delta s_I \delta s_J |_{\text{stat}} = F_{JI} \ . \end{aligned} \tag{115}$$

We note that there are no terms bilinear in the quantities $\{\delta \tilde{s}_I\}$ in (114), due to the linear way in which $\tilde{S}$ enters the expectation value (104). We may conveniently express (114) in the equivalent block-matrix form,

$$\bar{H} = E_0 + \frac{1}{2} (\delta x)^T \mathcal{H} \delta x + O(\delta x)^3 \ , \tag{116}$$

where

$$\delta x \equiv \begin{pmatrix} \delta s_I \\ \delta \tilde{s}_I \end{pmatrix} \ ; \quad \mathcal{H} \equiv \begin{pmatrix} F & E^T \\ E & 0 \end{pmatrix} = \mathcal{H}^T \ . \tag{117}$$

Linear Response to Static Perturbations: Many important physical parameters for real materials, such as polarisabilities and susceptibilities, cannot be calculated as ground-state expectation values of time-independent operators. An important tool in such cases is often *linear response theory.* Let us first consider *static linear response.* As above we subject the system Hamiltonian, to a small perturbation, $H \to H' = H + \lambda A$, and measure the consequent first-order change to the ground-state expectation value of another operator B,

$$\bar{B} \to \bar{B}' \equiv \bar{B} - \lambda R_{BA} + O(\lambda^2) \ , \tag{118}$$

where, by definition, R_{BA} is the associated *static response function*. Expanding about the stationary point of H, we have, correct to second order,

$$\bar{H}' \approx \bar{H}'|^0_{\text{stat}} + \frac{1}{2}(\delta x)^T \mathcal{H} \delta x + \lambda (\delta x)^T \frac{\delta \bar{A}}{\delta x} , \tag{119}$$

where $\delta \bar{A}/\delta x$ is the column vector formed from elements $\delta \bar{A}/\delta s_I$ and $\delta \bar{A}/\delta \tilde{s}_I$ evaluated at the stationary point of $\bar{H}$ alone. The stationarity condition, $\delta \bar{H}'/\delta x = 0$, for the perturbed Hamiltonian, then gives

$$\delta x = -\lambda \mathcal{H}^{-1} \left(\frac{\delta \bar{A}}{\delta x} \right) , \tag{120}$$

by making use of (119).

The average value of the operator B in the perturbed state is given by

$$\bar{B}(s_I + \delta s_I, \tilde{s}_I + \delta \tilde{s}_I) = \bar{B}(s_I, \tilde{s}_I) + \left(\frac{\delta \bar{B}}{\delta x} \right)^T \delta x + O(\lambda^2) . \tag{121}$$

Comparison of (118) and (121), together with (120), yields the result,

$$R_{BA} = \left(\frac{\delta \bar{B}}{\delta x} \right)^T \mathcal{H}^{-1} \left(\frac{\delta \bar{A}}{\delta x} \right) . \tag{122a}$$

By making use of the fact that $\mathcal{H}$ is a symmetric matrix, (122a) implies the symmetry relation $R_{BA} = R_{AB}$. Equation (122a) may also more explicitly be written as

$$R_{BA} = \sum_{I,J}{}' \left\{ (E^{-1})_{IJ} \left(\frac{\delta \bar{A}}{\delta s_I} \frac{\delta \bar{B}}{\delta \tilde{s}_J} + \frac{\delta \bar{B}}{\delta s_I} \frac{\delta \bar{A}}{\delta \tilde{s}_J} \right) - [(E^{-1})^T F E^{-1}]_{IJ} \frac{\delta \bar{A}}{\delta \tilde{s}_I} \frac{\delta \bar{B}}{\delta \tilde{s}_J} \right\} . \tag{122b}$$

We note that the perturbation λA often destroys a possible symmetry of the Hamiltonian. An example is the calculation of the density-density response function for a translationally-invariant system. In such cases the operators $\{S, \tilde{S}\}$ must be parametrised to allow for *symmetry breaking*. Thus, even though in the symmetric ground state the stationary values of those symmetry-breaking parameters will be zero, the second-order derivatives contained in $\mathcal{H}$ of $\bar{H}$ with respect to those parameters will generally not be zero. We also note that if the matrix $\mathcal{H}$ has an eigenvalue close to zero, the linear response to the perturbation A will be large. The approach of an eigenvalue to zero (i.e., a mode becoming soft), as some internal or external parameter is varied, often signifies the approach of a *phase transition* in the system, typically to a state which differs profoundly from the (naive vacuum or) current model state, $|\Phi_0\rangle$.

Dynamics and Dynamic Variational Principle: All of the above concepts relating to the stationary Schrödinger equation and static response may now be generalised to the dynamic (i.e., time-dependent) case. We define an action-like functional, $\mathcal{A}$,

$$\mathcal{A} = \mathcal{A}[|\Psi(t)\rangle; \langle\tilde{\Psi}(t)|] \equiv \int_{-\infty}^{\infty} \mathrm{d}t \, \langle\tilde{\Psi}(t)|(\mathrm{i}\partial/\partial t - H)|\Psi(t)\rangle \ . \tag{123}$$

The requirement that $\mathcal{A}$ be stationary, $\delta\mathcal{A} = 0$, with respect to arbitrary changes in $|\Psi(t)\rangle$ and $\langle\tilde{\Psi}(t)|$ independently is completely equivalent to the dynamic Schrödinger equations,

$$\delta\mathcal{A}/\delta\langle\tilde{\Psi}| = 0 \ \Rightarrow \ \mathrm{i}(\partial/\partial t)|\Psi(t)\rangle = H|\Psi(t)\rangle \ , \tag{124a}$$

$$\delta\mathcal{A}/\delta|\Psi\rangle = 0 \ \Rightarrow \ -\mathrm{i}(\partial/\partial t)\langle\tilde{\Psi}(t)| = \langle\tilde{\Psi}(t)|H \ . \tag{124b}$$

The time-dependent states are now parametrised in the NCCM, by complete analogy with their static counterparts in (93) and (101), as follows,

$$|\Psi(t)\rangle = \mathrm{e}^{k(t)}\mathrm{e}^{S(t)}|\Phi_0\rangle \ ; \ S(t) = \sum_I{}' s_I(t)C_I^{\dagger} \ , \tag{125a}$$

$$\langle\tilde{\Psi}(t)| = \mathrm{e}^{-k(t)}\langle\Phi_0|\tilde{S}(t)\mathrm{e}^{-S(t)} \ ; \ \tilde{S}(t) = 1\!\!1 + \sum_I{}' \tilde{s}_I(t)C_I \ . \tag{125b}$$

The c-number $k(t)$ is a necessary scale factor to describe the time-dependence of the amplitude of the wave functions, which remain manifestly normalised to each other at all times, however,

$$\langle\tilde{\Psi}(t)|\Psi(t)\rangle = \langle\Phi_0|\Phi_0\rangle = 1 \ , \ \forall t \ . \tag{126}$$

By making use of the relation (17) we may thus express $\mathcal{A}$ in the NCCM form,

$$\mathcal{A} = \int_{-\infty}^{\infty} \mathrm{d}t \left\{ \mathrm{i}\sum_I{}' \tilde{s}_I\dot{s}_I - \bar{H}(s_I, \tilde{s}_I) \right\} \ , \tag{127}$$

where $\dot{s}_I \equiv \mathrm{d}s_I/\mathrm{d}t$, and where $\bar{H}(s_I, \tilde{s}_I)$ is as specified in (104). Stationarity of $\mathcal{A}$ with respect to the complete set of NCCM parameters $\{s_I, \tilde{s}_I\}$ then re-expresses the *dynamic variational principle*, $\delta\mathcal{A} = 0$, in the form,

$$\frac{\delta\mathcal{A}}{\delta\tilde{s}_I} = 0 \ \Rightarrow \ \mathrm{i}\dot{s}_I = \frac{\delta\bar{H}}{\delta\tilde{s}_I} \ , \tag{128a}$$

$$\frac{\delta\mathcal{A}}{\delta s_I} = 0 \ \Rightarrow \ -\mathrm{i}\dot{\tilde{s}}_I = \frac{\delta\bar{H}}{\delta s_I} \ . \tag{128b}$$

We note that although the NCCM operators S and $\tilde{S}$ have seemingly entered the parametrisations (93) and (101), or (125a, b), in a very asymmetric fashion, their fundamental dynamics is now revealed in (128a, b) to be that of a *canonically conjugate pair*. Thus, $(s_I, \tilde{s}_I)$ are revealed to be canonically conjugate to each other in the usual terminology of classical Hamiltonian mechanics, in terms of a (classical) Hamiltonian functional $\bar{H} = \bar{H}(s_I, \tilde{s}_I)$. This conjugacy can be made absolutely precise by defining *generalised multi-configurational fields*, ϕ_I, and their *canonically conjugate multiconfigurational momenta*, π_I, as follows

$$\phi_I \equiv 2^{-\frac{1}{2}}(\tilde{s}_I + s_I) \ ; \quad \pi_I \equiv 2^{-\frac{1}{2}}\mathrm{i}(\tilde{s}_I - s_I) \ , \tag{129}$$

in terms of which the NCCM equations of motion ($128a, b$) can be written in the equivalent form,

$$\dot{\phi}_I = \delta\bar{H}/\delta\pi_I \ ; \quad \dot{\pi}_I = -\delta\bar{H}/\delta\phi_I \ , \tag{130}$$

which are precisely those of the classical Hamiltonian mechanics of a system described by a Hamiltonian $\bar{H} = \bar{H}(\phi_I, \pi_I)$.

By considering the dynamics of wave functions we have thus uncovered the (hitherto hidden) deep and fascinating result that the NCCM parametrisation represents an **exact** mapping of the general quantum many-body or field theory, described in terms of the complete set $\{|\Phi_0\rangle; C_I^\dagger\}$, onto the classical mechanics of a set of multiconfigurational classical fields $\{s_I, \tilde{s}_I\}$ interacting via the Hamiltonian, $\bar{H}(s_I, \tilde{s}_I) \equiv \langle\Phi_0|\tilde{S}\mathrm{e}^{-S}H\mathrm{e}^{S}|\Phi_0\rangle$. This classical Hamiltonian structure is now explored further below by considering the time-evolution of the expectation value of an arbitrary operator $A \in \mathcal{G}$.

Equation of Motion of an Arbitrary Expectation Value: We consider an arbitrary operator A with an expectation value, $\langle A\rangle = \bar{A} = \bar{A}(s_I, \tilde{s}_I; t)$, defined as in (104) but with the operators $S = S(t)$ and $\tilde{S} = \tilde{S}(t)$ now time-dependent. By making use of both the usual chain rule for partial differentiation and the equations of motion ($128a, b$), we readily prove that

$$\mathrm{d}\langle A\rangle/\mathrm{d}t = \langle\partial A/\partial t\rangle + \{\bar{A}, \bar{H}\} \ , \tag{131}$$

where the term $\langle\partial A/\partial t\rangle$ arises only from any intrinsic time-dependence of $A = A(t)$, and where we define a *generalised Poisson bracket*, $\{\bar{A}, \bar{B}\}$, as

$$\begin{aligned} \{\bar{A}, \bar{B}\} &= \frac{1}{\mathrm{i}}\sum_I{}' \left(\frac{\delta\bar{A}}{\delta s_I}\frac{\delta\bar{B}}{\delta\tilde{s}_I} - \frac{\delta\bar{B}}{\delta s_I}\frac{\delta\bar{A}}{\delta\tilde{s}_I}\right) \\ &= \sum_I{}' \left(\frac{\delta\bar{A}}{\delta\phi_I}\frac{\delta\bar{B}}{\delta\pi_I} - \frac{\delta\bar{B}}{\delta\phi_I}\frac{\delta\bar{A}}{\delta\pi_I}\right) \ . \end{aligned} \tag{132}$$

Thus, (131) represents the expected classical equation of motion in the canonical Hamiltonian formalism. We may complete the exploration of the classical Hamiltonian structure by considering the mapping of expectation values of commutators of operators in $\mathcal{G}$. We show below how these map onto precisely the generalised Poisson brackets (132) introduced above.

Expectation Values of Commutators: The NCCM expectation value (104) of the product of two operators may be written in the following form, by making use of the completeness relation (18),

$$\langle\tilde{\Psi}_0|AB|\Psi_0\rangle \equiv \overline{AB} = \sum_I \langle\Phi_0|\tilde{S}\mathrm{e}^{-S}A\mathrm{e}^{S}C_I^\dagger|\Phi_0\rangle\langle\Phi_0|C_I\mathrm{e}^{-S}B\mathrm{e}^{S}|\Phi_0\rangle \ . \quad (133)$$

By making use again of the definition (104) for expectation values, we readily prove the relations,

$$\delta\bar{A}/\delta s_I = \langle\Phi_0|\tilde{S}\mathrm{e}^{-S}(AC_I^\dagger - C_I^\dagger A)\mathrm{e}^{S}|\Phi_0\rangle \ ; \quad \forall I \neq 0 \ , \quad (134a)$$

$$\delta\bar{B}/\delta\tilde{s}_I = \langle\Phi_0|C_I\mathrm{e}^{-S}B\mathrm{e}^{S}|\Phi_0\rangle \ ; \quad \forall I \neq 0 \ . \quad (134b)$$

Equations (134*a*,*b*) may be used to rewrite (133) as follows,

$$\overline{AB} = \sum_I{}' \frac{\delta\bar{A}}{\delta s_I}\frac{\delta\bar{B}}{\delta\tilde{s}_I} + \sum_I \langle\Phi_0|\tilde{S}C_I^\dagger\mathrm{e}^{-S}A\mathrm{e}^{S}|\Phi_0\rangle\langle\Phi_0|C_I\mathrm{e}^{-S}B\mathrm{e}^{S}|\Phi_0\rangle \ . \quad (135)$$

By a judicious insertion of the completeness relation (18), we may rewrite (135) in the form

$$\overline{AB} = \sum_I{}' \frac{\delta\bar{A}}{\delta s_I}\frac{\delta\bar{B}}{\delta\tilde{s}_I} + \sum_{I,J} \langle\Phi_0|\tilde{S}C_I^\dagger C_J^\dagger|\Phi_0\rangle\langle\Phi_0|C_J\mathrm{e}^{-S}A\mathrm{e}^{S}|\Phi_0\rangle$$

$$\times\langle\Phi_0|C_I\mathrm{e}^{-S}B\mathrm{e}^{S}|\Phi_0\rangle \ . \quad (136)$$

We have rewritten the last term in (136) to show explicitly that it is symmetric under the interchange $A \rightleftharpoons B$, by making use of the commutation relation (14). Equation (136) thus yields the very important relation,

$$\langle\tilde{\Psi}_0|[A,B]|\Psi_0\rangle \equiv \overline{[A,B]} = \mathrm{i}\{\bar{A},\bar{B}\} \ . \quad (137)$$

In particular, we see that (131) and (137) are completely consistent with the usual quantum-mechanical Heisenberg equation of motion for an arbitrary operator $A \in \mathcal{G}$,

$$\frac{\mathrm{d}A}{\mathrm{d}t} = \frac{\partial A}{\partial t} + \frac{1}{\mathrm{i}}[A,H] \ . \quad (138)$$

Dynamic Linear Response: Let us now consider perturbing the ground state of our stationary system by a weak time-dependent perturbation, $H \to H'(t) = H + \lambda(t)A$, under which our ground-state NCCM correlation operators are perturbed as $S \to S'(t) = S + \delta S(t)$ and $\tilde{S} \to \tilde{S}'(t) = \tilde{S} + \delta\tilde{S}(t)$. Once again, we will work to leading order in λ and the associated small changes. If we now thus expand the equations of motion (128a,b) about the unperturbed ground-state stationary point, at which $\delta\bar{H}/\delta s_I = 0 = \delta\bar{H}/\delta\tilde{s}_I$, we readily find, in the same block-matrix notation introduced previously,

$$\mathrm{i}\frac{\mathrm{d}}{\mathrm{d}t}\begin{pmatrix}\delta s_I\\ -\delta\tilde{s}_I\end{pmatrix} \approx {\sum_J}' \begin{pmatrix}E_{IJ} & 0\\ F_{IJ} & E_{JI}\end{pmatrix}\begin{pmatrix}\delta s_J\\ \delta\tilde{s}_J\end{pmatrix} + \lambda(t)\begin{pmatrix}\delta\bar{A}/\delta\tilde{s}_I\\ \delta\bar{A}/\delta s_I\end{pmatrix} , \tag{139a}$$

where E_{IJ} and F_{IJ} are as defined in (115). In terms of the block column matrix δx and the block matrix $\mathcal{H}$ introduced in (117) above, we may rewrite (139a) in the equivalent form,

$$\mathrm{i}\frac{\mathrm{d}}{\mathrm{d}t}\delta x \approx J\mathcal{H}\delta x + \lambda(t)J\left(\frac{\delta\bar{A}}{\delta x}\right) , \tag{139b}$$

where we have introduced the block matrix, J,

$$J \equiv \begin{pmatrix}0 & \mathbb{1}\\ -\mathbb{1} & 0\end{pmatrix} . \tag{140}$$

In the linear response approximation, (139b) is thus a simple first-order differential equation for the column matrix $\delta x = \delta x(t)$. It is readily seen to have the following solution,

$$\delta x = -\mathrm{i}\int_{-\infty}^{t} \mathrm{d}t'\, \mathrm{e}^{-\mathrm{i}(t-t')J\mathcal{H}}\lambda(t')J\left(\frac{\delta\bar{A}}{\delta x}\right) , \tag{141}$$

where we have assumed $\lambda(t) \to 0$ as $t \to -\infty$.

As in the static case, to leading order, the change in expectation value $\bar{B}$ of any operator B, due to the perturbation $\lambda(t)A$, is as given by (121). The analogue of (118) for the dynamic case is now

$$\bar{B} \to \bar{B}'(t) \equiv \bar{B} - \int_{-\infty}^{t} \mathrm{d}t'\, R_{BA}(t-t')\lambda(t') + O(\lambda^2) , \tag{142}$$

where $R_{BA}(t)$ is the *dynamic response function.* A comparison of (141) and (142) gives the explicit solution,

$$R_{BA}(t) = \mathrm{i}\left(\frac{\delta\bar{B}}{\delta x}\right)^T \mathrm{e}^{-\mathrm{i}tJ\mathcal{H}}J\left(\frac{\delta\bar{A}}{\delta x}\right)\theta(t) , \tag{143}$$

where $\theta(x)$ is the usual unit-step function, equal to one if $x > 0$ and zero if $x < 0$. Equivalently, in the frequency domain, we may define the Fourier transform,

$$R_{BA}(\omega) = \int_0^\infty \mathrm{d}t\, \mathrm{e}^{\mathrm{i}\omega t} R_{BA}(t) \ . \tag{144}$$

So long as the *dynamic matrix*, H_D,

$$H_D \equiv J\mathcal{H} \ , \tag{145}$$

has no eigenvalues with a positive imaginary part (and see below), we may insert into (144) a factor $\mathrm{e}^{-\epsilon t}$ (or, equivalently, let $\omega \to \omega + i\epsilon$) with $\epsilon > 0$ a positive infinitesimal, to ensure convergence as $t \to \infty$, and thereby obtain the result

$$R_{BA}(\omega) = \left(\frac{\delta \bar{B}}{\delta x}\right)^T (J\mathcal{H} - \omega - \mathrm{i}\epsilon)^{-1} J \left(\frac{\delta \bar{A}}{\delta x}\right) \ . \tag{146}$$

A direct comparison of (122*a*) and (146) shows that $R_{BA}(\omega = 0) = R_{BA}$, viz., that the static limit is regained, as expected, in the limit of zero frequency. One may also show from (146) that $R_{BA}(\omega) = R_{AB}(\omega)$, although it is not entirely trivial to do so.

In view of the comment below (145), it is also interesting to note that the eigenvalues of the dynamic matrix H_D come in pairs, $\pm\omega_n$, as we now show. Since H_D is not Hermitian, we may solve either for the right or left eigenvectors, ζ_n or $\tilde{\zeta}_n$, respectively,

$$H_D \zeta_n = \omega_n \zeta_n \ ; \quad \tilde{\zeta}_n^T H_D = \omega_n \tilde{\zeta}_n^T \ . \tag{147}$$

By taking the transpose of the latter (left eigenvector) equation, and by premultiplying by the matrix J, we observe that

$$-J\mathcal{H}J\tilde{\zeta}_n = \omega_n J\tilde{\zeta}_n \ \Rightarrow \ H_D(J\tilde{\zeta}_n) = -\omega_n(J\tilde{\zeta}_n) \ , \tag{148}$$

where we have used the relations, $\mathcal{H}^T = \mathcal{H}$ and $J^T = -J$. Equation (148) thus shows that $J\tilde{\zeta}_n$ is a right eigenfunction with eigenvalue $-\omega_n$ if $\tilde{\zeta}_n^T$ is a left eigenfunction with eigenvalue $+\omega_n$. Our proof is thus complete. We note, furthermore, that the stationary point of $\bar{H}$ is therefore stable only if the dynamic matrix H_D has real eigenvalues. The appearance of complex eigenvalues as some internal or external parameter is varied through some critical value denotes the onset of an instability leading to a phase transition.

We note that such dynamic response functions, $R_{BA}(t)$, are extremely important in quantum many-body or quantum field theory. For example, for bosonic field theory, if A and B are single-particle field operators, $R_{BA}(t)$ becomes the *retarded single-particle Green's function*, from which one may also easily calculate the corresponding causal Green's function.

Excited States as Normal Modes of H_D: If we now consider the dynamics of the general system near its stationary ground state, $\delta \bar{H} = 0$, in the case where there is no external perturbation, small perturbations are described by (140) with $\lambda(t) \equiv 0$,

$$\mathrm{i}\frac{\mathrm{d}}{\mathrm{d}t}\delta x = H_D \delta x \ . \tag{149}$$

Just as in classical mechanics, we now seek *normal mode* solutions, $\delta x \to \zeta_n(\omega, t) = \zeta_n \mathrm{e}^{-\mathrm{i}\omega_n t}$, where ω_n and ζ_n are given by the eigenvalue equation (147). In more explicit block-matrix form, (147) may be written as follows,

$$\begin{pmatrix} E & 0 \\ -F & -E^T \end{pmatrix} \begin{pmatrix} \delta s_I \\ \delta \tilde{s}_I \end{pmatrix} = \omega_n \begin{pmatrix} \delta s_I \\ \delta \tilde{s}_I \end{pmatrix} \ . \tag{150}$$

Thus, the (positive) eigenvalues ω_n, which correspond to the excitation energies with respect to the ground-state energy, E_0, may be expressed purely in terms of the linked-cluster amplitudes $\{\delta s_I\}$,

$$\sum_J{}' \frac{\delta^2 \bar{H}}{\delta \tilde{s}_I \delta s_J} \delta s_J = \omega_n \delta s_I \ , \tag{151}$$

where we have made explicit use of the definition (115) of the matrix elements E_{IJ}. Using the definition (104) for the NCCM average value, $\bar{H}$, in (151) yields

$$\sum_J{}' \langle \Phi_0 | C_I \mathrm{e}^{-S} [H, C_J^\dagger] \mathrm{e}^S | \Phi_0 \rangle \delta s_J = \omega_n \delta s_I \ ; \ \forall I \neq 0 \ . \tag{152a}$$

By making use of the ground ket-state Schrödinger equation, (152a) may readily be rewritten in the equivalent form,

$$\sum_J{}' Q(\mathrm{e}^{-S} H \mathrm{e}^S - E_0) \delta s_J C_J^\dagger | \Phi_0 \rangle = \omega_n \sum_I{}' \delta s_I C_I^\dagger | \Phi_0 \rangle \ ; \ \forall I \neq 0 \ , \tag{152b}$$

where, as before, $Q \equiv 1\!\!1 - |\Phi_0\rangle\langle\Phi_0|$. Equation (152$b$) shows that the excitation energies, ω_n, of the system are obtained by diagonalisation of the same matrix, $Q(\mathrm{e}^{-S} H \mathrm{e}^S - E_0)Q$, whose inverse is required in the formal construction (107) of the NCCM bra-state operator $\tilde{S}$ from its ket-state counterpart S.

Ket Excited States: An alternative, but (as we shall see) equivalent, way of discussing excited states is via the time-independent Schrödinger equation,

$$H|\Psi_l\rangle - E_l|\Psi_l\rangle = (E_0 + \omega_l)|\Psi_l\rangle \ ; \ l \neq 0 \ . \tag{153}$$

The standard CCM description of excited states has been given, *inter alia*, by Emrich [107]. Strictly speaking, the NCCM parametrisation (93) describes not only the ground state of (62), but also any state with the same quantum numbers as the ground state and with nonzero overlap with the model state $|\Phi_0\rangle$. The possibility of obtaining multiple solutions to the nonlinear ground-state NCCM equations (96) has been discussed in some detail for a particular model case in Ref. [108].

For excited states (or, more generally, for states with zero overlap with $|\Phi_0\rangle$), we construct the respective ket wave functions $\{|\Psi_l\rangle\}$ in the NCCM in terms of a set of linear excitation operators $\{X^{(l)}\}$ which act on the corresponding exact ground-state ket wave function $|\Psi_0\rangle$,

$$|\Psi_l\rangle = X^{(l)}|\Psi_0\rangle = X^{(l)}\mathrm{e}^S|\Phi_0\rangle \ . \tag{154}$$

The operator $X^{(l)}$ is again decomposed wholly in terms of creation operators, $\{C_I^\dagger;\ I \neq 0\}$, with respect to $|\Phi_0\rangle$,

$$X^{(l)} = \sum_I{}' x_I^{(l)} C_I^\dagger \ . \tag{155}$$

Hence the operators $X^{(l)}$ and S commute; and the prime on the sum in (155) ensures that $\langle\Phi_0|\Psi_l\rangle = 0$. For extended systems with more than one phase, for example, the so-called ground-state parametrisation (93) will generally yield only the lowest state of a given symmetry imposed implicitly by the particular choice of model state. Indeed, phase transitions may be detected within this CCM description by observing, for fixed model state $|\Phi_0\rangle$, the onset of "excited" states of negative excitation energy (or, hence, "de-excited" states) from the so-called excited-state parametrisation (154)–(155), as some internal or external parameter is varied through some critical value [109].

By combining the ground- and excited-state Schrödinger equations (93) and (154), we readily find the NCCM eigenvalue equation,

$$\mathrm{e}^{-S}[H, X^{(l)}]\mathrm{e}^S|\Phi_0\rangle = \omega_l X^{(l)}|\Phi_0\rangle \ , \tag{156a}$$

or, equivalently,

$$(\mathrm{e}^{-S}H\mathrm{e}^S - E_0)X^{(l)}|\Phi_0\rangle = \omega_l X^{(l)}|\Phi_0\rangle \ , \tag{156b}$$

for the excitation energy, $\omega_l \equiv E_l - E_0$, directly. By taking the inner products of (156*b*) with each member of the set $\{C_I^\dagger|\Phi_0\rangle;\ I \neq 0\}$ we obtain a coupled set of homogeneous linear eigenvalue equations for the excited ket-state configuration coefficients $\{x_I^{(l)}\}$, which are identical to (152*a*). We note also that the left-hand side of (156*a*) may be developed as a nested commutator expansion, analogous to (97). In this way, we note that by making use of the commutativity relation, $[X^{(l)}, S] = 0$, the excited-state NCCM equations may, rather simply and very conveniently, be derived from their ground-state counterparts as follows. Thus, we replace each multinomial term in the coefficients $\{s_I\}$ arising from the expansion of the similarity transform in (94) with a corresponding set of terms in which each single coefficient s_J is replaced one at a time in turn by the corresponding factor $x_J^{(l)}$; and we simply drop the zeroth-order (inhomogeneous) term.

We also note that by projecting (156*b*) with the state $\langle\Phi_0|\tilde{S}$, and by making use of the bra-state Schrödinger equation (105), we readily prove the following relation,

$$\omega_l\langle\Phi_0|\tilde{S}X^{(l)}|\Phi_0\rangle = 0 \ . \tag{157}$$

Equation (157) immediately yields the required orthogonality relation,

$$\langle\tilde{\Psi}_0|\Psi_l\rangle = 0 \ , \ \text{unless } \omega_l = 0 \ . \tag{158}$$

Clearly, as $\omega_l \to 0$, the corresponding mode becomes soft. This mode may thus be easily excited into macroscopic occupation, and the corresponding ground state can undergo a phase transition into a state of markedly different character.

Hierarchical Approximation Schemes: In order to implement any of the above elements of the single-reference version of the NCCM, we need only to choose an appropriate model state or cyclic vector $|\Phi_0\rangle$, and then to approximate the corresponding sets of equations. The standard way to do this is to truncate the otherwise complete set of multiconfigurational set-indices $\{I\}$ to some appropriate finite or infinite subset. There are many ways of doing this, and in general one must be guided by the physics. One also needs in practice to develop hierarchical approximation schemes, in which at each increasing level one systematically incorporates more many-body correlations.

One of the simplest and intuitively most appealing such systematic hierarchical approximation schemes is the SUBn scheme for the ground state, or the more general SUB(m,n) scheme for excited states in the time-independent formalism. In the latter case, all amplitudes $\{x_I^{(l)}\}$ and $\{s_I\}$ which describe clusters of more than m and n particles (or particle/hole pairs in the case of number-conserving Fermi systems) respectively, are set to zero. The remaining equations, derived as described from (94) and (156) by taking their respective inner products with the wave functions $C_I^\dagger|\Phi_0\rangle$ of the configurations retained, are then solved without further approximation (except those introduced by the computational algorithms). In the static SUB(m,n) approximation the excitation energies $\{\omega_l\}$ are equivalently obtained by diagonalisation of the operator $Q(\mathrm{e}^{-S}H\mathrm{e}^{S} - E_0)Q$, where S is now the SUBn-approximated ground-state correlation operator, within some subspace of the multiconfigurational states defined by the truncation index m.

In general, there is no particular reason, at least, from within the static CCM formalism for ground and excited states, why the truncation indices m and n should not be chosen differently. Indeed, this freedom provides an additional degree of flexibility. Further work in this connection [110] has shown that by further embedding the theory of linear response within the CCM, each of the usual energy-weighted moment sum rules for the dynamic (liquid) structure function may, in principle, be exactly decomposed into an infinite cluster hierarchy of sub-sum-rules.

The above work [110] also provides a set of independent exact rules for the CCM cluster amplitudes, against which approximation schemes in both static ground- and excited-state formalisms can be tested for mutual compatibility. Furthermore, it provides a mechanism whereby any additional knowledge of the system (obtained either from experiment or from alternative theory or phenomenology) involving either the excitation spectrum or the ground-state correlations, may be used to extract information, from within the CCM, on the other. Finally, in this context, it is interesting to note that by making the simple approximation that the lowest members of the CCM sub-sum-rules are saturated by a single collective (or "giant resonance") state, we regain the important Bijl–Feynman relation [111] for the excitation spectrum in terms of the static structure function.

We note that for an infinite homogeneous system, translational invariance implies that the plane-wave single-particle basis is exact, and the ensuing momentum conservation then in turn implies that the one-body partition, S_1, of the correlation operator S, is identically zero. In such cases the lowest non-trivial SUBn approximation is the SUB2 scheme. This is already an extremely rich approximation. Amongst other terms it includes all of the two-body ladder diagrams and all of the two-body particle/hole ring diagrams [112].

We note that other CCM truncation schemes apart from the SUBn scheme have been used. In the case of pure hard-core potentials, for example, the SUBn scheme is not well defined, and for strongly repulsive (but not infinite hard-core) potentials the convergence is poor. A typical example of this case is the internucleon force in nuclear physics. The cure, however, is simple. It leads to the so-called (hard-core) HCSUBn scheme (otherwise known as the χ_n-truncation or Bochum truncation scheme). Thus, at a fixed ordinary SUBn level, one first identifies that subset of terms which when iterated together lead only to diagrams which are still contained in this original particular SUBn class when each bare interaction V is replaced by a ladder-summed G-matrix, and where the relative time orderings of the remaining interactions are kept fixed. The resulting HCSUBn approximation scheme [8,9,67] is closely related to the hole-line expansion of Bethe within the framework of time-independent perturbation theory.

For systems defined on a discrete spatial lattice, a localised approximation scheme called the LSUBn scheme has also been used to considerable advantage. In this scheme all possible many-body cluster configurations I are retained which occur in all different spatial locales of n contiguous sites on the lattice. Thus, to enumerate the fundamentally distinct LSUBn configurations, one must first evaluate the different *lattice animals* of size n which are inequivalent up to lattice symmetries (i.e., translations, rotations, and reflections) shared by the Hamiltonian. Then, on each such distinct lattice animal, one must consider all possible distinct sub-configurations (both connected and disconnected). Such LSUBn schemes have very successfully been applied to both quantum spin-lattice models (particularly of antiferromagnetism) [113–118] and lattice gauge theories [75–81].

All of the above schemes share the feature that the cluster amplitudes relating to configurations I not retained in the approximations are set to zero. For some purposes this may be too drastic an approximation. An alternative would be to attempt to approximate such amplitudes $\{s_I, \tilde{s}_I\}$ belonging to configurations I outside the basic subset retained by a given approximation, in terms of those retained. The invention and investigation of such schemes is at the forefront of current progress in the CCM [84].

3.6 Relationship of the CCM to TIPT and Goldstone Diagrams

For the sake of the simplest comparison with time-independent perturbation theory (TIPT) we consider a many-body Hamiltonian, $H = H_0 + V$, where H_0 is a one-body operator (e.g., kinetic energy) and V is a two-body interaction term. We also work in a single-particle basis, $|\alpha\rangle \equiv c_\alpha^\dagger|0\rangle$, in which H_0 is diagonal, so that in its second-quantised form,

$$H_0 = \sum_\alpha \epsilon_\alpha c_\alpha^\dagger c_\alpha \quad , \tag{159}$$

where $\{c_\alpha^\dagger\}$ are the generic single-particle creation operators (viz., the bosonic operators $b_\alpha^\dagger$ or the fermionic operators $a_\alpha^\dagger$ of Sect. 2). In the notation introduced in Sect. 2, we also work with a model state $|\Phi_0\rangle$ and a complete set of many-body states $C_I^\dagger|\Phi_0\rangle$ which are constructed to be eigenstates of H_0,

$$H_0|\Phi_0\rangle = \mathcal{E}_0|\Phi_0\rangle \quad ; \quad H_0 C_I^\dagger|\Phi_0\rangle = (\mathcal{E}_0 + e_I)C_I^\dagger|\Phi_0\rangle \quad . \tag{160}$$

Examples of such sets $\{|\Phi_0\rangle; C_I^\dagger\}$ have been given in Sects. 2.1–2.3.

We note that since H_0 is diagonal in the chosen basis, each term in (159) contains exactly one destruction operator with respect to $|\Phi_0\rangle$, since either one or other (but not both) of $c_\alpha^\dagger$ or c_α will destroy $|\Phi_0\rangle$. Thus, the nested commutator expansion (97) for $H \to H_0$ will terminate at the second term on the right-hand side,

$$\mathrm{e}^{-S} H_0 \mathrm{e}^{S} = H_0 + [H_0, S] \quad . \tag{161}$$

By making use of (160) and (161), we readily prove the following relation,

$$\langle\Phi_0|C_I \mathrm{e}^{-S} H_0 \mathrm{e}^{S}|\Phi_0\rangle = \mathcal{E}_0 \delta_{I0} + e_I s_I \quad , \tag{162}$$

where we have also made use of the properties (15) and (17) of the multiconfigurational creation operators $\{C_I^\dagger\}$. If we now insert (162) into the fundamental ket-state CCM equations (95) and (96), we find that the exact ground-state energy E_0 is given in terms of the unperturbed or model state energy $\mathcal{E}_0$ as

$$E_0 - \mathcal{E}_0 = \langle \Phi_0 | \mathrm{e}^{-S} V \mathrm{e}^{S} | \Phi_0 \rangle \ , \tag{163a}$$

in which the required ket-state cluster coefficients $\{s_I\}$ are given by the non-linear equations,

$$s_I = -\frac{1}{e_I} \langle \Phi_0 | C_I \mathrm{e}^{-S} V \mathrm{e}^{S} | \Phi_0 \rangle \ ; \ \forall I \neq 0 \ . \tag{163b}$$

It is now very straightforward to develop a diagrammatic representation of (163a, b) by complete analogy with the standard development of Goldstone (time-ordered) diagrams for TIPT (and see, e.g., Ref. [119]). Indeed, if one uses the nested commutator expansion (97) with $H \to V$ in (163b), and makes a straightforward iteration of the set of equations (163b) in powers of the potential V, one generates automatically the Goldstone diagrams for the linked-cluster amplitudes $\{s_I\}$. If no truncation is made, one generates in principle this way the complete set of diagrams. Alternatively, if the cluster operator S is truncated along the lines indicated above, a partial subset of diagrams is generated. Substitution of the resulting amplitudes into (163a) gives the corresponding expression for the energy in terms of a set of fully linked and closed (i.e., with no free ends) Goldstone diagrams.

We may equivalently derive (163b) and its counterpart for the bra-state cluster amplitudes $\{\tilde{s}_I\}$ from the static variational principle,

$$\frac{\delta \bar{H}}{\delta \tilde{s}_I} = 0 \ \Rightarrow \ s_I = -\frac{1}{e_I}\frac{\delta \bar{V}}{\delta \tilde{s}_I} = -\frac{1}{e_I} \langle \Phi_0 | C_I \mathrm{e}^{-S} V \mathrm{e}^{S} | \Phi_0 \rangle \ , \tag{164a}$$

$$\frac{\delta \bar{H}}{\delta s_I} = 0 \ \Rightarrow \ \tilde{s}_I = -\frac{1}{e_I}\frac{\delta \bar{V}}{\delta s_I} = -\frac{1}{e_I} \langle \Phi_0 | \tilde{S} \mathrm{e}^{-S} [V, C_I^\dagger] | \Phi_0 \rangle \ . \tag{164b}$$

Equations (164a, b) are precisely the *Dyson equations* for the NCCM amplitudes $\{s_I, \tilde{s}_I\}$, in which the unperturbed (or model) cluster excitation energies, e_I, appear in the familiar guise of the *energy denominators* of Rayleigh–Schrödinger perturbation theory. The iterative solutions of (164a, b) lead to a set of terms which can be put into one-to-one correspondence with classes of Goldstone diagrams. Whereas iteration of (164a) leads, as we have already discussed, automatically to a set of linked diagrams for the linked-cluster amplitudes $\{s_I\}$, iteration of (164b) is easily seen to lead to unlinked diagrams for the amplitudes $\{\tilde{s}_I\}$. Nevertheless, all expectation values $\bar{A} = \bar{A}(s_I, \tilde{s}_I)$ calculated by (104) for all such contributions from all diagrams for s_I and $\tilde{s}_I$ remain connected, as we have shown previously.

In the usual (physicist's) convention, one draws Goldstone diagrams with "time" axis running upwards (i.e., where matrix elements such as those in (164a, b) are read from right to left, and drawn, correspondingly, from bottom to top). Furthermore, by convention, $|\Phi_0\rangle$ plays the role of vacuum, so that only particle and hole lines which are excited from the model state are explicitly drawn. In this convention, the amplitude s_I is thus a "bottom amplitude" containing no legs from below and a set of particle/hole legs emanating from above corresponding to the configuration denoted by the set-index I. Correspondingly, $\tilde{s}_I$ is a "top amplitude".

Very related to the earlier discussion of the linked-cluster nature of the bottom amplitudes $\{s_I\}$ is the concept of *generalised time ordering* (GTO). This has traditionally been a rather useful formal device for classifying and combining certain classes of Goldstone diagrams [120]. It is based on a factorisation property of Goldstone diagrams containing disjoint (i.e., separately identifiable) sets I and J of particle/hole legs with unperturbed (or model) energies e_I, as given by (160). The simple identity

$$\frac{1}{e_I(e_I + e_J)} + \frac{1}{e_J(e_I + e_J)} = \frac{1}{e_I e_J} , \tag{165}$$

may then be used to factorise the individual energy denominators as on the right-hand side of (165), after the diagrams corresponding to all permitted time orderings which preserve the nature of the subclusters (i.e., do not change particles into holes, or vice versa) are combined together. It transpires [67,68] that the linked nature of the bottom amplitudes $\{s_I\}$ implies that the NCCM automatically generates diagrams (for the energy, for example) with such a GTO property in the "downward" direction (i.e., backwards in time). Conversely, the fact that the top amplitudes $\{\tilde{s}_I\}$ are generally unlinked implies the absence of a GTO property in the "upward" direction.

The upshot is that all of the diagrams for the energy obtained by iteration of the NCCM Dyson equations, as explained above, may thus be represented by what are termed *normal GTO tree diagrams* or NCCM trees [67]. Roughly speaking these are diagrams which "branch out" in the downward direction only (– and hence which represent the root system of a real tree, rather than its visible branch structure). It has been rather fully explained elsewhere [67] how each branch (– or root) of such a generalised diagram corresponds to a set of particle/hole lines labelled by a member of the complete set $\{I\}$. If such a (downward) closed NCCM tree diagram for the energy is divided into two open portions by cutting a single branch I, the bottom part constitutes a diagram for s_I, and will always itself be linked; whereas the top part constitutes a diagram for $\tilde{s}_I$, and may be unlinked.

We note again that whereas the NCCM equations *may* be iterated to yield terms corresponding to classes of Goldstone diagrams, we essentially never actually solve the equations this way in practice. In essence, the CCM performs a massive automatic resummation (to infinite order) of many (infinite) sets of Goldstone diagrams, and this is one of the underlying strengths of the method. By solving the equations as they stand (i.e., without iteration) at any level of truncation for the set of amplitudes $\{s_I, \tilde{s}_I\}$, we anticipate (and usually find) much better convergence properties in the hierarchical truncation scheme under consideration than the perturbation series.

Clearly, once we have made a particular choice of model state and multiconfigurational creation operators, and have chosen a particular truncation scheme, the practical implementation of the NCCM breaks into two tasks: (i) the algebraic evaluation of NCCM expectation values (104) and their functional derivatives; and (ii) the numerical solution of the resulting equations, e.g., (164a,b), for the truncated set $\{s_I, \tilde{s}_I\}$, in terms of which all physical quantities may be evaluated. We note that the underlying similarity transformation and the (usually) finite number of terms in the resulting nested commutator expansions (97), which lie at the heart of the CCM, make the first task well suited to the use of *computer-algebraic methods.* The interested reader is referred to Ref. [118], for example, for a thorough discussion of how this may be implemented very efficiently in practice for lattice Hamiltonian systems. The second task, namely the numerical solution of the equations so derived, normally presents no great difficulties, since the CCM equations are well-conditioned (unlike their CIM counterparts), and are also found in practice to be extremely robust in almost all applications.

3.7 The Multi-Reference Normal CCM

Physically, one might intuit that the single-reference version of the NCCM discussed above in Sects. 3.5 and 3.6 is most likely to be a good approach for "closed-shell" systems, for which a single model state $|\Phi_0\rangle$ suffices for a reasonable zeroth-order description. Conversely, we turn now to "open-shell" systems for which, *a priori*, a multi-reference approach appears to provide a more reasonable calculational framework. We argue by analogy with the single-reference ground-state NCCM approach discussed in Sect. 3.5. Thus, it is clear that the single-reference ground-state scheme may be viewed at a rather shallow level as simply providing, in any particular approximation scheme, a corresponding (and, evidently, a very sophisticated and clever) partial resummation of an infinite set of terms (or Goldstone diagrams) in the nondegenerate version of TIPT, as outlined in Sect. 3.6. As we have seen in Sect. 3.5, the basic CCM equations automatically embody the historically important linked cluster theorem of Goldstone [93] for the energy.

In the same spirit, Brandow [120] first showed that in the intermediate normalisation scheme there exists a similar linked diagram expansion, namely the *linked valence expansion,* in the degenerate version of TIPT. This was the first such formulation which was both size-extensive and size-consistent, where size-extensivity [52] is the property that the leading term in the energy of an N-particle system scales linearly with N, and where size-consistency [121] implies that a many-body wave function dissociates correctly into non-interacting fragments under infinite separation of the fragments. Brandow showed, in particular, how it was necessary for his formulation to be given in terms of a so-called *"complete model space."* This is defined to be one in which the multi-reference model space contained all possible N-body configurations (or Slater determinants for the usual number-conserving fermionic case to which the formalism is applied) that can be formed by distributing the valence particles among a selected set of valence orbitals. Brandow

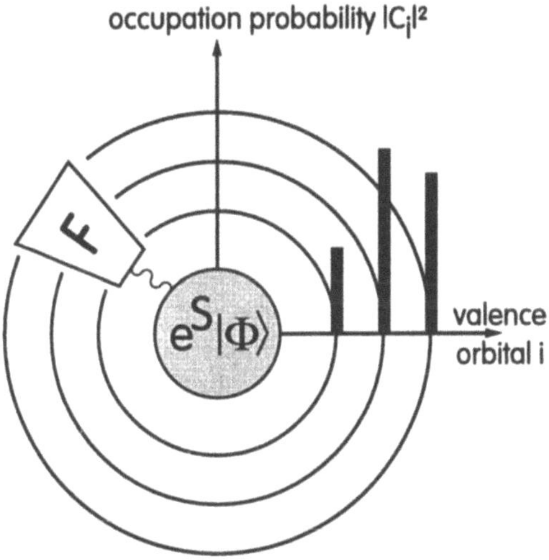

Fig. 1. A schematic representation of the essential elements of the multi-reference normal coupled cluster parametrisation of the ket-state energy eigenfunctions of an open-shell system.

further showed how his perturbative expansion could be expressed diagrammatically in terms of so-called *"folded diagrams."* Several authors have since described how to embody and extend this degenerate version of TIPT within the CCM framework, via a multi-reference approach. Perhaps the earliest within the physics (as opposed to chemistry) community were Kümmel and his co-workers [122]. Many other variants exist, most of which have been especially developed for use in quantum chemistry, but for present pedagogical purposes only this version is outlined below. Its essential ingredients are also indicated very schematically in Figure 1.

Thus, as in Sect. 3.1, let us consider for illustrative purposes, a closed-shell N-fermion system, whose CCM single-reference model state is now indicated as $|\Phi_0^N\rangle$, and whose exact ground-state energy is E_0^N. To this closed-shell system we imagine adding valence fermions (or holes) one at a time. In the notation of Sect. 2.3 we may now distinguish three sorts of single-fermion states $|\alpha_k\rangle$, namely: (i) orbitals occupied in $|\Phi_0^N\rangle$ (labelled $\alpha_k \to \nu_k$); (ii) valence orbitals (labelled $\alpha_k \to i_k$) partially occupied by the valence particles outside the core (and/or valence holes inside the core); and (iii) the remaining "unoccupied" orbitals (labelled $\alpha_k \to \rho_k$). The multi-reference CCM ansatz for the exact $(N+1)$-particle states is given as follows,

$$|\Psi_\lambda^{N+1}\rangle = \sum_{i\in\mathcal{V}} \mathrm{e}^S[1 + F^{(1)}]a_i^\dagger|\Phi_0^N\rangle C_i^\lambda \ , \tag{166}$$

where the correlation operator S is assumed known from the N-body "closed-shell" NCCM calculation, and where the sum on the index i runs over the set

$\mathcal{V}$ of valence orbitals considered as actually degenerate or quasi-degenerate. Thus, the states $\{a_i^\dagger|\Phi_0^N\rangle;\ i \in \mathcal{V}\}$ form a set of multi-reference $(N+1)$-body Slater determinants for the set of low-lying states (labelled λ) which we are attempting to construct.

Whereas the coefficients $\{C_i^\lambda\}$ determine the mixture of uncorrelated states in the multi-reference model space, the operator $F^{(1)}$ describes the dressing of the bare valence particle by its interaction with the core. Thus, we have the decomposition,

$$F^{(1)} = \sum_{n=1}^{N+1} F_n^{(1)} \ , \tag{167}$$

where, for example, $F_1^{(1)}$ describes the one-body (Hartree-Fock) part of the valence problem,

$$F_1^{(1)} = \sum_{\rho} \sum_{i \in \mathcal{V}} \langle \rho | F_1^{(1)} | i \rangle \, a_\rho^\dagger a_i \ , \tag{168a}$$

and $F_2^{(1)}$ describes the *"core polarisation"* terms which arise from the correlations between the valence particle and any one core particle,

$$F_2^{(1)} = \frac{1}{(2!)^2} \sum_{\eta_1,\eta_2} \sum_{\nu} \sum_{i \in \mathcal{V}} \langle \eta_1 \eta_2 | F_2^{(1)} | i \nu \rangle_A \, a_{\eta_1}^\dagger a_{\eta_2}^\dagger a_\nu a_i \ , \tag{168b}$$

where the labels η_k denote any extra-core state (i.e., valence or "unoccupied"), and hence run over i_k and ρ_k. We note that the CCM ansatz (166) is completely general, provided only that, just as in its single-reference counterpart (66*a*), the states $|\Psi_\lambda^{N+1}\rangle$ do not have zero inner product with all of the wave functions $\{a_i^\dagger|\Phi_0^N\rangle;\ i \in \mathcal{V}\}$ which now comprise the model space $\mathcal{M}$.

The corresponding ansatz for the two valence-particle $(N+2)$-fermion wave function is given as follows,

$$|\Psi_\mu^{N+2}\rangle = \sum_{i_1,i_2 \in \mathcal{V}} \mathrm{e}^S \left[1 + F^{(1)} + \frac{1}{2} : F^{(1)^2} : + F^{(2)} \right] a_{i_1}^\dagger a_{i_2}^\dagger |\Phi_0^N\rangle C_{i_1 i_2}^\mu \ , \tag{169}$$

where the factor of one-half in the quadratic $F^{(1)}$ term describing two "dressed" but uncorrelated valence particles is present, as usual, to avoid double-counting each excitation. We note that this term is also normal-ordered so as to avoid contractions (or links) between the two $F^{(1)}$ operators, since these are more properly contained in the genuine two-valence-particle-plus-core correlation operator, $F^{(2)}$,

$$F^{(2)} = \sum_{n=2}^{N+2} F_n^{(2)} \ . \tag{170}$$

If we proceed further in this fashion to add an arbitrary number of valence particles outside the core, we rapidly arrive at the normal-ordered exponential ansatz (where the normal ordering is always performed with respect to an appropriate closed-shell system or core) first written down explicitly by Lindgren [123], although the formulation of Ey [122] is completely equivalent.

By inserting the CCM parametrisations (166) and (169) into the respective $(N+1)$-body and $(N+2)$-body Schrödinger equations, and by premultiplying as usual by the factor e^{-S}, it is straightforward to derive equations for the energy eigenvalues E_λ^{N+1} and E_μ^{N+2}. Suitable projections onto the model space $\mathcal{M}$ thus lead to secular equations for the coefficients C_i^λ and $C_{i_1 i_2}^\mu$. These may be represented [8,9,122] as generalised eigenvalue equations for fully-linked one- and two-body *effective Hamiltonians,* respectively, which yield the *folded diagrams* of degenerate many-body perturbation theory. The corresponding eigenvalues are the respective excitation energies; for example, in the one-valence case, $\epsilon_\lambda \equiv E_\lambda^{N+1} - E_0^N$. Similarly, by projecting out of $\mathcal{M}$ onto "unoccupied" states, one derives the equations which determine the matrix elements of the operators $F^{(1)}$ and $F^{(2)}$. The interested reader is referred to the literature cited above, and to the article in the present volume by Kaldor, for further details.

In the original multi-reference formulation, the CCM and the associated linked diagrammatic expansions were restricted to complete model spaces and to the intermediate normalisation scheme. However, complete model spaces have prohibitively large dimensionalities, even when the number of active valence orbitals is restricted to relatively few. Mukherjee [124] first showed that for general incomplete model spaces, the condition that both the CCM cluster operators and the effective Hamiltonians are connected is normally incompatible with our previous choice of intermediate normalisation. Conversely, by abandoning this normalisation, these connectivity properties may be reinstated. A more detailed discussion of these points would, however, take us too far afield for present purposes, and the interested reader is referred to the literature [125].

3.8 Formal Elements of the Extended CCM

In the NCCM discussed so far, while all ground-state expectation values (104), $\bar{A} = \bar{A}(s_I, \tilde{s}_I)$, are linked-cluster quantities, and while the bottom amplitudes $\{s_I\}$ are also fully linked, the top amplitudes $\{\tilde{s}_I\}$ contain unlinked terms. For some purposes we might wish all of our basic amplitudes to be linked, and this is the basic motivation for the development of the *extended coupled cluster method.* (ECCM). We note firstly that the relations (14), (93), and (104) immediately imply that the NCCM top amplitudes are simply the ground-state expectation values of the multiconfigurational creation operators,

$$\tilde{s}_I = \langle C_I^\dagger \rangle \ . \tag{171}$$

If we now define the corresponding linked or connected quantities as

$$\tilde{\sigma}_I \equiv \langle C_I^\dagger \rangle_{\text{linked}} \equiv \langle C_I^\dagger \rangle_{\text{c}} \ , \tag{172}$$

our earlier discussion in Sect. 3.4 on general cumulant expansions immediately yields the basic ECCM parametrisation,

$$\tilde{S} = \mathrm{e}^{\tilde{\Sigma}} \; ; \quad \tilde{\Sigma} = \sum_I{}' \tilde{\sigma}_I C_I \; . \tag{173}$$

Although the cluster amplitudes $\{s_I, \tilde{\sigma}_I\}$ are a complete set of ECCM linked quantities, it is very convenient to define a new set of bottom ECCM amplitudes, $\sigma_I = \sigma_I(s_J, \tilde{\sigma}_J)$, such that the ECCM amplitudes $(\sigma_I, \tilde{\sigma}_I)$ again form a canonically conjugate pair, by complete analogy with the canonical conjugacy of the NCCM pairs $(s_I, \tilde{s}_I)$ expressed by (128a,b). This is very easy to achieve by writing the action-like functional (123) in the form,

$$\begin{aligned} \mathcal{A} &= \int_{-\infty}^{\infty} \mathrm{d}t \, \langle \Phi_0 | \mathrm{e}^{\tilde{\Sigma}(t)} \mathrm{e}^{-S(t)} (\mathrm{i}\partial/\partial t - H) \mathrm{e}^{S(t)} | \Phi_0 \rangle \\ &= \int_{-\infty}^{\infty} \mathrm{d}t \, \left[-\mathrm{i} \langle \Phi_0 | \dot{\tilde{\Sigma}}(t) \mathrm{e}^{\tilde{\Sigma}(t)} S(t) | \Phi_0 \rangle - \bar{H} \right] \; , \end{aligned} \tag{174}$$

where, in the second line of (174), we have integrated by parts. Insertion of a complete set (18) of multiconfigurational states into (174) then yields the ECCM form of the action-like functional,

$$\begin{aligned} \mathcal{A} &= \int_{-\infty}^{\infty} \mathrm{d}t \left\{ -\mathrm{i} \sum_I{}' \dot{\tilde{\sigma}}_I \sigma_I - \bar{H}(\sigma_I, \tilde{\sigma}_I) \right\} \\ &= \int_{-\infty}^{\infty} \mathrm{d}t \left\{ \mathrm{i} \sum_I{}' \tilde{\sigma}_I \dot{\sigma}_I - \bar{H}(\sigma_I, \tilde{\sigma}_I) \right\} \; , \end{aligned} \tag{175}$$

where the new ECCM bottom operator Σ is defined as

$$\Sigma | \Phi_0 \rangle \equiv Q \mathrm{e}^{\tilde{\Sigma}} S | \Phi_0 \rangle \; ; \quad \Sigma \equiv \sum_I{}' \sigma_I C_I^{\dagger} \; , \tag{176a}$$

in terms of $Q \equiv \mathbb{1} - |\Phi_0\rangle\langle\Phi_0|$, and amplitudes $\{\sigma_I\}$ given as follows,

$$\sigma_I = \langle \Phi_0 | C_I \mathrm{e}^{\tilde{\Sigma}} S | \Phi_0 \rangle \;\; \Leftrightarrow \;\; s_I = \langle \Phi_0 | C_I \mathrm{e}^{\tilde{\Sigma}} \Sigma | \Phi_0 \rangle \; . \tag{176b}$$

It is evident from the definition (176b) that the ECCM bottom amplitudes σ_I are fully linked, since S is a linked-cluster operator. Furthermore, stationarity of $\mathcal{A}$ from (175) now shows that the ECCM amplitudes $\{\sigma_I, \tilde{\sigma}_I\}$ form canonically conjugate pairs,

$$\frac{\delta \mathcal{A}}{\delta \tilde{\sigma}_I} = 0 \;\Rightarrow\; \mathrm{i}\dot{\sigma}_I = \frac{\delta \bar{H}}{\delta \tilde{\sigma}_I} \; , \tag{177a}$$

$$\frac{\delta \mathcal{A}}{\delta \sigma_I} = 0 \;\Rightarrow\; -\mathrm{i}\dot{\tilde{\sigma}}_I = \frac{\delta \bar{H}}{\delta \sigma_I} \; , \tag{177b}$$

in complete analogy with the canonical conjugacy of the NCCM pairs $\{s_I, \tilde{s}_I\}$ expressed via (128a,b).

Although we shall not go through the details here, we note that many of the NCCM results of Sect. 3.5 may now be generalised to the ECCM [67,68,101,126]. Many of the key distinguishing features of the ECCM follow from its *double* exponential structure and hence also its double similarity transform structure,

$$\begin{aligned}\bar{A} \equiv \langle\tilde{\Psi}_0|A|\Psi_0\rangle &= \langle\Phi_0|\mathrm{e}^{\tilde{\Sigma}}\mathrm{e}^{-S}A\mathrm{e}^{S}|\Phi_0\rangle \\ &= \langle\Phi_0|\mathrm{e}^{\tilde{\Sigma}}\mathrm{e}^{-S}A\mathrm{e}^{S}\mathrm{e}^{-\tilde{\Sigma}}|\Phi_0\rangle \\ &\equiv \bar{A}(\sigma_I, \tilde{\sigma}_I) \ . \qquad (178)\end{aligned}$$

In turn, these features lead to a well-defined *double linking* property of the corresponding ECCM diagrams [67]. By contrast to the CIM and NCCM, *all* of the basic ECCM amplitudes $\{\sigma_I, \tilde{\sigma}_I\}$ that now completely characterise the many-body system are linked-cluster quantities. In turn, they all therefore obey the important *cluster property,* namely that they approach zero asymptotically as any subset of the particles described by the configuration set-index I becomes far removed from the remainder. The entire system may thus be parametrised in terms of a complete set of ECCM multilocal, classical (i.e., c-number) amplitudes. Just as we have shown above for the NCCM, we can also show [101] how the ECCM maps an arbitrary quantum many-body or field-theoretic system with underlying Schrödinger dynamics onto a classical field theory.

Extensions of the ECCM have been made [68,126] to consider both excited states and general dynamical behaviour. It has thus been demonstrated how the amplitudes $\{\sigma_I, \tilde{\sigma}_I\}$ may also be viewed as *generalised (many-body) mean fields* or *quasilocal order parameters,* by considering their small-amplitude oscillations around a stationary equilibrium point, just as we did for the NCCM in Sect. 3.5. The overall structure of the ECCM is indicated very schematically in Figure 2, and the interested reader is referred to the literature [67–69,90,101,126–132] for further details.

We note that, to the best of our knowledge, the ECCM is unique as a formulation of quantum many-body/field theory in which *every* fundamental amplitude exactly obeys the cluster property at all levels of approximation. It is clear that only such formulations have the possibility to describe both the *local* properties of many-body systems and such *global* properties as their phase transitions, states of topological excitation or deformation, spontaneous symmetry breaking, general nonequilibrium behaviour, and nonlinear (large-amplitude) response.

3.9 Other Aspects of Coupled Cluster Theory

The range of individual components which are now contained under the general umbrella of coupled cluster theory or the CCM is large and varied, as

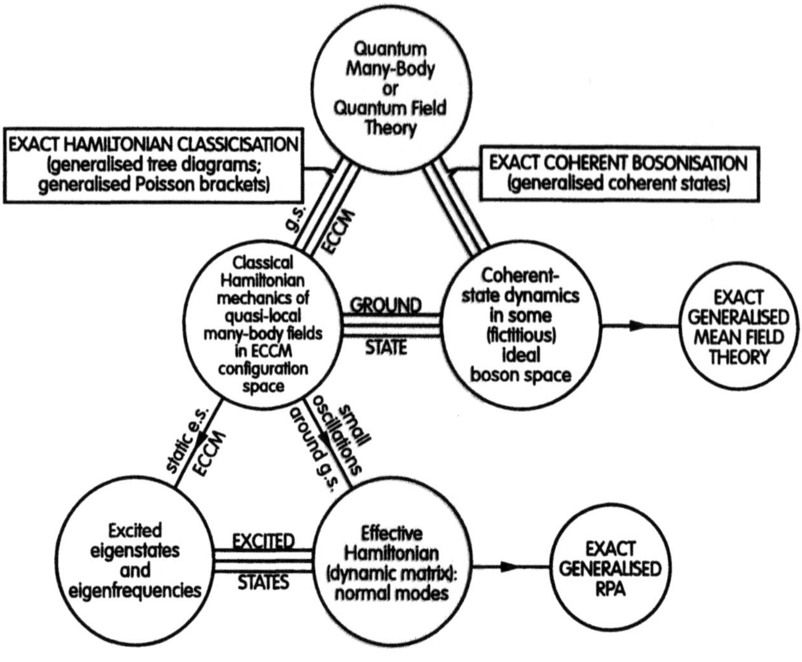

Fig. 2. A schematic representation of the hierarchical structure and the most general features of the ECCM.

we have attempted to describe above. In Figure 3 we have also attempted to make a very schematic overview of the hierarchical structure of the main elements of the CCM. We warn the reader that the schema shown is neither unique in its decomposition nor wholly rigorous or complete in the imposition of a logical structure on the resulting components. Nevertheless, we hope that Fig. 3 will serve, at least, as an *aide memoire* in surveying the formal aspects of what is, by now, a very wide-ranging formulation of microscopic quantum many-body theory.

We have already described, at least in outline, most of the CCM elements displayed in Fig. 3. We complete this Section with a very brief discussion of some additional items of coupled cluster theory, which largely reflect the work of the last few years or are the topics of current research.

CCM Parametrisation of Mixed States: Our entire discussion of the CCM up until this point has involved the parametrisation of quantum-mechanical pure states and, hence, is of relevance strictly to many-body systems at zero temperature. We note that an extension of the CCM to thermal mixed states, of relevance to systems at nonzero temperatures, was first given some ten years ago by Kümmel and his co-workers [133], in terms of the Bloch equation for the *statistical density operator*, $\hat{\rho} \equiv \mathrm{e}^{-\beta H} / \mathrm{Tr}(\mathrm{e}^{-\beta H})$,

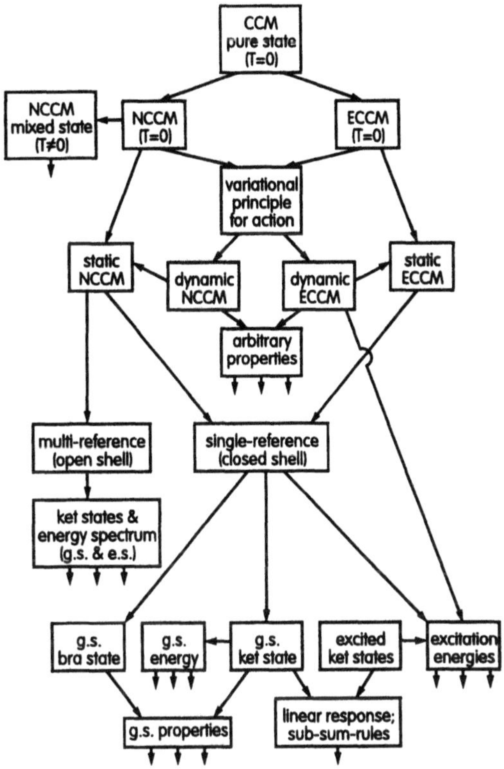

Fig. 3. A schematic diagram of the main ingredients of the CCM and their hierarchical structure.

where $\beta \equiv (k_{\mathrm{B}}T)^{-1}$ and T is the temperature. More recently, Mukherjee and his co-workers [134] have given a very natural and direct extension of the CCM to mixed states. In particular, they have provided a thermal cluster cumulant method for the grand partition function, $Z \equiv \mathrm{Tr}(\mathrm{e}^{-\beta(H-\mu\hat{N})})$, where μ is the chemical potential and $\hat{N}$ is the number operator. Unlike the earlier approach of Kümmel and his co-workers, the latter approach does not require a knowledge of the eigenspectrum of the many-body Hamiltonian H. For further details the reader is referred to the contribution by Mukherjee to the present volume.

Hermiticity-Preserving Versions of the CCM: As we have seen, both of the standard versions of coupled cluster theory (i.e., both NCCM and ECCM) are non-Hermitian by nature. The canonical coordinates, $\{s_I, \tilde{s}_I\}$ and $\{\sigma_I, \tilde{\sigma}_I\}$, which are intrinsically complex-valued for arbitrary wave functions, bear no simple relation to one another. In truncations to some subset of all available configurations $\{I\}$ the Hermiticity may be badly broken. Very recently, Arponen [135] has complemented both versions of the CCM by introducing the complex conjugates of the cluster amplitudes. The resulting extended phase spaces are complex manifolds of too large a dimensionality. Arponen has shown how a set of *constraint functions* may be introduced so as to return to the *physical submanifold* (or reduced phase space, or constraint surface) which corresponds precisely to the original Hilbert or Fock space, $\mathcal{G}$. He thereby eliminates the extra degrees of freedom, by the Dirac bracket method, leaving as independent coordinates a minimal set of (eventually real) multiconfigurational amplitudes. These intrinsically Hermitian versions of the NCCM and ECCM hold out the promise of making practicable Hermiticity-preserving truncations available for practical applications.

General Fermionic CCM: Most many-body Hamiltonians H contain only products of an even number of fermionic configuration operators multiplied by ordinary (complex) numbers. The same is true for the multiconfigurational creation operators $\{C_I^\dagger\}$ considered up to this point. An example is provided by Sect. 2.3. However, for the purposes of calculating odd fermionic quantities such as Green's functions, correlation functions, or excited states, it is often convenient to break the fermionic parity of H, e.g., by the addition of *source terms* of the form $\sum_\alpha(\tilde{\mu}_\alpha a_\alpha + a_\alpha^\dagger \mu_\alpha)$. In such cases it is necessary to enlarge the commutation rules (14) to *graded commutation rules,* and to enlarge the corresponding set of CCM amplitudes, e.g., $\{s_I, \tilde{s}_I\}$ and $\{\sigma_I, \tilde{\sigma}_I\}$ from ordinary c-numbers to (complex) *Grassmann numbers* of both even and odd parity [119,136]. The procedure to do so is relatively straightforward, as has been demonstrated by Arponen [131].

Functional Form of the CCM: As we have seen above, the standard CCM parametrisations are operatorial in nature, where the ground-state wave functions, for example, are described in terms of correlation operators $(S, \tilde{S})$ or $(\Sigma, \tilde{\Sigma})$. However, as we have already mentioned in Sect. 1, similar functional forms of the CCM have also been employed recently [75–80,82], in which the ground-state ket wave function, for example, is specified in the exponential form, $\langle\{r\}|\Psi_0\rangle = \exp[S(\{r\})]$, where $|\{r\}\rangle$ is some suitably chosen complete set of wave functions. At corresponding levels of truncation the functional and operatorial versions are not identical, as has been discussed in detail in Ref. [82], to which the interested reader is referred for further details.

4 Applications of the CCM

Although our aim in the present article has been to review the general aspects of the coupled cluster method itself, we close with a very brief summary of some of the more important applications of the method that have so far been made. For further details of each application we refer the interested reader to the references cited in each case, as well as to other reviews of the CCM [8,9,66–69,76,101,110,112,118,126,131,132,135,137–142].

Without any attempt to be exhaustive, we note that some of the most important applications of the CCM have been to the following systems:

- **Atomic Nuclei:** The NCCM has very successfully been applied to the ground and excited states of both closed-shell nuclei (e.g., ^{4}He, ^{16}O, ^{40}Ca) and to the open-shell nuclei (e.g., ^{15}N, ^{17}O, ^{14}C, ^{18}O, ^{18}F) formed from the addition to the closed-shell nuclei of one or two valence particles or holes. Calculations have been performed [8,9,70–75,100,107,122,143] for a variety of phenomenological two-body (and three-body) internucleon potentials, using the HCSUBn truncation scheme essentially up to the $n = 4$ level. Numerical convergence has been demonstrated at this level; and no other *ab initio* technique has ever bettered (or even equalled) these results, most of which are now about twenty years old.
- **Nuclear Matter:** Very similar calculations have also been performed [144] for nuclear matter, with similarly converged results for the binding energy per nucleon and the saturation density, for a range of phenomenological internucleonic forces.
- **Atoms and Molecules:** The CCM has become the method of choice in quantum chemistry wherever high accuracy is required. A huge range of calculations on such quantities as electron affinities, excitation energies, and ionisation potentials has been performed (see, e.g., Refs. [10,52,99, 105,106,123–125,137,139,140]) for a wide variety of atoms and molecules with up to about 100 active electrons. Especially for the heavier atoms, calculations have been performed with relativistic and QED corrections incorporated. The interested reader is referred also to recent reviews [145], and to the article by Kaldor in the present volume.[2]
- **The Electron Gas:** The NCCM has been applied to the electron gas in the high-density (RPA) limit, in the metallic density range, and in the low-density (Wigner crystal) regime. Results [112,146,147] in the important metallic density range for the correlation energy, for example, are accurate to better than 1 mH per electron (or $< 1\%$) by comparison to the essentially exact GFMC results [148] of Ceperley and Alder. This accuracy has never been equalled in any other *ab initio* calculation of what is still one of the best studied of all quantum many-body systems.
- **Charged Impurity in a Polarisable Medium:** The technique of allowing low-energy positrons to annihilate inside metals, alloys, and other forms of condensed matter has become an important experimental tool. Such systems comprising a charged impurity in a polarisable medium are prototypes of field-theoretical one-body problems. By casting the polarisation degrees of freedom as *internal gauge fields*, it has been shown [128]

[2] Issues 2–6 of Theoretica Chimica Acta **80** (1991) and a forthcoming issue of Molecular Physics (1998) are also wholly devoted to articles on coupled cluster theory and its applications.

how the ECCM can provide a powerful microscopic description of such systems.

- **Quantum Fluid Mechanics:** By coupling to *external gauge fields* (viz., the U(1) scalar and vector potentials), which serve to set the system into its most general state of motion, it has been shown [127] how the ECCM can provide a very complete, microscopic description of the zero-temperature quantum hydrodynamics of a strongly-interacting condensed Bose fluid.
- **Quantum Spin Lattice Models:** Extensive and very successful application of the CCM have been made in recent years to a wide variety of quantum spin lattice models. Examples include the solid phases of ^{3}He [149]; the isotropic (Heisenberg) and anisotropic XXZ models in one dimension and on the two-dimensional square lattice, both for spin-half systems [113–115,150] and higher-spin systems [116,151]; the spin-one Heisenberg-biquadratic model [152]; and to such *frustrated* spin models as the spin-half J_1-J_2 model [92,153], and the two-dimensional (anisotropic) Heisenberg antiferromagnet on a triangular lattice [117,118]. The CCM has also been applied to the spin-one easy-plane ferromagnet [154].
- **Electron Lattice Models:** The CCM has been applied [155] to such lattice models of strongly interacting electrons as the Hubbard model, both precisely at half-filling and with the further removal of one or two electrons.
- **Anharmonic Oscillators:** Both the NCCM and ECCM have been intensively applied to anharmonic oscillator systems [69,90,108,130–132,156] as $(0+1)$-dimensional models of nonlinear quantum field theories. The structure of the NCCM and ECCM has been particularly explored by applications to this model of the powerful holomorphic (or Bargmann) representation. Analogous anharmonic spin models have also been studied [90].
- **Continuum Quantum Field Theory:** Kümmel and his co-workers have very successfully applied the NCCM to several continuum quantum field theories. These include: (i) Φ^4 field theory [157] in $1+1$ (and $2+1$) dimensions, in the vacuum (ground state) and one-particle (mass gap) sectors, and in the soliton sector; (ii) $\Phi^4 + \Phi^6$ field theory [158] in $1+1$ dimensions, in the vacuum, one- and two-particle sectors; (iii) the sine-Gordon model in $1+1$ dimensions [159] in the vacuum, one-particle, and soliton sectors; and (iv) the vacuum, one-particle (physical nucleon mass), and two-particle (deuteron) sectors of a model $(3+1)$-dimensional field theory of pions and nucleons [160] interacting via the usual isospin-invariant pseudoscalar coupling. We also note that a preliminary application of the ECCM to quantum field theory has also been made [161], using the sine-Gordon model as an example.
- **Lattice Quantum Field Theory:** As discussed in Sect. 1, the NCCM has been widely applied to various lattice gauge field theories. These

include the discrete Z(2) model [75], and such continuous groups as the Abelian U(1) case [75-80] and the non-Abelian SU(2) case [81]. Both the operatorial and functional forms of the NCCM have also been applied very recently [82] to a latticised O(4) nonlinear sigma model of chiral meson field theory.

- **Quantum Optics:** A very recent application of the NCCM has been made to one of the prototypical models of quantum optics, namely the Rabi Hamiltonian (or Jaynes–Cummings model without the rotating wave approximation) [162].
- **Solid-State Optoelectronics:** A similar application of the NCCM has been made to the linear *E-e* Jahn–Teller effect [163].
- **Other Model Problems:** Various other applications of the CCM have been made to such important model problems as the Lipkin–Meshkov–Glick model referred to [89] in Sect. 2.4 [67,107,129,164]; and the one-dimensional Lieb model [86,87] of bosons interacting via repulsive delta-function potentials [84].

References

1. F. Coester, Nucl. Phys. **7**, 421 (1958).
2. F. Coester and H. Kümmel, Nucl. Phys. **17**, 477 (1960).
3. B.D. Day, Rev. Mod. Phys. **50**, 495 (1978); Nucl. Phys. **A328**, 1 (1979).
4. C. Mahaux, Nucl. Phys. **A328**, 24 (1979).
5. B.D. Day, in *From Nuclei to Particles*, edited by A. Molinari (North-Holland, Amsterdam, 1981).
6. D. Bohm and D. Pines, Phys. Rev. **82**, 625 (1951); **92**, 609 (1953); D. Pines, *ibid.* **92**, 626 (1953).
7. M. Gell-Mann and K.A. Brueckner, Phys. Rev. **106**, 364 (1957).
8. H. Kümmel, K.H. Lührmann, and J.G. Zabolitzky, Phys. Rep. **36C**, 1 (1978).
9. H.G. Kümmel, in *Nucleon-Nucleon Interaction and Nuclear Many-Body Problems,* edited by S.S. Wu and T.T.S. Kuo (World Scientific, Singapore, 1984), p. 46.
10. V. Kvasnička, V. Laurinc, and S. Biskupič, Chem. Phys. Lett. **73**, 81 (1979); M.F. Guest and S. Wilson, *ibid.* **73**, 607 (1980); M.J. Frisch, R. Krishnan, and J.A. Pople, *ibid.* **75**, 66 (1980); R.J. Bartlett, H. Sekino, and G.D. Purvis, *ibid.* **98**, 66 (1983).
11. J. Linderberg and Y. Öhrn, *Propagators in Quantum Chemistry* (Academic, London, 1973).
12. Y. Öhrn, in *Aspects of Many-Body Effects in Molecules and Extended Systems* (Lecture Notes in Chemistry, Vol. **50**), edited by D. Mukherjee (Springer, Berlin, 1989), p. 187; J. Oddershede and P.W. Sengeløv, in *ibid.*, p. 207.
13. I.T. Diatlov, V.V. Sudakhov, and K.A. Ter-Matirosian, Sov. Phys.–JETP **5**, 631 (1957).
14. R. Roulet, J. Gavoret, and P. Nozières, Phys. Rev. **178**, 1072 (1969).
15. G. Ripka, Phys. Rep. **56**, 1 (1979).
16. A.D. Jackson, A. Lande, and R.A. Smith, Phys. Rep. **86**, 55 (1982).

17. A.D. Jackson, A. Lande, and R.A. Smith, Phys. Rev. Lett. **54**, 1469 (1985); E. Krotscheck, R.A. Smith, and A.D. Jackson, Phys. Rev. A **33**, 3535 (1986).
18. A. Lande and R.A. Smith, Phys. Lett. **131B**, 253 (1983).
19. R.A. Smith and A.D. Jackson, in *Recent Progress in Many-Body Theories*, Vol. **1**, edited by A.J. Kallio, E. Pajanne, and R.F. Bishop (Plenum, New York, 1988), p. 327; A. Lande and R.A. Smith, in *ibid.*, p. 335.
20. R.A. Smith and A. Lande, in *Condensed Matter Theories,* Vol. **3**, edited by J.S. Arponen, R.F. Bishop, and M. Manninen (Plenum, New York, 1988), p. 1.
21. A.D. Jackson, A. Lande, R.W. Guitink, and R.A. Smith, Phys. Rev. B **31**, 403 (1985).
22. G. Baym and L.P. Kadanoff, Phys. Rev. **124**, 287 (1961); G. Baym, *ibid.* **127**, 1391 (1962).
23. R.A. Smith, in *Condensed Matter Theories,* Vol. **4**, edited by J. Keller (Plenum, New York, 1989), p. 129.
24. R. Jastrow, Phys. Rev. **98**, 1479 (1955).
25. F. Iwamoto and M. Yamada, Prog. Theor. Phys. **17**, 543 (1957).
26. J.W. Clark and P. Westhaus, J. Math. Phys. **9**, 131 (1968); J.W. Clark, *ibid.* **9**, 149 (1968).
27. M. Gaudin, J. Gillespie, and G. Ripka, Nucl. Phys. **A176**, 237 (1971); S. Fantoni and S. Rosati, Nuovo Cim. **20A**, 179 (1974).
28. G. Ripka, Nucl. Phys. **A314**, 115 (1979).
29. E. Feenberg, *Theory of Quantum Fluids* (Academic, New York, 1969).
30. S. Fantoni and S. Rosati, Lett. Nuovo Cim, **10**, 545 (1974); Nuovo Cim. **25A**, 593 (1975).
31. E. Krotscheck and M.L. Ristig, Phys. Lett. **48A**, 17 (1974); Nucl. Phys. **A242**, 389 (1975).
32. J.W. Clark, in *Progress in Particle and Nuclear Physics,* Vol. **2**, edited by D.H. Wilkinson (Pergamon, Oxford, 1979), p. 89.
33. V.R. Pandharipande and R.B. Wiringa, Rev. Mod. Phys. **51**, 821 (1979); R.B. Wiringa and V.R. Pandharipande, Nucl. Phys. **A317**, 1 (1979); J.C. Owen, Phys. Lett. **82B**, 23 (1979); L.J. Lantto and P.J. Siemens, Nucl. Phys. **A317**, 55 (1979); J.G. Zabolitzky, in *Advances in Nuclear Physics,* Vol. **12**, edited by J.W. Negele and E. Vogt (Plenum, New York, 1980); S. Fantoni and V.R. Pandharipande, Nucl. Phys. **A427**, 473 (1984).
34. S. Rosati and S. Fantoni, in *The Many-Body Problem: Jastrow Correlations Versus Brueckner Theory* (Lecture Notes in Physics, Vol. **138**), edited by R. Guardiola and J. Ros (Springer, Berlin, 1981), p. 1.
35. S. Rosati and M. Viviani, in *First International Course on Condensed Matter* (ACIF Series, Vol. **8**, edited by D. Prosperi, S. Rosati, and G. Violini (World Scientific, Singapore, 1988), p. 231.
36. C. Ciofi degli Atti, in *Perspectives on Theoretical Nuclear Physics* (Proc. of the Primo Convegno su Problemi di Fisica Nucleare Teorica, Cortona, Italy, 1985), edited by L. Bracci *et al.* (ETS Editrice, Pisa, 1986), p. 1.
37. J.W. Clark, Nucl. Phys. **A328**, 587 (1979).
38. J.W. Clark and E. Feenberg, Phys. Rev. **113**, 388 (1959); H.W. Jackson and E. Feenberg, Ann. Phys. (NY) **15**, 266 (1961); E. Feenberg, and C.W. Woo, Phys. Rev. **137**, A391 (1965); J.W. Clark and P. Westhaus, Phys. Rev. **141**, 833 (1966); **149**, 990 (1966).

39. J.W. Clark, L.R. Mead, E. Krotscheck, K.E. Kürten, and M.L. Ristig, Nucl. Phys. **A328**, 45 (1979).
40. E. Krotscheck and J.W. Clark, Nucl. Phys. **A328**, 73 (1979).
41. J.W. Clark, in *The Many-Body Problem: Jastrow Correlations Versus Brueckner Theory* (Lecture Notes in Physics, Vol. **138**), edited by R. Guardiola and J. Ros (Springer, Berlin, 1981), p. 184.
42. J.M.C. Chen, J.W. Clark, and D.G. Sandler, Z. Phys. A **305**, 223, (1982); **305**, 367 (1982).
43. E. Krotscheck, R.A. Smith, J.W. Clark, and R.M. Panoff, Phys. Rev. B **24**, 6383 (1981); M.F. Flynn, J.W. Clark, E. Krotscheck, R.A. Smith, and R.M. Panoff, *ibid.* **32**, 2945 (1985).
44. E. Krotscheck, J.W. Clark, and A.D. Jackson, Phys. Rev. B **28**, 5088 (1983).
45. J.W. Clark, E. Krotscheck, and B. Schwesinger, Phys. Lett. **143B**, 287 (1984); Anales Fisica **A81**, 116 (1985).
46. R.F. Bishop, in *Recent Progress in Many-Body Theories*, Vol. **1**, edited by A.J. Kallio, E. Pajanne, and R.F. Bishop (Plenum, New York, 1988), p. 385.
47. J.W. Clark and M.L. Ristig, Phys. Rev. C **7**, 1792 (1973); M.L. Ristig and J.W. Clark, Nucl. Phys. **A199**, 351 (1973).
48. L.R. Mead and J.W. Clark, Phys. Lett. **90B**, 331 (1980).
49. E Krotscheck, H. Kümmel, and J.G. Zabolitzky, Phys. Rev. A **22**, 1243 (1980); E. Krotscheck and J.W. Clark, in *The Many-Body Problem: Jastrow Correlations Versus Brueckner Theory* (Lecture Notes in Physics, Vol. **138**), edited by R. Guardiola and J. Ros (Springer, Berlin, 1981), p. 356.
50. R.K. Nesbet, Phys. Rev. **109**, 1632 (1958).
51. A. Cauchy, *Cours d'Analyse de l'École Polytechnique,* Oeuvres Complète Vols. **2**,**3**; J.K.L. Macdonald, Phys. Rev. **43**, 830 (1933).
52. R.J. Bartlett and G.D. Purvis, Int. J. Quantum Chem. **14**, 561 (1978); Phys. Scripta **21**, 251 (1980).
53. H. Primas, in *Modern Quantum Chemistry,* Vol. **II**, edited by O. Sinanoglu (Academic, New York, 1965), p. 45.
54. K.A. Brueckner, Phys. Rev. **97**, 1353 (1955); **100**, 36 (1955).
55. P.M. Lam, J.W. Clark, and M.L. Ristig, Phys. Rev. B **16**, 222 (1977); J.W. Clark, P.M. Lam, J.G. Zabolitzky, and M.L. Ristig, *ibid.* **17**, 1147 (1978); M.L. Ristig, K.E. Kürten, and J.W. Clark, *ibid.* **19**, 3539 (1979); M.F. Flynn, J.W. Clark, R.M. Panoff, O. Bohigas, and S. Stringari, Nucl. Phys. **A427**, 253 (1984).
56. E. Krotscheck, Nucl. Phys. **A465**, 461 (1987).
57. G. Co', A. Fabrocini, S. Fantoni, and I.E. Lagaris, Nucl. Phys. **A549**, 439 (1992); F. Arias de Saavedra, G. Co', A. Fabrocini, and S. Fantoni, *ibid.* **A605**, 359 (1996).
58. J.M.C. Chen, J.W. Clark, E. Krotscheck, and R.A. Smith, Nucl. Phys. **A451**, 509 (1986); V.A. Khodel, V.V. Khodel, and J.W. Clark, *ibid.* **A598**, 390 (1996).
59. M. Saarela, Phys. Rev. B **33**, 4596 (1986).
60. E. Krotscheck and M. Saarela, Phys. Rep. **232**, 1 (1993); B.E. Clements, J.L. Epstein, E. Krotscheck, and M. Saarela, Phys. Rev. B **48**, 7450 (1993).
61. J.W. Clark and E. Krotscheck, in *Recent Progress in Many-Body Theories* (Lecture Notes in Physics, Vol. **198**), edited by H. Kümmel and M.L. Ristig (Springer, Berlin, 1984), p. 127.

62. S. Fantoni, X. Wang, E. Tosatti, and Lu Yu, Physica C **153–155**, 1255 (1988); X.Q. Wang, S. Fantoni, E. Tosatti, and Lu Yu, in *Condensed Matter Theories,* Vol. **5**, edited by V.C. Aguilera-Navarro (Plenum, New York, 1990), p. 203; M.L. Ristig, Z. Phys. B **79**, 351 (1990).
63. C.E. Campbell, T. Pang, and E. Krotscheck, in *Condensed Matter Theories,* Vol. **5**, edited by V.C. Aguilera-Navarro (Plenum, New York, 1990), p. 265.
64. M.L. Ristig, J.W. Kim, and R. Mehlmann, in *Condensed Matter Theories,* Vol. **11**, edited by E.V. Ludeña, P. Vashishta, and R.F. Bishop (Nova Science, Commack, New York, 1996), p. 113; M.L. Ristig, J.W. Kim, and J.W. Clark, Phys. Rev. B **57**, 56 (1998).
65. A. Dabringhaus, M.L. Ristig, and J.W. Clark, Phys. Rev. D **43**, 1978 (1991).
66. F. Coester, in *Lectures in Theoretical Physics: Quantum Fluids and Nuclear Matter,* Vol. **XI B** (Gordon and Breach, New York, 1969).
67. J. Arponen, Ann. Phys. (NY) **151**, 311 (1983).
68. R.F. Bishop, J. Arponen, and E. Pajanne, in *Aspects of Many-Body Effects in Molecules and Extended Systems,* (Lecture Notes in Chemistry, Vol. **50**), edited by D. Mukherjee (Springer, Berlin, 1989), p. 79; R.F. Bishop and J.S. Arponen, Int. J. Quantum Chem.: Quantum Chem. Symp. **24**, 197 (1990).
69. J.S. Arponen and R.F. Bishop, Ann. Phys. (NY) **207**, 171 (1991).
70. R.F. Bishop, M.F. Flynn, M.C. Boscá, E. Buendía, and R. Guardiola, J. Phys. G **16**, L61 (1990); in *The Nuclear Equation of State, Part A: Discovery of Nuclear Shock Waves and the EOS,* edited by W. Greiner and H. Stöcker (Plenum, New York, 1990), p. 605; in *Condensed Matter Theories,* Vol. **5**, edited by V.C. Aguilera-Navarro (Plenum, New York, 1990), p. 253.
71. R.F. Bishop, M.F. Flynn, M.C. Boscá, E. Buendía, and R. Guardiola, Phys. Rev. C **42**, 1341 (1990).
72. R.F. Bishop, E. Buendía, M.F. Flynn, and R. Guardiola, in *Condensed Matter Theories,* Vol. **6**, edited by S. Fantoni and S. Rosati (Plenum, New York, 1991), p. 405.
73. R.F. Bishop, E. Buendía, M.F. Flynn, and R. Guardiola, J. Phys. G **17**, 857 (1991); **18**, 1157 (1992); **19**, 1163 (1993).
74. R. Guardiola, P.I. Moliner, J. Navarro, R.F. Bishop, A. Puente, and N.R. Walet, Nucl. Phys. **A609**, 218 (1996).
75. R.F. Bishop, A.S. Kendall, L.Y. Wong, and Y. Xian, Phys. Rev. D **48**, 887 (1993); in *Condensed Matter Theories,* Vol. **8**, edited by L. Blum and F.B. Malik (Plenum, New York, 1993), p. 269.
76. R.F. Bishop, in *Many-Body Physics,* edited by C. Fiolhais, M. Fiolhais, C. Sousa, and J.N. Urbano (World Scientific, Singapore, 1994), p. 3.
77. R.F. Bishop and Y. Xian, in *Condensed Matter Theories,* Vol. **9**, edited by J.W. Clark, K.A. Shoaib, and A. Sadiq (Nova Science, Commack, New York, 1994), p. 433.
78. N.J. Davidson and R.F. Bishop, Nucl. Phys. B (Proc. Suppl.) **42**, 817 (1995).
79. R.F. Bishop, N.J. Davidson, and Y. Xian, in *Recent Progress in Many-Body Theories,* Vol. **4**, edited by E. Schachinger, H. Mitter, and H. Sormann (Plenum, New York, 1995), p. 237.
80. S.J. Baker, R.F. Bishop and N.J. Davidson, Phys. Rev. D **53**, 2610 (1996); Nucl. Phys. B (Proc. Suppl.) **53**, 834 (1997).
81. C.H. Llewellyn Smith and N.J. Watson, Phys. Lett. **B302**, 463 (1993).

82. N.F. Ligterink, N.R. Walet, and R.F. Bishop, "A Coupled-Cluster Formulation of Hamiltonian Lattice Field Theory: The Nonlinear Sigma Model," UMIST preprint (UMIST/Phys/TP/97-11), (UMIST, Manchester, 1997).
83. R.G. McGarry, PhD Thesis (University of Manchester, 1990).
84. R.F. Bishop, in *Condensed Matter Theories,* Vol. **10**, edited by M. Casas, M. de Llano, J. Navarro, and A. Polls (Nova Science, Commack, New York, 1995), p. 483.
85. H.A. Bethe, Z. Phys. **71**, 205 (1931); M. Gaudin, *La Fonction d'Onde de Bethe* (Masson, Paris, 1963); V.E. Korepin, N.M. Bogoliubov, and A.G. Izergin, *Quantum Inverse Scattering Method and Correlation Functions* (Cambridge Univ. Press, 1993), p. 3.
86. E.H. Lieb and W. Liniger, Phys. Rev. **130**, 1605 (1963); E.H. Lieb, *ibid.* **130**, 1616 (1963).
87. J.B. McGuire, J. Math. Phys. **5**, 622 (1964); C.N. Yang, Phys. Rev. Lett. **19**, 1312 (1967).
88. J. Bardeen, L.N. Cooper, and J.R. Schrieffer, Phys. Rev. **106**, 162 (1957); **108**, 1175 (1957).
89. H.J. Lipkin, N. Meshkov, and A.J. Glick, Nucl. Phys. **62**, 188 (1965); **62**, 199 (1965); **62**, 211 (1965).
90. E. Aalto, J.S. Arponen, and R.F. Bishop, in *Condensed Matter Theories,* Vol. **5**, edited by V.C. Aguilera-Navarro (Plenum, New York, 1990), p. 295.
91. J.B. Parkinson, J. Phys. C **12**, 2873 (1979).
92. Y. Xian, J. Phys.: Condens. Matter **5**, 7489 (1993); **6**, 5965 (1994); in *Condensed Matter Theories,* Vol. **10**, edited by M. Casas, M. de Llano, J. Navarro, and A. Polls (Nova Science, Commack, New York, 1995), p. 541.
93. J. Goldstone, Proc. Roy. Soc. (London) **A239**, 267 (1957).
94. J. Hubbard, Proc. Roy. Soc. (London) **A240**, 539 (1957).
95. M. Gell-Mann and F. Low, Phys. Rev. **84**, 350 (1951).
96. D.J. Thouless, *The Quantum Mechanics of Many-Body Systems* (Academic, New York, 1961).
97. R. Kubo, J. Phys. Soc. Japan **17**, 1100 (1962).
98. J.E. Mayer and M.G. Mayer, *Statistical Mechanics,* 2nd ed. (Wiley, New York, 1977).
99. J. Čižek, J. Chem. Phys. **45**, 4256 (1966); Advan. Chem. Phys. **14**, 35 (1969).
100. M. Fink, Nucl. Phys. **A221**, 163 (1974)
101. J.S. Arponen, R.F. Bishop, and E. Pajanne, Phys. Rev. A **36**, 2519 (1987); in *Condensed Matter Theories,* Vol. **2**, edited by P. Vashishta, R.K. Kalia, and R.F. Bishop (Plenum, New York, 1987), p. 357.
102. H. Hellmann, Acta Physicochimica USSR **I(6)**, 513 (1935); R.P. Feynman, Phys. Rev. **56**, 340 (1939).
103. H.G. Kümmel, Int. J. Quantum Chem. **24**, 79 (1983).
104. H.J. Monkhorst, Int. J. Quantum Chem.: Quantum Chem. Symp. **11**, 421 (1977).
105. R.J. Bartlett, in *Geometrical Derivatives of Energy Surfaces and Molecular Properties,* edited by P. Jørgensen and J. Simons (Reidel, Dordrecht, 1986), p. 35; E.A. Salter, G.W. Trucks, and R.J. Bartlett, J. Chem. Phys. **90**, 1752 (1989).

106. A.C. Scheiner, G.E. Scuseria, J.E. Rice, T.J. Lee, and H.F. Schaefer, J. Chem. Phys. **87**, 5365 (1987).
107. K. Emrich, Nucl. Phys. **A351**, 379 (1981); **A351**, 397 (1981); K. Emrich and J.G. Zabolitzky, *ibid.* **A351**, 439 (1981).
108. R.F. Bishop, M.C. Boscá, and M.F. Flynn, Phys. Lett. **A132**, 440 (1988); Phys. Rev. A **40**, 3484 (1989).
109. R.F. Bishop, Anales Fisica **A89**, 9 (1985); R.F. Bishop, W. Piechocki, and G.A. Stevens, Few-Body Systems **4**, 161 (1988); **4**, 179 (1988).
110. R.F. Bishop, in *Recent Progress in Many-Body Theories* (Lecture Notes in Physics, Vol. **198**), edited by H. Kümmel and M.L. Ristig (Springer, Berlin, 1984), p. 310; in *Nucleon-Nucleon Interaction and Nuclear Many-Body Problems,* edited by S.S. Wu and T.T.S. Kuo (World Scientific, Singapore, 1984), p. 604.
111. A. Bijl, Physica (Utrecht) **7**, 869 (1940); R.P. Feynman, Phys. Rev. **94**, 262 (1954).
112. R.F. Bishop and K.H. Lührmann, Phys. Rev. B **17**, 3757 (1978); **26**, 5523 (1982).
113. R.F. Bishop, J.B. Parkinson, and Y. Xian, Phys. Rev. B **43**, 13782 (1991); Theor. Chim. Acta **80**, 181 (1991); Phys. Rev. B **44**, 9425 (1991); in *Recent Progress in Many-Body Theories,* Vol. **3**, edited by T.L. Ainsworth, C.E. Campbell, B.E. Clements, and E. Krotscheck (Plenum, New York, 1992), p. 117.
114. R.F. Bishop, J.B. Parkinson, and Y. Xian, J. Phys.: Condens. Matter **4**, 5783 (1992).
115. R.F. Bishop, R.G. Hale, and Y. Xian, Phys. Rev. Lett. **73**, 3157 (1994).
116. R.F. Bishop, J.B. Parkinson, and Y. Xian, Phys. Rev. B **46**, 880 (1992).
117. C. Zeng and R.F. Bishop, in *Coherent Approaches to Fluctuations,* edited by M. Suzuki and N. Kawashima (World Scientific, Singapore, 1996), p. 296.
118. C. Zeng, D.J.J. Farnell, and R.F. Bishop, J. Stat. Phys. **90**, 327 (1998).
119. A.L. Fetter and J.D. Walecka, *Quantum Theory of Many-Particle Systems* (McGraw-Hill, New York, 1971); J.P. Blaizot and G. Ripka, *Quantum Theory of Finite Systems* (MIT Press, Cambridge, Mass., 1986); J.W. Negele and H. Orland, *Quantum Many-Particle Systems* (Addison-Wesley, Redwood City, Calif., 1988).
120. B.H. Brandow, Rev. Mod. Phys. **39**, 771 (1967); Ann. Phys. (NY) **57**, 214 (1970).
121. J.A. Pople, J.S. Brinkley, and R. Seeger, Int. J. Quantum Chem.: Quantum Chem. Symp. **10**, 1 (1976).
122. R. Offermann, W. Ey, and H. Kümmel, Nucl. Phys. **A273**, 349 (1976); R. Offermann, *ibid.* **A273**, 368 (1976); W. Ey, *ibid.* **A296**, 189 (1978).
123. I. Lindgren, Int. J. Quantum Chem.: Quantum Chem. Symp. **12**, 33 (1978).
124. D. Mukherjee, Chem. Phys. Lett. **125**, 207 (1986); Int. J. Quantum Chem.: Quantum Chem. Symp. **20**, 409 (1986).
125. I. Lindgren and D. Mukherjee, Phys. Rep. **151**, 93 (1987); R. Chowdhuri, D. Mukherjee, and M.D. Prasad, in *Aspects of Many-Body Effects in Molecules and Extended Systems,* (Lecture Notes in Chemistry, Vol. **50**), edited by D. Mukherjee (Springer, Berlin, 1989), p. 3.
126. J.S. Arponen, R.F. Bishop, and E. Pajanne, Phys. Rev. A **36**, 2539 (1987).

127. J. Arponen, R.F. Bishop, E. Pajanne, and N.I. Robinson, Phys. Rev. A **37**, 1065 (1988); in *Condensed Matter Theories,* Vol. **3**, edited by J.S. Arponen, R.F. Bishop, and M. Manninen (Plenum, New York, 1988), p. 51; in *Aspects of Many-Body Effects in Molecules and Extended Systems,* (Lecture Notes in Chemistry, Vol. **50**), edited by D. Mukherjee (Springer, Berlin, 1989), p. 241.
128. J. Arponen, R.F. Bishop, and E. Pajanne, in *Condensed Matter Theories,* Vol. **2**, edited by P. Vashishta, R.K. Kalia, and R.F. Bishop (Plenum, New York, 1987), p. 173.
129. N.I. Robinson, R.F. Bishop, and J. Arponen, Phys. Rev. A **40**, 4256 (1989).
130. J.S. Arponen and R.F. Bishop, Phys. Rev. Lett. **64**, 111 (1990); in *Recent Progress in Many-Body Theories,* Vol. **2**, edited by Y. Avishai (Plenum, New York, 1990), p. 193; Theor. Chim. Acta **80**, 289 (1991).
131. J.S. Arponen, Theor. Chim. Acta **80**, 149 (1991).
132. J.S. Arponen and R.F. Bishop, Ann. Phys. (NY) **227**, 275 (1993); **227**, 334 (1993).
133. M. Altenbokum, K. Emrich, H. Kümmel, and J.G. Zabolitzky, in *Condensed Matter Theories,* Vol. **2**, edited by P. Vashishta, R.K. Kalia, and R.F. Bishop (Plenum, New York, 1987), p. 389.
134. G. Sanyal, S.H. Mandal, and D. Mukherjee, Chem. Phys. Lett. **192**, 55 (1992); G. Sanyal, S.H. Mandal, S. Guha, and D. Mukherjee, Phys. Rev. E **48**, 3373 (1993); G. Sanyal, S.H. Mandal, and D. Mukherjee, Proc. Indian Acad. Sci. (Chem. Sci.) **106**, 407 (1994).
135. J. Arponen, Phys. Rev. A **55**, 2686 (1997).
136. M. Nakahara, *Geometry, Topology and Physics* (Adam Hilger, Bristol, 1990).
137. R.J. Bartlett, Ann. Rev. Phys. Chem. **32**, 359 (1981); J. Phys. Chem. **93**, 1697 (1989); Theor. Chim. Acta **80**, 71 (1991).
138. R.F. Bishop and H.G. Kümmel, Phys. Today **40(3)**, 52 (1987).
139. V. Kvasnička, V. Laurinc, and S. Biskupič, Phys. Rep. **90C**, 160 (1982).
140. K. Szalewicz, J.G. Zabolitzky, B. Jeziorski, and H.J. Monkhorst, J. Chem. Phys. **81**, 2723 (1984).
141. R.F. Bishop, in *Dirkfest '92 — A Symposium in Honor of J. Dirk Walecka's Sixtieth Birthday,* edited by W.W. Buck, K.M. Maung, and B.D. Serot (World Scientific, Singapore, 1992), p. 21.
142. R.F. Bishop, in *Recent Progress in Many-Body Theories,* Vol. **4**, edited by E Schachinger, H. Mitter, and H. Sormann (Plenum, New York, 1995), p. 237.
143. H. Kümmel and J.G. Zabolitzky, Phys. Rev. C **7**, 547 (1973); J.G. Zabolitzky, Nucl. Phys. **A228**, 272 (1974); **A228**, 285 (1974); J.G. Zabolitzky and W. Ey, Phys. Lett. **76B**, 527 (1978); Nucl. Phys. **A328**, 507 (1979); J.G. Zabolitzky, Phys. Lett. **100B**, 5 (1981).
144. B.D. Day, Phys. Rev. Lett. **47**, 226 (1981); B.D. Day and J.G. Zabolitzky, Nucl. Phys. **A366**, 221 (1981); B.D. Day, Phys. Rev. C **24**, 1203 (1981); P. Bodden, Nucl. Phys. **A384**, 449 (1982).
145. R.J. Bartlett and J.F. Stanton, in *Reviews in Computational Chemistry,* Vol. **V** (VCH Publishers, New York, 1994); *Recent Advances in Coupled-Cluster Methods,* edited by R.J. Bartlett (World Scientific, Singapore, 1997).
146. D.L. Freeman, Phys. Rev. B **15**, 5512 (1977); J. Arponen and E. Pajanne, J. Phys. C **15**, 2665 (1982); **15**, 2683 (1982).
147. K. Emrich and J.G. Zabolitzky, Phys. Rev. B **30**, 2049 (1984).

148. D.M. Ceperley and B.J. Alder, Phys. Rev. Lett. **45**, 566 (1977); S.H. Vosko, L. Wilk, and M. Nusair, Can. J. Phys. **58**, 1200 (1980).
149. M. Roger and J.H. Hetherington, Phys. Rev. B **41**, 200 (1990).
150. F.E. Harris, Phys. Rev. B **47**, 7903 (1993); F. Cornu, Th. Jolicoeur, and J.C. Le Guillou, *ibid.* **49**, 9548 (1994).
151. W.H. Wong, C.F. Lo, and Y.L. Wang, Phys. Rev. B **50**, 6126 (1994).
152. R.F. Bishop, J.B. Parkinson, and Y. Xian, J. Phys.: Condens. Matter **5**, 9169 (1993).
153. D.J.J. Farnell and J.B. Parkinson, J. Phys.: Condens. Matter **6**, 5521 (1994); R. Bursill, G.A. Gehring, D.J.J. Farnell, J.B. Parkinson, T. Xiang, and C. Zeng, *ibid.* **7**, 8605 (1995).
154. C.F. Lo, K.K. Pang, and Y.L. Wang, J. Appl. Phys. **70**, 6080 (1991).
155. M. Roger and J.H. Hetherington, Europhys. Lett. **11**, 255 (1990); C.F. Lo, E. Manousakis, and Y.L. Wang, Phys. Lett. **A156**, 42 (1991); F. Petit and M. Roger, Phys. Rev. B **49**, 3453 (1994); R.F. Bishop, Y. Xian, and C. Zeng, Int. J. Quantum Chem. **55**, 181 (1995).
156. C.S. Hsue and J.L. Chern, Phys. Rev. D **29**, 643 (1984); U.B. Kaulfuss and M. Altenbokum, *ibid.* **33**, 3658 (1986); H. Kümmel, in *Condensed Matter Theories,* Vol. **1**, edited by F.B. Malik (Plenum, New York, 1986), p. 33; in *Condensed Matter Theories,* Vol. **3**, edited by J.S. Arponen, R.F. Bishop, and M. Manninen (Plenum, New York, 1988), p. 21; R.F. Bishop and M.F. Flynn, Phys. Rev. A **38**, 2211 (1988).
157. U. Kaulfuss, Phys. Rev. D **32**, 1421 (1985); C.S. Hsue, H. Kümmel, and P. Ueberholz, *ibid.* **32**, 1435 (1985); M. Altenbokum and H. Kümmel, *ibid.* **32**, 2014 (1985); M. Funke, U. Kaulfuss, and H. Kümmel, *ibid.* **35**, 621 (1987).
158. M. Funke and H. Kümmel, Phys. Rev. D **50**, 991 (1994).
159. H. Kümmel, Ruhr-Universität Bochum preprints (1997).
160. H. Kümmel, Phys. Rev. C 27, 765 (1983); G. Hasberg and H. Kümmel, *ibid.* **33**, 1367 (1986).
161. U. Kaulfuss and M. Altenbokum, Phys. Rev. D **35**, 609 (1987).
162. R.F. Bishop, N.J. Davidson, R.M. Quick, and D.M. van der Walt, Phys. Rev. A **54**, R4657 (1996).
163. W.H. Wong and C.F. Lo, Phys. Rev. B **56**, 17615 (1994).
164. K.H. Lührmann, Ann. Phys. (NY) **103**, 253 (1977); H.G. Kümmel, Nucl. Phys. **A317**, 199 (1979); J. Arponen, J. Phys. G **8**, L129 (1982); J. Arponen and J. Rantakivi, Nucl. Phys. **A407**, 141 (1983).

Atomic and Molecular Applications of the Coupled Cluster Method

Uzi Kaldor

School of Chemistry, Tel Aviv University, 69978 Tel Aviv, Israel

Abstract. Representative applications of the coupled cluster method to atomic and molecular systems are described. Properties calculated for molecules include potential energy surfaces, geometries, ionization potentials and excitation energies. For atomic systems, emphasis is put on heavy atoms, where the relativistic coupled cluster scheme yields atomic transition energies with unprecedented accuracy. The nature of the ground state is investigated for some superheavy (Z > 100) atoms; strong relativistic effects produce in many cases ground states different from those of lighter elements in the same group of the periodic table.

1 Introduction

The coupled cluster (CC) method, discussed in detail elsewhere in the present volume [1], has become one of the most effective and widely used schemes for atoms and small molecules. We start with a brief discussion of certain relevant aspects of the method, and devote most of the chapter to the description of selected applications, both in the nonrelativistic and relativistic regimes. Only a few representative applications are presented here; most (but not all) of them are taken from our research. The reader is referred to recent reviews [2] and workshop proceedings [3, 4], where many more examples may be found.

2 Aspects of the Coupled Cluster Method

2.1 Single Reference CC

The wave function of many atomic and molecular states may be written approximately as a single determinant. This is true, e.g., for closed-shell systems or for high-spin open shells (where all open-shell electrons have the same spin). The "best" such determinant (the one with the lowest energy) is the Hartree-Fock function Φ_0. The exact eigenfunction Ψ is obtained by applying the wave operator Ω,

$$\Psi = \Omega\Phi_0 \ . \tag{1}$$

The distinctive feature of the coupled cluster method is the *exponential* wave operator,

$$\Omega = \exp(S) \ ; \tag{2}$$

S is the excitation operator, the sum of virtual *connected* excitations of one, two, ..., N electrons (N is the total number of electrons in the system):

$$S = S_1 + S_2 + \cdots + S_N = \sum_{ij} \{a_i^\dagger a_j\} s_j^i + \frac{1}{2} \sum_{ijkl} \{a_i^\dagger a_j^\dagger a_l a_k\} s_{kl}^{ij} + \cdots \quad . \tag{3}$$

The right-hand side of Eq. (3) is in the widely-used second quantized notation, with $a_i^\dagger$ and a_i the creation and annihilation operators, respectively, and the braces denote normal order with respect to Φ_0 [5].

Keeping all terms in the expansion (3) leads to the exact wave function Ψ. This is impractical in all but the simplest cases, and one must resort to truncated schemes. The simplest such scheme includes just single and double excitations,

$$S \simeq S_1 + S_2 \ , \tag{4}$$

and is called CCSD by chemists and SUB2 by physicists. More accurate schemes are CCSDT or SUB3,

$$S \simeq S_1 + S_2 + S_3 \ , \tag{5}$$

and CCSDTQ or SUB4,

$$S \simeq S_1 + S_2 + S_3 + S_4 \ . \tag{6}$$

The computational effort increases rapidly with each additional term. Thus, while the cost of CCSD calculations scales as n^6, where n is the size of the system, CCSDT scales as n^8, and CCSDTQ goes up with n^{10}. There are also intermediate methods, where S_1 and S_2 are calculated exactly, while S_3 is included approximately, either as a single n^7 calculation from the converged CCSD amplitudes [the CCSD(T) method] or by an iterative n^7 scheme with simplified S_3 equations (CCSDT-1). A full discussion of these and other schemes may be found in the review of Bartlett and Stanton [2] and in references quoted therein.

It is interesting to compare the CC method with the oldest and most straightforward strategy for correlation calculations, the configuration interaction (CI) method. The CI wave function is a sum of all single, double, ..., excitations from Φ_0,

$$\Psi = (1 + \hat{C}_1 + \hat{C}_2 + \cdots + \hat{C}_N)\Phi_0 \ . \tag{7}$$

The operators $\hat{C}_i$, which include connected as well as disconnected excitations, may be expressed in terms of the connected CC operators S_i,

$$\begin{aligned}
\hat{C}_1 &= S_1 \\
\hat{C}_2 &= S_2 + \frac{1}{2!} S_1^2 \\
\hat{C}_3 &= S_3 + S_2 S_1 + \frac{1}{3!} S_1^3 \\
\hat{C}_4 &= S_4 + S_3 S_1 + \frac{1}{2!} S_2^2 + \frac{1}{2!} S_2 S_1^2 + \frac{1}{4!} S_1^4 \ .
\end{aligned} \tag{8}$$

The full, non-truncated operator in (7) is equivalent to the full expansion in (3), leading in both cases to the exact Ψ; this is called "full CI" or basis set limit, and can be applied only to very small systems. Truncated CC is, however, superior to similarly truncated CI, since it incorporates more excitations. For example, the rather large term S_2^2 is accounted for in the CCSD expansion, whereas CI will have to go to four-electron excitations. This is the reason that CC expansions converge faster than CI. An example is shown in Table 1. Differences from full CI energies in a rather small basis are presented for many-body perturbation theory (MBPT) of orders 2–5, as well as for CI and CC calculations in a number of truncation schemes [2]. Convergence is much faster for CC than for the other methods. Note the importance of the CI quadruple excitations, due largely to the S_2^2 term in $\hat{C}_4$ [see Eq. (8)].

Table 1. Errors relative to full CI in many-body perturbation theory and truncated CI and CC calculations (millihartree). H_2O at equilibrium (R_e) and stretched ($1.5R_e$ and $2.0R_e$) geometries [2].

Method	R_e	$1.5R_e$	$2.0R_e$
MBPT(2)	13.0	23.3	53.7
MBPT(3)	7.22	26.4	74.6
MBPT(4)	0.92	5.76	14.9
MBPT(5)	0.70	4.98	17.0
CISD	12.9	30.4	75.6
CISDT	10.6	23.5	60.3
CISDTQ	0.40	1.55	6.29
CCSD	4.12	10.2	21.4
CCSD(T)	0.72	2.09	4.63
CCSDT	0.53	1.78	−2.47
CC5SD(TQ*)	0.19	0.13	−1.96
CCSDTQ	0.02	0.14	−0.02

2.2 Fock-Space Coupled Cluster

Many atomic and molecular open-shell states cannot be described to a good approximation by a single determinant. Multireference methods are then required. The function space is partitioned into a small P (also called model space), which includes the most important determinants, and a large Q, comprising the rest of the space. An effective Hamiltonian in the P space is constructed, where the effect of Q is included approximately; H_{eff} is then diagonalized to yield the corrected energies,

$$H_{\text{eff}}\Psi_0^a = E^a\Psi_0^a \ . \tag{9}$$

The effective Hamiltonian in the CC method is given by

$$H_{\text{eff}} = PH\Omega P \ . \tag{10}$$

There are many varieties of multireference coupled cluster methods; a review may be found in Ref. [6]. Only the Fock-space method [7] will be presented here. One starts from a closed-shell reference state;[1] the states of interest are reached from the reference by adding and/or removing electrons. The CC excitation operator S may be written in terms of section excitation operators $S^{(m,n)}$, where the (m,n) sector includes states reached by removing m and adding n electrons to the reference. The practical limit is $m+n \leq 2$, although higher sectors have been investigated [8]. The CC equations for $S^{(m,n)}$ involve only $S^{(k,l)}$ with $k \leq m$ and $l \leq n$ [9]; the equations are therefore solved hierarchically, first iterating the $(0,0)$ sector to convergence, then moving to the $(0,1)$ or $(1,0)$ sectors, and continuing as needed.

Intruder States and Incomplete Model Spaces. The P space is usually selected by designating certain orbitals as valence orbitals, and including in P all possible distributions of the valence electrons in the valence orbitals. The linked-cluster theorem was proved for multireference many-body methods [10] with this kind of space, nowadays called a *complete* model space (CMS). A simple example is provided by the first excited ^1S state of the He atom. The valence orbitals will be $1s$ and $2s$, and P will include the four determinants $1s\alpha 2s\beta$, $1s\beta 2s\alpha$, $1s\alpha 1s\beta$, and $2s\alpha 2s\beta$. Obviously, the last two are not particularly important for the state, and their inclusion needlessly complicates the calculation. More seriously, the energy range of P will be large and encompass such Q states as $1s3s$. This situation often leads to *intruder states* [11], which slow or destroy the convergence of the CC iterations. The problem may be solved in many cases by resorting to an *incomplete* model space (IMS) [12, 13], where P is constructed by selecting determinants rather than orbitals. In the He example above, only the first two determinants will go into P. The selection of P determinants may often be carried out on the basis of orbital energies alone, as in the He example and in many of the cases described below. This is not always true. Consider the Be atom, where $2s$ (orbital energy -0.67 au) and $2p$ (-0.52 au) seem well separated from $1s$ (-5.67 au) below and from $3s$ (-0.27 au) and $3p$ (-0.23 au) above. A CMS calculation with $2s$ and $2p$ as valence orbitals is, however, extraordinarily difficult to converge, requiring some 200 iterations, and exhibits an extremely large excitation amplitude (> 5) for $2p^2 \to 2s3s$ [14]. The resulting energies are in good agreement with experiment, except for the $2p^2\ {}^1S$ energy, which in fact reproduces the experimental $2s3s\ {}^1S$ state (see Table 2). The latter

[1] An open-shell reference may also be used; this leads to complications and is not employed in the applications described below.

state is the intruder, which appears unexpectedly within the P space energy range. An incomplete model space, formed by adding the $2s3s$ determinant, gives faster convergence and better agreement with experiment (last column of Table 2) [14].

Table 2. Ionization potentials (IP) and excitation energies (EE) of Be in complete and incomplete model spaces (eV) [14]

		Expt.	CMS	IMS
IP	$2s^2 \rightarrow 2s$	9.320	9.283	9.283
EE	$2s^2 \rightarrow 2s2p\ ^3P$	2.725	2.721	2.721
	$2s^2 \rightarrow 2s2p\ ^1P$	5.277	5.323	5.323
	$2s^2 \rightarrow 2s3s\ ^3S$	6.457	...	6.432
	$2s^2 \rightarrow 2s3s\ ^1S$	6.779	...	6.765
	$2s^2 \rightarrow 2p^2\ ^1D$	6.997	7.075	7.075
	$2s^2 \rightarrow 2p^2\ ^3P$	7.402	7.403	7.403
	$2s^2 \rightarrow 2p^2\ ^1S$	9.447	(6.802)	9.520

3 Applications in the Nonrelativistic Regime

Calculations of various atomic and molecular properties will now be described. Light atoms and molecules made of them are treated by the nonrelativistic Hamiltonian; some of these applications are presented in this section. The next section describes applications of the relativistic coupled cluster method, which is necessary for systems containing heavy atoms. It should be remembered that all molecular calculations (and a large part of atomic calculations) are carried out in a finite basis, which is, of necessity, incomplete. The sources of error in CC results are, therefore, basis set incompleteness and truncation of the S expansion.

3.1 Geometry and Harmonic Frequencies of Water

The potential energy surfaces of molecules, or their energy as function of nuclear geometry, are of utmost interest to chemists; they determine, among other things, the molecular equilibrium configuration, vibrational frequencies, and binding energy. The water molecule has been studied extensively, since it is small enough for inexpensive experimentation and its physical constants have been measured with high accuracy. Errors in the values of geometrical parameters and vibrational frequencies calculated by the Hartree-Fock, CI, MBPT and CC methods are collected in Table 3, compiled by Bartlett and

Stanton [2]. The coupled cluster method clearly shows better agreement with experiment than the other computational approaches.

Table 3. Geometrical parameters R (O–H distance) and θ (HOH angle) and vibrational frequencies ω of H_2O [2]

		Errors relative to experiment				
	Expt.	HF	CISD	MBPT2	CCSD	CCSD(T)
R (Å)	.957		−.013	.006	−.001	.002
θ (deg)	104.5		1.8	−0.3	0.0	−0.3
ω_1 (cm^{-1})	3832	332	135	77	80	64
ω_2 (cm^{-1})	1649	102	44	15	34	28
ω_3 (cm^{-1})	3943	344	147	112	98	83

3.2 Electronic Excitations in the Nitrogen Molecule

The three highest occupied orbitals of N_2 are rather close in energy, and the molecule has a number of low-lying unoccupied orbitals. This situation gives rise to a large number of electronic excitations within a relatively narrow energy range, and complicates the analysis of experimental spectra. A Fock-space CC calculation of low-lying excitations has been carried out [15]. The ground state of the molecule was selected as reference, and the (1,1) sector included all excitations from the three highest occupied orbitals (with energies between −0.8 and −0.6 hartree) to thirteen unoccupied orbitals with Hartree-Fock energies below 0.3 hartree. Ninety six excited states resulted. The higher states are probably not meaningful, since they may require contributions from even higher orbitals for proper description; only the 35 excitations below 15 eV are therefore considered. The sixteen energies known experimentally are collected in Table 4 and compared with values calculated by different methods. These methods include time-dependent Hartree-Fock (TDHF) [16], second-order polarization propagator (SOPPA) [17], many-body perturbation theory (MBPT) [18], and multireference double-excitation CI (MRDCI) [19]. Average errors are compared for the lowest eight states, as these are available in all calculations. The CC method again shows better results than all its rivals. The average CCSD error for all 16 states is 0.13 eV. Very good ionization potentials are also obtained: 15.44 eV for the $3\sigma_g$ orbital (experimental 15.6 eV), 17.11 eV for $1\pi_u$ (experimental 17.2 eV), and 18.66 eV for $2\sigma_u$ (experimental 18.7 eV).

3.3 Potential Functions – Nine States of Li_2

Li_2 is an interesting molecule, with a large number of low-lying states. When applying the Fock-space coupled cluster method, the first question is the se-

Table 4. Vertical excitation energies in N_2 [15]

		Expt.	CCSD	TDHF	SOPPA	MBPT	MRDCI
A	$^3\Sigma_u$	7.75	7.56	7.64	7.91	7.49	7.96
B	$^3\Pi_g$	8.04	8.05	8.17	7.87	8.39	8.27
W	$^3\Delta_u$	8.88	8.93	8.86	9.01	8.85	9.25
a	$^1\Pi_g$	9.31	9.27	9.60	9.32	9.90	9.63
B'	$^3\Sigma_u$	9.67	9.86	10.07	9.99	9.48	9.98
a'	$^1\Sigma_u$	9.92	10.09	10.40	10.05	10.02	10.56
w	$^1\Delta_u$	10.27	10.54	10.76	10.54	10.26	10.57
C	$^3\Pi_u$	11.19	11.19	11.43	11.10		11.19
E	$^3\Sigma_g$	12.00	11.75			11.63	11.59
a"	$^1\Sigma_g$	12.20	12.20			12.08	11.78
c	$^1\Pi_u$	12.90	12.84		12.11		
C'	$^1\Sigma_u$	12.98	12.82		12.30		
b	$^1\Pi_u$	13.24	13.61		13.68		13.77
o	$^1\Pi_u$	13.63	13.71		13.83		13.23
b'	$^1\Sigma_u$	14.25	14.31		14.30		14.74
e'	$^1\Sigma_u$	14.48	14.65				14.66
Average error (8 states)			0.11	0.27	0.14	0.22	0.29

lection of the reference state. It is well known that Hartree-Fock functions lead to improper dissociation; thus, the H_2 function dissociates in the covalent ($H + H$) and ionic ($H^+ + H^-$) channels with equal probability, which is far from the real situation. Correct dissociation occurs only when all products have closed-shell structure. It is therefore convenient to choose Li_2^{2+} (isoelectronic with He_2) as the reference, so that correct dissociation is ensured. The Li_2 states of interest are then obtained by adding two electrons, one at the time, using the (0,2) sector of the Fock space [20].

Li_2^{2+} has the electronic configuration $1\sigma_g^2$. To get the Li_2 states, electrons are added in the $2\sigma_g$, $3\sigma_g$, $2\sigma_u$, $1\pi_u$, and $1\pi_g$ orbitals. Adding two electrons in some of these orbitals creates states with rather high energies, so that the P space has a broad energy span and intruders come in; this is why no convergence could be obtained with a complete model space. An incomplete space was therefore used, selected on the basis of zero-order energies, and convergence was easy. The nine Li_2 states are shown in Fig. 1, and the molecular constants are collected and compared with experiment in Table 5. Excellent agreement with experiment is obtained, within 0.01 Å for equilibrium internuclear separations, 0.02 eV for dissociation and adiabatic excitation energies, and 3 cm^{-1} for vibrational frequencies.

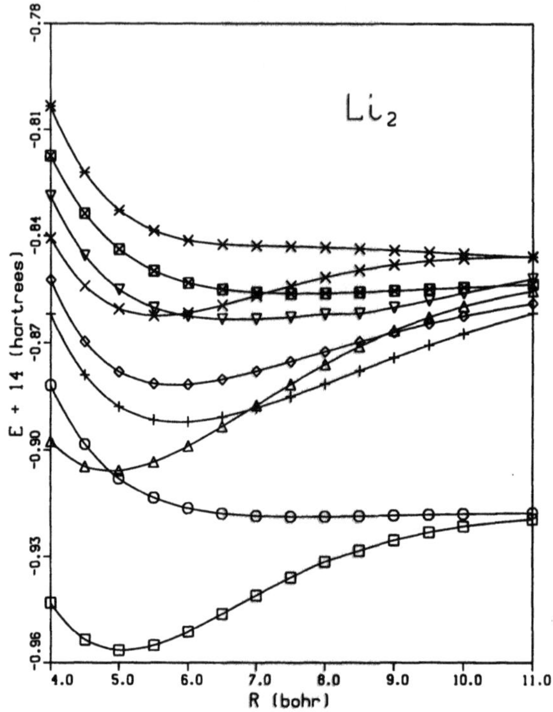

Fig. 1. Potential functions of Li_2 [20]

Table 5. Molecular constants of Li_2: equilibrium distances R_e, dissociation energies D_e, adiabatic excitation energies T_e, and vibrational frequencies ω_e [20]

		R_e(Å)	$T_e(eV)$	$D_e(eV)$	$\omega_e(cm^{-1})$
$X^1\Sigma_g^+$	Expt	2.673	—	1.056	351.4
	CCSD	2.67	—	1.061	351
$a^3\Sigma_u^+$	CCSD	4.05	1.023	0.038	75
$b^3\Pi_u$	Expt	2.591	1.394	1.510	345.7
	CCSD	2.58	1.375	1.533	349
$A^1\Sigma_u^+$	Expt	3.108	1.744	1.160	255.5
	CCSD	3.10	1.750	1.158	257
${}^3\Sigma_g^+$	CCSD	3.06	2.032	0.876	252
${}^1\Sigma_g^+$	CCSD	3.56	2.530	0.378	137
$B^1\Pi_u$	Expt	2.936	2.534	0.370	270.7
	CCSD	2.94	2.557	0.350	269
${}^1\Pi_g$	CCSD	4.02	2.726	0.181	97

3.4 The Charge Transfer Complex $NH_3^+Cl^-$

Experimental studies of the photoelectron spectrum of the NH_3Cl^- cluster revealed the expected line at 4.0 eV, corresponding to the detachment of the extra chlorine electron to leave the neutral radical NH_3Cl, and an additional broad band at 6.4 eV, which at higher resolution showed a long series of lines. It was conjectured that the second band involved electron detachment from the NH_3 moiety of the anion; this would leave the charge transfer (CT) complex $NH_3^+Cl^-$, the first gas-phase CT complex to be identified. To test this conjecture, a CC calculation of the system was undertaken [21]. First, the geometry of the parent NH_3Cl^- anion was optimized at the CCSD level. Then, the ionization potentials of the anion were calculated by going to the (1,0) sector of the Fock space and allowing electron ionization from the four highest occupied orbitals. The lowest IP (actually three nearly degenerate IPs, since the three $3p$ orbitals of chlorine are not exactly degenerate in the molecular environment) was 3.88 eV, and the next IP was 6.36 eV; these agree very well with experimental values. On analyzing the orbitals, it was found that the lowest IP corresponded to ionization from the Cl^- part of the anion, whereas the second IP described ionization from the NH_3 moiety, confirming the identification of a gas-phase CT complex. The reason for the diffuse structure of the second band was found by optimizing the geometry of that complex. Contrary to the spatial structure of the parent anion, where NH_3 is pyramidal, the CT complex is planar (as is the free NH_3^+ anion). The vertical transition from the equilibrium geometry of the parent anion leaves a CT complex at a highly distorted geometry, leading to a long series of vibrational lines in the photoelectron spectrum. The spacing between these lines is indeed close to the vibrational spacing in NH_3^+.

4 Relativistic Coupled Cluster

4.1 The Relativistic Hamiltonian

The relativistic many-electron Hamiltonian cannot be written in closed form; it may be derived perturbatively from quantum electrodynamics [22]. The simplest form is the Dirac-Coulomb (DC) Hamiltonian, where the nonrelativistic one-electron terms in the Schrödinger equation are replaced by the one-electron Dirac operator h_D,

$$H_{\mathrm{DC}} = \sum_i h_D(i) + \sum_{i<j} 1/r_{ij} \ , \tag{11}$$

with

$$h_D = c\boldsymbol{\alpha} \cdot \boldsymbol{p} + \beta c^2 + V_{\mathrm{nuc}}. \tag{12}$$

$\boldsymbol{\alpha}$ and β are the four-dimensional Dirac matrices, and V_{nuc} is the nuclear attraction operator, with the nucleus modeled as a point or finite-size charge.

The uniformly-charged nucleus model is used in the applications presented below. A nucleus of mass A and charge Z is described by a sphere of radius $R = 2.2677 \times 10^{-5} A^{1/3}$, with the potential

$$V_{\mathrm{nuc}}(r) = \begin{cases} -Z/r & \text{for } r > R \\ -(Z/2R)(3 - r^2/R^2) & \text{for } r \leq R \; . \end{cases} \tag{13}$$

Only the one-electron terms in the DC Hamiltonian include relativistic effects, and the two-electron repulsion remains in the nonrelativistic form. The lowest-order correction to the two-electron repulsion is the Breit [23] operator

$$B_{12} = -\frac{1}{2}[\boldsymbol{\alpha}_1 \cdot \boldsymbol{\alpha}_2 + (\boldsymbol{\alpha}_1 \cdot \boldsymbol{r}_{12}) \cdot (\boldsymbol{\alpha}_2 \cdot \boldsymbol{r}_{12})/r_{12}^2]/r_{12} \; , \tag{14}$$

yielding the Dirac-Coulomb-Breit (DCB) Hamiltonian

$$H_{\mathrm{DCB}} = \sum_i h_D(i) + \sum_{i<j} (1/r_{ij} + B_{ij}) \; . \tag{15}$$

All equations are in atomic units.

Neither the DC nor the DCB Hamiltonians are appropriate starting points for accurate many-body calculations. The reason is the admixture of the negative-energy eigenstates of the Dirac Hamiltonian by the two-body terms in an erroneous way [24, 25]. The no-virtual-pair approximation [26, 27] is invoked to correct this problem: the negative-energy states are eliminated by the projection operator Λ^+, leading to the projected Hamiltonians

$$H_{\mathrm{DC}}^+ = \Lambda^+ H_{\mathrm{DC}} \Lambda^+ \tag{16}$$

or

$$H_{\mathrm{DCB}}^+ = \Lambda^+ H_{\mathrm{DCB}} \Lambda^+ \; . \tag{17}$$

H_{DCB}^+ is correct to second order in the fine-structure constant α, and is expected to be highly accurate for all neutral and weakly-ionized atoms [28]. Higher quantum electrodynamic (QED) terms are required for strongly-ionized species; these are outside the scope of this chapter. A comprehensive discussion of higher QED effects and other aspects of relativistic atomic physics may be found in the proceedings of the 1988 Santa Barbara program [29].

Since the Dirac Hamiltonian is not bound from below, failure to observe correct boundary conditions leads to "variational collapse", where admixture of negative-energy solutions may yield energies much below experimental. To avoid this failure, the basis sets used for expanding the large and small components must maintain "kinetic balance" [30, 31]. In the nonrelativistic limit ($c \rightarrow \infty$), the small component is related to the large component by [32]

$$Q_{n\kappa}(r) = (2c)^{-1} \left(\mathrm{d}/\mathrm{d}r + \kappa/r\right) P_{n\kappa}(r) \; . \tag{18}$$

The simplest way to obtain kinetic balance is to derive the small-component basis functions from those used to span the large component by

$$\chi^S_{\kappa j} = (\mathrm{d}/\mathrm{d}r + \kappa/r)\,\chi^L_{\kappa j} \;. \tag{19}$$

Ishikawa and coworkers [33, 34] have shown that G-spinors, with functions spanned in Gaussian-type functions (GTF) chosen according to (19), satisfy kinetic balance for finite c values if the nucleus is modeled as a uniformly-charged sphere.

4.2 Application to Atoms

Different ways of implementing the relativistic coupled cluster (RCC) method are known [35]. A numerical procedure for solving the pair equation has been developed by Lindgren and coworkers [36] and applied to two-electron atomic systems [37]. Other approaches use discrete basis sets of local or global functions. This makes the application of the projection operators onto the positive-energy space much easier than in the numerical scheme; one simply ignores the negative-energy branch of the one-electron spectrum. A technique based on local splines was developed by Blundell et al. [38], while the Göteborg group introduced another type of local basis, obtained by discretizing the radial space [39]. The first relativistic coupled cluster calculation in a global basis [40] appeared in 1990, but was limited to s orbitals only, both in the occupied and virtual space. A more general and sustained implementation started two years later, with pilot calculations for light atoms in closed-shell [41] and open-shell [42] states. The method has since been applied to many heavy atoms, where relativistic effects are crucial to the correct description of atomic structure. Calculated properties include ionization potentials, excitation energies, electron affinities, fine-structure splittings, and for super-heavy elements — the nature of the ground state. The additivity of relativistic and correlation effects was also studied. Systems investigated include the gold atom [43], few-electron ions [44], the alkali-metal atoms Li–Fr [45], the Xe atom [46], the f^2 shells of Pr^{3+} and U^{4+} [47], the ytterbium [48], lutetium [48], mercury [49], barium [50], radium [50], thallium [51], and bismuth [52] atoms, and the super-heavy elements lawrencium [48], rutherfordium [53], 111 [54], 112 [49], 113 [51], 115 [52], and 118 [55]. Representative applications are described below.

The spherical symmetry of atoms, which leads to angular decomposition of the wave function and coupled-cluster equations, is used at both the Dirac-Fock-Breit [33] and RCC [43, 45] stages of the calculation. The energy integrals and CC amplitudes which appear in the Goldstone-type diagrams defining the CC equations are decomposed in terms of vector-coupling coefficients, expressed by angular-momentum diagrams, and reduced Coulomb-Breit or S matrix elements, respectively. The reduced equations for single and double excitation amplitudes are derived using the Jucys-Levinson-Vanagas theorem [5] and solved iteratively. This technique makes possible the use of large basis sets with high l values, as a basis orbital gives rise to two functions at most, with $j = l \pm 1/2$, whereas in Cartesian coordinates the number of functions increases rapidly with l. Typically we go up to h ($l = 5$) or i ($l = 6$) orbitals. To account for core-polarization effects, which may be important for many systems, we correlate at least the two outer shells, usually 20–40 electrons. Finally, uncontracted Gaussians are used, since contraction leads to problems in satisfying kinetic balance and correctly representing the small components. On the other hand, it has been found that high-energy virtual orbitals have little effect on the transition energies we calculate, since these orbitals have nodes in the inner regions of the atom and correlate mostly the inner-shell electrons, which we do not correlate anyway. These virtual orbitals, with energies above 80 or 100 hartree, are therefore eliminated from the RCC calculation.

4.3 Gold and Ekagold (Element 111)

The gold atom exhibits very large relativistic effects on its chemical and physical properties, due to the contraction and stabilization of the 6s orbital. Nonrelativistic calculations lead to large errors, including the reversal of the two lowest excited states [56, 57]. Gold was therefore selected as the first heavy atom to be treated by the RCC method [43]. Two closed-shell states can be used as starting points for the Fock-space treatment, defining the (0,0) sector, namely Au^+ or Au^-. Electrons are then added or removed according to the schemes

$$Au^+(0,0) \rightarrow Au(0,1) \rightarrow Au^-(0,2), \tag{20}$$

or

$$Au^-(0,0) \rightarrow Au(1,0) \rightarrow Au^+(2,0). \tag{21}$$

The basis consisted of 21s17p11d7f Gaussian spinors [57]; the 4$spdf$5spd6s electrons were correlated. The various transition energies of the gold atom and its ions are shown and compared with experiment [58] in Table 6. The nonrelativistic results have errors of several eV. The RCC values, on the other hand, are highly accurate, with an average error of 0.06 eV. The inclusion of the Breit effect does not change the result by much, except for some improvement of the fine-structure splittings.

A major relativistic effect in the gold atom is the stabilization of the 6s orbital. This is manifested by the energy separation between the $5d^{10}6s$ ^2S ground state and the $5d^9 6s^2$ ^2D excited state. Looking at the group 11 (or coinage metal) atoms, the ^2D excitation energies of Cu are 1.389 ($J = 5/2$) and 1.642 ($J = 3/2$) eV, increasing to 3.749 and 4.304 eV for Ag [58]. Were it not for relativity, one would expect even higher energies for Au. Indeed, nonrelativistic CCSD (Table 6) puts the ^2D energy at 5.301 eV above the ^2S ground state, in line with expectations. Relativistic effects reduce this value radically, giving 1.150 and 2.669 eV for the excited ^2D sublevels, within 0.015

Table 6. CCSD transition energies in Au (eV). IP is the ionization potential, EA denotes electron affinity, and EE — excitation energy relative to the ground state. FS denotes fine-structure splittings

			NR	DC	DCB	Expt. [58]
IP	$5d^{10}6s$	$^2S_{1/2}$	6.981	9.101	9.086	9.22
EE	$5d^96s^2$	$^2D_{3/2}$	5.301	2.661	2.669	2.658
	$5d^96s^2$	$^2D_{5/2}$	5.301	1.115	1.150	1.136
	$5d^{10}6p_{1/2}$	$^2P_{1/2}$	3.313	4.723	4.720	4.632
	$5d^{10}6p_{3/2}$	$^2P_{3/2}$	3.313	5.193	5.184	5.105
FS		^2D	0	1.546	1.519	1.522
		^2P	0	0.470	0.466	0.473
EA	$5d^{10}6s^2$	$^1S_0$	1.267	2.278	2.269	2.31

eV of experiment [58]. An even larger stabilization may be expected for the next member of the group, element 111. The question arose whether this stabilization would be sufficient to push the ^2D level below the ^2S and make it the ground state of the atom.

The calculations were carried out as for gold above, starting with the 111^- anion as reference [54]. As expected, very large relativistic effects are observed. The $7s$ orbital energy of the anion goes down from -0.018 to -0.136 hartree, while the $6d$ goes up from -0.355 to -0.186 ($j = 3/2$) and -0.080 ($j = 5/2$) hartree. The s and p orbitals undergo very large contraction (see figure 2). Atomic transition energies also show dramatic changes (Table 7). Of particular interest to us is the $6d^97s^2$ $^2D_{5/2}$ state, predicted by nonrelativistic CCSD to lie 5.43 eV above the $6d^{10}7s$ ^2S state, which is reduced

Table 7. CCSD excitation energies (EE), electron affinity (EA) and ionization potentials (IP) of element 111 (eV)

	Transition	DC	DCB	NR
EE	$6d^97s^2\ ^2D_{5/2} \rightarrow 6d^97s^2\ ^2D_{3/2}$	2.719	2.687	0
	$6d^97s^2\ ^2D_{5/2} \rightarrow 6d^{10}7s\ ^2S_{1/2}$	3.006	2.953	−5.430
EA	$6d^97s^2\ ^2D_{5/2} \rightarrow 6d^{10}7s^2\ ^1S_0$	1.542	1.565	6.484
IP	$6d^97s^2\ ^2D_{5/2} \rightarrow 6d^87s^2\ ^3D_4$	10.57	10.60	22.98
	$6d^97s^2\ ^2D_{5/2} \rightarrow 6d^97s\ ^3D_3$	12.36	12.33	0.92
	$6d^97s^2\ ^2D_{5/2} \rightarrow 6d^{10}\ ^1S_0$	15.30	15.23	−0.44

relativistically to 3 eV *below* the ^2S, thus becoming the ground state. Ionization potentials of the atom show relativistic effects of 12–15 eV!

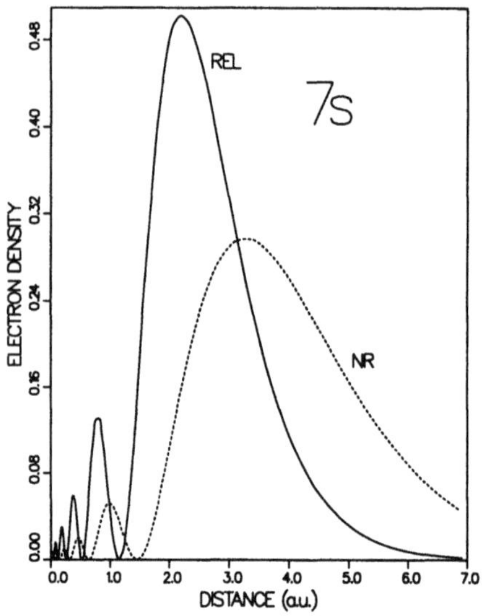

Fig. 2. Relativistic and nonrelativistic densities of element 111 $7s$ orbital

4.4 Rutherfordium — Role of Dynamic Correlation

The nature of the rutherfordium ground state has been a subject of interest for a long time. Rutherfordium is the first atom after the actinide series, and in analogy with the lighter group 4 elements it should have the ground-state configuration $[Rn]5f^{14}6d^27s^2$. Keller [59] suggested that the relativistic stabilization of the $7p_{1/2}$ orbital would yield a $7s^27p_{1/2}^2$ ground state. Recent multiconfiguration Dirac-Fock (MCDF) calculations [60, 61] found that the $7p^2$ state was rather high; they indicate a $6d7s^27p$ ground state, with the lowest state of the $6d^27s^2$ configuration higher by 0.5 [60] or 0.24 eV [61]. The two calculations are similar, using numerical MCDF in a space including all possible distributions of the four external electrons in the $6d$, $7s$ and $7p$ orbitals, and the difference may be due to the different programs used or to minor computational details. These MCDF calculations take into account nondynamic correlation only, which is due to near-degeneracy effects and can be included by using a small number of configurations. A similar approach by Desclaux and Fricke [62] gave errors of 0.4–0.5 eV for the energy differences between $(n-1)d$ and np configurations of Y, La and Lu, with the calculated np energy being too low. Desclaux and Fricke corrected the corresponding energy difference in Lr by a similar amount [62]. If a shift of similar magnitude is applied to the MCDF results for Rf, the order of the two lowest states may be reversed. It should also be noted that dynamic correlation, largely omitted from MCDF, has been shown to play a significant role in determining atomic excitation energies, reducing the average error in calculating Pr^{3+} excitation energies by a factor of four relative to MCDF results [47]. The RCC method has therefore been applied to Rf [53]. Starting from Rf^{2+} with the closed-shell configuration $[Rn]5f^{14}7s^2$, two electrons were added, one at a time, in the $6d$ and $7p$ orbitals, to form the low-lying states of Rf^+ and Rf. A large basis set of $34s24p19d13f8g5h4i$ Gaussian spinors was used, and the external 36 electrons were correlated, leaving only the $[Xe]4f^{14}$ core uncorrelated. A series of calculations, with increasing l values in the virtual space, was performed to assess the convergence of the results. Some of the calculated transition energies are shown in Table 8. Others may be found in the original publication [53].

Table 8. Transition energies in Rf^+ and Rf (eV)

	Rf^+		Rf	
	$7s^26d_{3/2}$ $^2D_{3/2}$ IP	$7s^26d_{5/2}$ $^2D_{5/2}$ EE	$7s^26d^2$ $^3F_2$ IP	$7s^27p6d$ $^3D_2$ EE
MCDF [61]	13.47		5.30	−0.24
MCDF [60]				−0.50
RCCSD				
$l \leq 2$[a]	13.37	0.79	5.15	−0.60
$l \leq 3$[a]	13.95	0.82	5.65	−0.11
$l \leq 3$	14.05	0.87	5.76	0.03
$l \leq 4$	14.20	0.90	5.90	0.17
$l \leq 5$	14.34	0.92	5.99	0.25
$l \leq 6$	14.37	0.92	6.01	0.27
$l \leq 5$[b]	14.34	0.87	5.99	0.27

[a] $5f$ electrons not correlated
[b] With Breit interaction.

The salient feature of the calculated transition energies is their monotonic behavior with the amount of correlation accounted for. The correlation of the $5f$ electrons and the gradual inclusion of higher l spaces all increase the four transition energies in Table 8, as well as those not shown here. The MCDF results fall invariably between the d and f limits. This makes sense, since the MCDF function optimizes the orbitals and CI coefficients in a space including configuration state functions which correspond to all possible distributions of the four external electrons in the $6d$, $7s$ and $7p$ orbitals. Nondynamic correlation, resulting from interactions of configurations relatively close in energy, is thus described very well; dynamic correlation, which is more difficult to include, requiring many thousands of configurations, is not described as well, leading to an error in the identification of the Rf ground state. The latter is determined by the sign of the excitation energy in the last column of Table 8. A negative energy means that the $7s^27p6d$ configuration is lower than the $7s^26d^2$, and is therefore the ground state. From the calculations reported, we estimate the CCSD converged value for this energy at 0.30–0.35 eV, making the $7s^26d^2$ state the ground state of atomic Rf. Recent state-of-the-art experiments with Rf confirmed [63] that the chemistry of the atom is similar to that of Hf, which has a $6s^25d^2$ ground state.

This example shows the intricate interplay of relativity and correlation. It is well known that relativity stabilizes s vs. d orbitals, and correlation has the opposite effect. When both effects are important and the result not obvious *a priori*, one must apply methods, such as RCC, which treat relativity and correlation simultaneously to high order.

4.5 Element 118 — a Rare Gas with Electron Affinity

One of the most dramatic effects of relativity is the contraction and concomitant stabilization of s orbitals (see Fig. 2). An intriguing question is whether the $8s$ orbital of element 118, the next rare gas, would be stabilized sufficiently to give the atom a positive electron affinity. Using the neutral atom Dirac-Fock orbitals as a starting point raises a problem, since the $8s$ orbital has positive energy and tends to "escape" to the most diffuse basis functions. This may be avoided by calculating the unoccupied orbitals in artificial fields, obtained by assigning partial charges to some of the occupied shells. The unphysical fields are compensated by including an appropriate correction in the perturbation operator. A series of calculations with a variety of fields gave electron affinities differing by a few wave numbers. The correlated relativistic electron affinity is 0.056 eV, with an estimated error of 0.01 eV. Inclusion of both relativity and correlation is required to yield a positive EA. It should be noted that similar calculations did not give ^{2}S bound states for Rn^-. Further details may be found in the original publication [55].

4.6 Application to Molecules – One-Component

Molecules are always more difficult to treat accurately than atoms, because of the reduced symmetry. An additional complication arises in relativistic calculations: the Dirac-Fock(-Breit) orbitals will in general be complex. One way to circumvent this difficulty is by the Douglas-Kroll-Hess transformation [64], which yields a one-component function with computational effort essentially equal to that of a nonrelativistic calculation.

Before implementing the Douglas-Kroll-Hess transformation to molecules, we tested its accuracy by applying it to the gold atom [65] and comparing with the full four-component results [43] and with experiment [58]. Since no spin-orbit interaction is present, only term energies (averaged over spin components) can be compared. All transition energies were within a few hundredths of an eV of the corresponding four-component values. It seems therefore safe to apply the one-component method to states where spin-orbit effects are not significant. Calculations were carried out for AuH and some of its ions [65]. The first set of calculations started from the closed-shell AuH^{2+} ion, and added two electrons to obtain states of the monocation and neutral species. Figure 3 shows very large relativistic effects on the potential of the dication. The species is not bound, but it has a sizeable metastable well and barrier to dissociation. The shape of the potential comes from the crossing of two curves dissociating to $Au^{2+} + H$ and to $Au^{+} + H^{+}$, respectively. The depth of the well is more than doubled by relativity. Similar effects occur in the AuH potential (see figure 4). Relativity shortens the bond length of the molecule by 15% and increases the bond energy by 60%. The relativistic potential of the anion is shown in figure 5, and the spectroscopic constants and compared with experiment [66] in Table 9.

Table 9. Molecular constants of AuH

	Nonrel	Rel	Expt
R_e(Å)	1.749	1.525	1.524
D_e(eV)	1.79	2.92	3.22
ω_e(cm^{-1})	1565	2288	2305

Similar calculations were carried out for the hydride of element 111 [67]. As expected, realtivistic effects were even larger than for AuH. The equilibrium separation is reduced by relativity from 1.92 to 1.51Å, and the bond energy increases from 1.83 to 3.56 eV. Preliminary work on Au_2 has also been carried out [68], giving molecular constants in good agreement with experiment: R_e is 2.487Å compared to the experimental 2.472Å, and the harmonic frequency is 2330 cm^{-1} compared to 2305 cm^{-1}.

4.7 Four-Component Application: SnH_4

The Dirac-Fock part of the MOLFDIR program [69] has been interfaced with our Fock-space coupled-cluster package. The first molecule investigated was SnH_4 [70]. The results, collected and compared with experiment [71] in Table 10, show the relativistic and correlation effects on the energy and some spectroscopic constants of the molecule. The relativistic CC results show very good agreement with experiment.

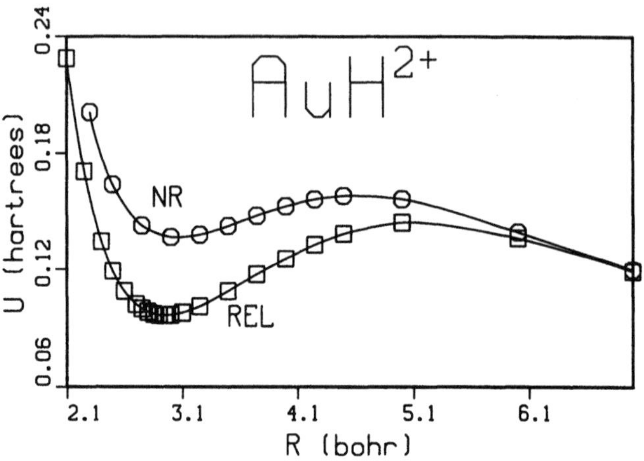

Fig. 3. Relativistic and nonrelativistic potentials of AuH^{2+}

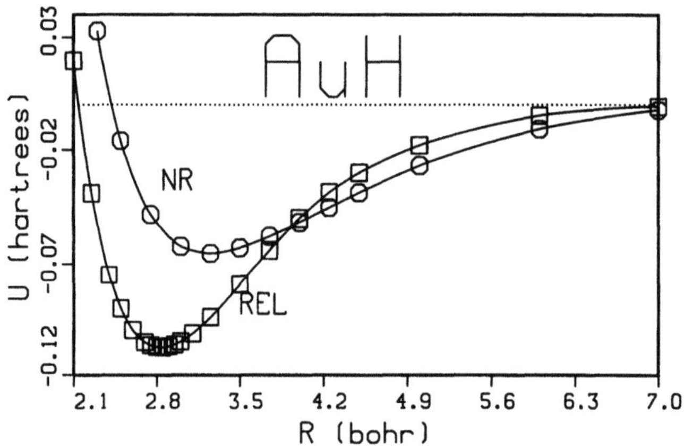

Fig. 4. Relativistic and nonrelativistic potentials of AuH

5 Summary and Conclusion

Representative applications of the coupled cluster method to atomic and molecular systems have been described. Whenever applicable, the method yields accurate predictions of such properties as transition energies, potential surfaces, and molecular geometries. Of particular interest are recent applications to heavy and superheavy atoms, where the CC method was applied in the framework of the four-component Dirac-Coulomb-Breit Hamiltonian, which includes relativistic terms through second order in the fine-structure constant α. The spherical symmetry of the atoms allows the use of large basis sets, leading to calculation of large numbers of heavy-atom states with

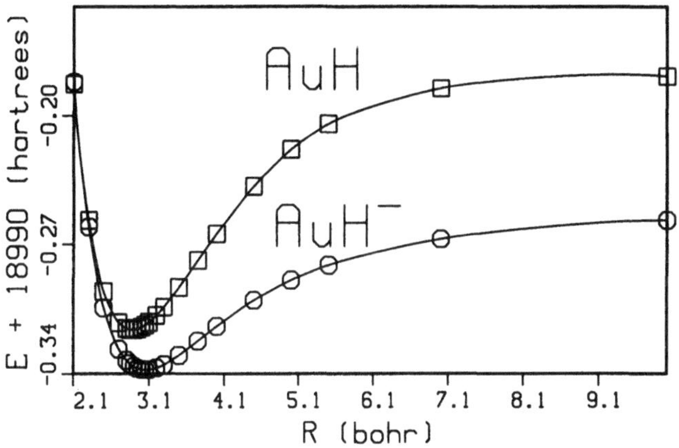

Fig. 5. Relativistic potentials of AuH and AuH^-

Table 10. SCF, second-order perturbation theory, and coupled-cluster values of the equilibrium bond length R_e, energy E_e at R_e, and harmonic frequency ω_e of the breathing mode of SnH_4.

Method	R_e(Å)	E_e(a.u.)	$\omega_e(cm^{-1})$
Nonrelativistic			
SCF	1.727	−6025.21370	2067
PT2	1.710	−6025.39218	2030
CCSD	1.717	−6025.41979	2025
Relativistic			
DF	1.711	−6172.71153	2073
PT2	1.694	−6172.89890	2034
CCSD	1.702	−6172.92553	2019
Expt. [71]	1.701		1955

unprecedented accuracy, and giving reliable predictions for superheavy elements. The largest remaining source of error is probably the omission of triple virtual excitations. Molecules containing heavy atoms present a more difficult challenge, and few four-component RCC applications have been reported to date. The one-component Douglas-Kroll-Hess transformation provides an inexpensive, accurate alternative when spin-orbit interaction may be neglected.

Acknowledgments

Most of the research reported here was done in collaboration with Ephraim Eliav of Tel Aviv University. Yasuyuki Ishikawa of the University of Puerto Rico took part in the heavy atom work, and Bernd Hess of Bonn University in the molecular Douglas-Kroll-Hess investigations. Support was provided by the U.S.-Israel Binational Science Foundation, the German-Israeli Foundation, and the Ministry of Science.

References

1. R.F. Bishop, contribution to this volume
2. R.J. Bartlett and J.F. Stanton, in Rev. Comput. Chem., Vol. V, edited by K.B. Lipkowitz and D.R. Boyd (VCH Publishers, New York, 1994) p. 65ff
3. Proceedings of the Workshop held in Cambridge, Mass., 1990, published in Theor. Chim. Acta **80**, issues 2–6 (1991).
4. Proceedings of the Workshop held in Cedar Key, Florida, 1997, to be published in Mol. Phys.
5. I. Lindgren and J. Morrison, *Atomic Many-Body Theory*, 2nd ed. (Springer-Verlag, Berlin, 1986)
6. D. Mukherjee and S. Pal, Adv. Quantum Chem. **20**, 292 (1989)
7. U. Kaldor, Theor. Chim. Acta **80**, 427 (1991) and references therein
8. S.R. Hughes and U. Kaldor, Phys. Rev. A **47**, 4705 (1993); Chem. Phys. Lett. **194**, 99 (1992); ibid. **204**, 339 (1993); Int. J. Quantum Chem. **55**, 127 (1995)
9. A. Haque, Ph.D. Thesis, Calcutta University (1983); A. Haque and D. Mukherjee, J. Chem. Phys. **80**, 5058 (1984)
10. B.H. Brandow, Rev. Mod. Phys. **39**, 771 (1967)
11. T. H. Schucan and H.A. Weidenmüller, Ann. Phys. (NY) **73**, 108 (1972); ibid. **76**, 486 (1973)
12. G. Hose and U. Kaldor, J. Phys. B **12**, 3827 (1979); Phys. Scr. **21**, 357 (1980); J. Chem. Phys. **86**, 2133 (1984)
13. B. Jeziorski and H.J. Monkhorst, Phys. Rev. A **24**, 1668 (1981)
14. U. Kaldor, Phys. Rev. A **38**, 6013 (1988)
15. S. Berkovic Ben-Shlomo and U. Kaldor, J. Chem. Phys. **92**, 3680 (1990)
16. M. Jaszunski, A. Rizzo, and D.L. Yaeger, Chem. Phys. **136**, 385 (1989)
17. J. Oddershede, Adv. Chem. Phys. **69**, 201 (1987)
18. U. Kaldor, J. Chem. Phys. **81**, 2406 (1984)
19. S.K. Shih, W. Butscher, R.J. Buenker, and S.D. Peyerimhoff, Chem. Phys. **29**, 241 (1978)
20. U. Kaldor, Chem. Phys. **140**, 1 (1990)
21. G. Markovich, O. Cheshnovsky, and U. Kaldor, J. Chem. Phys. **99**, 6201 (1993)
22. See, e.g., J. Sucher, in *Relativistic, Quantum Electrodynamic, and Weak Interaction Effects in Atoms*, edited by W. Johnson, P. Mohr, and J. Sucher (American Institute of Physics, New York, 1989), p. 28
23. G. Breit, Phys. Rev. **B34**, 553 (1929); *ibid.* **36**, 383 (1930); *ibid.* **39**, 616 (1932)
24. G.E. Brown and D.G. Ravenhall, Proc. Roy. Soc. A **208**, 552 (1951)

25. H. A. Bethe and E. E. Salpeter, *Quantum Mechanics of One- and Two-Electron Atoms* (Springer-Verlag, Berlin, 1957)
26. J. Sucher, Phys. Rev. A **22**, 348 (1980); Phys. Scr. **36**, 271 (1987)
27. W. Buchmüller and K. Dietz, Z. Phys. C **5**, 45 (1980)
28. I. Lindgren, in *Many-Body Methods in Quantum Chemistry*, edited by U. Kaldor, Lecture Notes in Chemistry Vol. 52 (Springer-Verlag, Heidelberg, 1989) p. 293; Nucl. Instrum. Methods **B31**, 102 (1988)
29. *Relativistic, Quantum Electrodynamic, and Weak Interaction Effects in Atoms*, edited by W. Johnson, P. Mohr, and J. Sucher (AIP, New York, 1989)
30. R.E. Stanton and S. Havriliak, J. Chem. Phys. **81**, 1910 (1984)
31. Y. Ishikawa, R. Baretty, and R.C. Binning, Intern. J. Quantum Chem. Symp. **19**, 285 (1985); Y. Ishikawa and H. Sekino, Chem. Phys. Lett. **165**, 243 (1990)
32. Y.S. Lee and A.D. McLean, J. Chem. Phys. **76**, 735 (1982)
33. Y. Ishikawa, R.C. Binning, and H. Sekino, Chem. Phys. Lett. **160**, 206 (1989); Y. Ishikawa, Phys. Rev. A **42**, 1142 (1990); Chem. Phys. Lett. **166**, 321 (1990); Y. Ishikawa and H.M. Quiney, Phys. Rev. A **47**, 1732 (1993); Y. Ishikawa and K. Koc, Phys. Rev. A **50**, 4733 (1994).
34. Y. Ishikawa, R.C. Binning, and K.M. Sando, Chem. Phys. Lett. **101**, 111 (1983); *ibid.* **105**, 189 (1984); *ibid.* **117**, 444 (1985); Y. Ishikawa, R. Baretty, and R.C. Binning, Chem. Phys. Lett. **121**, 130 (1985); Y. Ishikawa and H.M. Quiney, Intern. J. Quantum Chem. Symp. **21**, 523 (1987)
35. For recent reviews see Y. Ishikawa and U. Kaldor, in *Computational Chemistry: Review of Current Trends*, edited by J. Leszczynski, (World Scientific, Singapore, 1996), vol. I, p. 1; U. Kaldor, in *Recent Advances in Coupled Cluster Theory*, edited by R.J. Bartlett (World Scientific, Singapore, 1997), p. 175
36. S. Salomonson, I. Lindgren, and A.M. Mårtensson, Phys. Scr. **21**, 351 (1980)
37. E. Lindroth, Phys. Rev. A **37**, 316 (1988)
38. S.A. Blundell, W.R. Johnson, Z. W. Liu, and J. Sapirstein, Phys. Rev. A **39**, 3768 (1989); *ibid.* **40**, 2233 (1989); S.A. Blundell, W.R. Johnson, and J. Sapirstein, Phys. Rev. Lett. **65**, 1411 (1990); Phys. Rev. A **43**, 3407 (1991); Z.W. Liu and H.P. Kelly, *ibid.* **43**, 3305 (1991)
39. S. Salomonson and P. Öster, Phys. Rev. A **40**, 5548 (1989); E. Lindroth and S. Salomonson, *ibid.* **41**, 4659 (1990); A.C. Hartly, E. Lindroth and A.M. Mårtensson-Pendrill, J. Phys. B **23**, 1990 (1990); A.C. Hartly and A.M. Mårtensson-Pendrill, Z. Phys. D **15**, 309 (1990); J. Phys. B **24**, 1193 (1991); A.M. Mårtensson-Pendrill, S.A. Alexander, L. Adamowicz, N. Oliphant, J. Olsen, P. Öster, H.M. Quiney, S. Salomonson, and D. Sundholm, Phys. Rev. A **43**, 3355 (1991); E. Lindroth, H. Persson, S. Salomonson and A.M. Mårtensson-Pendrill, Phys. Rev. A **45**, 1493 (1992); E. Lindroth and J. Hvarfner, *ibid.* **45**, 2771 (1992)
40. H. Sekino and R.J. Bartlett, Intern. J. Quantum Chem. Symp. **24**, 241 (1990)
41. E. Ilyabaev and U. Kaldor, Chem. Phys. Lett. **194**, 95 (1992)
42. E. Ilyabaev and U. Kaldor, J. Chem. Phys. **97**, 8455 (1994); Phys. Rev. A **47**, 137 (1993)
43. E. Eliav, U. Kaldor, and Y. Ishikawa, Phys. Rev. A **49**, 1724 (1994)
44. E. Eliav (Ilyabaev), U. Kaldor, and Y. Ishikawa, Chem. Phys. Lett. **222**, 82 (1994)
45. E. Eliav, U. Kaldor, and Y. Ishikawa, Phys. Rev. A **50**, 1121 (1994)

46. E. Eliav, U. Kaldor, and Y. Ishikawa, Intern. J. Quantum Chem. Symp. **28**, 205 (1994)
47. E. Eliav, U. Kaldor, and Y. Ishikawa, Phys. Rev. A **51**, 225 (1995)
48. E. Eliav, U. Kaldor, and Y. Ishikawa, Phys. Rev. A **52**, 291 (1995)
49. E. Eliav, U. Kaldor, and Y. Ishikawa, Phys. Rev. A **52**, 2765 (1995)
50. E. Eliav, U. Kaldor, and Y. Ishikawa, Phys. Rev. A **53**, 3050 (1996)
51. E. Eliav, U. Kaldor, Y. Ishikawa, M. Seth, and P. Pyykkö, Phys. Rev. A **53**, 3926 (1996)
52. E. Eliav, U. Kaldor, and Y. Ishikawa, to be published
53. E. Eliav, U. Kaldor, and Y. Ishikawa, Phys. Rev. Lett. **74**, 1079 (1995)
54. E. Eliav, U. Kaldor, P. Schwerdtfeger, B.A. Heß, and Y. Ishikawa, Phys. Rev. Lett. **73**, 3203 (1994)
55. E. Eliav, U. Kaldor, Y. Ishikawa, and P. Pyykkö, Phys. Rev. Lett. **77**, 5350 (1996)
56. P. J. Hay, W.R. Wadt, L.R. Kahn, and F.W. Bobrowicz, J. Chem. Phys. **69**, 984 (1978)
57. A. Pizlo, G. Jansen, and B.A. Heß, J. Chem. Phys. **98**, 3945 (1993)
58. C.E. Moore, *Atomic Energy Levels*, Natl. Bur. of Stand. (U.S.) Circ. No. 467 (U.S. GPO, Washington, DC, 1948)
59. O.L. Keller, Radiochim. Acta **37**, 169 (1984)
60. V.A. Glebov, L. Kasztura, V.S. Nefedov, and B.L. Zhuikov, Radiochim. Acta **46**, 117 (1989)
61. E. Johnson, B. Fricke, O.L. Keller, C.W. Nestor Jr., and T.C. Tucker, J. Chem. Phys. **93**, 8041 (1990)
62. J.P. Desclaux and B. Fricke, J. Phys. **41**, 943 (1980)
63. M. Schädel, Radiochim. Acta **70/71**, 207 (1995)
64. M. Douglas and N.M. Kroll, Ann. Phys. (NY) **82**, 89 (1974); B.A. Heß, Phys. Rev. A **32**, 756 (1982); *ibid.* **33**, 3742 (1986); G. Jansen and B.A. Heß, Phys. Rev. A **39**, 6016 (1989); R. Samzow, B.A. Heß, and G. Jansen, J. Chem. Phys. **96**, 1227 (1992); B.A. Heß, R.J. Buenker, and P. Chandra, Intern. J. Quantum Chem. **29**, 737 (1986); R. Samzow and B.A. Heß, Chem. Phys. Lett. **184**, 491 (1991)
65. U. Kaldor and B.A. Heß, Chem. Phys. Lett. **230**, 1 (1994)
66. K.P. Huber and G. Herzberg, *Molecular Spectra and Molecular Structure Constants of Diatomic Molecules*, (Van Nostrand, New York, 1979)
67. M. Seth, P. Schwerdtfeger, M. Dolg, K. Fægri, B.A. Heß, and U. Kaldor, Chem. Phys. Lett. **250**, 461 (1996)
68. U. Kaldor and B.A. Heß, unpublished results.
69. P. J.C. Aerts, O. Visser, L. Visscher, H. Merenga, W. A. de Jong, and W.C. Nieuwpoort, MOLFDIR, University of Groningen, The Netherlands
70. E. Eliav and U. Kaldor, Chem. Phys. Lett. **248**, 405 (1996)
71. H.W. Kattenberg and A. Oskam, J. Mol. Spectr. **51**, 377 (1974); Y. Oshima, Y. Matsumoto, M. Takami, S. Yamamoto, and K. Kuchitsu, J. Chem. Phys. **87**, 5141 (1987); M. Halonen, H. Bürger, and S. Sommer, J. Phys. Chem. **94**, 5222 (1990)

A Thermal Cluster-Cumulant Theory

Sheikh Hannan Mandal, Goutam Sanyal and Debashis Mukherjee

Department of Physical Chemistry
Indian Association for the Cultivation of Science
Calcutta 700-032, India

Abstract. The preeminent success of the coupled cluster theory for treating correlation effects to high accuracy has prompted attempts in recent years to extend this approach to encompass temperature-dependent situations. We shall discuss here one such nonperturbative finite-temperature version, to be called a cluster-cumulant formalism, which shares the best features of a zero-temperature coupled cluster theory. The key theoretical ingredients of the method are :(a) the notion of thermal normal ordering and Wick-like expansion theorem to simplify the imaginary-time ordered exponential representation of the Boltzmann operator, and (b) a representation of the Boltzmann operator as an exponential of a cluster-cumulant in the thermal normal order. The grand partition function is generated systematically as a thermal trace of the Boltzmann operator. Generalization of the concept of the normal ordering to cover imaginary-time path-integral based methods for partition functions offers an access to systematic nonperturbative extensions of the usual Feynman-Kleinert approximations.

1 Introduction

Computation of equilibrium thermal averages of quantum mechanical observables in an efficient, compact and systematic manner continues to be of major theoretical interest in statistical physics. Thermal averaging is usually required in two distinct physical situations: (i) to compute equilibrium thermal properties of quantum many-body systems, and (ii) to study dynamical behaviour or transport properties of systems in contact with a robust thermal bath. Thermal averaging thus has wide applicability in various branches of physics and chemistry [1, 2]. There is a vast literature on the subject and we mention only a few major methodologies of immediate relevance to us. These are the Matsubara formalism as used in the context of the temperature Green's function [3, 4], the finite temperature perturbation theory [5, 6], the complex time-path method [7] and the related path-integral techniques [8, 9], and the recently emerging formalism of the thermofield dynamics [10]. Most of these methods exploit in one form or another the striking algebraic similarity of the Boltzmann statistical operator for the equilibrium partition function and the Heisenberg time-evolution operator. This allows one to treat the Boltzmann operator as an evolution operator in imaginary-time.

The most commonly used pathway for treating thermal averages is to invoke a perturbative treatment of the Boltzmann operator by partitioning the Hamiltonian into a convenient unperturbed component and a perturbation. A naive truncation of the thermal trace at a given order of perturbation would violate the multiplicative separability of the partition function, Z_q and consequently would violate the additive separability of extensive properties such as the free energy. In the many-body parlance this manifests itself as the appearances of disconnected diagrams in the perturbation series for free energy. This difficulty is obviated via the introduction of cumulants [5, 6, 11, 12]. The basic idea is to write Z_q as an exponential of an entity and perform a perturbation series for this entity. The exponent then remains connected at any given order of perturbation so that the logarithm of Z_q now generates an extensive free energy. The entity in the exponent is called a cumulant [11]. The Fox-Kubo operator cumulant expansion [11, 12] and the conceptually related Bloch-Balian-deDominicis thermodynamic perturbation theory [5, 6] are both developed using a perturbative expansion of the cumulant. The path-integral based techniques of Feynman, Kleinert and others [8, 9] also use a perturbative treatment beyond the mean field model of the cumulant. While a perturbation theory may work well for weak perturbations, it is inadequate when the perturbation is strong. Although various resummation strategies have been developed to include certain classes of diagrams to all order [6, 13], there does not exist systematic resummation methods to handle arbitrary classes of diagrams and one must look for nonperturbative methods. While there are special variational [14] and related techniques [15] to tackle specific problems, there are no systematic nonperturbative procedures to compute the cumulant.

We would present here a nonperturbative cluster expansion approach for treating the Boltzmann operator. Our intention is to represent the Boltzmann operator as a manifestly multiplicatively separable manner. This we would achieve by positing on it an exponential type of ansatz, involving in the exponent both operators and a number. We will call the operators in the exponent as cluster-cumulant operators and the number component as the zero-body component of the cluster-cumulant. We would invoke a special type of ordering in the exponential such that the operators appearing in the exponential do not survive the thermal averaging procedure. In that case only the number component survives, which then provides a nonperturbative access to free energy. Just as a normal ordered exponential of a sum of operators and a number retains the exponential of only the number on taking vacuum expectation value in a zero-temperature field theory, our ansatz may be viewed as a finite temperature version of an analogous normal ordered exponential. This is why we would call this ordering as a thermal normal ordering [16, 17, 19]. The operators commute (anticommute) under the thermal normal ordering for bosons (fermions) just as in the zero-temperature case, and the thermal trace of a normal ordered product of operators vanishes. An important aspect of our formulation is the realization that an analogous concept may be invoked to formulate a thermally normal ordered exponential representation of the Boltzmann operator in the path-integral based formulation as well [18], which we shall also describe here. It turns out that a factorization property of the thermal trace of an ordered product of field operators in imaginary-time discovered by Matsubara [3], Bloch [5, 6] and Gaudin [20] is at the root of our being able to define the thermal ordering. With a partition of the Hamiltonian into a harmonic unperturbed component and a perturbation, the same factorization property holds good for the thermal trace (with respect to the unperturbed Hamiltonian) of an ordered product of dynamical variables such as positions and momenta in imaginary-time in the path-integral representation. Once this is realized, we may extend our formulation to generate a thermally normal ordered exponential representation of the Boltzmann operator in the path-integral form.

2 Towards a Nonperturbative Cluster-Cumulant Formalism for Grand Partition Function (GPF)

2.1 Theoretical Preliminaries

We start with the definition of the grand partition function (gpf) Z_q of a quantum system

$$Z_q = \mathrm{Tr}[e^{-\beta K}] \text{ and } K = H - \mu N \ , \tag{1}$$

where H is the Hamiltonian of the system, μ is the chemical potential and N is the number operator. If we split H into an unperturbed part, H_0 and a perturbation, V as

$$H = H_0 + V \ , \tag{2}$$

and define an operator $U_I(\beta)$ via

$$U_I(\beta) = e^{\beta K_0} e^{-\beta K} \ , \tag{3}$$

where $K_0 = H_0 - \mu N$, then $U_I(\beta)$ can be computed by solving the differential equation

$$-\frac{\partial U_I(\tau)}{\partial \tau} = V_I(\tau) U_I(\tau) \ , \tag{4}$$

where

$$V_I(\tau) = e^{\tau K_0} V e^{-\tau K_0} \ . \tag{5}$$

The equation above may be viewed as the Heisenberg's evolution equation in imaginary-time τ, in an 'interaction representation'. Just like the Dyson-form of solution for the evolution operator for real-time in the interaction picture, the solution of eqn.(4) for $U_I(\tau)$, the imaginary-time evolution operator at 'time' β, is given by

$$U_I(\beta) = T \exp\left(- \int_0^\beta V_I(\tau) d\tau \right) \ , \tag{6}$$

where T is the ordering operator in imaginary-time.

Let us define the unperturbed partition function, Z_{0q} as

$$Z_{0q} = \mathrm{Tr}\,[e^{-\beta K_0}] \ . \tag{7}$$

Using Eqs.(1), (3) and (6), we obtain

$$Z_q = \mathrm{Tr}\,[e^{-\beta K_0} U_I(\beta)] = \mathrm{Tr}\left[e^{-\beta K_0}\left[T \exp\left(-\int_0^\beta V_I(\tau) d\tau\right)\right]\right] \ . \tag{8}$$

Then

$$\frac{Z_q}{Z_{0q}} = \frac{\mathrm{Tr}\,[e^{-\beta K_0}[T \exp{(-\int_0^\beta V_I(\tau) d\tau)}]]}{\mathrm{Tr}\,[e^{-\beta K_0}]}$$
$$= \left\langle T \exp\left(-\int_0^\beta V_I(\tau) d\tau\right)\right\rangle = \langle U_I(\beta)\rangle \ , \tag{9}$$

where $\langle \cdots \rangle$ denotes thermal averaging with respect to the unperturbed K_0. As we have emphasized in Sect.1, we may look upon the process of taking the thermal trace $\langle \cdots \rangle$ as a finite temperature analogue of taking the vacuum expectation value. Evaluation of the ratio Z_q/Z_{0q} obviously involves expanding the T-ordered exponential above and evaluating the thermal traces of such products. This would generate a perturbation series of the ratio. We want to formulate a nonperturbative theory, and in particular, we want to respect extensivity of the free energy.

Bloch [5, 6], and Matsubara [3] have proved a remarkable factorization theorem. They have shown that a thermal trace with respect to K_0 of a T-ordered product of field-operators in the interaction picture in the imaginary-time can be written as sum of products of thermal traces of all the pairs of operators. For bosons, there are no phase factors, while for fermions there will be a phase factor in each term of the sum depending on the parity of permutation needed to bring the operators in the pairs side by side. Again, we may view this as a finite-temperature generalization of the Wick's theorem for the vacuum expectation value. We now want to go one step further and try to generate a Wick-like theorem which reorders the T-ordered products of operators in terms of thermal normal ordered products mentioned in Sect. 1. The thermal normal products will have vanishing traces by definition. If we can have such a theorem, the Bloch-Matsubara theorem will then follow naturally on taking the thermal trace of the T-ordered product and using the 'thermal' Wick's theorem. Moreover, this theorem will simplify the procedure of taking the thermal trace enormously. Assuming for the moment that a thermal Wick's theorem exists, we want to posit a thermal normal ordered exponential ansatz for the operator $U_I(\tau)$

$$U_I(\tau) = \{\exp[S_q(\tau) + x_q(\tau)]\}_\beta \ , \tag{10}$$

where $S_q(\tau)$ is the operator part of the exponent (i.e., a cluster-cumulant operator, introduced in Sect.1), and $x_q(\tau)$ is its number component. The curly bracket $\{\cdots\}_\beta$ signifies the thermal normal ordering. Clearly, if the thermal Wick's theorem is valid, we have

$$Z_q/Z_{0q} = \langle U_I(\beta)\rangle = \exp[x_q(\beta)] \ . \tag{11}$$

The Thermal Cluster Cumulant (TCC) equations determining the cluster amplitudes of S_q and x_q are obtained if we substitute the ansatz Eq. (10) in Eq. (4), and generate the differential equations for S_q and x_q. It turns out that the equations for S_q are decoupled from that for x_q, so the amplitudes of S_q are determined first and these are used next to get x_q's. The shift ΔF of the free energy with respect to the unperturbed value F_0 is related to $x_q(\beta)$ by

$$\Delta F = F - F_0 = -\frac{x_q(\beta)}{\beta} \text{ and } F_0 = -\frac{\ln[Z_{0q}(\beta)]}{\beta} \ . \tag{12}$$

For a path-integral treatment of the gpf for perturbed harmonic (bosonic) systems, we start from a path-integral representation [8] of Z_q as

$$Z_q = \oint D[x] e^{-\int_0^\beta \left[\frac{1}{2}\dot{x}^2(\tau)+\frac{1}{2}\Omega^2 x^2(\tau)+V[x(\tau)]\right]d\tau} , \tag{13}$$

where $\oint$ denotes path-integration over all cyclical paths and $x(\tau)$ denotes quantum paths periodic in β. Clearly, with Z_{0q} defined as

$$Z_{0q} = \oint D[x] e^{-\int_0^\beta \left[\frac{1}{2}\dot{x}^2(\tau)+\frac{1}{2}\Omega^2 x^2(\tau)\right]d\tau} \tag{14}$$

one can write Z_q/Z_{0q} as an average

$$\frac{Z_q}{Z_{0q}} = \left\langle e^{-\int_0^\beta V[x(\tau)]d\tau} \right\rangle , \tag{15}$$

where the average has to be interpreted as the ratio obtained from Eqs. (13) and (14). The ordering of $V[x(\tau)]$'s in the exponentials in the path-integral is automatically in T-ordering. The path-integrals for a general V cannot be evaluated analytically in closed form. A perturbation expansion, however, can be generated by expanding the exponential in Eq. (15) and the resulting series can be computed term by term by using the property of gaussian integrals. This becomes possible since the T-ordered products of x's satisfy the gaussian factorization property similar to the Bloch-Matsubara theorem when the weight factor in the path-integral is a gaussian, as in Eq. (15). Then we can envisage representing the T-ordered exponential in Eq. (15) as a normal ordered exponential to formulate a path-integral based cluster-cumulant theory.

For the discussion above, it implies that the notion of normal ordering and reordering theorems are of pretty general validity provided a gaussian factorization property holds good. In the next subsection, we shall explore this.

2.2 The Notion of Normal Ordering and Wick-Like Reordering Theorem for Quantum Variables with Gaussian Factorization Property

Let us assume that we have several gaussian variables ϕ_i's which are distinguished from each other either because they are different or they differ in their time-arguments or both. We consider bosons first. The case for fermions can be generated from that of bosons via Grassmann variables [21]. The variables display their gaussian nature with respect to a prescribed averaging:

$$\langle \phi_i(t) \rangle = 0 \ , \tag{16}$$

$$\langle T[\phi_i(t_1)\phi_j(t_2)] \rangle = g_{ij}(t_1, t_2) \ , \tag{17}$$

$$\langle T[\phi_i(t_1) \cdots \phi_m(t_{2n})] \rangle = \sum_{\text{all pairs } (ij)} \prod \langle T[\phi_i(t_i)\phi_j(t_j)] \rangle \ , \tag{18}$$

$$\langle T[\phi_i(t_1) \cdots \phi_m(t_{2n+1})] \rangle = 0 \ , \tag{19}$$

where $\langle \cdots \rangle$ denotes the averaging process.

We define the *operator derivative* $\delta A / \delta \phi_i$ of a T-ordered product A of several ϕ_i's as an operator where we commute ϕ_i in A to bring it as the leftmost entry and delete it. Thus, with $A = T[\phi_1(t_1)\phi_2(t_2)\phi_3(t_3)]$

$$\frac{\delta A}{\delta \phi_1(t_1)} = T[\phi_2(t_2)\phi_3(t_3)] \tag{20}$$

$$\frac{\delta A}{\delta \phi_2(t_2)} = T[\phi_1(t_1)\phi_3(t_3)] \tag{21}$$

$$\frac{\delta A}{\delta \phi_3(t_3)} = T[\phi_1(t_1)\phi_2(t_2)] \ . \tag{22}$$

Owing to the presence of the T-ordering operator, the operator derivatives behave like ordinary derivatives for bosons. Taken by themselves, though, the ϕ_i's do not generally commute, and operator derivatives of ordinary products would be rather complicated.

Let us now define the following T-ordered exponential operator C:

$$C = T \exp \left[\sum_i \alpha_i \phi_i \right] \ , \tag{23}$$

where α_i's are complex numbers. It follows from the property of the operator-derivatives that

$$\frac{\delta C}{\delta \phi_i} = \alpha_i C \ . \tag{24}$$

Our intention is to define a *normal product* of operators ϕ_i, which will also satisfy the same operator differentiation property in the sense that the operator derivative with respect to a ϕ_i of a normal product of n operators will generate a normal product of $(n-1)$ operators with the ϕ_i deleted. In addition, we want the average of a normal product to be zero unless the product has no ϕ-operators. Denoting this normal ordering by a curly bracket, we introduce an operator D via

$$D = \left\{ \exp \left[\sum_i \alpha_i \phi_i \right] \right\} \quad . \tag{25}$$

D is an exponential-like series in $\sum_i \alpha_i \phi_i$ in normal ordering. Since we want to preserve the same operator-derivative property of normal ordered products as in T-ordered products, we want $\delta D / \delta \phi_i$ to satisfy

$$\frac{\delta D}{\delta \phi_i} = \alpha_i D \tag{26}$$

in exact analogy to Eq. (24) for the T-ordered exponentials. Since D contains all powers $\sum_i \alpha_i \phi_i$, we may express D in terms of T-ordered products of ϕ_i's:

$$D = \sum_{k_1 \cdots k_n} d_{k_1 \cdots k_n} \, T \left[\phi_1^{k_1} \cdots \phi_n^{k_n} \right] \quad . \tag{27}$$

Using Eq. (26), we find that

$$\sum_{k_1 \cdots k_n} d_{k_1 \cdots k_n} \, k_i \, T \left[\phi_1^{k_1} \cdots \phi_i^{k_i - 1} \cdots \phi_n^{k_n} \right] = \alpha_i \sum_{k_1 \cdots k_n} T \left[\phi_1^{k_1} \cdots \phi_i^{k_i} \cdots \phi_n^{k_n} \right] \tag{28}$$

Equating the coefficients of $T \left[\phi_1^{k_1} \cdots \phi_i^{k_i - 1} \cdots \phi_n^{k_n} \right]$ on both sides, we have

$$k_i \, d_{k_1 \cdots k_i \cdots k_n} = \alpha_i \, d_{k_1 \cdots k_i - 1 \cdots k_n} \quad . \tag{29}$$

Starting with $k_i = 1$ for each k_i, and going upwards to $k_i = 2, 3, \ldots$, etc, we find that

$$d_{k_1 \cdots k_i \cdots k_n} = d_{0,0,0,\cdots,0} \frac{\alpha_1^{k_1} \cdots \alpha_n^{k_n}}{k_1! \cdots k_n!} \equiv D_0 \, \frac{\alpha_1^{k_1} \cdots \alpha_n^{k_n}}{k_1! \cdots k_n!} \quad . \tag{30}$$

Thus, D can be written as

$$D = D_0 \sum_{k_1 \cdots k_n} \frac{\alpha_1^{k_1} \cdots \alpha_n^{k_n}}{k_1! \cdots k_n!} T \left[\phi_1^{k_1} \cdots \phi_i^{k_i} \cdots \phi_n^{k_n} \right] \equiv D_0 T \exp \left(\sum_i \alpha_i \phi_i \right) \quad . \tag{31}$$

Since the average of a normal product $T \left[\phi_1^{k_1} \cdots \phi_i^{k_i} \cdots \phi_n^{k_n} \right]$ is zero unless $k_1 = k_2 = \cdots = k_n = 0$, we have, on taking average of Eq. (31), the relation

$$1 = D_0 \left\langle T \exp \left(\sum_i \alpha_i \phi_i \right) \right\rangle \quad . \tag{32}$$

From the gaussian factorization property, Eq. (18), it follows that

$$\left\langle T\exp\left(\sum_i \alpha_i\phi_i\right)\right\rangle = \exp\left(\frac{1}{2}\sum_{ij}\alpha_i\alpha_j\langle T[\phi_i\phi_j]\rangle\right) . \tag{33}$$

Hence, from Eqs. (25) and (32), we deduce

$$T\exp\left(\sum_i \alpha_i\phi_i\right) = \left\{\exp\left(\sum_i \alpha_i\phi_i\right)\right\}\exp\left(\frac{1}{2}\sum_{ij}\alpha_i\alpha_j\langle T[\phi_i\phi_j]\rangle\right) . \tag{34}$$

Expanding both sides of Eq. (34), and equating first power of each of $\alpha_1, \cdots, \alpha_n$, we have

$$\begin{aligned} T[\phi_1 \cdots \phi_n] &= \{\phi_1\cdots\phi_n\} + \sum_{i>j}\langle T[\phi_i\phi_j]\rangle\{\phi_1\cdots(\phi_i)\cdots(\phi_j)\cdots\phi_n\} \\ &+ \sum_{\substack{i>j\\k>l}}\langle T[\phi_i\phi_j]\rangle\langle T[\phi_k\phi_l]\rangle\{\phi_1\cdots(\phi_i)\cdots(\phi_j)\cdots(\phi_k)\cdots(\phi_l)\cdots\phi_n\} \\ &+ \cdots . \end{aligned} \tag{35}$$

where $\{\phi_1\cdots(\phi_i)\cdots(\phi_j)\cdots\phi_n\}$ etc. denote normal products with ϕ_i and ϕ_j missing. Introducing the contraction $\overline{\phi_i\phi_j} \equiv \langle T[\phi_i\phi_j]\rangle$, and the notation $\{\phi_1\cdots\overline{\phi_i\cdots\phi_j}\cdots\phi_n\} = \langle T[\phi_i\phi_j]\rangle\{\phi_1\cdots(\phi_i)\cdots(\phi_j)\cdots\phi_n\}$ etc., we find that

$$\begin{aligned} T[\phi_1\cdots\phi_n] &= \{\phi_1\cdots\phi_n\} \\ &+ \sum_{i>j}\{\phi_1\cdots\overline{\phi_i\cdots\phi_j}\cdots\phi_n\} \\ &+ \sum_{\substack{i>j\\k>l}}\{\phi_1\cdots\overline{\phi_i\cdots\phi_j}\cdots\overline{\phi_k\cdots\phi_l}\cdots\phi_n\} + \cdots . \end{aligned} \tag{36}$$

which shows how one can write a T-ordered product in terms of normal products with all possible contractions. Henceforth we shall refer to Eq. (36) as a 'Wick's theorem', defined with respect to a normal ordering corresponding to the averaging $\langle\cdots\rangle$. We note that we need not explicitly indicate the nature of the averaging process. The only requirement to arrive at Eq. (36) is the gaussian factorization property of the products of ϕ_i's with respect to the averaging, viz. the validity of Eq. (18).

From Eq. (34), we can also find out a generalized Wick's theorem, which shows how to rewrite an T-ordered product of two strings of operators which are themselves in normal order:

$$\begin{aligned} T[\{\phi_1\cdots\phi_m\}\{\psi_1 \cdots \psi_n\}] &= \{\phi_1\cdots\phi_m\,\psi_1\cdots\psi_n\} \\ &+ \sum_{ik}\{\phi_1\cdots\overline{\phi_i\cdots\phi_m\,\psi_1\cdots\psi_k}\cdots\psi_n\} \end{aligned}$$

$$+ \sum_{ijkl} \left\{ \phi_1 \cdots \overline{\phi_i \cdots \overline{\phi_j \cdots \phi_m \, \psi_1 \cdots \psi_k} \cdots \psi_l} \cdots \psi_n \right\}$$

$$+ \text{ all possible cross} - \text{contractions} \ . \tag{37}$$

Thus the T-ordered product of normal ordered products of operators can be written as normal products of all the operators plus normal products with only cross-contractions, excluding contractions between operators which are already in normal order. This relation can be generalized to a T-ordered product of more than two strings of normal ordered products of operators. We shall make extensive use of this theorem to generate the equations for the cluster amplitudes in our cluster-cumulant theory in various different situations.

The normal ordering introduced by us above pertains to boson variables which commute under T. Fermions anticommute under permutation under the T-ordering. This can be taken care of by writing fermion variables a_i in terms of Grassmann numbers [21] and some auxiliary boson variables ϕ_i:

$$\phi_i = \gamma_i a_i \ , \tag{38}$$

where γ_i's are Grassmann numbers which anticommute among themselves and also with the fermion variables. Then, using the anticommuting properties of the Grassmann numbers, we can move all the γ's in Eq. (36) to the left, and get rid of them by Berezin integration [21]. The result is a Wick's theorem for fermions, looking exactly like Eq. (36), but now the fermions anticommute under the normal ordering, and there is a nontrivial phase factor present in the terms in normal order with contractions:

$$\{a_1 \cdots \overline{a_i \cdots a_j} \cdots a_n\} = (-1)^\eta \, \overline{a_i a_j} \, \{a_1 \cdots (a_i) \cdots (a_j) \cdots a_n\} \ , \tag{39}$$

with η as the parity of permutations to bring the contracted pair side by side. Although we have studied femionic many-body systems such as Lipkin model with our formalism, we shall not consider applications to fermions in this paper.

3 A Thermal Cluster-Cumulant Theory (TCC) for GPF: Formalism and Applications

3.1 General Considerations

To develop the Thermal Cluster Cumulant theory for bosons/fermions, we start from the expression Eq. (9) for Z_q/Z_{0q}. As discussed in Sects. 1 and 2, we would utilize the gaussian factorization property of the T-ordered products of creation/annihilation operators to posit on U_I a thermal normal ordered exponential ansatz, Eq. (10). For bosons, S_q operators commute under the normal ordering. For fermions, there are always even number of creation / annihilation operators in S_q, and thus they commute again. If we substitute the ansatz, Eq. (10), in the Heisenberg's equation for imaginary-time, Eq. (4), and utilize the fact that the cluster operators S_q commute under the normal ordering, we derive

$$-\left\{\left[\frac{\partial S_q(\tau)}{\partial\tau}+\frac{\partial x_q(\tau)}{\partial\tau}\right]\; e^{(S_q(\tau)+x_q(\tau))}\right\}_\beta = V_I(\tau)\;\left\{e^{(S_q(\tau)+X_q(\tau))}\right\}_\beta \;. \quad (40)$$

The advantage of using a thermal normal order ansatz for U_I is evident, since we have a rather simple expression for the 'time-derivative' of U_I. Using again the generalized thermal Wick's theorem Eq. (37) on the right side of Eq. (40), we have

$$-\left\{\left[\frac{\partial S_q(\tau)}{\partial\tau}+\frac{\partial x_q(\tau)}{\partial\tau}\right]\quad e^{(S_q(\tau)+x_q(\tau))}\right\}_\beta$$

$$=\left\{\overline{V_I(\tau)\;\{e^{(S_q(\tau)+x_q(\tau))}\}}\;\{e^{(S_q(\tau)+x_q(\tau))}\}\right\}_\beta \quad (41)$$

where $\overline{V_I\;\exp(S_q+x_q)}$ involves thermal contractions between V_I and various powers of S_q and x_q from the exponential, omitting contractions among S_q and x_q. But x_q is just a number and hence cannot contract with V_I. Since the various components of S_q and the x_q are linearly independent, it follows that [16, 17]

$$-\frac{\partial S_q(\tau)}{\partial\tau}=\left\{\overline{V_I\;e^{S_q(\tau)}}\right\}_{ex}\;, \quad (42)$$

$$-\frac{\partial x_q(\tau)}{\partial\tau}=\left\{\overline{V_I\;e^{S_q(\tau)}}\right\}_{cl}\;, \quad (43)$$

where the suffices *ex* and *cl* denote the operator (excitation) and the number (closed, completely contracted) components of the composites.

Equations (42) and (43) are the principal working equations of our thermal cluster-cumulant (TCC) theory for interacting many-particle systems. The connectedness of the right-hand side of the equations indicates that $S_q(\beta)$ and $x_q(\beta)$ will be connected operators, justifying the term 'cluster-cumulant' used to denote them. Any truncation of the right side of Eqs. (42) and (43) would still preserve the connectedness of the S_q and x_q. Since Z_q/Z_{0q} is the thermal average of U_I, only the zero-body (number) component of U_I will servive, owing to thermal normal ordering present in the expression of U_I. The advantage of thermal normal ordering is again evident since otherwise we would have contractions from the operators S_q and x_q in Z_q/Z_{0q}.

Since x_q is a connected quantity-even when computed via some truncated series of Eqs. (42) and (43), it follows that ΔF is also a connected quantity and hence is an extensive quantity. The chemical potential, μ should be determined from the condition that

$$\bar{N}=\frac{\langle NU_I\rangle}{\langle U_I\rangle} \quad (44)$$

has a preassigned value. Since there are both exponentially growing and decaying terms in U_I, a direct solution of Eqs. (42) and (43) might entail potential 'stiffness' and hence numerical instability. To obviate this, it seems better (unlike in the real-time theories) to revert to the 'Schrödinger picture', and solve the corresponding cluster operators $S(\tau)$, defined as

$$S(\tau) = e^{-\tau(H_0-\mu N)} S_q(\tau) e^{\tau(H_0-\mu N)} \ . \tag{45}$$

The cluster-cumulant equations for $S(\tau)$ and $x_q(\tau)$ are given by

$$-\frac{\partial S(\tau)}{\partial \tau} = [(H_0 - \mu N), S(\tau)] + \left\{\overline{V\ \exp(S)}\right\}_{ex} \ . \tag{46}$$

$$-\frac{\partial x_q(\tau)}{\partial \tau} = \left\{\overline{V\ \exp(S)}\right\}_{cl} \ . \tag{47}$$

In our numerical applications, we shall solve for S first from Eq. (46), then x_q from Eq. (47) and finally compute Z_q/Z_{0q} from Eq. (11). The free energy can now be obtained by using Eqs. (12).

If we expand the cluster-cumulants in Eqs. (42) and (43) in order of V_I, we generate in a straightforward manner the variant of the thermodynamic perturbation theory of Refs. [5, 6]. Bloch, Balian and de Dominicis [5, 6] had formulated another resummed version of the perturbative formulation, which involves 'perturbed' or exact occupation probabilities. This is very useful for treating translationally invariant systems. Our present cluster-cumulant formulation does not naturally generate this perturbation series. By introducing a suitable self-consistent one particle potential, we may, however, generate such a resummed version.

We should mention that Altenbokum *et al.* [22] had suggested earlier a coupled cluster approach for canonical partition functions, which required the knowledge of the eigenspectra of H, and is thus structurally very different from our more direct formulation.

We emphasize here that the thermal field theory developed by us in Sect. 2 and Sect. 3.1, unlike in the traditional thermofield dynamics [10] never explicitly used the concept of a thermal vacuum nor introduced auxiliary variables for computing the thermal averages. The thermal Wick's theorem derived by us requires only the knowledge of contractions.

The right sides of Eqs. (42) and (43) are finite power series expansion of S_q. This follows from the normal ordered ansatz for $U_I(\beta)$, since all the S_q's in $\left\{\overline{V\ \exp[S_q(\tau)]}\right\}_\beta$ have then to be joined to V_I, and V_I has only a finite number of creation/annihilation operators. Thus our TCC equations share the best features of the zero-temperature coupled cluster (CC) theory where the CC equations are also finite power series in the cluster operators.

In the zero-temperature limit, the TCC equations reduce to the stationary solutions of the single reference coupled cluster theory in the same truncation scheme [23, 24]. In the high temperature limit, TCC reduces to a truncated version of the classical partition function. But since the computation of the exact classical partition function is much easier to perform, it seems natural to seek for an extention of our TCC where the classical component is incorporated fully. This we undertake below.

3.2 Complete Inclusion of the Classical Component in GPF

To see how we can incorporate the exact classical contribution in the TCC theory, we look upon the values of S_q and x_q as being generated as perturbative series in $V_I(\tau)$. Any typical term will be a linked term joining various V_I's, with contractions as propagators between the various vertices. We illustrate our development with bosonic systems with a number nonconserving H, so that $\mu = 0$.

A propagator $G_q(\tau_1, \tau_2)$ is given by

$$\begin{aligned} G_q(\tau_1, \tau_2) &= \left\langle T[b^\dagger(\tau_1) b(\tau_2)] \right\rangle_0 , \quad \forall \ \tau \\ &= G_q^+(\tau_1, \tau_2)\, \theta(\tau_1 - \tau_2) + G_q^-(\tau_1, \tau_2)\, \theta(\tau_2 - \tau_1) \ . \end{aligned} \tag{48}$$

where $b^\dagger/b$ are the creation/annihilation operators. The propagator $G_q(\tau_1, \tau_2)$ for $\tau_1 > \tau_2$, $G_q^+(\tau_1, \tau_2)$, is the *forward* component, and is given by

$$G_q^+(\tau_1, \tau_2) = \left\langle b^\dagger b \right\rangle_0 \, e^{\omega_0 (\tau_1 - \tau_2)} = n_q \, e^{\omega_0 (\tau_1 - \tau_2)} \ , \tag{49}$$

where

$$n_q = \frac{1}{e^{\beta \omega_0} - 1} \ , \tag{50}$$

and ω_0 is the unperturbed frequency.

Similarly, $G_q(\tau_1, \tau_2)$ for $\tau_1 \leq \tau_2$, $G_q^-(\tau_1, \tau_2)$ is the *backward* component, and is given by

$$G_q^-(\tau_1, \tau_2) = \left\langle b b^\dagger \right\rangle_0 \, e^{\omega_0 (\tau_1 - \tau_2)} = (n_q + 1)\, e^{\omega_0 (\tau_1 - \tau_2)} \ , \tag{51}$$

where

$$n_q + 1 = \frac{e^{\beta \omega_0}}{e^{\beta \omega_0} - 1} \ . \tag{52}$$

It is well-known [3, 20] that $G_q^\pm(\tau_1, \tau_2)$ are periodic functions in β. Hence, they can be written as Fourier sums over the harmonics of a fundamental frequency, $2\pi/\beta$. For G_q^+, we have

$$G_q^+(\tau_1, \tau_2) = \frac{1}{\beta} \sum_{m=0}^{\infty} \frac{e^{i\Omega_m (\tau_1 - \tau_2)}}{(\omega_0 + i\Omega_m)} = G_c^+ + G_f^+(\tau_1, \tau_2) \ , \tag{53}$$

where $\Omega_m (= 2\pi m/\beta)$ are called the Matsubara frequencies. We are particularly interested in the $m = 0$ component of the propagator. It is easy to see that in the high temperature limit, $\beta \to 0$ and the only contribution to the propagator comes from the $m = 0$ Matsubara frequency. This would then correspond to the *classical* component of the propagator G_q^+, and let us denote it by $G_c^+(=1/\beta\omega_0)$. The rest, denoted by G_f^+, is given by,

$$G_f^+(\tau_1,\tau_2) = \frac{1}{\beta}\sum_{m=1}^{\infty}\frac{e^{i\Omega_m(\tau_1-\tau_2)}}{(\omega_0 + i\Omega_m)} , \tag{54}$$

and is the *fluctuation* part of the propagator.

Similarly, the corresponding Fourier representation for the backward propagator is given by

$$G_q^-(\tau_1,\tau_2) = \frac{1}{\beta}\sum_{m=0}^{\infty}\frac{e^{i\Omega_m(\tau_1-\tau_2)}}{(\omega_0 - i\Omega_m)} = G_c^- + G_f^-(\tau_1,\tau_2) , \tag{55}$$

where $G_c^-(=1/\beta\omega_0)$ is again the classical counter part; and

$$G_f^-(\tau_1,\tau_2) = \frac{1}{\beta}\sum_{m=1}^{\infty}\frac{e^{i\Omega_m(\tau_1-\tau_2)}}{(\omega_0 - i\Omega_m)} . \tag{56}$$

We should note that in the classical limit, viz. at high temperature ($\beta \to 0$), the values of the forward and the backward propagators become equal and converge to $G_c^+ = G_c^- = 1/\beta\omega_0$. In the high temperature limit, $G_q \to G_c$ and hence one should land up with just the exact classical partition function, Z_c for an untruncated calculation with

$$\frac{Z_c}{Z_{0c}} = \frac{\int dq\, e^{-\beta[\frac{1}{2}\Omega^2 q^2 + V(q)]}}{\int dq\, e^{-\beta[\frac{1}{2}\Omega^2 q^2]}} . \tag{57}$$

But in practice, we invoke a truncation in the rank of S_q. This would then produce an approximate partition function, Z_q^a. Any inclusion of a class of cluster operators contributing to the free energy has a built-in 'classical' component, when all contractions are classical, i.e., evaluated by propagators $G_c^\pm$, and the rest which constitute the quantum (fluctuation) correction. Hence, the classical limit of Z_q^a (i.e., in the truncated calculation) will include only the subset of terms generated by the S_q included in the truncated calculation, but evaluated using the classical propagators. This would yield an approximate classical partition function, Z_c^a. The corresponding approximate harmonic classical partition function Z_{0c} is the classical counterpart of Z_{0q} of Eq. (7). The ratio Z_c^a/Z_{0c} in Eq. (57) is the classical limit of Z_q^a/Z_{0q}. Now Z_q^a/Z_{0q} can be written as $\exp(x_q^a)$, involving the approximate x_q, according to the traditional TCC. Hence the classical limit of Z_q^a/Z_{0q} is given by $\exp(x_c^a)$ where x_c^a is the classical counterpart of x_q^a. Thus,

$$Z_c^a = Z_{0c}\, e^{x_c^a} . \tag{58}$$

We want now to replace the contribution of the approximate classical partition function by the *exact* classical value, Z_c such that our computed Z_q

exactly includes the full Z_c. To do this, we compute x_c^a by the same TCC, but replacing G_q by G_c. The difference of x_q^a and x_c^a is the contribution of the fluctuation part; we call it x_f^a. When no approximation is made, $x_q^a = x_q$; $x_c^a = x_c$; $x_f^a = x_f$. Using these considerations, one can factorize Z_q as

$$Z_q = Z_{0c} Z_{0f} \, e^{x_c + x_f} \ , \tag{59}$$

where $Z_{0c} (= 1/\beta\omega_0)$ is the classical limit of the unperturbed partition function, Z_{0q}.

Similarly, for the approximate calculation one can also have

$$Z_q^a = Z_{0c} Z_{0f} \, e^{x_c^a + x_f^a} = Z_c^a Z_{0f} \, e^{x_f^a} \ , \tag{60}$$

where

$$x_f^a = x_q^a - x_c^a \ . \tag{61}$$

By replacing the approximate classical partition function, Z_c^a by the exact one, Z_c, we may obtain a much better approximation to the quantum partition function, Z_q; we call it $\overline{Z}_q^a$. Thus $\overline{Z}_q^a$ is given by

$$\overline{Z}_q^a = Z_c Z_{0f} \, e^{x_q^a - x_c^a} = \overline{Z}_{0q} \, e^{x_q^a - x_c^a} \ , \tag{62}$$

where

$$\overline{Z}_{0q} = Z_c \frac{(\beta\omega_0/2)}{\sinh(\beta\omega_0/2)} \ , \tag{63}$$

where ω_0 is the frequency of the zeroth order unperturbed Hamiltonian.

Since we have separated the quantum propagator $G_q(\tau_1, \tau_2)$ into a purely classical component G_c, and a fluctuation part $G_f(\tau_1, \tau_2)$, this automatically separates the cluster amplitudes S_q into a classical part S_c, and a fluctuation part S_f. We obtain the purely *classical component* of the TCC equations as

$$-\frac{\partial S_c(\tau)}{\partial \tau} = \left\{ \overline{V_c(\tau) \exp\left[S_c(\tau)\right]} \right\}_\beta^{ex} \ , \tag{64}$$

$$-\frac{\partial x_c(\tau)}{\partial \tau} = \left\{ \overline{V_c(\tau) \exp\left[S_c(\tau)\right]} \right\}_\beta^{cl} \ , \tag{65}$$

where $V_c(\tau)$ is the classical limit of the thermally normal ordered potential $V(\tau)$, and all contractions in Eqs. (64) and (65) are evaluated in terms of G_c. By solving Eqs. (64) and (65) for x_c and using Eq. (59), we obtain

$$Z_c = Z_{0c} \, e^{x_c} \ . \tag{66}$$

Again, for a truncated calculation, we obtain an approximate x_c^a, yielding an approximate Z_c^a:

$$Z_c^a = Z_{0c} \, e^{x_c^a} \ . \tag{67}$$

Finally, one can obtain the partition function, $\overline{Z}_q^a$ from Eq. (62) which incorporates the classical component of partition function *exactly*. We shall illustrate the formalism by taking specific example in Sect. 4.3.

4 A Path-Integral-Based Cluster-Cumulant Theory for GPF: Formalism and Applications

4.1 A Path-Integral Inspired Transcription of TCC

Our starting point is the path-integral expression for the pf given in Eq. (13). Introducing the unperturbed harmonic pf via

$$Z_{0q} = \oint D[x] \exp\left(-\int_0^\beta \left[\frac{\dot{x}^2(\tau)}{2} + \frac{1}{2}\Omega^2 x^2(\tau)\right] d\tau\right) \tag{68}$$

we can write Z_q/Z_{0q} as

$$\frac{Z_q}{Z_{0q}} = \frac{\oint D[x]\, e^{-\int_0^\beta H_0[x(\tau)]d\tau}\, e^{-\int_0^\beta [V[x(\tau)]+\frac{1}{2}(\omega^2-\Omega^2)x^2(\tau)]d\tau}}{\oint D[x]\, e^{-\int_0^\beta H_0[x(\tau)]d\tau}} , \tag{69}$$

where

$$H_0[x(\tau)] = \frac{\dot{x}^2(\tau)}{2} + \frac{1}{2}\Omega^2 x^2(\tau) . \tag{70}$$

Z_q/Z_{0q} can then be expressed as

$$\frac{Z_q}{Z_{0q}} = \left\langle \exp\left(-\int_0^\beta \left[V[x(\tau)] + \frac{1}{2}\omega^2 x^2(\tau) - \frac{1}{2}\Omega^2 x^2(\tau)\right] d\tau\right)\right\rangle , \tag{71}$$

where the symbol $\langle\cdots\rangle$ denotes the average with respect to the harmonic 'reference' path-integral corresponding to the frequency Ω with its weight measure. Following Feynman [8], we write the cyclical path $x(\tau)$ as a Fourier series involving the Fourier components x_m :

$$x(\tau) = x_0 + \sum_{m=1}^{\infty}(x_m e^{i\omega_m\tau} + c.c.) , \tag{72}$$

$$= x_0 + y_+(\tau) + y_-(\tau) , \tag{73}$$

where $y_\pm(\tau)$ denote the cyclical paths describing the quantum fluctuations around the centroid, x_0, of each point of the quantum path $x(\tau)$. x_0 is given by

$$x_0 = \frac{1}{\beta}\int_0^\beta x(\tau)d\tau . \tag{74}$$

ω_m in Eq. (94) is the m-th Fourier harmonic of the fundamental Matsubara frequency ω_0.

As we emphasized in Sect. 2, $x(\tau)$'s satisfy the gaussian factorization property with respect to this average. Hence all the averages for a function of $x(\tau)$ can be evaluated in terms of the two-time correlation function $\langle x(\tau_1)x(\tau_2)\rangle \equiv C(\tau_1, \tau_2)$. Using the periodicity of $x(\tau)$, we have

$$C(\tau_1,\tau_2) = \frac{1}{\beta\Omega^2} + \frac{1}{2\beta\Omega}\left[\sum_{m\neq 0}\frac{e^{i\omega_m(\tau_1-\tau_2)}}{\omega - i\omega_m} + \sum_{m\neq 0}\frac{e^{-i\omega_m(\tau_1-\tau_2)}}{\omega - i\omega_m}\right] \quad (75)$$

$$\equiv \langle x_0^2\rangle_{x_0} + \langle y_+(\tau_1)y_-(\tau_2)\rangle_y + \langle y_-(\tau_1)y_+(\tau_2)\rangle_y \ , \quad (76)$$

where $\langle\cdots\rangle_{x_0}$ and $\langle\cdots\rangle_y$ indicate averaging with respect to x_0 and y respectively. Introducing variables $x_+(\tau)$ and $x_-(\tau)$ by the relations

$$x(\tau) = x_+(\tau) + x_-(\tau) \ , \quad (77)$$

with

$$x_\pm(\tau) = x_{0\pm} + y_\pm(\tau) \ , \quad (78)$$

where x_{0+} and x_{0-} are assumed to have the following non-zero averages:

$$\langle x_{0+}x_{0-}\rangle_{x_0} = \langle x_{0-}x_{0+}\rangle_{x_0} = \frac{1}{2}\langle x_0^2\rangle_{x_0} \ , \quad (79)$$

we can write the correlation function as

$$\langle x(\tau_1)x(\tau_2)\rangle = \langle x_+(\tau_1)x_-(\tau_2)\rangle + \langle x_-(\tau_1)x_+(\tau_2)\rangle \ . \quad (80)$$

Thus we can generate the traditional TCC in the path-integral framework by treating $x_+(\tau)$ and $x_-(\tau)$ as the gaussian variables and writing the exponential in the path-integral expression for Z_q/Z_{0q} in terms of a normal ordered exponential representation involving cluster operators. The normal ordering is obviously with respect to the path-integral measure. The connection with the traditional field-theoretic treatment becomes complete if we can identify $x_+(\tau)$ and $x_-(\tau)$ as $\frac{1}{\sqrt{2\Omega}}a_I^\dagger(\tau)$ and $\frac{1}{\sqrt{2\Omega}}a_I(\tau)$ respectively. Using suitable integral representation of $C(\tau_1,\tau_2)$, one can verify that they are indeed so.

Our traditional TCC theory can thus be exactly transcribed into a path-integral based theory, where x_+ and x_- play the same role as creation and annihilation field-operators, and the path-integral average plays the role of the thermal trace.

We note that both the present and the traditional TCC method provide a 'global' description of the various points on the classical path, since the coordinate x_0 is also integrated over in the propagators. We now look for a path-integral based TCC where the integral over x_0 will be performed only after a cluster expansion summation of the fluctuation is performed. In that case, we separate the contribution of the fluctuation component for each point of the path-centroid variable, and finally integrate over all points of the centroid path. This would then provide a method of choosing a different reference point for each point of the centroid path, and would add more flexibility to the formalism. This will also be a 'local' description.

4.2 A Path-Integral-Based Cluster-Cumulant Formalism: A Local Inclusion of the Classical Component in GPF

Since at a high enough temperature, $\beta \to 0$, the terms containing $y(\tau)$ makes vanishing contribution to the correlation function and hence to the path-integral, the only surviving term is from the centroid path coordinate x_0. x_0 can thus be identified as the *classical path.* In terms of x_0 and y, the partition function can be expressed as

$$
\begin{aligned}
Z_q &= \oint D[x] \exp\left(-\int_0^\beta \left[\frac{\dot{x}^2}{2} + \frac{1}{2}\omega^2 x^2 + V(x)\right] d\tau\right) \\
&= \frac{1}{\sqrt{2\pi\beta}} \int_{-\infty}^{\infty} dx_0 \, \exp\left(-\beta\left[\frac{1}{2}\omega^2 x_0^2 + V(x_0)\right]\right) \\
&\times \oint D[y] \exp\left(-\int_0^\beta \left[\frac{\dot{y}^2}{2} + \frac{1}{2}\omega^2 y^2 + V(x_0+y) - V(x_0)\right] d\tau\right) \qquad (81)
\end{aligned}
$$

Let us define the zeroth-order 'local' pf for the fluctuation component as

$$
\zeta_0 = \oint D[y] \exp\left(-\int_0^\beta \left[\frac{\dot{y}^2}{2} + \frac{1}{2}\Omega^2(x_0) y^2\right] d\tau\right) , \qquad (82)
$$

and the perturbed counterpart as

$$
\begin{aligned}
\zeta = \oint D[y] \exp\Bigg(-\int_0^\beta \Bigg[&\left(\frac{\dot{y}^2}{2} + \frac{1}{2}\Omega^2(x_0) y^2\right) \\
&+ \left(\frac{1}{2}\omega^2 y^2 - \frac{1}{2}\Omega^2(x_0) y^2 + V(x_0+y) - V(x_0)\right)\Bigg] d\tau\Bigg) . \qquad (83)
\end{aligned}
$$

Then the ratio ζ/ζ_0 is given by

$$
\begin{aligned}
\frac{\zeta}{\zeta_0} &= \left\langle \exp\left(-\int_0^\beta \left[\frac{1}{2}\omega^2 y^2 - \frac{1}{2}\Omega^2(x_0) y^2 + V(x_0+y) - V(x_0)\right] d\tau\right)\right\rangle_y \\
&\equiv \langle U(\beta)\rangle_y , \qquad (84)
\end{aligned}
$$

where $\langle\cdots\rangle_y$ denotes average with respect to the fluctuation. Our description is local because we have the flexibility to use a different reference frequency $\Omega(x_0)$ at each centroid point x_0. We now note that the product of $y(\tau)$'s also satisfies the gaussian factorization property with respect to the y-average. Hence we can again define normal order of the y-products with respect to the y-average and use Wick's expansion in normal order to generate $U(\beta)$. Our intention is to write the exponential, symbolized as $U(\beta)$ in Eq. (84) as an exponential type operator in normal order. Unfortunately, unlike in the case of creation/annihilation operators, the contractions for the pairs $(y_+ y_-)$ etc. do not generate simple exponentials with 'time' differences. This would preclude developing a TCC formalism, since the TCC formalism is predicated by the factorizability of the propagators with two arguments as products of two factors with its own argument. To bypass this difficulty, we rewrite the fluctuation propagators as differences between a quantum propagator and a classical propagator:

$$g_+(T-\tau_1) = G_q^+(T-\tau_1) - G_c^+ \ , \tag{85}$$

$$g_-(\tau_2-T) = G_q^-(\tau_2-T) - G_c^- \ . \tag{86}$$

The values of the classical propagators G_c^+ and G_c^- are both equal to $1/\beta\Omega(x_0)$, which we call n_c from now on. We note the similarity of Eqs. (85) and (86) to Eqs. (53) and (55), with the major difference now that the frequency $\Omega(x_0)$ has a dependence on the centroid point x_0. This implies the manifest locality of our description here.

We can write y_+/y_- as differences of the 'quantum' variables A_+/A_- and the 'classical' variables A_{0+}/A_{0-} as

$$y_\pm = A_\pm - A_{0\pm} \ . \tag{87}$$

Let us posit the following nonzero averages (i.e., contractions) of the variables A:

$$\langle A_+(\tau_1)A_-(\tau_2)\rangle = n_q \, e^{\Omega(\tau_1-\tau_2)} \ , \tag{88}$$

$$\langle A_-(\tau_1)A_+(\tau_2)\rangle = (n_q+1) \, e^{-\Omega(\tau_2-\tau_1)} \ , \tag{89}$$

$$\langle A_{0+}(\tau_1)A_{0-}(\tau_2)\rangle \equiv \langle A_{0+}A_{0-}\rangle = -n_c \ , \tag{90}$$

$$\langle A_{0-}(\tau_1)A_{0+}(\tau_2)\rangle \equiv \langle A_{0-}A_{0+}\rangle = -n_c \ . \tag{91}$$

We now rewrite the entity $\langle U(\beta)\rangle_y$ in terms of a normal ordered exponential representation of $U(\beta)$. To this end, we introduce $U(\tau)$ as a normal ordered exponential:

$$U(\tau) = \{\exp(\sigma(\tau) + \chi(\tau))\}_y \ , \tag{92}$$

where

$$\sigma = \sum_{m,n,k,l} \{A_+^m A_-^n A_{0+}^k A_{0-}^l\}_y \, \sigma_{m,n}^{k,l} \tag{93}$$

is the operator part of U and χ is the number part. The powers of σ in $U(\tau)$ are to be interpreted as

$$\frac{1}{2}\{\sigma^2\} = \frac{1}{2} \sum_{\substack{m1,n1\\k1,l1\\m2,n2\\k2,l2}} \left\{A_+^{m1+m2} A_-^{n1+n2} A_{0+}^{(k1+k2)} A_{0-}^{(l1+l2)}\right\}_y \sigma_{m1,n1}^{k1,l1} \sigma_{m2,n2}^{k2,l2} \ . \tag{94}$$

The curly bracket with the suffix y denotes the normal ordering corresponding to the average with respect to y (or equivalently to A's).

We can now set up the TCC equations for determining σ and χ:

$$-\frac{\partial \sigma_{m,n}^{k,l}[x_0,\tau]}{\partial \tau} = \left\{\overline{V(\tau)\exp(\sigma)}\right\}_{m,n}^{k,l} , \tag{95}$$

$$-\frac{\partial \chi[x_0,\tau]}{\partial \tau} = \left\langle \overline{V(\tau)\exp(\sigma)} \right\rangle , \tag{96}$$

where the right side of Eq. (96) is the completely contracted component of $\langle V(\tau)\exp(\sigma)\rangle$.

From this, we get

$$\frac{\zeta}{\zeta_0} = e^{\chi[x_0,\beta]} , \tag{97}$$

where χ depends on x_0 via the Hamiltonian, appearing in the exponential of Eq. (84), which depends on x_0. Thus, having solved for $\chi[x_0,\beta]$, we get Z_q as

$$Z_q = \frac{1}{\sqrt{2\pi\beta}} \int_{-\infty}^{\infty} dx_0 \; e^{-\beta\left[\frac{1}{2}\omega^2 x_0^2 + V(x_0)\right]} \; \zeta_0 \; e^{\chi[x_0,\beta]} . \tag{98}$$

Defining a nonperturbative effective potential $V_{eff}(x_0)$ via

$$V_{eff}(x_0) = \frac{1}{2}\omega^2 x_0^2 + V(x_0) - \frac{1}{\beta}\ln\zeta_0 - \frac{1}{\beta}\chi[x_0,\beta] , \tag{99}$$

we have

$$Z_q = \frac{1}{\sqrt{2\pi\beta}} \int_{-\infty}^{\infty} dx_0 \; e^{-\beta V_{eff}(x_0)} . \tag{100}$$

Since V_{eff} contains the complete classical component V of the potential, our formulation automatically subsumes the total classical partition function. In practice, we first generate an optimal reference ζ_0 by defining a suitable frequency $\Omega(x_0)$ at each x_0 by optimizing the free energy. We shall illustrate this choice by taking the concrete example of the anharmonic oscillator in the next section.

5 Applications: The Anharmonic Oscillator with Cubic Plus Quartic Perturbations

In this section, we shall apply our two TCC formalisms described in Sects. 3 and 4 :

5.1 The Optimal Zeroth-Order 'Global' Description

We shall apply our extended TCC formalism using Eq. (62), to the model one-dimensional anharmonic oscillator Hamiltonian:

$$H = \frac{p^2}{2} + \frac{x^2}{2} + \gamma\, x^3 + \lambda\, x^4 = \frac{p^2}{2} + V(x) , \tag{101}$$

where γ and λ are the coupling constants for cubic and quartic perturbations respectively. In terms of the creation/annihilation operators, $a^\dagger/a$ the Hamiltonian becomes

$$H = (a^\dagger a + \frac{1}{2}) + \frac{\gamma}{2\sqrt{2}}(a^\dagger + a)^3 + \frac{\lambda}{4}(a^\dagger + a)^4 \ . \tag{102}$$

In ref. [17], we have presented in detail the procedure of computing the approximate quantum partition function, Z_q^a for this model Hamiltonian. Since the harmonic component of this Hamiltonian would be a poor starting point when the anharmonic coupling is strong, we induce a Bogoliubov transformation of the creation/annihilation operators to generate an optimal starting function in a certain sense. Inducing the transformation given by

$$b = (1 - t^2)^{-\frac{1}{2}}(a - ta^\dagger - d) \ , \tag{103}$$

$$b^\dagger = (1 - t^2)^{-\frac{1}{2}}(a^\dagger - ta - d) \ , \tag{104}$$

the Hamiltonian H can be written in thermal normal order in terms of $b/b^\dagger$. Using the Gibbs-Bogoliubov inequality [8] $Z_q \geq Z_{0q}$ where the Hamiltonian is in thermal normal order, we get $F_q \leq F_{0q}$. The optimal transformation can then be obtained by minimizing the unperturbed free energy with respect to the parameters t and d. We have computed x_q and x_c by solving Eqs. (43) and (65) by including in our calculation cluster operators up to rank four. The partition function, $\overline{Z}_q^a$ can now be evaluated from Eq. (62) using the value of x_q and x_c and also the free energy, $\overline{F}_q^a$. For an assesment of the importance of the inclusion of the classical component, we have also computed the traditional free energy, F_q^a, using Eq. (12).

5.2 An Optimal Zeroth-Order 'Local' Description

To apply our path-integral based TCC formalism, Eq. (98), we first generate a fluctuation part, $\mathcal{H}$ corresponding to the Hamiltonian given in Eq. (101) as

$$\mathcal{H} = \frac{\dot{y}^2(\tau)}{2} + \frac{1}{2}\, W^2(x_0)\, y^2(\tau) + \gamma(x_0)\, y^3(\tau) + \lambda\, y^4(\tau) \ , \tag{105}$$

where $W(x_0) = 1 + 6\gamma\, x_0 + 12\lambda\, x_0^2$ and $\gamma(x_0) = \gamma + 4\lambda\, x_0$. The corresponding classical component is given by

$$V(x_0) = \frac{1}{2}x_0^2 + \gamma\, x_0^3 + \lambda\, x_0^4 \ . \tag{106}$$

Hereafter we shall omit the imaginary-time argument 'τ' for notational convenience. Since the $y_\pm$ variables have properties similar to $1/\sqrt{2\Omega}$ times creation/annihilation operators denoted by $a^\dagger/a$, we have

$$\mathcal{H} = W(x_0)\left[a^\dagger a + \frac{1}{2}\right] + \overline{\gamma}(x_0)\left(a + a^\dagger\right)^3 + \overline{\lambda}(x_0)\left(a + a^\dagger\right)^4 \ , \tag{107}$$

where $\overline{\gamma}(x_0) = \gamma(x_0)[2W(x_0)]^{-\frac{3}{2}}$ and $\overline{\lambda}(x_0) = \lambda[2W(x_0)]^{-2}$.

In the strong coupling regime, the unperturbed harmonic-like component of $\mathcal{H}$ forms a poor starting point, and just as in Sect. 3, we should look for an optimal zeroth-order description. Thus we induce a Bogoliubov transformation [18], of the variables $a^\dagger/a$ to a new set of variables $b^\dagger/b$ such that

$$b^\dagger = (1-t^2)^{-\frac{1}{2}}(a^\dagger - ta) \ . \tag{108}$$

$$b = (1-t^2)^{-\frac{1}{2}}(a - ta^\dagger) \ . \tag{109}$$

Since y is a cyclical variable we do not need to introduce any shift parameter in the Bogoliubov transformation. Inducing the Bogoliubov transformation through Eqs. (108) and (109) in Eq. (107) we can rewrite $\mathcal{H}$ in normal order in terms of $b/b^\dagger$. The optimal value of t is obtained by minimizing the 'local' free energy $[-\ln \zeta_0/\beta]$. With this optimal t, the Hamiltonian $\mathcal{H}$ takes the form

$$\begin{aligned}\mathcal{H} &= \Omega(x_0)\left(b^\dagger b + \frac{1}{2}\right) + \langle \mathcal{H}\rangle_y - \Omega(x_0)\left(n_f + \frac{1}{2}\right) \\ &+ \Gamma(x_0)\{\frac{1}{3!}\left(b+b^\dagger\right)^3\}_y + \Lambda(x_0)\{\frac{1}{4!}\left(b+b^\dagger\right)^4\}_y \ ,\end{aligned} \tag{110}$$

where n_f is the contraction $\langle b^\dagger b\rangle_y$, given by

$$n_f = \frac{1}{2}\coth\left(\frac{1}{2}\beta\Omega(x_0)\right) - \frac{1}{\beta\Omega(x_0)} - \frac{1}{2} \ . \tag{111}$$

The unperturbed Hamiltonian, $\mathcal{H}_0$ is given by

$$\mathcal{H}_0 = \Omega(x_0)\left(b^\dagger b + \frac{1}{2}\right) + \langle \mathcal{V}\rangle_y \ , \tag{112}$$

where $\langle \mathcal{V}\rangle_y = \langle \mathcal{H}\rangle_y - \Omega(x_0)\left(n_f + \frac{1}{2}\right)$.

If we substitute just the value of ζ_0 obtained from above to the expression, Eq. (98) for Z_q, and ignore the perturbation part of the fluctuation, then we obtain the total, centroid-integrated, unperturbed partition function

$$Z_{0q} = \frac{1}{\sqrt{2\pi\beta}}\int_{-\infty}^{\infty} dx_0\, e^{-\beta V(x_0)}\, \frac{\beta\Omega(x_0)/2}{\sinh(\beta\Omega(x_0)/2)}\, e^{-\beta\langle \mathcal{V}\rangle_y} \ . \tag{113}$$

We now evaluate the quantity $\chi(x_0,\beta)$ in the Eq. (98) by applying our local TCC strategy. To do this we express our renormalized TCC operator, $\sigma^{k,l}_{m,n}(x_0,\beta)$ in Eq. (95) as a sum of normal ordered operators up to rank 4, i.e., we include in $\sigma^{k,l}_{m,n}$ all operators whose total rank $m+n+k+l \leq 4$. $\chi(x_0,\beta)$ is the totally contracted number part as stated earlier.

We have constructed the path-integral based TCC equations by truncating the exponential of the right hand side of the exact Eqs. (95) and (96) up to linear terms. To achieve a better numerical stability, the TCC equations thus obtained are converted to the Schrödinger picture of the cluster amplitudes ($\overline{\sigma}^{k,l}_{m,n}$):

$$\sigma^{k,l}_{m,n}(x_0,\tau)\{b^{\dagger m}b^n\}_\beta = e^{[-(m-n)\Omega\tau]}\,\overline{\sigma}^{k,l}_{m,n}(x_0,\tau)\{b^{\dagger m}b^n\}_\beta\ . \tag{114}$$

Having known the cluster amplitude $\chi(x_0,\beta)$, one can evaluate the quantum

Table 1. Comparison of the results of free energy, F^a_q, $\overline{F}^a_q$, F^a_{picc} and F_{exact} for the 'cubic plus quartic' anharmonic oscillator with the various values of γ, λ and T

γ	λ	T	F^a_q	$\overline{F}^a_q$	F^a_{picc}	F_{exact}
		0.8	1.49685	1.50060	1.49379	1.49355
1	10	1.2	1.45518	1.45181	1.45090	1.45047
		1.4	1.41483	1.40738	1.40914	1.40870
		0.8	1.37659	1.38170	1.37232	1.37203
5	10	1.0	1.35949	1.35896	1.35471	1.35421
		1.4	1.28524	1.27316	1.27722	1.27662
		1.5	2.44088	2.48733	2.42390	2.42407
50	100	2.0	2.40157	2.40144	2.38352	2.38200
		2.5	2.32240	2.29156	2.29948	2.29721
		4.0	6.54974	6.56491	6.53163	6.53003
100	1000	5.0	6.44703	6.43756	6.42551	6.42322
		6.0	6.27263	6.23730	6.24404	6.24154
		8.0	12.04509	12.12220	11.97973	11.97635
2000	10000	10.0	11.84340	11.80978	11.76971	11.76251
		12.0	11.50186	11.37422	11.40876	11.39980

F^a_q are the free energies computed from the traditional TCC theories.
$\overline{F}^a_q$ are the free energies with the extended scheme.
F^a_{picc} are the results of our path-integral based cluster-cumulant approach.
F_{exact} are the exact results.

partition function, Z^a_{picc} in the linear approximation by

$$Z^a_{picc} = \frac{1}{\sqrt{2\pi\beta}}\int_{-\infty}^{\infty} dx_0\, e^{-\beta V(x_0)}\,\frac{\beta\Omega(x_0)/2}{\sinh(\beta\Omega(x_0)/2)}\, e^{-\beta\langle V\rangle_v}\, e^{\chi(x_0,\beta)} \tag{115}$$

We have also verified that the second order perturbative cumulant expansion can be generated from our TCC equations by expanding σ and χ in powers of the perturbation. Our formulation may thus be viewed as a nonperturbative generalization of the Feynman-Kleinert strategy [9].

5.3 Results and Discussions

We present the results of free energy computed by the traditional TCC, the extended TCC scheme and the path-integral based TCC scheme in Table 1 for the various values of the coupling constants, γ, λ and the temperature, $T = 1/\beta$. We also list the exact values for assessing the performance of our methods. Our path-integral results are very close to the exact results, even at very low temperatures, where the quantum fluctuation is high. Of the other two results, i.e., F_q^a and $\overline{F}_q^a$, at the low temperature F_q^a is closer to the exact one but at the high temperature $\overline{F}_q^a$ is the closer one, as it should be.

6 Concluding Remarks

We have developed in this paper a nonperturbative cluster expansion formalism for computing thermally averaged equilibrium properties of quantum many-body systems. We have shown how to express the Boltzmann operator in an exponential type of cluster ansatz. It is shown that when the quantum variables satisfy the gaussian factorization property with respect to the thermal average, we can invoke in a general manner the notion of normal ordering of the variables with respect to the averaging such that the thermal average of an ordered product of the variables in this normal order is zero. For boson/fermion creation/annihilation operators, the Bloch-Matsubara theorem ensures this gaussian factorization property. For a path-integral based treatment, the path-integral average with a gaussian (harmonic) weight of an ordered product of coordinates also satisfy the factorization property. We can thus invoke the same notions of normal ordering in the path-integral framework. Using the concept of normal ordering, we have expressed the Boltzmann operator as a normal ordered exponential of extensive clusters. There are both operators and numbers in the cluster, which we have termed as cluster-cumulants. We have formulated the cluster-cumulant equations for both the operator and the number component which are coupled first order differential equations in the imaginary-time. Owing to the normal ordering, there is a great simplification in the expression of the time derivative of the exponential Boltzmann operator, and also in the fact that, in the thermal trace of the normal ordered exponential, only the number component survives. This is a generic feature of our formalism, and we have exploited it for both the traditional and the path-integral based formulation of the partition function. In general, it has found to be better to induce a Bogoliubov transformation of the variables to have an optimal starting point. We have taken the optimal mean field description to be the one with the best mean field free energy. We have shown how we can incorporate the entire classical component of the partition function in a global manner in the formalism to ensure a good high-temperature behaviour of the partition function.

In contrast, in the path-integral treatment, we have introduced the best mean-field potential around the centroid of the quantum path by invoking 'local' thermal mean field condition [18] at each point of the centroid path. We have then gone beyond the mean-field description in a very systematic and compact way by adopting our Thermal Cluster Cumulant strategy [16, 17, 19]. This approach thus includes in a local manner the entire classical component at each point of the centroid path. Our development also makes manifest the deep algebraic homomorphy between the traditional and the path-integral based TCC approaches.

Acknowledgements

The authors thank the department of Science and Technology (New Delhi), the University Grants Commision (New Delhi) and the Council for Scientific and Industrial Research (New Delhi) for financial support. We also thank S. Ray, S. Mukhopadhyay and S. Roy Chowdhury for help and technical advice.

References

1. A. L. Fetter and J. D. Walecka, *Quantum Theory of Many-Particle Systems* (McGraw-Hill, New York, 1971)
2. G. Parisi, *Statistical Field Theory* (Addison-Wesley, Palo Alto, CA, 1988); J. W. Negele and H. Orland, *Quantum Many Particle Systems* (Addison-Wesley, Palo Alto, CA, 1988)
3. T. Matsubara, Prog. Theor. Phys. **14**, 351, 628 (1955)
4. D.J. Thouless, Phys. Rev. **107**, 1162 (1957)
5. C. Bloch, in *Studies in Statistical Mechanics* Vol. **3**, edited by J. de Boer and G. E. Uhlenbeck (North-Holland, Amsterdam, 1965)
6. C. Bloch, Nucl. Phys. **7**, 451, 459 (1958); C. Bloch, in *Lectures on the Many-Body Problem* Vol. **1**, edited by E.R. Caianiello (Academic Press, New York, 1962), p. 31; R. Balian, *ibidem*, p.139; C. de Dominicis, *ibidem*, p. 163
7. P.C. Martin and J. Schwinger, Phys. Rev. **115**, 1342 (1959); E.S. Fradkin, Nucl. Phys. **12**, 465 (1959); J. Schwinger, J. Math. Phys. **2**, 407 (1961)
8. R. P. Feynman and A. R. Hibbs, *Quantum Mechanics and Path-Integrals* (McGraw-Hill, New York, 1965); H. Kleinert, *Path-integrals in Quantum Mechanics, Statistics and Polymer Physics* (World Scientific, Singapore, 1990), and references therein.
9. R. Giachetti and V. Tognetti, Phys. Rev. Lett. **55**, 912 (1985); Phys. Rev. B **33**, 7647 (1986); R. P. Feynman and H. Kleinert, Phys. Rev. **A34**, 5080 (1986); H. Büttner and N. Flytzanis, Phys. Rev. A **36**, 3443 (1987); S.K. Lee, K.L. Liu and K. Young, Phys. Rev. A **44**, 7951 (1991); H. Kleinert, Phys. Lett. **B280**, 251 (1992); Phys. Lett. **A173**, 332 (1993)
10. L. Leplae, H. Umezawa, and F. Mancini, Phys. Rep. **10**, 151 (1974); H. Umezawa, H. Matsumoto and M. Tachiki, *Thermofield Dynamics and Condensed States* (North-Holland, Amsterdam, 1982)
11. R. Kubo, J. Phys. Soc. Japan **12**, 570 (1957); **17**, 1100 (1962)
12. R.F. Fox, Phys. Rep. **48**, 179 (1978)

13. W. Kohn and J.M. Luttinger, Phys. Rev. **118**, 41 (1960); J.M. Luttinger and J.C. Ward, Phys. Rev. **118**, 1417 (1960)
14. N.N. Bogoliubov, D.B. Zubarev, and I.A. Tserkovnikov, Sov. Phys. Dokl. **2**, 535 (1957)
15. C. De Dominicis, J. Math. Phys. **3**, 983 (1962); *ibidem*, **4**, 255 (1963); R. Balian and M. Veneroni, Ann. Phys. (NY) **164**, 334 (1985); *ibidem* **187**, 29 (1988)
16. G. Sanyal, Sh.H. Mandal and D. Mukherjee, Chem. Phys. Letts. **192**, 55 (1992)
17. G. Sanyal, Sh.H. Mandal, S. Guha and D. Mukherjee, Phys. Rev. E **48**, 3373 (1993)
18. G. Sanyal, Sh.H. Mandal and D. Mukherjee. Proc. Indian. Acad. Sci. (Chem. Sci.) **106**, 407 (1994)
19. Sh.H. Mandal, G. Sanyal, R. Ghosh and D. Mukherjee, in *Condensed Matter Theories*, Vol. **9**, edited by J.W. Clark, K.A. Shoaib and A. Sadiq (Nova Science Publishers, New York, 1994)
20. M. Gaudin, Nucl. Phys. **15**, 89 (1960)
21. F.A. Berezin, *The Method of Second Quantization* (Academic Press, New York, 1966)
22. M. Altenbokum, K. Emrich, H.Kümmel and J.G. Zabolitzky, in *Condensed Matter Theories*, Vol. **2**, edited by P. Vashista, R. K. Kalia and R.F. Bishop (Plenum, New York, 1987)
23. F. Coester, Nucl. Phys. **7**, 421 (1958); F. Coester and H. Kümmel, Nucl. Phys. **17**, 477 (1960)
24. J. Čižek, J. Chem. Phys. **45**, 4256 (1966); Adv. Chem. Phys. **14**, 35 (1969)

Correlated Basis Function Theory for Fermion Systems

Stefano Fantoni[1,2] and Adelchi Fabrocini[3,4]

[1] International School for Advanced Studies, SISSA, Via Beirut 2/4, I–34014, Trieste, Italy
[2] International Centre for Theoretical Physics, ICTP, Strada Costiera 11, I–34014, Trieste,Italy
[3] Department of Physics, University of Pisa, Piazza Torricelli,2, I–56100, Pisa, Italy
[4] Istituto Nazionale di Fisica Nucleare, Sezione di Pisa, Pisa, Italy

Abstract. A description of correlated basis function theory for Fermi systems is given with the perspective of discussing some of the recent developments and its range of applications. The actual status of variational calculations in liquid helium, strongly correlated electrons and nuclear matter is presented, by discussing the merits and limitations of the most realistic trial wave functions. Correlated basis function perturbation theory is described as a tool to go beyond variational calculations. The not–orthogonal and the newly developed orthogonal versions are discussed and compared. A brief review of the one–body Green's function calculation in nuclear matter is presented, as one of the main applications of the orthogonal version of the perturbation theory. We describe a generalized Fermi Hyper–Netted Chain scheme, showing that it constitutes a powerful method to compute the hamiltonian and identity operator matrix elements, which enter the correlated basis theory, starting from its zeroth order or variational theory. Such scheme sums up both reducible and irreducible cluster terms, whereas the original FHNC one takes care of the irreducible terms only, and it is particularly suitable for calculations with state dependent correlations, as well as for finite systems, like nuclei and helium droplets.

1 Introduction

Nearly all of physics is many-body physics at the most microscopic level of understanding, appropriate to the energy scale of the particular branch of physics under consideration. Thus, the subject of quantum many-body theory (QMBT) can fairly be said to virtually strengthen from beneath all of modern physics. The fundamentally many-particle aspects of nuclei, atoms, molecules, solids, and fluids are all manifestly apparent, but also the single nucleon problem is itself rapidly becoming a multiparticle problem.

Physical applications range from condensed matter (e.g., antiferromagnets, quantum liquids and solids, plasmas, metals, superconductors), nuclei and high energy physics (e.g., nuclear matter, finite nuclei, quark–gluon plasmas), dense matter astrophysics (e.g., neutron stars, white dwarfs), atoms and molecules (e.g., electron correlations, reactions, quantum chemistry), and elementary particles (e.g., quantum field theory, lattice gauge field models).

Most of the many–body systems of interest are characterized by strong correlations between the microscopic constituents, which cannot be properly taken into account by effective one–body treatments, like the Landau theory of Fermi liquids, the polarization potential method of Pines and co–workers or density functional theories. These are the systems which are addressed in these lecture notes, and they require use of the fully QMBT methods.

In **quantum liquids and solids** there is a number of important issues for the ground state, low–lying excitations, finite temperature properties, where the real predictive power of QMBT is still to be fully exploited. In particular, the accuracy in the description of Fermion *ground states* is in general far behind the level reached for Bose systems. Not to say that, after more than thirty years from the discovery of superfluidity in ^{3}He [1], a microscopic theory for such fascinating Fermi system is still lacking. Bridging this gap is a fundamental methodological issue. In addition, extending our knowledge of the ground state properties will provide new *phenomenological* inputs for further developments within density functional and polarization potential theories.

The role played by short–range correlations in **strongly interacting electrons** is an open and still very much debated subject. The strong short–range Coulomb repulsion, which has been actualized in real electronic systems, can produce a number of interesting and competing instabilities, as superconductivity, charge and spin density waves, phase separation, lattice instabilities and Fractional Quantum Hall effect. Moreover, many–particle correlation effects in the electron plasma in confined geometry may be responsible of a variety of excitations, which cannot be understood on the basis of bulk electron fluid properties.

In **nuclear physics**, problems such as hadron propagation in nuclear matter, quark–gluon plasma transition and the equation of state of neutron matter are subjects of intense theoretical and experimental study. Experimental data coming out from on–going and future accelerator facilities require a quantitative understanding of the structure of the hadronic matter under normal and extreme non equilibrium conditions, as well as, of the nature of the vacuum of the underlying field theory, and how it is altered across the quark–gluon plasma transition. The neutron star physics asks for fully microscopic calculations of asymmetric nuclear matter, in the limit of the neutron matter, both for the normal and the superfluid phases.

A **variational theory** directly or indirectly underlies almost all the more modern and efficient QMBT techniques, which heavily rely on it. The knowledge of realistic trial functions for the ground and the excited states is a fundamental requirement to get, first of all, a realistic picture of the system, and (more important) to proceed any further towards the *exact* solutions of the many–body problem under consideration.

The most efficient techniques employed in variational calculations are the Fermi–Hyper–Netted–Chain (FHNC) theory [2, 3] and the Variational MonteCarlo (VMC) method [4, 5]. FHNC is a widely applied integral equation method which sums up series of cluster diagrams associated with the distribution functions of the many–body wave function, and allows for calculations both in the thermodynamical limit and in finite, inhomogeneous systems. In spite of being inherently not exact, because there is a class of cluster diagrams (the so–called elementary diagrams) which cannot be included in any known closed form, FHNC provides extremely accurate results for a variety of strongly interacting many–fermion systems. VMC is a well known stochastic method, which allows for exactly evaluating (in the statistical sense) the expectation values of various observables, but it is limited to simulations with a finite number of particles. Finite size effects, however, can be controlled in most of the cases. FHNC and VMC should not be viewed as competing methodologies, but rather as complementary ones.

The full optimization of the trial functions is more natural in the case of FHNC–type calculations, where it leads to Schrödinger–like equations to be solved in a self–consistent way [6, 7] and may provide unique information about possible instabilities. Recently, semi–full optimizations have become possible also in VMC calculations [8, 9] by using the re-weighting method [5].

Two main strategies are followed, at present, to go **beyond variational calculations**. One is based upon integral equation approaches, such as Correlated Basis Function (CBF) theory [10, 11, 12], which makes use of the FHNC technology, or Coupled Cluster (CC) theory [13]. They basically provide various levels of approximation schemes, hopefully converging to exact solutions. The other strategy relies on numerical simulation techniques and fall into the category of Quantum MonteCarlo methods [5, 14, 15, 16, 17].

In these lecture notes we will focus the attention on the CBF theory for Fermi systems at zero temperature, and its recent advances in (i) quantum liquids and solids, (ii) strongly correlated electrons and (iii) nuclear matter. CBF developments for finite nuclei will be treated elsewhere in this volume [18]. Other interesting fields of applications of CBF theory, such as the electron plasma [19, 20] or Ising spin models [21] will not be discussed here. Some of the presented material have been previously reviewed [10, 22, 23, 24, 25, 26]. The aim, here, is to give a description of the broad range of applications of this theory, together with the most recent developments.

The initial step of any CBF calculation is to search for a realistic wave function for the ground state of the system, or, differently phrased, the zeroth order of CBF theory is the variational theory. Therefore, in the following pages we will give a brief review of the trial wave functions used in variational calculations, focusing on those which are more convenient for CBF theory, and discuss their merits and limitations. Then we will describe the CBF scheme and outline the calculation of the one–body Green function of symmetrical nuclear matter. This has been, together with the dynamical response of nuclear matter [18], one of the major recent developments of CBF theory, with several important applications in inclusive and semi exclusive e–scattering reactions off nuclei at intermediate energies. The cluster expansion and the FHNC techniques which are needed to compute the diagonal and not–diagonal matrix elements required by CBF theory will be also presented.

2 Hamiltonians and variational calculations

Strongly correlated systems are characterized by short–range correlations between the constituent particles. A well known effect of such correlations is to deplete the population of single–particle states, and to spread out the missing strength over an extremely wide range of excitation energy [27]. A clear picture of this feature is shown in Fig. 1, where the momentum distribution $n(k)$ of liquid ^{3}He and of nuclear matter are displayed. It is much more apparent for the denser liquid helium system, but also in nuclear matter is very pronounced, certainly much more than it was believed in the past. The figure provides indications for the spreading of the missing strength. The integral over the missing energy of the nuclear matter hole spectral function $P_h(k, E)$ (i.e. the probability density of removing a particle with momentum k from the ground state, leaving the residual system with an excitation energy E) gives the momentum distribution. Integrating $P_h(k > k_F, E)$ up to 100 MeV one misses about two order of magnitude in the tail of $n(k)$ (see the inset of the figure).

As a consequence, many–body theories based upon single–particle (or mean field, MF) basis $\{\Psi^{(i)}_{MF}\}$ have hard time in including the effects of the short–range correlations. To overcome this problem one can use a set of correlated basis functions [28]

$$\Psi^{(i)}_{CBF}(R, \Xi) = \hat{F}(1, \cdots, N)\Psi^{(i)}_{MF}(R, \Xi), \tag{1}$$

with R and Ξ representing the set of spatial and spin–isospin variables of the N particles, where the correlation operator $\hat{F}$ incorporates the main features of the short– and long–range correlations. This is commonly achieved by performing a variational calculation on the ground state energy, which provides an *optimal* $\hat{F}$. The hope is that perturbation theory built on such correlated basis functions will take care of the *small* differences between $\{\Psi^{(i)}_{CBF}\}$ and $\{\Psi^{(i)}_{exact}\}$, or eventually manifests phase instabilities.

A useful scheme to generate the ground state trial functions of a Fermi system is the following. One starts with a mean field wave function $\Psi^{(0)}_{MF}$, describing the *uncorrelated* system. For instance, in a homogeneous liquid of particles with spin degeneracy d, at a given density ρ, $\Psi^{(0)}_{MF}$ is given by a Slater determinant of plane waves:

$$\Psi_{MF}(R, \Xi) = \mathrm{Det}_{ij}\left[\exp\left(i\mathbf{k}_i \cdot \mathbf{r}_j\right)\chi_i\left(j\right)\right], \tag{2}$$

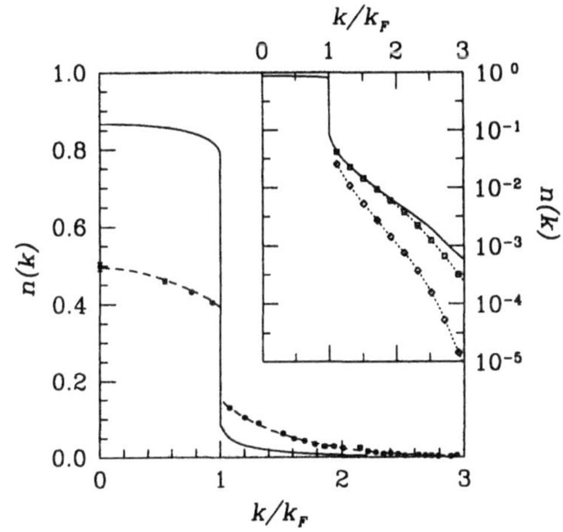

Fig. 1. Momentum distribution $n(k)$ of liquid ^{3}He (dashed line) [29] and of nuclear matter (solid line) [30] at equilibrium densities. The inset displays the integral of the nuclear matter hole spectral function over the missing energy up to ∞ (solid line) or to 100 MeV and 300 MeV (dotted lines) [31].

where $\chi_i(j)$ is the spin state $|i\rangle$[1] for particle j and k is lesser or equal to the Fermi momentum $k_F = (6\pi^2\rho/d)^{1/3}$. In a second step one formally applies the projection operator $e^{-\beta H}$ to $\Psi^{(0)}_{MF}$. In the limit $\beta \to \infty$, the lowest eigenstate not–orthogonal to $\Psi^{(0)}_{MF}$ is projected out. Different approximations of the projection operator lead to different forms and systematic expansions of $\hat{F}$.

2.1 Strongly correlated atoms

Liquid and solid helium are the paradigmatic examples of strongly correlated atoms. Helium atoms can be considered as structureless, spherical particles interacting through a two–body potential, whose attractive part causes the gas to condense into a liquid phase at a critical temperature $T_0 = 3.2$ K for ^{3}He and $T_0 = 4.2$ K for ^{4}He at normal pressure. Decreasing the temperature below T_0, helium does not solidify unless a pressure of 30 bar is applied. The strong quantum effects due to the large zero point motion and the weakness of the attractive part of the interaction, lead to the occurrence of several fascinating phenomena, like superfluidity.

It is believed that Bose and Fermi liquid helium at relatively low pressure are very well described by the hamiltonian

[1] In nuclear matter $i = \sigma(\uparrow)\tau(\uparrow), \sigma(\downarrow)\tau(\uparrow), \sigma(\uparrow)\tau(\downarrow), \sigma(\downarrow)\tau(\downarrow)$ and $d = 4$, in helium–3 $i = (\uparrow, \downarrow)$ and $d = 2$, in spinless fermions $d = 1$.

$$H = -\frac{\hbar^2}{2m}\sum_{i=1}^{N} \Delta_i + \sum_{i<j=1}^{N} v(r_{ij}), \tag{3}$$

where $v(r)$ is the HFDHE2 potential of Aziz *et al.* [32, 33]. The HFDHE2 potential reproduces extremely well the equation of state of ^{4}He, both for the liquid and the solid phase. However, the possibility of a three–body interaction is currently taken under consideration, mostly in connection with ^{3}He physics.

Feenberg wave functions. The correlation operator $\hat{F}$ is commonly taken, in CBF calculations, as a product of a state dependent part $\hat{F}_{SD}$ and a Feenberg scalar part $F_{Fee}(R)$,

$$\hat{F}(1,\cdots,N) = \hat{F}_{SD}(1,\cdots,N)F_{Fee}(R), \tag{4}$$

where $F_{Fee}(R)$ is expressed as a series of n–body pseudopotentials,

$$F_{Fee}(R) = \exp\left[-\frac{1}{2}\sum_{i<j=1}^{N} u_2(r_{ij}) - \frac{1}{6}\sum_{i<j<k=1}^{N} u_3(i,j,k) - \cdots\right]. \tag{5}$$

The simplest form of $F_{Fee}(R)$, denoted as *Jastrow* ansatz, is obtained by setting $\hat{F}_{SD} = 1$ and keeping only the **two–body pseudopotential**. The McMillan form (M) $(b/r)^m$, with b and m as variational parameters, provides a reasonable description of $u_2(r)$ and has been used in a number of variational calculations. A full optimization of $u_2(r)$ leads to a more structured pseudopotential (optimal Jastrow (OJ)), less repulsive at intermediate range ($r \sim 3 - 4\mathring{A}$), with a wiggle at about $5\mathring{A}$ (Fig. 2)[2]. The energy gain from (M) to (OJ) is about 0.2 K, coming almost completely from the potential energy expectation value.

The **three–body pseudopotential** is very well described by the so–called Feynman form, given by

$$u_3(i,j,k) = \sum_{cycl.} \xi(r_{ij})\xi(r_{ik})P_1(\hat{\mathbf{r}}_{ij}\cdot\hat{\mathbf{r}}_{ik}), \tag{6}$$

where P_1 is the first order Legendre polynomial and *cycl.* means summation over the permutations of i, j, k. This form is generated by the kinetic energy operator acting on the two–body pseudopotential $(\nabla_i u_2(r_{ij})\cdot\nabla_i u_2(r_{ik}))$, and

[2] The optimal pair pseudopotential has a long range behavior of the type α/r^2, where the strength α is related to the sound velocity in the liquid. This tail is very important for the collective excitation of the liquid, because it determines the long wavelength behavior of the static structure function. However, it gives an extremely small contribution to the energy, so that it cannot be obtained by VMC calculations. When needed it can be included using either Euler–FHNC calculations [6] or independent information on the equation of state.

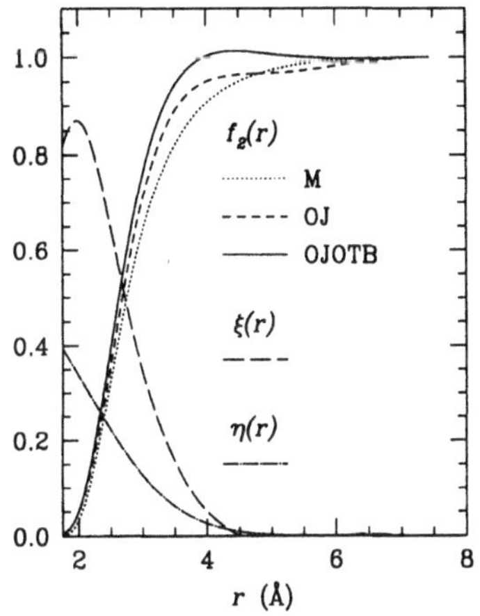

Fig. 2. Pair correlation $f_2(r) = \exp[-u_2(r)/2]$, triplet correlation $\xi(r)$ and backflow correlation $\eta(r)$ of liquid ^{3}He at equilibrium density [9]. Various optimization levels are shown for f_2.

has been used in several calculations [34, 35]. It provides an energy lowering of about 0.5 K. Full optimization of u_3 [9, 17] shows that the Feynman form is indeed nearly optimal. The function $\xi(r)$ should have a long range tail of the type β/r^3 from the projection operator argument.

Another very important effect, caused by the kinetic energy operator acting on both the two–body pseudopotential and the Slater determinant of plane waves $(\nabla_i u_2(r_{ij}) \cdot \nabla_i \exp(i\mathbf{k} \cdot \mathbf{r}_i))$, is the **backflow correlation.** This is a velocity dependent correlation, due to the flow induced by a moving atom [36], and which increases the effective mass. It can be incorporated in $\Psi^{(0)}_{MF}$ through the transformation

$$\exp[i\mathbf{k}_i \cdot \mathbf{r}_j] \;\rightarrow\; \exp[i\mathbf{k}_i \cdot (\mathbf{r}_j + \sum_{k \neq j} \eta(r_{kj})\mathbf{r}_{kj})]. \tag{7}$$

The shape of $\eta(r)$ is very similar to that of $\xi(r)$, in accordance with the projection operator argument, and it is very well represented by

$$\eta(r) = \lambda \exp\left[-\left(\frac{r - r_B}{\omega_B}\right)^2\right]\left(\frac{2r - L}{L}\right)^3, \tag{8}$$

where λ, r_B and ω_B are variational parameters and L is the side of the simulation box. The backflow correlation accounts for about 0.4 K of the binding energy (see Table 1).

Transformation (7) cannot be fully exploited in FHNC calculations. In this case, backflow correlations are included in the correlation part $\hat{F}_{SD}$ of

Table 1. Energy E and kinetic energy T (in K) of ^{3}He at the experimental equilibrium density $\rho_{eq} = 0.01635\,\text{Å}^{-3}$. VMC results [9] are fully optimized for the Jastrow and Jastrow + triplet parts. FHNC results use full optimized Jastrow and a three parameter function for the triplet and the backflow correlations. FHNC/s reports the results obtained using the *scaling approximation* for the elementary diagrams [35], whereas FHNC/α refers to the *interpolating approximation* estimates [37]. Fixed-node DMC, performed by using the OJOTB trial function as sampling function, gives $E = -2.37$ K and $T = 12.02$ K [9].

	T(FHNC/s)	E(FHNC/s)	T(FHNC/α)	E(FHNC/α)	T(VMC)	E(VMC)
J	13.48	-1.28	13.25	-1.31	12.85	-1.30
J+T	12.97	-1.83	12.73	-1.80	12.54	-1.78
J+T+B	12.28	-2.36	12.02	-2.31	12.27	-2.16

the wave function, by means of momentum dependent two–body correlations [35].

As far as the present status of variational calculations on helium–3 is concerned, the following comments are in order.

1. The FHNC technology is not advanced enough to get completely reliable estimates of the energy expectation value. Better approximations than the existing ones are needed to include the effect of the elementary diagrams (see section 4). Table 1 shows a comparison between VMC and FHNC results at equilibrium density of liquid ^{3}He. When triplet and/or backflow correlations are taken into account, the agreement is not fully satisfactory[3]. Recently, the various types of FHNC distribution functions have been calculated [41] by Monte Carlo in a liquid ^{3}He test case. It may be possible to use such results to improve upon the scaling [42] or the interpolating [43] approximations.
2. The Feenberg's series in (5) is not rapidly converging, as shown in Table 1 and Fig. 3. The VMC results listed in the table unambiguously indicate that n–body pseudopotentials, with $n > 3$, are not negligible. The problem becomes more serious at higher densities. Furthermore, in the solid phase one needs to include in $\Psi^{(0)}_{MF}$ a one–body Nosanow term $\exp\{-(\mathbf{r}_i - \mathbf{R}_i)^2\}$, which localizes the particles at the lattice sites $\mathbf{R}_i$ to get sensible variational estimates.

Shadow wave functions. A new trial function for Bose systems has been recently proposed: the so–called Shadow wave function [44, 45]. It corresponds

[3] The VMC energy E(J+T) for a triplet correlation with ξ of the of the form (8), and therefore not fully optimized (as in the FHNC calculations reported in Table 1), is -1.73 K .

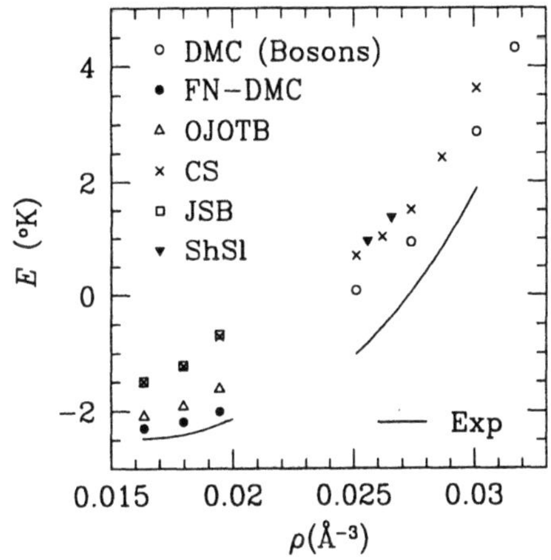

Fig. 3. Equation of state of ^{3}He at $T = 0$. For the liquid phase the results obtained with a coherent–state (CS) with McMillan correlations [38] are compared with those provided by a Jastrow–Slater with McMillan and backflow correlations (JSB), an optimized Jastrow+triplet+backflow (OJOTB) trial function, and the fixed node Diffusion Monte Carlo results with backflow nodes (FN–DMC) [9]. For the solid phase the CS data (obtained with a Nosanow factor in the single particle orbitals) [39] are compared with a Shadow wave function calculation (ShSl) [40] and the DMC energy for mass–3 bosons.

to a correlation operator which may be viewed as a one–step discretized path integral representation of the projection operator, using the primitive approximation (consisting in assuming that the kinetic and potential energy operators commute in $e^{-\beta H}$ [16]). It has been widely used for microscopic calculations on liquid and solid ^{4}He at zero temperature [17].

Auxiliary (shadow) variables $\mathbf{s}_i$ are introduced, representing centers of motion of the particles, with coordinates $\mathbf{r}_i$:

$$F_{SWF}(R) = \exp\left[-\sum_{i<j=1}^{N} u_R(r_{ij})\right] \times \int dS K(R-S) \exp\left[-\sum_{i<j=1}^{N} u_S(s_{ij})\right], \quad (9)$$

where S is the whole set of shadow variables $(\mathbf{s}_1, \mathbf{s}_2, \cdots, \mathbf{s}_N)$ and u_R and u_S are taken of the Jastrow form. The kernel K has the form $K(R-S) = \exp[-\sum_{i=1}^{N} u_{RS}(|\mathbf{r}_i - \mathbf{s}_i|)]$, where $u_{RS}(r)$ is a confining pseudopotential (the simplest approximation is the harmonic oscillator potential). Concerning the comparison of $\Psi_{Fee}^{(0)}$ with $\Psi_{SWF}^{(0)}$, it has to be noticed that: (i) F_{SWF} includes

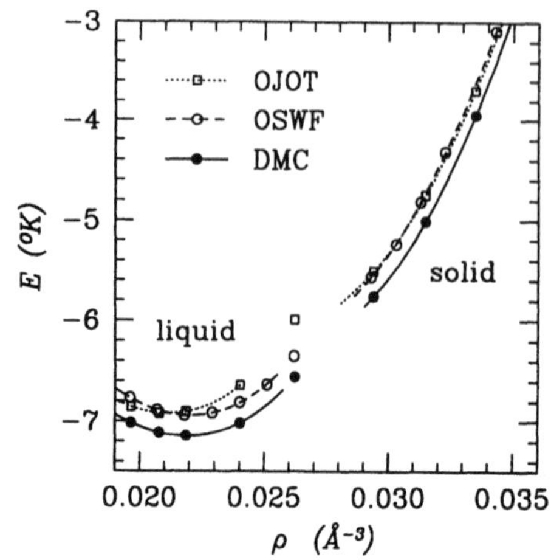

Fig. 4. Equation of state of ^{4}He at $T = 0$. Results are obtained from variational calculations with an optimal Jastrow+triplet Ψ_{Fee} (OJOT) and with an optimized Ψ_{SWF} (OSWF), as well as from Diffusion Monte Carlo simulation (DMC) [9, 47]. Lines are cubic fits to the Monte Carlo data. For each set of data lines are broken across the coexistence region. The experimental equation of state, not shown, would be barely distinguishable from the DMC result.

correlations at any order (pair, triplet, quadruples, etc.) through the shadow correlations, after integration over S; (ii) the shadow variables can localize the particles with no cost in kinetic energy. Therefore the same *translationally invariant* wave function can be used to describe both the liquid and the solid phases, with no need of a Nosanow factor. This makes Ψ_{SWF} unique to perform microscopic calculations on the liquid–solid coexistence region to study the liquid–solid phase transition and interface [46]; and (iii) the price to be paid when introducing F_{SWF} is that, for each particle i, we now have three vector coordinate variables, $\mathbf{r}_i$, $\mathbf{s}_i^{right}$ (coming from the ket in $\langle H \rangle$) and $\mathbf{s}_i^{left}$ (coming from the bra).

VMC calculations performed with fully optimized Ψ_{SWF} show an uniformly good equation of state for ^{4}He for both the liquid and the solid phases, and compare rather well with the corresponding results obtained with Ψ_{Fee} for the liquid phase (see Fig. 4).

VMC calculations for ^{3}He using (1), with F_{SWF} as correlation operator, have been recently carried out [40] with results indicating that it is variationally equivalent to F_{Fee} for the liquid phase, but, in addition, it is able to provide a stable crystalline order even starting from a liquid–like configuration of the atoms. As in the Bose case, the many–body correlations of the shadow type, at sufficiently high densities, introduce a spontaneously broken translational symmetry.

FHNC calculations have not been done yet with the shadow wave functions. The auxiliary variables introduce new classes of elementary diagrams which are expected to be not negligible in the limit of strong shadow–shadow correlations, and this may be difficult to handle. However, an FHNC treatment for the fermionic case would surely be of interest.

The present status of variational calculations in normal liquid and solid ^{3}He at zero temperature is summarized in Fig. 3. They underestimate the binding energy of 0.3-0.5 K in the liquid phase and of about 1.7 K in the solid one (the variational calculations in the latter case have not been fully optimized). Diffusion Monte Carlo calculations [17] performed in the Fixed–Node (FN–DMC) approximation for the liquid phase are also displayed for comparison. They are much closer to the experimental data. The remaining underbinding, 0.1-0.2 K, may still be due to the Fixed–Node approximation. However, the fact that the DMC results for the mass–3 boson in the solid phase (shown in Fig. 3) are much higher (about 1 K) than the experimental energies is disturbing and unexpected, and it may point to an inadequacy of the Aziz model hamiltonian in ^{3}He.

2.2 Strongly correlated electrons

The interest in strongly correlated electron models, raised by phenomena like high temperature superconductivity and the quantum Hall effect, does not give signals of becoming less intense. Several theoretical schools have blossomed, each of them focused on some specific hamiltonian and approach. These hamiltonians, however, have proven to be so difficult to solve to become a challenge by themselves, quite independently on their original motivation.

It is not the scope of this section to give a review of the immense work done in this field, but rather to briefly describe the advances which have recently been made by CBF theory in connection with the fractional quantum Hall effect (FQHE) [48] and with the Hubbard model for high T_c superconductivity [49].

The fractional quantum Hall effect. The fractional quantum Hall effect stems from a strongly correlated incompressible fluid state [50, 51], formed at special densities, of a two–dimensional electronic system in the extreme quantum limit of a strong perpendicular magnetic field ($B > 5$ T), low temperature ($T < 2$ K) and high mobility of electrons ($\mu > 10^5$ cm^2/Vs). For fully spin–polarized (spinless) electrons, the most pronounced states occur when filling of the lowest Landau level (LLL) is $\nu = 1/m$, where m is an odd integer [52]. There are other interesting electronic states, the so–called higher order FQHE states at $\nu = p/q$, whose origin is, however, less clear. In particular, the filling $1/2$ state (the simplest filling factor with even denominator) has recently received much attention. Extrapolation of exact calculations with few electrons to the thermodynamic limit is not totally unambiguous. Therefore,

CBF theory could play an important role with respect to this challenging problem.

The many–electron system is described by the hamiltonian $H = T + V$ where T is the kinetic energy operator

$$T = \frac{1}{2m_e} \sum_{i=1}^{N} [-\imath\hbar\nabla_i + e\mathbf{A}(r_i)]^2, \tag{10}$$

and

$$V = \sum_{i<j=1}^{N} \frac{e^2}{\epsilon |z_i - z_j|} + \sum_{i=1}^{N} v_{eb}(z_i) + V_{bb}. \tag{11}$$

Here, z_i is the complex coordinate $z_i = x_i + \imath y_i$, $v_{eb}(z_i)$ the electron–neutralizing background and V_{bb} the background–background interaction energy, with ϵ, e and m_e being the background dielectric constant, the charge and the mass of the electron.

When considering the symmetric gauge vector potential, $\mathbf{A} = \frac{1}{2}B(y\hat{x} - x\hat{y})$, it is convenient to regard the $x-y$ plane as a complex one. The eigenvalue of the kinetic energy operator (the *Landau levels*) are those of a harmonic oscillator, $E_n = (n + \frac{1}{2})\omega_c$, where $\omega_c = eB/mc$ is the cyclotron frequency of a classical particle in the magnetic field B. Each Landau level is infinitely degenerate. For the LLL ($n = 1$) the corresponding eigenfunctions $|n = 1, m\rangle$ are also eigenfunctions of the orbital angular momentum.

The area covered by a single electron in state $|1, m\rangle$ is proportional to m, and given by $\pi\langle r^2 \rangle = 2\pi(m + 1)l_0^2$, where $l_0 = (\hbar c/eB)^{1/2}$ is the magnetic length. In a finite cell of area A, the maximum number of allowed states is $N_s = A/(2\pi l_0^2)$, which corresponds to the ratio of the magnetic flux $\Phi = AB$ with the flux quantum $\Phi_0 = hc/e$. Thus, the filling factor $\nu = N/N_s$, where N is the number of electrons in the cell, is given by $\nu = 2\pi l_0^2 \rho$, and is proportional to the electronic density ρ and inversely proportional to the magnetic field.

The ground state wave function lies in the LLL, and, therefore, is a linear combination of states $|1, m\rangle$. Due to the absence of an energy scale, like a bandwith of a periodic solid or the Fermi energy of an electron liquid in zero magnetic field, the Coulomb interaction induces such strong correlations amongst the electrons that cannot be accounted for in a perturbative way. The variational principle is applied here to the expectation value of the Coulomb potential (plus the neutralizing background), the kinetic energy and V_{bb} providing a constant shift only.

For states with filling factor $1/m$ an extremely good trial function is given by [52]:

$$\Psi_L^{(0)}(R) = \prod_{i,j=1}^{N} (z_i - z_j)^m \prod_{i=1}^{N} \exp[-|z_i|^2/4l_0^2] \,. \tag{12}$$

This wave function obeys Fermi statistics because m is an odd integer, and is entirely built with states in the LLL. One could view it [53] as if every electron position acts as a center of m vortices, each carrying a flux $m\Phi_0$. Since $Nm\Phi_0$ must be equal to the total flux Φ, the filling factor is precisely given by $1/m$. Therefore we have no variational parameters in the trial wave function.

For $m = 1$ (filled Landau level), the polynomial $\prod_{i,j=1}^{N}(z_i - z_j)^m$ is the Vandermond determinant of order N, and one can easily prove that, in the thermodynamic limit, the particle density tends to a constant value $\rho = 1/(2\pi l_0^2)$. A similar feature occurs for $m > 1$. This can be understood by noticing that the modulus square of $\Psi_L^{(0)}(R)$ can be mapped onto a charge neutral two–dimensional classical plasma [54] with $e^2 = 2m_e$ and $\rho = 1/(2m\pi l_0^2)$.

One can use Hyper–Netted–Chain (HNC) theory to compute the correlation energy separating a short and a long range part in the corresponding *Jastrow* pseudopotential $u_2(r)$ [55]

$$\begin{aligned} u_2^{short}(r) &= -2mK_0(Qr) \\ u_2^{long}(r) &= 2m[\ln r + K_0(Qr)], \end{aligned} \tag{13}$$

where K_0 is the modified Bessel function, and Q a cutoff parameter of the order unity. Results obtained with the *scaling* approximation are in extremely good agreement with Monte Carlo data [56].

More recently, a remarkably simple picture has been proposed [57] to understand the origin of the FQHE for states at $n = p/q$. It introduces the idea of a new kind of particle, called *composite fermion* (CF), which is an electron carrying an even number of vortices of the wave function. The fundamental property of the CF system is that each CF experiences an effective field $B^\star = B - q_e\Phi_0\rho$, where q_e is the number of vortices and it is an even integer. Thus, the liquid of strongly correlated electrons at B is equivalent to a liquid of weakly interacting CF–s at $B^\star$, or, alternatively, the FQHE is regarded as the integer quantum Hall effect of novel composite fermions.

The stable fractional fillings obtained in this way are: $\nu = p/(q_e p \pm 1)$, where q_e is the even number of vortices attached to each electron and $p = 1, 2, \cdots$ is the corresponding number of CF Landau levels. The electronic trial function for the ground state at $\nu = p/(q_e p + 1)$ is given by

$$\Psi_{CF}^{(0)}(R) = \hat{P}_{LLL} \prod_{i,j=1}^{N} (z_i - z_j)^{q_e} \Psi_{MF}^{(0)}(R), \tag{14}$$

where $\hat{P}_{LLL}$ is the lowest–Landau–level projection operator, and $\Psi_{MF}^{(0)}(R)$ is the Slater determinant built with the first p Landau levels. For the particular case of $p = 1$, the CF wave function is identical to the Laughlin wave function.

Table 2. Correlation energies per particle expressed in units $e^2/\epsilon l_0$, for unprojected CF wavefunctions at various fillings, computed by using FHNC/0 and HNC/s [60]. The Monte Carlo data obtained for projected CF wavefunctions in the spherical geometry [61] are also reported for comparison.

	1/5	3/13	1/3	2/5	4/9	8/17
MC	-0.327	-0.348	-0.410	-0.433	-0.447	
FHNC/0	-0.324	-0.344	-0.406	-0.453	-0.458	-0.466
HNC/s	-0.327		-0.410			

There is evidence for the validity of the CF theory from several numerical studies performed mainly on few electron systems [58, 59].

FHNC theory has been recently applied to unprojected CF wave functions [62]. Some of the results are shown in Table 2, and one can see that there is very good agreement with the Monte Carlo data obtained for few electron systems [61]. The state at filling 1/3 and 1/5 can be treated with both HNC and FHNC. The HNC results include the contribution from elementary diagrams (about 1%) by using the *scaling* approximation of [42] (HNC/s) (see section 4 for a discussion on this approximation), whereas the FHNC/0 data have been obtained by totally neglecting them. For these states there is no problem of projection onto the LLL, because only the first Landau level is included in the Slater determinant. It appears that, in general, the projection is not particularly effective for the ground state correlation energy. However, it becomes crucial when studying the excitations. For instance, the energy gaps for the unprojected wave functions are about a factor three larger than that for the projected ones [60]. How to include the projection operator effects in FHNC–type calculations is an open problem.

The Hubbard model. A second important example of applications of CBF theory to strongly interacting electrons, is the Hubbard lattice model. It was proposed in [49], in its strong on–site Coulomb repulsion limit, to understand the high T_c superconductivity discovered in copper–oxide materials [63, 64]. For a review on the subject see [65, 66].

The Hubbard model is defined by the lattice hamiltonian

$$H = -t \sum_{<i,j>,\sigma} c^\dagger_{i\sigma} c_{j\sigma} + U \sum_i n_{i\uparrow} n_{i\downarrow} , \tag{15}$$

where $c^\dagger_{i\sigma}$ ($c_{j\sigma}$) is the creation (annihilation) operator of an electron with spin $\sigma = \pm\frac{1}{2}$ on site i. t and U represent the nearest–neighbor hopping energy and the on–site Coulomb repulsive interaction with opposite spin, respectively.

Typical values of the coupling constant U/t range from 1 to 10. In the strong coupling limit the Hubbard hamiltonian becomes the $t - J$ model.

While an exact solution is only available in $1D$ [67], all common methods of solution for $D > 1$ are strictly numerical, and eventually limited by finite size problems. In particular, for $D = 2$ no consensus has been reached on whether superconductivity is present or not within the model. CBF theory may provide methods of solution which are capable of dealing with the thermodynamic limit from the outset and which do not become prohibitively difficult as the space dimensionality is increased.

The variational ground state energy can be expressed in the form [68] $E_0 = \langle T \rangle + \langle V \rangle$, where

$$\begin{aligned} \langle T \rangle / N &= -z\rho t n(\mathbf{r}_{11'} = \mathbf{a}), \\ \langle U \rangle / N &= \frac{\nu^2}{2} U g(r_{12} = 0), \end{aligned} \tag{16}$$

where $\mathbf{a} = \sum_i^D \mathbf{l}_i \hat{\mathbf{e}}_i$ is the unit vector for the lattice, z is the coordination number in the D–dimensional lattice system, and the density ρ is twice the filling factor. The functions $n(\mathbf{r}_{11'})$ and $g(r_{12})$ are the one–body density matrix and the pair function, respectively (see sections 3 and 4 for their definitions). As seen from (16), we need to know the latter at $r_{ij} = 0$ (on–site repulsion) and the former at $\mathbf{r}_{1,1'} = \mathbf{a}$ (first neighbor hopping).

The trial wave functions are of the form (1), where the correlation operator $\hat{F}$ may contain spin–dependent correlations as in (22) with $p = 1, 2$. The most common one is the so–called *Gutzwiller ansatz*, which contains strictly on–site two–body correlations:

$$f_2(r_{ij}) = \exp[-\frac{1}{2} u_2(r_{ij})] = 1 - (1 - g)\delta_{r_{i\uparrow}, r_{j\downarrow}}. \tag{17}$$

For the 1D case, the variational problem has been solved exactly [69]. In spite of this, the 1D Hubbard model has been investigated [70, 71], at the variational level of CBF, with correlation operators $\hat{F}$ of increasing reachness, beginning from Gutzwiller, to Jastrow (first and second neighbor sites correlations), to include the so–called *e–d* correlations by means of quartet correlations, to spin correlations of the form $f_\sigma(i,j) = 1 + \eta(r_{ij})\sigma_i \cdot \sigma_j$. Various mean field wave functions $\Psi_{MF}^{(0)}$, with and without a Fermi surface, have been considered, starting from the paramagnetic Slater function of plane waves, to a spin density wave (SDW) Slater determinant, to a BCS wave function. The main scope of such calculations has been not to uncover any new physics related to the model, which is rather well studied, but to assess the applicability and limitations of the method, and to obtain precise indications planning future calculations aimed to extracting new physics for the Hubbard or related models in $D > 1$ dimensions. Under this respect, the following comments are in order.

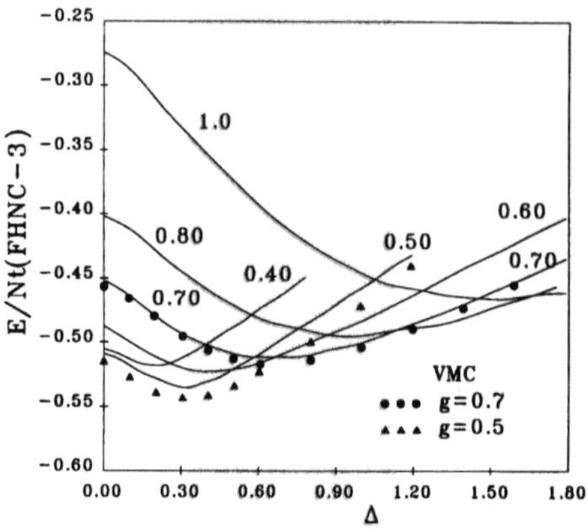

Fig. 5. The variational energy E/Nt calculated with FHNC [71]) and VMC [73]). FHNC calculations include the elementary diagrams following the scheme FHNC–n discussed in section 4. g and Δ are the Gutzwiller and SDW variational parameters.

1. The accuracy of the FHNC calculations is generally rather satisfactory. The results obtained are in very good with the existing Monte Carlo data or exact solutions. The contribution from elementary diagrams is not at all negligible, particularly for the correlated SDW states.
2. The Gutzwiller ansatz, acting on the bare Fermi sea, describes very well the existing correlations. Adding second neighboring site correlations or spin-dependent correlations, practically does not change the variational energy. On the contrary, the quartet correlations are relatively important. They mimic the feature that double occupied and empty sites, which the Gutzwiller wave function leaves mutually uncorrelated, prefer to sit on neighboring sites in the true ground state [72], at least near filling 1/2. They yield a ground state energy which is very close to the exact result. Yet, the jump at the Fermi surface is still quite large. This is not correct, since it is know that the 1D Hubbard model is a true insulator at $\rho = 1$ and only a marginal metal, with pole residue Z of the single particle Green's function equal to zero. Implicitly this demonstrates that the departure from Fermi–liquid behavior is probably connected with an extremely small and delicate energetic improvement, at least for $\rho \neq 1$.
3. There is no indication that correlated BCS (which is a wave function without a Fermi jump) is energetically preferred to the paramagnetic Slater. This is partially expected, because the BCS correlations are not favorable for the 1D repulsive Hubbard model. The Tomonaga mapping, for example [74], indicates that the pairing susceptibility does not diverge in this regime. However, in the 2D $U = \infty$, Anderson has indicated that a Gutzwiller projected BCS state may be a reasonable representative for his RVB state. In fact, in 2D there are indications that such state with a d–wave pairing function has quite a good energy [75, 76].

4. Correlated spin–density–wave function is a second type of wave function without a Fermi jump. It has been recently reconsidered by [77] in connection with the spin–bag model of high–T_c superconductivity. FHNC results for the variational energy, the gap in the excitation energy and the staggered magnetization are in very good agreement with the available VMC calculations [73] as it can be seen from Fig. 5.

From the above discussion it follows that perturbative CBF calculations would be very valuable in this field of research.

2.3 Strongly correlated nucleons

In spite of the fact that the first excitation of the nucleon (the Δ resonance) is at $\simeq$ 400 MeV, it is believed that a microscopic description of nuclear structure is reasonably well achieved by regarding the nucleus as a collection of bare nucleons interacting through the non–relativistic hamiltonian

$$H = -\frac{\hbar^2}{2m}\sum_{i=1}^{N} \Delta_i + \sum_{i<j=1}^{N} v_{ij} + \sum_{i<j<k=1}^{N} V_{ijk}. \tag{18}$$

The interactions v_{ij} and V_{ijk} are not well known as in the helium case, but v_{ij} is rather well determined by the nucleon–nucleon scattering and the deuteron data, and binding energies of light nuclei and nuclear matter properties put stringent constraints on V_{ijk}. The interparticle short–range behavior is influenced by the repulsive core (as in ^{3}He) and the tensor components of the two–body potential. They are contained in most realistic models of v_{ij}, such as the Reid [78], Paris [79], Urbana [80] and Argonne v_{18} [81], which all seem to predict similar structures.

The two–body interaction in Reid, Urbana and Argonne potentials can be expressed as

$$v_{ij} = \sum_{l}\sum_{i<j=1}^{N} v_l(r_{ij})\hat{O}^l(i,j), \tag{19}$$

where the firsts fourteen operators are given by

$$\begin{aligned} \hat{O}^{2l-1}(i,j) &= 1; \sigma_i \cdot \sigma_j; S(ij); \mathbf{L}\cdot\mathbf{S}; L^2; L^2\sigma_i\cdot\sigma_j; (\mathbf{L}\cdot\mathbf{S})^2 \\ \hat{O}^{2l}(i,j) &= \hat{O}^{2l-1}(i,j) \times \tau_i \cdot \tau_j,\ (l = 1,7), \end{aligned} \tag{20}$$

where $S(ij) = (3r_{ij}^{\alpha}r_{ij}^{\beta} - \delta_{\alpha\beta}r_{ij}^2)\sigma_i^{\alpha}\sigma_j^{\beta}$ is the tensor operator, and $\mathbf{L}$ and $\mathbf{S}$ are respectively the total angular momentum and spin of the pair (ij). The remaining 4 components in the Argonne v_{18} potential are charge and charge–independence symmetry breaking terms. The Paris potential uses the ∇^2 operator in place of L^2, which introduces a strong momentum dependence, and for this reason it has been used very little in CBF calculations.

The static part of the potential (the first six components, or v_6 model) is the most important one and characterizes the main features of the ground state wave function. The eight components depending on the angular momentum have, however, a not negligible effect on the variational parameters of $\hat{F}$, therefore the v_{14} model should be considered as a basic model for variational and realistic calculations in nuclei. The remaining charge and charge–independence symmetry breaking components can safely be treated in first order perturbation theory[4].

At interparticle distance $r > 1.5$ fm, the v^{stat} components are dominated by the one–pion exchange potential, while at smaller r they are significantly model dependent. However, much of the model dependence is counterbalanced by the momentum–dependent terms in the models. It is interesting to

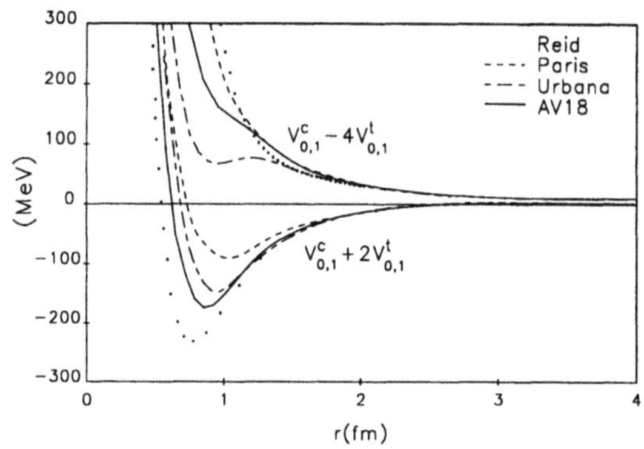

Fig. 6. Spin dependence of v^{sta} in the deuteron channel. The upper four lines display $\langle T = 0, S = 1, M_S | v^{stat} | T = 0, S = 1, M_S \rangle$ for $M_S = 0$ and $\theta = 0$, and the lower for lines for $M_S = 0$ and $\theta = \pi/2$ or equivalently for $M_S = \pm 1$ and $\theta = 0$. The results for $M_S = \pm 1$ and $\theta = \pi/2$ (not shown) are half–way in between.

analyze the spin dependence of the static part of the interaction (v_6) and its effect on the two–body density distribution, for the various spin–isospin channels ($T, S = 0, 1$) and spin projection $M_S = 0, \pm 1$. This has been recently done in light nuclei [82] and in nuclear matter [83], confirming in part predictions of classical Skyrme effective Lagrangians, related to QCD in the limit of infinite colors.

Fig. 6 displays $\langle T = 0, S = 1, M_S | v^{stat} | T = 0, S = 1, M_S \rangle$, as a function of M_S, the interparticle distance r and the polar angle θ between $\mathbf{r}$ and the quantization axis $\hat{\mathbf{z}}$. Outside the repulsive core, the $M_S = 0$ expectation value has a strong attraction for $\theta = \pi/2$ and a repulsion for $\theta = 0, \pi$. Therefore, in

[4] In first order perturbation theory, they do not contribute to symmetrical nuclear matter because it is a $T = 0$ state

this state the np pairs form a toroidal shape in the xy plane. On the contrary, the $M_S = \pm 1$ states are characterized by an attraction for $\theta = 0, \pi$ and a repulsion $\theta = \pi/2$, with a dumbell density distribution shape.

The three-body force is largely phenomenological and it is commonly fitted to reproduce the observed energy of ^{3}H nucleus, $E(^3\mathrm{H}) = -8.48$ MeV and the equilibrium density of nuclear matter, $\rho_0 = 0.16\ \mathrm{fm}^{-3}$. The most used form in CBF calculations is the so-called Urbana model [84], made up of two isoscalar terms:

$$V_{ijk} = V_{ijk}^{2\pi} + V_{ijk}^{R}. \tag{21}$$

The first term is a two-pion exchange interaction [85], with a strength $A_{2\pi}$ as a free parameter. The second one is a purely phenomenological interaction, of the form $V_{ijk}^R = U_0 \sum_{cyc} T_\pi^2(r_{ij}) T_\pi^2(r_{ik})$, where $T_\pi^2(r_{ij})$ is the radial shape of the one-pion exchange tensor. It is repulsive and counterbalances the attraction of $V_{ijk}^{2\pi}$. The latest fits [86] have led to the values $A_{2\pi} = -0.0293$ MeV and $U_0 = 0.0048$ MeV (Model IX), not too different from those of the earlier Model VII (-0.0028 MeV and 0.005 MeV respectively). The three-body potential provides a quite small contribution to the potential energy (less than 5%), but because of the large cancelation between kinetic and potential energies it cannot be disregarded. Table 3 displays the latest results for the binding energy of light nuclei and nuclear matter, used to fit the three body potential [87]. There is a very good agreement among the results obtained for the binding energies of s–shell nuclei, by means fo different techniques. For instance, the pair-correlated hyperspherical harmonic (PHH) method [88]

Table 3. Contributions to the energy E (in MeV) of light nuclei and to the energy per particle of symmetrical nuclear matter (SMN) calculated with the Argonne v_{18} plus the Model IX three–body potential. Light nuclei have been obtained by using Green Function Monte Carlo [87], whereas SMN results refer to variational calculations [83] in FHNC/SOC approximation. (A CBF perturbative correction of –1.91 MeV has also been estimated for a net binding of 11.96 MeV).

	$^3\mathrm{H}(\frac{1}{2})$	$^4\mathrm{He}(0)$	$^5\mathrm{He}(\frac{3}{2})$	$^5\mathrm{He}(\frac{1}{2})$	SMN
T	50.0	118.0	122.0	117.0	42.26
$\langle v_{14} \rangle$	-58.0	-142.0	-145.0	-140.0	-55.09
$\langle v_{18} \rangle$	-57.3	-139.8	-141.5	-136.3	-55.09
$\langle V \rangle$	-1.20	-6.5	-7.0	-6.4	2.78
$E(cal)$	-8.47	-28.3	-27.2	-25.8	-10.05
$E(exp)$	-8.48	-28.3	-27.2	-25.8	-16.0

provides a binding energy of –8.49 MeV for ^{3}H [89]. For nuclear matter we are not at the same level of accuracy as for light nuclei, and a more extensive discussion on this point will be given in the next sections.

The many–body correlation operator must include spin- and isospin–dependent correlations, induced by the nucleon–nucleon potential. This is commonly done [22] by taking the following operator[5]

$$\hat{F} = Sim \prod_{i,j=1}^{N} \sum_{p=1,8} f^p(r_{ij}) O^p(ij), \tag{22}$$

where Sim is the symmetrization operator. The corresponding trial function is not any more of the factorized form, and the two body correlations do not commute between themselves ($[\hat{O}^p(ij), \hat{O}^q(ik)] \neq 0$). This is the reason for introducing the symmetrizer in (22). In dealing with medium–heavy nuclei or nuclear matter, such non-commutativity poses serious problems, which have been tackled only under ad–hoc approximations (the most used one is the FHNC/SOC [22], which basically consists in retaining only the leading terms in the spin–isospin–dependent correlations for each cluster diagrams, as discussed in section 4). Nuclear matter is a relatively lower density system than liquid helium, and, therefore, triplet correlations can be probably neglected.

The correlation functions $f^p(r)$ are usually obtained through a partial optimization which will be described in section 4. The strong spin–isospin dependence of the short–range correlations in SMN is shown in Fig. 7, where the static components ($p = 1, 6$) of the two–body density operator, given by

$$\rho_2^p(r) = \frac{2}{N} < \Psi^{(0)} | \sum_{i,j=1}^{N} \delta(r - r_{ij}) \hat{O}^p(i,j) | \Psi^{(0)} \rangle, \tag{23}$$

are displayed.

Spin–coherent wave functions. A new type of trial wave functions, denoted as spin–coherent wave function (CS), has been recently proposed [38]. It includes spin–dependent correlations in such a way that the aforementioned non–commutativity problem is overcome. The wave function is constructed using spin–coherent states to represent the spin–states of particles. On this basis, the states are parameterized by a continuous variable that looks like a classical spin vector. Thus, spin–operators can be represented as simple integrals over these *spin variables*.

In this formulation, the number of integration variables increases. However, these integrals can be evaluated using standard numerical techniques. With this approach, VMC calculations for nuclei and larger systems may

[5] Another variationally equivalent formulation can be given by using the exponential form $\hat{F} = \exp[-1/2 \sum_{\{ij\},p} u^p(r_{ij}) O^p(ij)]$. The relationship between the u's and the f's is not trivial due to the non commutativity problem

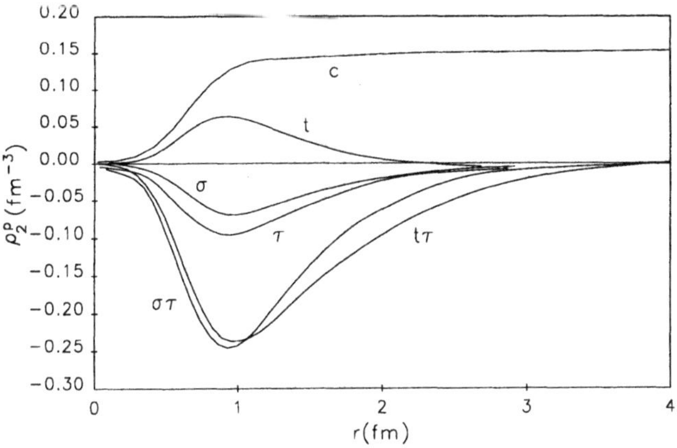

Fig. 7. $\rho_2^{p=1,6}$ for symmetrical nuclear matter with the Argonne v_{18} plus the Model IX three–body model hamiltonian.

become feasible. Additionally, since these classical spin variables commute, commutator terms in the FHNC equations do not appear and it is possible to go beyond the FHNC/SOC approximation.

In the coherent state formulation, the spin state of a given particle is expressed on the following basis:

$$|\hat{\omega}\rangle \equiv |\theta,\phi\rangle \equiv \exp[-i\phi(\hat{S}_z - S)]\exp[-i\theta\hat{S}_y]\;|S\rangle, \tag{24}$$

where $|\hat{\omega}\rangle$ is a unity vector on a sphere having coordinates $|1,\theta,\phi\rangle$, $0 \leq \theta \leq \pi$ and $0 \leq \phi \leq 2\pi$, and the state $|S\rangle$ is the eigenstate of $\hat{S}_z$ with the largest possible eigenstate. There is a one–to–one correspondence between the coherent states and points on the unity sphere. For spin 1/2 one has the simple relation

$$|\hat{\omega}\rangle = \cos(\frac{\theta}{2})\exp(i\frac{\phi}{2})\mid\uparrow\rangle + \sin(\frac{\theta}{2})\exp(-i\frac{\phi}{2})\mid\downarrow\rangle. \tag{25}$$

The individual coherent states are not orthogonal and the set $\{|\hat{\omega}\rangle\}$ is over complete. Any spin–operator $\hat{O}$ can be represented in a form that looks like a diagonal operator,

$$\hat{O} = \int \frac{d\hat{\omega}}{2\pi}|\hat{\omega}\rangle f(\hat{\omega}) < \hat{\omega}|, \tag{26}$$

where $f(\hat{\omega})$ is, in general, a non–unique, c–valued function of the angles (θ,ϕ). For example, for the Pauli matrices σ_i, $f(\hat{\omega}) = 3\hat{\omega}_i$, and, in general, if the operator $\hat{O}$ is linear in σ, then the corresponding $f(\hat{\omega})$ is also linear in $\hat{\omega}$

The coherent–spin representation of the correlation operator $\hat{F}$ of (22) (or of that in the exponential form given in the footnote) is formally obtained by inserting the resolution of the identity $I = \int d\hat{\Omega}/(2\pi)^N |\hat{\Omega}\rangle\langle\hat{\Omega}|$ twice, where $\int d\hat{\Omega}$ represents the product $\int \prod_{i=1}^{N} d\hat{\omega}_i$, and $|\hat{\Omega}\rangle$ corresponds to the inner product of the N single–particle coherent states $\{|\hat{\omega}_i\rangle\}$.

In analogy with explicit coherent state representation for propagators of quantum spin models constructed using path integral methods [90], the spin–coherent trial wave function is taken of the form:

$$\begin{aligned} \Psi_{CS}^{(0)}(R) &= F_R(R) \int d\hat{\Omega}_\sigma d\hat{\Omega}_\tau |\hat{\Omega}_\sigma \hat{\Omega}_\tau\rangle \\ &\times \exp\left[-\frac{1}{2} \sum_{i<j=1}^{N} \sum_{p=1}^{8} u^p(r_{ij}) \hat{O}_\omega^p(i,j) \right] \langle \hat{\Omega}_\sigma \hat{\Omega}_\tau | \Psi_{MF}^{(0)} \rangle, \end{aligned} \tag{27}$$

where $\hat{\Omega}_\sigma$ and $\hat{\Omega}_\tau$ refer to the spin and isospin variables, respectively. The operators $\hat{O}_\omega^p(i,j)$ are obtained from the corresponding $\hat{O}_p(i,j)$ of (20) by replacing σ_i with $\hat{\omega}_\sigma(i)$ and τ_i with $\hat{\omega}_\tau(i)$. The function $F_R(R)$ can be taken of the Jastrow form, or one can add triplet correlations, if needed.

The overlap of the spin wave functions with the corresponding single–particle coherent state replaces the spin wave functions with the corresponding matrix elements $\langle\hat{\omega}_\sigma(i)\hat{\omega}_\tau(i)|\chi_j(i)\rangle$. The full overlap with the coherent–state–spin wave function is expressed as a determinant of the single particle orbitals $\mathrm{Det}(M_{ij})$, where $M_{ij} = \exp(i\mathbf{k}_j \cdot \mathbf{r}_i)\langle\hat{\omega}_\sigma(i)\hat{\omega}_\tau(i)|\chi_j(i)\rangle$. The ground state expectation value of the Hamiltonian and other operators are many–body integrals, and they can be evaluated by either FHNC or Monte Carlo methods.

VMC calculations with $\Psi_{CS}^{(0)}$ present technical difficulties due to the fact that the product $\mathrm{Det}(M_{ij}^{left}) \times \mathrm{Det}(M_{ij}^{right})$ is not positive definite, and, therefore one is facing with the sign–problem. Recent VMC calculations [38, 39] performed for liquid and solid ^{3}He, with only the $p = 2$ component in (27), have provided extremely promising results (see Fig. 3).

In FHNC calculations, the $\hat{\Omega}$ integrations can be made easier by using a discrete set of points (see section 4). Work is in progress to find an efficient algorithm to increase the number of particles in VMC calculations, and in solving the FHNC equations for $\Psi_{CS}^{(0)}$.

3 Correlated basis function theories

Correlated basis function theory uses a set of correlated basis functions, instead of mean–field states. There is a direct correspondence between a mean field state $\Psi_{MF}^{(i)}$ and a CBF state $\Psi_{CBF}^{(i)}$ through the correlation operator $\hat{F}$ (see eq.(1)):

$$\Psi_{MF}^{(i)} \rightarrow \Psi_{CBF}^{(i)} = \hat{F}\Psi_{MF}^{(i)} \equiv |n_i, m_i\rangle, \tag{28}$$

where we have used the notation $|n_i, m_i\rangle$ to indicate that the original mean–field state has n particles and m holes, namely, n single particle states above the Fermi and m missing states below the Fermi sea. The subindex i specifies the momenta and the spin–isospin assignments of the particle and hole states. Often we will use for brevity $|i\rangle$ to denote $|n_i, m_i\rangle$

As discussed in the previous sections, the correlation operator is determined by a variational calculation of the ground state energy. The diagonal matrix elements of the hamiltonian on the CBF states

$$E_i^v = \frac{\langle i|H|i\rangle}{\langle i|i\rangle} = \textit{variational energies} \tag{29}$$

are generally called *variational energies*, although only E_0^v has been variationally estimated. However, E_i^v are of order N, whereas the excitation energies $E_i^v - E_0^v$ are of order 1, and cannot be minimized variationally.

The diagonal matrix elements define the unperturbed hamiltonian:

$$H = H_0 + H_I, \tag{30}$$

$$\langle i|H_0|j\rangle = \delta_{i,j} E_i, \tag{31}$$

$$\langle i|H_I|j\rangle = (1 - \delta_{i,j}) \tfrac{\langle i|H|j\rangle}{\langle i|i\rangle}, \tag{32}$$

so that there are no first order terms in the perturbation expansion.

A major problem of CBF theory is that the basis $|i\rangle$ is not orthogonal,

$$N_{ij} = \frac{\langle i|j\rangle}{\langle i|i\rangle^{1/2}\langle j|j\rangle^{1/2}} \neq \delta_{i,j}. \tag{33}$$

Originally, *non–orthogonal* perturbation theory has been used to deal directly with the states $|i\rangle$. The problem with orthogonalizing the states $|i\rangle$ among themselves is that the estimates of the hamiltonian diagonal matrix elements in a fully orthogonal basis set are worst than the variational ones [10].

More recently, a new orthogonalization procedure has been devised [12], which preserves the variational diagonal matrix elements of the hamiltonian and allows for using ordinary orthogonal perturbation theory in zero temperature calculations.

Although, formally, the two scheme are equivalent, in practice *orthogonal* CBF theory is more efficient and flexible, particularly in dealing with quantities other than the energy eigenvalues. A more detailed comparison of the two schemes will be given in the remaining of the section.

3.1 Non–orthogonal CBF scheme

In the case of a non–orthogonal basis, the perturbative correction to E_n^v is given by [91]

$$\Delta E_n = E_n - E_n^v = \sum_{p \neq n} \frac{W_{np}(n) W_{pn}(n)}{E_n - E_p^v} \tag{34}$$

$$+ \sum_{q \neq p \neq n} \frac{W_{np}(n) W_{pq}(n) W_{qn}(n)}{(E_n - E_p^v)(E_n - E_q^v)} + \cdots, \tag{35}$$

where

$$W_{ij}(n) = H_{ij} - E_n N_{ij}. \tag{36}$$

The correction at various orders, $\Delta E_n^{(i)}$, of the perturbative series can be obtained by expanding eq. (35) around $E_n = E_n^v$ and then inserting ΔE_n back in the right–hand side repeatedly. Correspondingly,

$$W_{ij}(n) = (H_{ij} - E_n^v N_{ij}) - \Delta E_n N_{ij} = W_{ij}^v(n) - \Delta E_n N_{ij} \, , \tag{37}$$

$$\frac{1}{E_n - E_p} = \frac{1}{(E_n^v - E_p^v) - \Delta E_n} = \frac{1}{E_n^v - E_p^v} \sum_{i=0}^{\infty} (-\frac{\Delta E_n}{E_n^v - E_p^v})^i. \tag{38}$$

As in ordinary perturbation theory, each $\Delta E_n^{(i)}$ has a catastrophic dependence on the particle number N. However it has been proved, in an analogous way, that the divergent terms appearing at different orders cancel each other [92].

CBF perturbation theory has two major features, absent in ordinary perturbation theory:

1. there is an energy dependence carried by the *interaction* $W(n)$, besides the one coming from energy denominators;
2. due to correlations, $W^v(n)$ is a many body operator built up with the two–body dressed hamiltonian $f_2(1,2)H(1,2)f_2(1,2)$, acting between the interacting particles and the pseudopotential $u_2(\alpha,\beta) = -\frac{1}{2}\ln(f_2(\alpha,\beta))$ acting between the particles of the medium [6].

These features imply an unavoidable interplay between the structure of the CBF perturbation series and that of W and of N in terms of the pseudopotentials included in the $\hat{F}$. This has been studied by means of *cluster expansions* of H_{ij} with respect to the pseudopotentials (see the following section).

Let us analyze, through an example, the two aforementioned problems. The diagram shown in Fig. 8 is a representation of a typical term generated by eq. (35) for ΔE_0. The intermediate states are $|p\rangle = |p_1, p_2; h_1, h_2\rangle$, $|q\rangle = |p_1, \cdots, p_4; h_1, \cdots, h_4\rangle$, and again $|p\rangle$. Correspondingly, the matrix elements $W_{ij}^v(0)$ are characterized by CBF states $|i\rangle$ and $|j\rangle$ which differ in two orbitals, namely $d_{ij} = 2$. However the matrix element $W_{pq}^v(0)$ also depends upon the

[6] One may also have three–body pseudopotential, as discussed in section 2. Moreover, spin–isospin dependent correlations bring spin–isospin dependent pseudopotentials. In this case f_2 is an operator in the spin–isospin space.

orbitals $p_1, p_2; h_1, h_2$ (the dependence on the other occupied orbitals below the Fermi sea is integrated out and therefore is a passive one). This brings in an *off-shell* energy dependence of $W^v_{pq}(0)$.

The many–body structure of W_{ij}, embedded in the horizontal boxes, has the consequence that it may depend on any number of orbitals (W_{pq} in the figure depends upon eight orbitals). In fact, the W–box implies an infinite series of cluster diagrams (see next section) built with one fHf interaction line and pseudopotentials lines, correlating the medium particles among themselves. Similarly, N_{ij} is characterized by cluster diagrams, in which, however, the fHf line is missing.

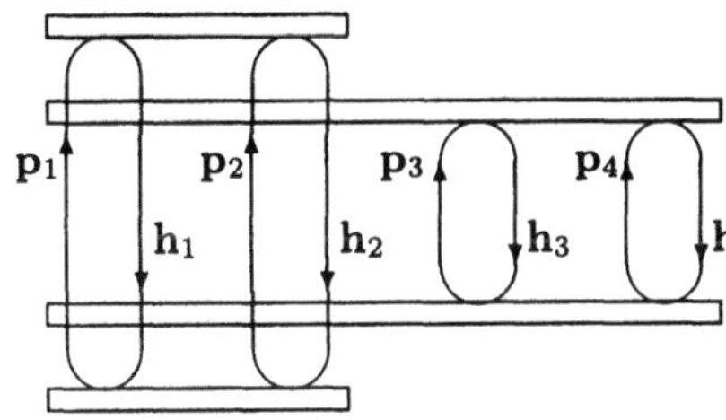

Fig. 8. Typical CBF diagram contributing to ΔE_0. The horizontal boxes represent the interaction $W(0)$. Up going and down going lines denote particle and hole states respectively.

In order to keep track of the energy dependence of $W^v_{ij}(n)$, it has been found convenient to split W into two parts:

$$W^v_{ij}(n) = W^v_{ij}(i) + (E^v_i - E^v_n)N_{ij}, \tag{39}$$

where $W^v_{ij}(i) = H_{ij} - E^v_i N_{ij}$ is the matrix element of the hamiltonian between $|i\rangle$ and the state $\overline{|j\rangle} = |j\rangle - N_{ij}|i\rangle$, orthogonal to $|i\rangle$. From cluster expansion arguments [93, 94, 92] it follows that, at leading orders in the number of particles N, both $W^v_{ij}(i)$ and N_{ij} have no off-shell dependence. The whole off–shell dependence of $W^v_{ij}(n)$ is left to the factor $(E^v_i - E^v_n)$. Equation (35) can be recast in the form:

$$\Delta E_n = \sum_{p\neq n}(W^v_{np}(n) - \Delta E_n N_{np})(\frac{W^v_{pn}(p)}{E_n - E^v_p} - N_{pn})$$
$$+ \sum_{q\neq p\neq n}(W^v_{np}(n) - \Delta E_n N_{np})(\frac{W^v_{pq}(p)}{E_n - E^v_p} - N_{pq})$$

$$\times \left(\frac{W^v_{qn}(q)}{E_n - E^v_q} - N_{qn}\right) + \cdots, \tag{40}$$

where the energy denominators are expanded as in (38). According to (40), CBF diagrams have two types of interaction boxes, one representing $W^v_{ij}(i)$ and the other denoting N_{ij}.

$W^v_{ij}(i)$ and N_{ij} are matrix elements of many–body operators, and,for $d_{ij} \geq 4$ they are of the order Ω^l, with a l–value too large ($> 1 - d_{ij}$). So, they lead to divergencies in the thermodynamic limit. In fact, the leading terms of W_{ij} and N_{ij} with $d_{ij} \geq 4$ can be factorized in two or more parts, or, equivalently, the corresponding cluster diagrams are unlinked (a cluster diagram with a number n_l of linked portions is of the order $\Omega^{(n_l-d)}$).

Working out eq.(40) iteratively, it can be proved [92] that ΔE_n is given by the sum of all linked Goldstone–type diagrams built up with $W^v_{ij}(i)$ and N_{ij} interaction boxes. The interaction boxes represent sums of only linked cluster diagrams. Therefore, the CBF perturbation series is well behaved in the thermodynamic limit.

The topmost box of a CBF diagram must be of the W–type. Each N–box carries a minus sign and no energy denominator is present between an N–box and an interaction box immediately above it.

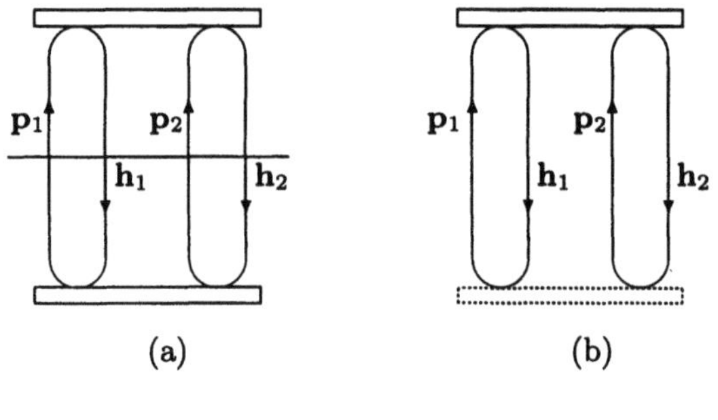

Fig. 9. Second order CBF diagrams contributing to ΔE_0. The horizontal boxes with solid lines represent the linked terms of $W^v_{ij}(i)$. The horizontal box with dashed lines denotes the linked terms of N_{ij}. The horizontal bar in diagram (a) represents the energy denominator $1/(E^v_0 - E^v_p)$.

The number of interaction boxes has been generally considered to characterize the order of the perturbative correction. Indeed, the perturbative order should be determined by the number of the W–boxes, because the interaction of type N are related to orthogonality corrections.

Typical calculations of the second order CBF perturbative corrections to the ground state energy have included the diagrams with two interaction boxes displayed in Fig. 9. The corresponding expression is

$$\Delta E_0 = \sum_{p\neq 0} W^v_{0p}(0)\left(\frac{W^v_{p0}(p)}{E^v_0 - E^v_p} - N_{p0}\right)$$
$$= \sum_{p\neq 0} \frac{W^v_{0p}(0)W^v_{p0}(0)}{E^v_0 - E^v_p}. \tag{41}$$

Non orthogonal CBF perturbation theory has been extensively applied to nuclear matter and much less to other quantum liquids. In the earlier calculations, the correlation operator was taken of the Jastrow form [10, 95, 96, 97]. However, the second order corrections to the binding energy, single–particle spectrum, etc. are large in Jastrow CBF. The correlation operator of eq.(22) provides much smaller corrections and, therefore, is expected to significantly improve the convergence. Jastrow CBF is much more justified in liquid ^{3}He or electron gas calculations, where the interaction is purely radial. Attempts in this direction may be found in [99, 100, 101, 26] for ^{3}He and in [20] for electron plasma.

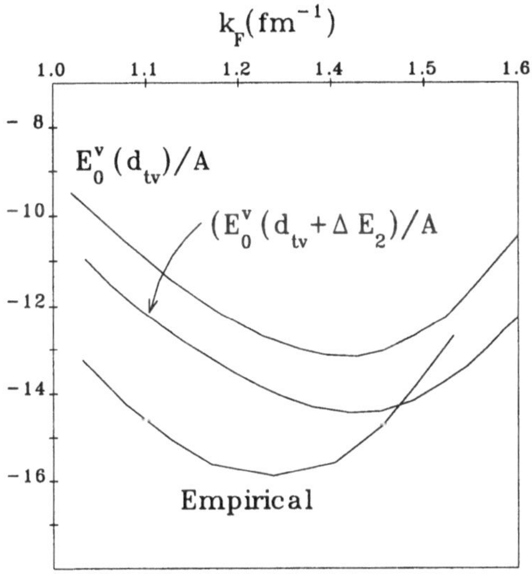

Fig. 10. The energy per particle in nuclear matter obtained with the *Urbana* v_{14} plus DDI is compared with the empirical equation of state [104].

Perturbative CBF calculations have been limited to the second order terms, eq. (41) for the binding energy.

We shall discuss, in the following, a few nuclear matter calculations of quantities of physical interest, like binding energy, optical potential and momentum distribution. They are based on correlated basis with the correlation operator (22).

Extensive CBF perturbative calculations have been done for the *Urbana* v_{14} [80] and the *Argonne* v_{14} [102] models of nuclear matter. The underlying variational calculations of the binding energy and the optimal variational parameters of the correlation operator, have been carried out including also the three–nucleon potential.

V_{ijk} has been approximated by a density dependent interaction (DDI), originally proposed by [103], in several calculations. DDI is made up of two terms: a repulsive (TNR) and an attractive (TNA) one, approximating V_{ijk}^R and the two–pion exchange interaction $V_{ijk}^{2\pi}$, respectively. TNR is obtained by multiplying the intermediate–range part of v_{14} by a factor $\exp(-\gamma_1\rho) - 1$, whereas TNA is a Skyrme–like term, given by $3\gamma_2\rho^2\exp(-\gamma_3\rho)$. The DDI parameters are fitted so that the $v_{14} + DDI$ model reproduces the nuclear matter saturation density $\rho_0 = 0.16\text{fm}^{-3}$, binding energy $E_0/N = 16$ MeV and compressibility $K = 240$ MeV.

DDI cannot be used in finite nuclei, where the density is not constant, unless ρ is substituted by a local density operator, as done in the ^{4}He liquid–solid coexistence calculations with shadow wave functions [46]. DDI is not well suited for perturbative calculations either; in fact, TNA is independent on the wave function, and the use of TNR in the calculation of $W_{ij}^v(i)$, as a simple modification of the two–body part of the interaction, is not fully consistent.

However, the three–nucleon interaction provides less than 5% contribution of the potential energy, and, therefore, it is expected to have little effect on the perturbative corrections. For this reason, the three-nucleon interaction has never been explicitly included in the calculation of $W_{ij}^v(i)$, so far.

Equation of state of nuclear matter. Typical estimates of second order perturbative correction to the binding energy at the equilibrium density of nuclear matter range from ~ -1 to -3 MeV, depending on the model interaction used. Fig. 10 shows the results obtained in [104] for second order estimates of ΔE_0 (labeled in the figure with ΔE_2) for the *Urbana* v_{14} potential, added to the variational estimates E_0^v obtained with the *Urbana* v_{14} plus model VII interaction.

More recent CBF calculations [105] have found $\Delta E_0^{(2)} = -1.82$ MeV for the *Urbana* $v_{14} + DDI$ model and $\Delta E_0^{(2)} = -3.12$ MeV for *Argonne* $v_{14} + DDI$ model, at ρ_0. Tensor correlations are quite important for perturbative corrections. *Argonne*–potential has a stronger tensor component than *Urbana*–potential and this explains the quite different results obtained for $\Delta E_0^{(2)}$.

Very recently, $\Delta E_0^{(2)}$ has been estimated for the *Argonne* v_{18}+*Urbana* IX interaction [83]. The resulting value, at saturation density, is -1.89 MeV, with a variational energy $E_0^v = -11.96$ MeV.

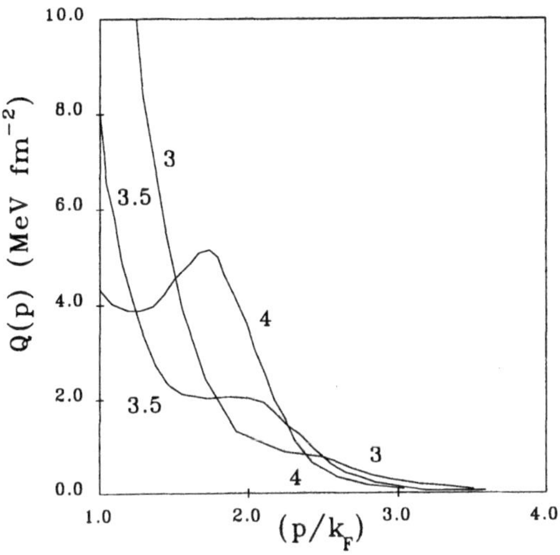

Fig. 11. The function $Q(p)$ calculated with the *Urbana* v_6 interaction is shown at three values of the variational parameter $d_t = 3, 3.5$ and $4r_0$ [104].

A poorly adapted feature, which deserves some discussion, is that $E_0^v + \Delta E_0^{(2)}$ is not, as one would expect, more independent on the variational parameters than E_0^v itself. Actually, $E_0^v + \Delta E_0^{(2)}$ shows well pronounced maxima as a function of d_t, where d_t is the *healing* distance of the tensor correlations (see section 4 for a discussion of the variational parameters). The reason lies in the small contribution of the low–lying eigenstates of the hamiltonian to E_0^v. This is because $E_0^v = E_0 + \sum_{i\neq 0} \alpha_{0i}^2 (E_i - E_0)$, and, even if α_{0i} are large, the factor $(E_i - E_0)$ cuts off their contribution. The same does not hold for the perturbative correction. Fig. 11 shows the function $Q(p)$, defined as

$$Q(p) = p^2 \int_{-\infty}^{e^v(k_F)} dE \frac{Im\Sigma(p,E)}{e^v(p) - E}, \tag{42}$$

calculated for the *Urbana* v_{16} model. $\Sigma(p,E)$ is the self–energy (defined below in the section) and $e^v(p) = E^v(p) - E_0^v$ is the excitation energy of the single particle state with momentum p. In terms of Q, the perturbative correction $\Delta E_0^{(2)}$ turns out to be

$$\Delta E_0^{(2)} = -\frac{N}{\pi^3 \rho} \int_{k_F}^{\infty} dp Q(p), \tag{43}$$

therefore $Q(p)$ is a measure of the contribution of states of momentum p to the second order perturbative correction. One can see that $Q(p = k_F)$ is large for values of d_t close to the optimal variational value $d_{tv} \sim 3r_0$, with $r_0 = (3/4\pi\rho)^{1/3}$, and decreases for increasing values of d_t. Therefore $d_t = d_{tv}$ is not necessarily a good choice for CBF calculations in normal Fermi liquids. It is plausible that this is a manifestation of instabilities due to pairing phenomena, taking place in the region of low–lying states.

The perturbative corrections are calculated at values $d_t = d_{tp}$ which minimize the quantity k_0 defined as [7]

$$
\begin{aligned}
k_0 &= \frac{1}{4N} \sum \frac{|\langle 0|(H - E_0^v)|\mathbf{p}_1, \mathbf{p}_2; \mathbf{h}_1, \mathbf{h}_2\rangle|^2}{(e^v(h_1) + e^v(h_2) - e^v(p_1) - e^v(p_2))^2} \\
&= \frac{1}{N} \sum_{i \neq 0} \alpha_{0i}^2 + \mathcal{O}(\alpha^3),
\end{aligned}
\tag{44}
$$

and which provides an unbiased measure of the magnitudes of α_{0i}. The resulting values of d_{tp} are slightly larger than d_{tv}. ΔE_0^v is taken as

$$
\Delta E_0^v = E_0^v(d_{tp}) + \Delta E_0(d_{tp}) - E_0^v(d_{tv}). \tag{45}
$$

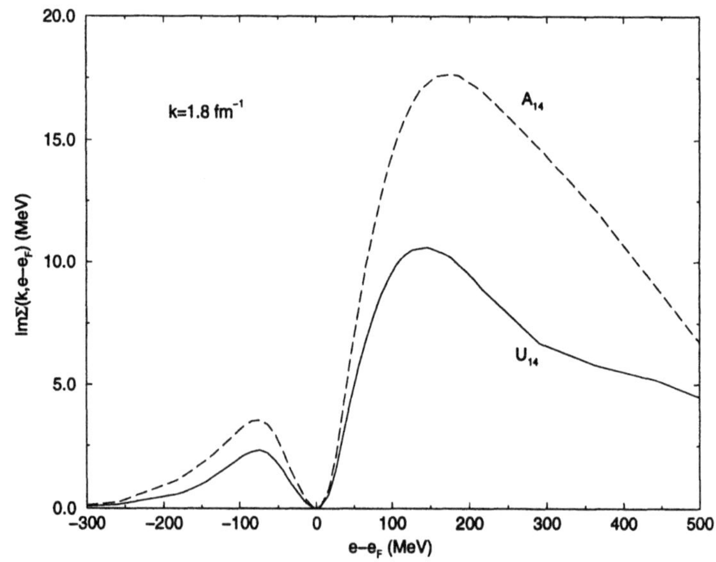

Fig. 12. Self–energy of the Urbana and Argonne models of nuclear matter [105].

[7] Note that, in the thermodynamic limit, the energy denominator $E_0^v - E^v(\mathbf{p}_1, \mathbf{p}_2; \mathbf{h}_1, \mathbf{h}_2)$ can be expressed in terms of single particle excitation energies.

Self–energy of nuclear matter. A very important quantity, which enters most of the CBF perturbative calculations, is the self–energy $\Sigma(\mathbf{k}, E)$, usually splitted into a *correlation* plus a *polarization* term. The imaginary part of Σ^{CO}, in second order perturbation theory, is defined as [106]

$$Im\Sigma^{CO}(\mathbf{p}, E < e^v(k_F)) = \frac{\pi}{2} \sum |\langle 0|H - E_0^v|\mathbf{p}, \mathbf{p}_2; \mathbf{h}_1, \mathbf{h}_2\rangle|^2 \times \delta(E + e^v(p_2) - e^v(h_1) - e^v(h_2)), \tag{46}$$

for particle states, and

$$Im\Sigma^{CO}(\mathbf{h}, E > e^v(k_F)) = \frac{\pi}{2} \sum |\langle 0|H - E_0^v|\mathbf{p}_1, \mathbf{p}_2; \mathbf{h}, \mathbf{h}_2\rangle|^2 \times \delta(E + e^v(h_2) - e^v(p_1) - e^v(p_2)), \tag{47}$$

for hole states. Similarly the polarization term is defined as

$$Im\Sigma^{PO}(\mathbf{p}, E > e^v(k_F)) = \frac{\pi}{2} \sum |\langle \mathbf{p}|H - E^v(p)|\mathbf{p}_1; \mathbf{p}_2, \mathbf{h}_2\rangle|^2 \times \delta(E + e^v(h_2) - e^v(p_1) - e^v(p_2)), \tag{48}$$

for particle states, and

$$Im\Sigma^{PO}(\mathbf{h}, E < e^v(k_F)) = \frac{\pi}{2} \sum |\langle \mathbf{h}|H - E^v(h)|\mathbf{p}_2; \mathbf{h}_1, \mathbf{h}_2\rangle|^2 \times \delta(E + e^v(p_2) - e^v(h_1) - e^v(h_2)), \tag{49}$$

for hole states. Detailed expressions to compute $Im\Sigma(\mathbf{k}, E)$ can be found in [100]. Fig. 12 shows the calculated $Im\Sigma$ at $k = 1.8\text{fm}^{-1}$ for the *Urbana* (U_{14}) and the *Argonne* (A_{14}) models. One can see that the two typical maxima of $Im\Sigma$ are $\sim 50\%$ larger for A_{14}, as a consequence of its stronger tensor part. Similar shapes are found at any other value of the momentum.

The imaginary part of the self–energy can be used to evaluate the perturbative correction to the ground state energy and the optical potential $U_0 + iW_0$ of a nucleon in nuclear matter. Two equivalent expressions can been obtained for $\Delta E_0^{(2)}$:

$$\Delta E_0^{(2)} = \frac{1}{2\pi} \int_{e^v(k_f)}^{\infty} dE \frac{Im\Sigma^{CO}(\mathbf{h}, E)}{e^v(h) - E} = \frac{1}{2\pi} \int_{-\infty}^{e^v(k_f)} dE \frac{Im\Sigma^{CO}(\mathbf{p}, E)}{E - e^v(p)}. \tag{50}$$

A comparison of the two expressions provides a useful test for the accuracy of the numerical techniques used in the calculation.

The perturbative corrections to the single particle excitation energies are given by

$$\Delta e^{(2)}(k) = \Delta E(k)^{(2)} - \Delta E_0^{(2)} = \frac{\mathcal{P}}{\pi} \int_{-\infty}^{\infty} dE \frac{Im\Sigma(k, E)}{e^v(k) - E}, \tag{51}$$

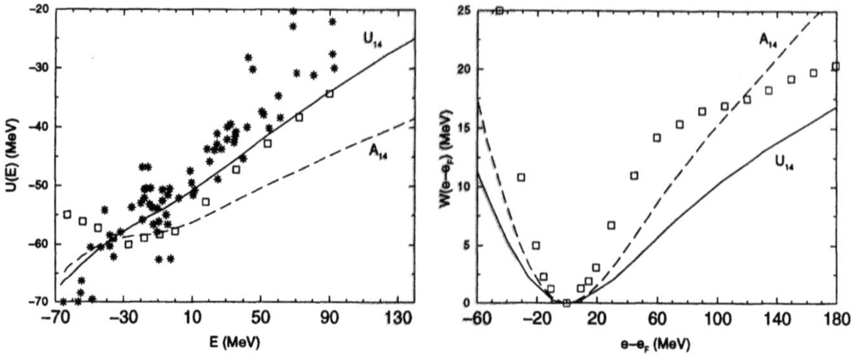

Fig. 13. The real (left panel) and imaginary (right panel) parts of the nuclear matter optical potential for the Urbana+*DDI* (solid line) and Argonne+*DDI* (dashed line) models [105]. The squares correspond to the Brueckner–Hartree–Fock (BHF_2) results [107]. The stars are the finite nuclei experimental estimates from [108].

where $\mathcal{P}$ means principal part of the integral. The above equation neglects the two *rearrangement* terms, each one of them being of order Ω and whose sum is of order 1 [109]. Their contribution is quite small in nuclear matter, and, in any case, it does not vary rapidly with k in the region of $k \sim k_F$. One can verify [104] that $\Delta e^{(2)}(k)$ of eq.(51) satisfies the continuity at $k = k_F$, and check whether the neglected rearrangement terms are small or not through the equation

$$\Delta e^{(2)}(k) = \frac{\Delta E_0^{(2)}(\rho)}{N} + \frac{\rho}{N}\partial \Delta E_0^{(2)}(\rho)/\partial\rho, \tag{52}$$

The real part of the optical potential,

$$e(k) = \frac{\hbar^2}{2m}k^2 + U_0(\rho, e), \tag{53}$$

is shown in Fig. 13 for $U_{14} + DDI$ and $A_{14} + DDI$, and compared with the Brueckner–Hartree–Fock estimates. The $U_{14} + DDI$ results are in slightly better agreement with the experimental data. The contribution from perturbative corrections is at most $\sim 10\%$. The relevant feature is the enhancement of the effective mass $m^\star(k)/m = \hbar^2 k(mde/dk)^{-1}$ around $k = k_F$, which is totally absent in the variational estimate, and is produced by the wiggle of U around $E \sim -16$ MeV. The perturbative correction to $m^\star(k_f)/m$ is 0.16 for $U_{14} + DDI$ and 0.30 for $A_{14} + DDI$.

The imaginary part of the optical potential W_0 is directly related to the imaginary part of the on–shell self–energy through the equation

$$W_0(e) = \frac{m^\star(k)}{m} Im\Sigma(k(e), e), \tag{54}$$

where the factor $m^\star(k)/m$ takes into account the effect of non locality of the real part of the potential [110, 111]. CBF results for the $U_{14}+DDI$ and $A_{14}+DDI$ models of nuclear matter are displayed in Fig. 13.

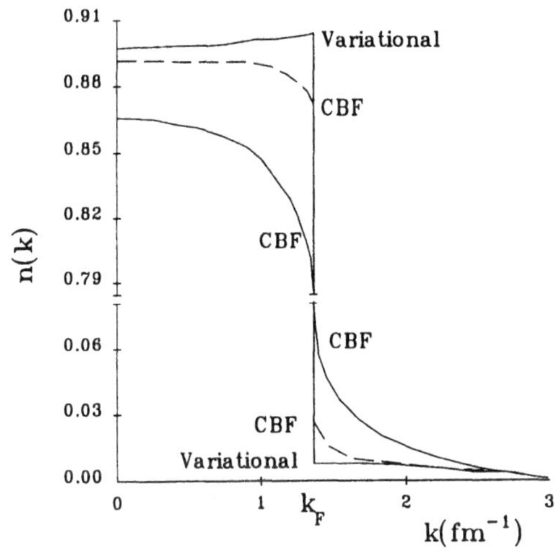

Fig. 14. Momentum distribution $n(k)$ of a nucleon in nuclear matter [30]. The solid line is the full CBF calculation, whereas the dashed line is the CBF calculation without tensor correlations.

Momentum distribution of nucleons in nuclear matter. Non orthogonal CBF theory has also been used to compute the momentum distribution of nuclear matter [30]. The second order perturbative correction to $n^v(k)$ has been obtained from the expectation value of $a^\dagger_{\mathbf{k}}a_{\mathbf{k}}$ on the state

$$|0_2\rangle = |0\rangle + \frac{1}{4}\sum \alpha(\mathbf{p}_1, \mathbf{p}_2; \mathbf{h}_1, \mathbf{h}_2)|\mathbf{p}_1, \mathbf{p}_2; \mathbf{h}_1, \mathbf{h}_2\rangle, \tag{55}$$

where

$$\alpha(\mathbf{p}_1, \mathbf{p}_2; \mathbf{h}_1, \mathbf{h}_2) = \frac{\langle 0|H - E_0^v|\mathbf{p}_1, \mathbf{p}_2; \mathbf{h}_1, \mathbf{h}_2\rangle}{(e^v(h_1) + e^v(h_2) - e^v(p_1) - e^v(p_2))}, \tag{56}$$

and keeping only the terms of order α^2.

The perturbative corrections $\Delta n^{(2)}(k)$ recover the property that $n(k)$ has infinite slopes as $|k - k_F| \to 0$ [112, 113], namely

$$n(k \sim k_F) = n(k_F^\pm) \pm A \ln \frac{|k - k_F|}{k_F}, \tag{57}$$

where the upper (lower) sign is for particle (hole) state. This singularity is missing at the variational level, and comes entirely on account of the second order perturbative correction. More explicitly, it derives from the well–known second order contribution to $n(k)$ in ordinary (i.e. orthogonal basis) perturbation theory,

$$\begin{aligned}\left(\Delta n^{(2)}(k)\right)_{OB} &= \pm\frac{1}{2}\sum \alpha^2(\mathbf{p}_1,\mathbf{p}_2;\mathbf{h}_1,\mathbf{h}_2) \\ &= \frac{1}{\pi}\int_{-\infty}^{\infty} dE\,\Theta(\pm E \mp e^v(k_F))\frac{Im\Sigma(k,E)}{(e^v(k)-E)^2}, \end{aligned} \tag{58}$$

where the step function $\Theta(x)$ is one or zero whether x is negative or positive. At $k \sim k_F$, $Im\Sigma(k \sim k_F, E) = A_0(E - e(k_F))^2$ at small values of $E - e(k_F)$ due to phase space limitations. Therefore, the coefficient A in eq.(57) is given by

$$A = \frac{2A_0 k_F}{\pi}\frac{\partial e^v(k_F)}{\partial k}. \tag{59}$$

From the calculated $Im\Sigma$, $A_0 \sim 0.0023\mathrm{MeV}^{-1}$, and $A \sim 0.2$.

The momentum distribution for the $U_{14} + DDI$ model hamiltonian, displayed in Fig. 14, shows the importance of the perturbative corrections around $k = k_F$ and the large effect of the tensor correlations. The quenching of the occupation probability for states below the Fermi sea and the magnitude of the discontinuity of $n(k)$ at $k = k_F$ (~ 0.7, giving the renormalization constant Z of the Green's function), are in fair agreement with recent inclusive and semi exclusive electron scattering data [114, 27]. Such data, together with the interpretation provided by CBF calculations, represent the clearest and cleanest signature of the importance of N–N correlations in a nuclear medium.

More recently, the momentum distribution has been calculated within the orthogonal CBF scheme [115], as a byproduct of the hole spectral function. The results obtained are in very good agreement with those presented here and they will be given in a subsequent section.

3.2 Orthogonal CBF scheme

A clear analysis of the convergence properties of the non–orthogonal CBF perturbation theory has not been done yet. Actually, the truncation of the series at some perturbative order (characterized by the number of W or N interaction lines) leads to non–orthogonality spuriousities, whose effects may not be negligible. For instance, in the calculation of $W^v_{0i}(0)$ the non–orthogonality between $|i\rangle$ and the ground state $|0\rangle$ is properly taken into account. However, corrections due to the non–orthogonality of the various intermediate states $|i\rangle$ among themselves come only at higher order of the perturbation theory.

Moreover, the properly orthogonalized eigenvectors cannot be easily extracted out from nonorthogonal CBF theory. This is a major problem in calculating quantities other than the eigenvalues of the hamiltonian, like the dynamical structure function $S(q,\omega)$ or the Green's function $G(k,E)$.

It is possible to orthogonalize a set of states by using the Löwdin transformation [116]. However, the resulting orthogonal states are worst then the original correlated states under several respects. For instance, the expectation value of the hamiltonian in the ground state of the orthogonal set is higher than E_0^v. The Löwdin orthogonalization moves up the energy of the low–lying states, and the perturbative corrections move them down, both effect being larger than the net displacement [28, 93]. In non–orthogonal basis theory only the net displacement is calculated.

We describe in the following a two–step orthogonalization procedure, denoted as SL orthogonalization, having the property that the diagonal matrix elements of the hamiltonian and of the identity operator in the SL states are equal to those of correlated states in the thermodynamic limit. Equivalently, the unperturbed hamiltonian H_0 in SL orthogonalized basis is the same as in non–orthogonal basis.

Correlated orthogonal (CO) states. The main point of SL orthogonalization is that a given $|m_i,n_i\rangle$ state is orthogonalized only to the states $|m_j,n_j\rangle$ with $m_j \le m_i$ and $n_j \le n_i$. This is achieved by a two–step procedure, based on Schmidt (as in $W_{ij}^v(i)$) and Löwdin orthogonalizations (this is the reason of the SL acronym).

Let us consider $|0\rangle$ as the reference state, namely $|SL;0\rangle \equiv |0\rangle$. The SL procedure to orthogonalize the intermediate states needed for the perturbative correction to the ground state energy proceeds as follows:

1. Schmidt orthogonalize $|2_i,2_i\rangle$ to $|SL;0\rangle$ by
$$|S;2_i,2_i\rangle = |2_i,2_i\rangle - |0\rangle\langle 0|2_i,2_i\rangle .$$
The overlap matrix $\langle 0|2_i,2_i\rangle$ coincides with the matrix element N_{0i};
2. Löwdin orthogonalize $|S;2_i,2_i\rangle$ among themselves by
$$|SL;2_i,2_i\rangle = |S;2_i,2_i\rangle - \frac{1}{2}\sum_{j\neq i}|S;2_j,2_j\rangle\langle 2_j,2_j;S|S;2_i,2_i\rangle$$
$$+\frac{3}{8}\sum_{j\neq k\neq i}|S;2_k,2_k\rangle\langle 2_k,2_k;S|S;2_j,2_j\rangle\langle 2_j,2_j;S|S;2_i,2_i\rangle+\cdots$$
The coefficients $+1,-1/2,+3/8,\cdots$ are those occurring in the expansion of $(1+x)^{-1/2}$;
3. perform step 1 for $|3,3\rangle$ with all $|SL;2,2\rangle$ and with $|SL;0\rangle$;
4. perform step 2 for $|S;3,3\rangle$;
5. iterate the process for the various $|m_i,n_i\rangle$ states.

The procedure can be used with any other reference state and for any other application of the CBF theory. In the above scheme, we have omitted the one particle–hole states because they do not enter as intermediate state in the calculation of E_0. In general they should be included. Obviously, they do not need to be Schmidt orthogonalized to $|0\rangle$, since they are already orthogonal to $|0\rangle$, due to momentum conservation.

In [12], it is proved that

$$\langle SL; i|H|i; SL\rangle = \langle i|H|i\rangle + \mathcal{O}(\frac{1}{N}), \tag{60}$$

$$\langle SL; i|i; SL\rangle = \langle i|i\rangle + \mathcal{O}(\frac{1}{N}). \tag{61}$$

The proof is very technical, and it will not be reported here.

Perturbative scheme. Standard perturbation theory may be formally used with SL states. As far as the divergences coming from the matrix elements of many–body interaction are concerned, arguments similar to those for the non–orthogonal case can be used. The only exceptions are that there are no more N–interaction lines and that non diagonal matrix elements of the hamiltonian with SL states include only linked cluster diagrams, and have no off–shell energy dependence.

Properly orthogonalized states can be obtained by using the formalism of degenerate perturbation theory [117]). Let us consider the exact ground state $|\overline{0}\rangle$, and its expansion in terms of SL states (for simplicity, we will omit, in the following, the label SL):

$$|\overline{0}\rangle = \sum_i a_i |i\rangle. \tag{62}$$

One has the following equations

$$(H - E_0)|\overline{0}\rangle = 0, \tag{63}$$

$$(H_0 - E_i^v)|i\rangle = 0, \tag{64}$$

which imply

$$(E_0 - E_i^v)a_i = \langle i|H_I|\overline{0}\rangle. \tag{65}$$

If we select $|0\rangle$ as reference state and take the Green's function

$$G = \sum_{i\neq 0} \frac{|i\rangle\langle i|}{E_0 - E_i^v} = \frac{Q}{E_0 - H_0}, \tag{66}$$

then it follows that

$$|\overline{0}\rangle = |0\rangle + GH_I|\overline{0}\rangle. \tag{67}$$

Now, we can iterate the above equation and expand E_0 around E_0^v, using eq. (38), with the result

$$|\bar{0}\rangle = \sum_n \left(\frac{1}{E_0^v - H_0} (H_I - \Delta E_0) \right)^n |0\rangle. \tag{68}$$

The linked cluster theorem states that

$$\frac{\langle\bar{0}|H|\bar{0}\rangle}{\langle\bar{0}|\bar{0}\rangle} = \frac{\langle\bar{0}|\bar{0}\rangle \times \text{Linked terms}}{\langle\bar{0}|\bar{0}\rangle}. \tag{69}$$

A similar result holds for other static quantities, as the static structure function and the momentum distribution.

Orthogonal CBF theory has been successfully applied to compute the structure function $S(q,\omega)$ and the hole spectral function $P_h(k,E)$ of nuclear matter [118, 31]. They are proportional to the imaginary part of the density–density correlation function $D(\mathbf{q},\omega)$ and the hole Green's function $G_h(\mathbf{k},E)$, respectively, and are defined as

$$S(q,\omega) = \frac{1}{\pi} Im \frac{\langle\bar{0}|\rho_{\mathbf{q}}^\dagger (H - e_0 - \omega - i\eta)^{-1} \rho_{\mathbf{q}}|\bar{0}\rangle}{< \bar{0}|\bar{0}\rangle}, \tag{70}$$

$$P_h(k,E) = \frac{1}{\pi} Im \frac{\langle\bar{0}|a_{\mathbf{k}}^\dagger (H - e_0 - E - i\eta)^{-1} a_{\mathbf{k}}|\bar{0}\rangle}{\langle\bar{0}|\bar{0}\rangle}, \tag{71}$$

where $\rho_{\mathbf{q}} = \sum_i \exp(i\mathbf{q}\cdot\mathbf{r}_i)$ is the density fluctuation operator. They satisfy the sum rules

$$\int d\omega S(q,\omega) = \langle \rho_{\mathbf{q}}^\dagger \rho_{\mathbf{q}} \rangle = S(q), \tag{72}$$

$$\int dE P_h(k,E) = \langle a_{\mathbf{k}}^\dagger a_{\mathbf{k}} \rangle = n(k), \tag{73}$$

and play a fundamental role in the studies of inclusive and semi exclusive electron scattering reactions on nuclei.

In the calculation, an expansion of E_0 around E_0^v in the propagator, similar to that discussed for the energy (see eq. (68)), is needed, namely

$$\frac{1}{H - E_0 - E - i\eta} = \frac{1}{H_0 - E_0^v - E - i\eta} \times \sum_{n=0}^{\infty} \left(-(H_I - \Delta E_0) \frac{1}{H_0 - E_0^v - E - i\eta} \right)^n. \tag{74}$$

The corresponding perturbative series are obtained by explicitly expanding the energy denominators in $|\bar{0}\rangle$ and the propagator, and keeping track of the ΔE_0 insertions. The resulting terms are calculated by inserting $\sum |n_i, m_i\rangle \langle n_i, m_i|$ between the contiguous operators, $\rho_{\mathbf{q}}$, $(H_I - \Delta E_0)$ or $(H_0 - E_0^v -$

$E - i\eta)^{-1}$. This is more conveniently done in a diagrammatic approach, as in ordinary many–body perturbation theory, and as it has been previously discussed in non–orthogonal CBF theory. The ΔE_0 insertions and the unlinked Goldstone–like diagrams cancel exactly, and the perturbative series results to be well behaved in the thermodynamic limit.

Actual evaluations are performed by truncating the number of intermediate states at the leading level, (typically two particle–hole states for the ground state energy and for the structure function or (two hole)–(one particle) states for the hole spectral function) and the perturbative series to two interaction lines. In orthogonal CBF theory this implies a re–summation at all orders on the N–lines through the SL orthogonalization process.

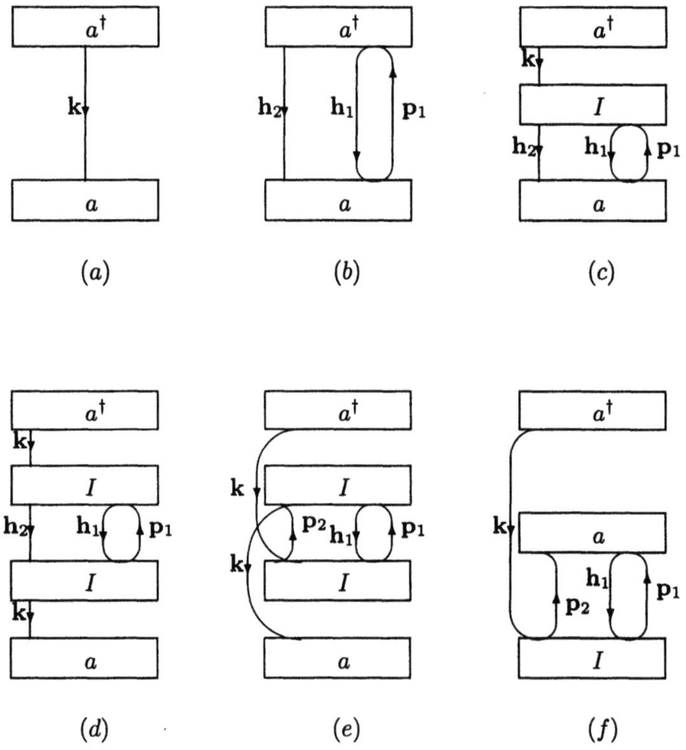

Fig. 15. A few examples of orthogonal CBF diagrams contributing to $P_h(k, E)$. The notation is different from that of [31]: the circles are substituted with a boxes and the solid points by interaction boxes. The lines representing energy denominators are not drawn in this figure. Diagrams (a) and (b) are zeroth CBF order, (c) and (f) first order, and (d),(e) second order. Diagrams (c), (d) and (e) have self–energy insertions. See the text for a detailed discussion.

A few examples of orthogonal CBF diagrams contributing to $P_h(k,E)$ are shown in Fig. 15. They are classified according to the number of excitations of the intermediate states. In lowest–order, the one hole state $|\mathbf{k}\rangle$ is the only intermediate state (diagram (a)). The resulting $P_h(k,E)$ of a Fermi liquid is zero for $k > k_F$ and for $E \neq e(k)$, with a strength at $E = -e(k)$ which is quenched with respect to 1 of $|\langle 0|a^\dagger_{\mathbf{k}}|\mathbf{k}\rangle|^2$.

In next order approximation, the set of intermediate states is enlarged to include (two hole)–(one particle) and (two–hole)–(two particle) states. They account for the following physical effects: (i) the width of the state $|\mathbf{k}\rangle$ gets the main contribution from its coupling with (two hole)–(one particle) states, $|\mathbf{p}_1;\mathbf{h}_1,\mathbf{h}_2\rangle$. Self–energy insertions, as in diagrams (c) and (d) account for this effect; (ii) (two hole)–(one particle) states contribute to $P_h(k,E)$ through two–particle emission processes. Diagram (b) provides the leading contribution to this process; (iii) (two–hole)–(two particle) admixture to $|0\rangle$ is the most important perturbative correction to $|0\rangle$. Diagrams (e) and (f) (together with other second order CBF diagrams, omitted in the figure but considered in nuclear matter calculations [31]) include such corrections. From a more methodological point of view, (a) and (b) are zeroth order, (c) and (f) are first order and the others are second order CBF diagrams. Diagrams (c),(d) and (e) are obtained by developing the propagator, and (f) by expanding eq. (68). The *twisted* diagram (f) is a remnant from the cancelation process between unlinked CBF diagrams and ΔE_0 insertions. The energy denominators between interaction boxes are on–shell (they have no E dependence) except for the peculiar twisted self–energy insertion of diagram (f) (see [31] for a detailed discussion on this diagram). On the contrary, the energy denominators connected with the a boxes do depend on E.

The hole spectral function will be discussed later, and the dynamical structure function is treated elsewhere in this book [18]).

Let us mention a feature which shows the effects of orthogonality spuriousities in non orthogonal CBF theory. The most important intermediate states in the calculation of $S(q,\omega)$ are the one particle–one hole states. It is easy to verify that such states completely saturate the sum rule (72), namely

$$S(q) = \sum \langle 0|\rho^\dagger_{\mathbf{q}}|\mathbf{p},\mathbf{h}\rangle\langle \mathbf{p},\mathbf{h}|\rho_{\mathbf{q}}|0\rangle. \tag{75}$$

The reason is that $\rho_{\mathbf{q}}|0\rangle$ is a combination of one particle–hole states, if $\rho_{\mathbf{q}}$ commutes with the correlation operator [8]. Non orthogonal CBF does not satisfy the above sum rule at any finite order of the perturbative series, because the one particle–hole states need to be orthogonalized among themselves.

Let us conclude this section by mentioning that CBF perturbation theories have been mostly applied to nuclear matter and finite nuclei, and not so much to liquid helium or strongly correlated electrons. A possible reason is that it heavily relies on FHNC technology, which is somewhat biased by

[8] If $\hat{F}$ contains spin–orbit correlations the above equation is no longer valid.

the intrinsic approximation due to the elementary diagrams. This approximation is less severe for nuclear systems, where the density is relatively low with respect to the range of the repulsion of the interaction. In denser atomic systems, the CBF perturbative scheme has been pushed to higher orders to study the static and dynamical properties of liquid ^{4}He [98], ^{3}He [99], and ^{3}He-^{4}He mixtures [119, 120]. As a valid alternative, it has been considered preferable to go beyond variational estimates by using Quantum Monte Carlo techniques. However, this might be prohibitive to study phases other than the normal Fermi fluid, because of finite size problems. For instance, the superfluid phase of liquid ^{3}He requires enormous simulation boxes in Monte Carlo calculations, due to the large correlation length of the pairs. Obviously, the study of phase transitions is a formidable task for CBF theory too. One certainly needs to change the symmetry of the underlying correlated basis, when dealing with phases characterized by an off–diagonal long range order. However, one expects that the occurrence of a phase transition is manifested by divergences in the perturbative calculations.

4 Cluster Expansion Technique and FHNC Summation Scheme

We shall devote this section to the more technical issue of discussing the methodology needed to compute the quantities entering CBF perturbation theories. In essence, we will unfold the interaction boxes in terms of the bare and the pseudopotential interactions. In doing so, we will introduce *passive* single particle lines, not carrying any label, because a sum on all the occupied states is always implied.

From the definition of the correlated states of eq.(1), it follows that the diagonal and not–diagonal matrix elements of the hamiltonian, as well as the overlap matrix elements N_{ij}, are built from the matrix elements of $\hat{F}^\dagger H \hat{F}$ and of $\hat{F}^\dagger \hat{F}$ on the Fermi gas states.

The operators $\hat{F}^\dagger H \hat{F}$ and $\hat{F}^\dagger \hat{F}$ are expanded in terms of the correlation interaction $\hat{h}(i,j)$ (CI), defined as

$$h(r_{ij}) = \exp(-u_2(r_{ij})) - 1, \tag{76}$$

$$h_3(i,j,k) = \exp(-u_3(i,j,k)) - 1, \tag{77}$$

for the scalar correlations, and of

$$\hat{h}(i,j) = \sum f^p(r_{ij}) \hat{O}^p(i,j), \tag{78}$$

for the spin–isospin ones. In the last case one has to take care of the commutators of the CI's between themselves and with the hamiltonian.

The terms of the resulting expansions are characterized by integrals containing a given number of CI's, the hamiltonian and $\exp(i\mathbf{k} \cdot r_{ij})$ factors joining the correlated particles. They are usually called *cluster terms* and are conveniently represented by *cluster diagrams*. A few examples are displayed in Figs. 16 and 17.

We will limit ourselves to the simplest (and still realistic) case of Jastrow correlations. Therefore, $h(r_{ij})$ is the only CI we will consider. This is certainly not adequate to describe nuclear matter (mainly because it neglects the tensor forces), but it will make (hopefully) our discussion less technical and more transparent.

Whereas the inclusion of triplets or higher order scalar correlations is straightforward, that of spin–isospin correlations is not at all trivial, mainly because one needs to keep track of the orderings of the CI's. As a matter of fact, we are not yet able to treat such correlations at the same level of accuracy (FHNC) as the scalar ones. In actual calculations, only the leading cluster terms with spin–isospin CI are included. We will briefly discuss at the end of this section what *leading orders* means, and we refer to the review paper [22] for technical details.

The problem concerning the order of the spin dependent correlation does not arise if spin–coherent states are used (see section 2.3). In fact, spin–coherent wave function can be treated at the full FHNC level.

A second limitation is that we will apply the cluster expansion technique to the pair distribution function only. We will describe a sort of generalized FHNC summation scheme, which is used in the more recent applications in nuclear systems and in liquid ^{3}He. A direct application of this scheme is the explicit calculation of the CBF matrix elements, which however is beyond the scope of these lecture notes, and can be found in the original papers and in [26, 18].

4.1 Pair distribution function

The pair distribution function is given by

$$\begin{aligned} g(1,2) &= \frac{N-1}{\rho}\langle 0|\delta(r-r_{12})|0\rangle \\ &= \frac{N(N-1)}{\rho^2}\frac{Tr(\Xi)\int d\mathbf{r}_1,\cdots,d\mathbf{r}_N|\Psi^{(0)}_{CBF}(R,\Xi)|^2}{Tr(\Xi)\int dR|\Psi^{(0)}_{CBF}(R,\Xi)|^2}, \end{aligned} \tag{79}$$

where $Tr(\Xi)$ means trace on the spin–isospin variables. The knowledge of $g(r_{12})$ is necessary for calculating the ground state expectation value of the hamiltonian . For instance, the mean value per nucleon of the two–body scalar potential is

$$\langle v_1\rangle/N = \frac{1}{2}\rho\int dr_{12}v_1(r_{12})g(r_{12}). \tag{80}$$

Particle 1 and 2 are the *interacting* particles, and, therefore, are distinguished from the particles of the medium.

Cluster decomposition of the pair distribution function. The main steps of the cluster decomposition are the following.

Step 1 : decompose $\hat{F}^\dagger \hat{F}$ *in the numerator of eq. (79).* One obtains:

$$\hat{F}^\dagger \hat{F} = f^2(r_{12}) \left(1 + \sum_{i \neq 1,2} X^{(3)}(12;i) + \sum_{i<j \neq 1,2} X^{(4)}(12;ij) + \cdots \right) . \quad (81)$$

The generic cluster term $f^2(r_{12})X^{(n)}$ correlates the two interacting particles and $n-2$ medium particles, and should be considered as an n–body operator. For instance, $X^{(3)}(12;i) = h_{1i} + h_{2i} + h_{1i}h_{2i}$.

Step 2 : perform the cluster expansion of the numerator. Let us calculate the expectation values of the $f^2(r_{12})X^{(n)}$ on the MF wave function Ψ_{MF}. They can be expressed in terms of the MF n–body distribution function $g_{MF}(1, \cdots, n)$:

$$\begin{aligned} numerator \; &= f^2(r_{12})(g_{MF}(1,2) \\ &+ \sum_{n=3}^{\infty} \frac{\rho^{n-2}}{(n-2)!} \int d\mathbf{r}_3 \cdots d\mathbf{r}_n X^{(n)}(12;3\cdots n) g_{MF}(1\cdots n)). \end{aligned} \quad (82)$$

$g_{MF}(1 \cdots n)$ can be written in terms of the MF one–body density matrix:

$$l(1,1') = \frac{1}{N} \sum_{n \in \epsilon_F} \phi_n^\star(1)\phi_n(1'), \quad (83)$$

where ϕ_n are the single particle orbitals in Ψ_{MF} and the sum is restricted to the occupied states. In the case of a normal Fermi liquid, like nuclear matter, $\phi_n(i) = \exp(i\mathbf{k}_n \cdot \mathbf{r}_i)$, and $l(1,1') \equiv l(x) = \frac{3}{x^3}(\sin(x) - x\cos(x))$, with $x = k_F r_{11'}$.

The structure of $g_{MF}(1 \cdots n)$ reflects the property of the antisymmetrization operator, which can be written in terms of two–body exchange operators P_{ij} only:

$$\begin{aligned} A \rightarrow 1 - \sum P_{ij} + \sum \overbrace{(P_{ij}P_{jk} + P_{jk}P_{ki})}^{(ijk-loop)} \\ + \sum (ij - loop) \times (kl - loop) - \sum (ijkl - loop) + \cdots \end{aligned} \quad (84)$$

The exchange of spatial variables leads to the one–body density matrix. The exchange of spin–isospin variables implies that, in the case of a scalar CI (like for Jastrow), the spin–isospin assignments of all the particles in a loop must be the same. For this reason, each l brings a $1/d$ factor, where d is the spin degeneracy of the system ($d = 4$ for nuclear matter). Each exchange loop carries an extra factor $2d$, where d sums over the possible spin assignments of the loop and the factor 2 is due to the two possible directions of the loop (clockwise and anticlockwise). The factor 2 is missing for the two–particle loops, where the two directions coincide. For instance,

$$g_{MF}(1,2) = 1 - \frac{1}{d} l^2(1,2), \tag{85}$$

$$g_{MF}(1,2,3) = 1 - \frac{1}{d} l^2(1,2) - \frac{1}{d} l^2(1,3) - \frac{1}{d} l^2(2,3) + \frac{2}{d^2} l(1,2) l(2,3) l(3,1). \tag{86}$$

In general, there are loops and products of separate loops of $(-\frac{1}{d}l)$–functions, each loop carrying a $-2d$ factor, except for the two–particle loop, which carries a factor $-d$.

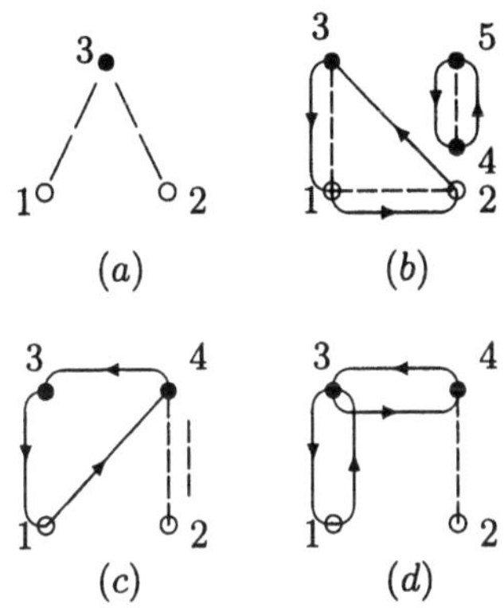

Fig. 16. Examples of cluster diagrams. Diagrams (c) and (d) are not allowed. Diagram (b) is unlinked.

We are now ready to give the general structure of the cluster decomposition of the numerator of eq. (79). This is better done in terms of cluster diagrams $\overline{X}^{(n)}(12;\alpha)$:

1. The diagrams are built with points, oriented lines and dashed lines. The points (vertices) of a diagram represent the coordinate $\mathbf{r}_i$. Solid points (internal points) refer to particles in the medium, and imply an integration over the corresponding coordinates times a factor ρ; the circles (external points) refer to the interacting particles and represent coordinates over which there is no integration. The dashed lines are CI–lines and they cannot be superimposed[9].

[9] CI corresponding to triplet correlations are generally represented with black triangles, and spin–isospin CI's are represented with wavy lines. Two wavy lines can be superimposed, representing the CI's coming from the bra and from the ket.

An oriented lines going from i to j denotes a $(-\frac{1}{d}l)$–function, meaning that particles i and j have been exchanged in the ket (particle i in the bra is in the same state of particle j in the ket). There is no label on the oriented lines, because a sum over all the occupied state has been done. In cluster diagrams of not–diagonal matrix elements we will have to add a **k**–label to the oriented line corresponding to an excited state and distinguish the *passive* oriented lines (with no labels) from the *active* ones.

2. Each solid point must be reached by at least one CI–line. This is not necessary for the circles, because they are already correlated by the operator, whose matrix elements are being computed. The exchange lines may or may not connect the points of a diagram.
3. The exchange lines must form closed loops and different loops cannot have common points.

Fig. 16 shows four cluster diagrams. Diagrams (c) and (d) are not allowed, the first because the solid point 3 is not reached by a CI–line (moreover, between points 2 and 4 there are two superimposed dashed lines), the second because there are two touching exchange loops. The corresponding expressions of diagrams (a) and (b) are

$$Diagram\ (a) = \rho \int d\mathbf{r}_3 h(r_{13}) h(r_{23}), \tag{87}$$

$$\begin{aligned} Diagram\ (b) &= (-2d)(-\frac{l(1,2)}{d})h(r_{12})\rho^3 \\ &\times \int d\mathbf{r}_3 d\mathbf{r}_4 d\mathbf{r}_5 (-\frac{l(2,3)}{d})(-\frac{l(3,1)}{d})(-\frac{l^2(4,5)}{d})h(r_{13})h(r_{45}). \end{aligned} \tag{88}$$

The cluster terms have no specific prefactor, except those coming from the exchange rules and a ρ factor for each integration. One might wonder where the factor $1/(n-2)!$ of eq. (82) ended up. The factor is due to the counting of the permutations of the internal points and it is automatically taken into account by the fact that the labels of the solid points in the cluster diagrams are dummy indices. The only remnant of that factor is the inverse of what is usually called *symmetry factor*, s, *i.e.* the number of permutations of the solid point labels leaving unchanged the cluster term. For instance, a four point diagram fully connected with CI's has $s = 2$, because interchanging the point 3 with 4 leaves unchanged the corresponding integral, which, therefore, has to be multiplied by an extra factor $1/2$.

Cluster diagrams may be irreducible (diagrams (a), (c) and (d)), reducible and unlinked (diagram (b)). The integrals corresponding to irreducible diagrams cannot be factorized. The reducible diagrams are still linked, but they can be separated into two parts having only one point in common, and, as a consequence, the corresponding integrals can be factorized. The unlinked diagrams have at least two parts with no common points.

The order of magnitude of a cluster diagram is Ω^{N_u-1}, where N_u is the number of its unlinked parts. In the calculation of an expectation value of a given two–body operator, there are two extra integrations, one on $\mathbf{r}_{12}$ and the other, say, on $\mathbf{r}_1$. The latter brings one more Ω factor, and the final order of magnitude is the well known Ω^{N_u}. Therefore, unlinked diagrams are divergent in the thermodynamic limit.

Step 3 : repeat steps 1 and 2 for the denominator. All the previous discussion extends to the denominator of eq. (79), with only one exception: there are no interacting particles. Therefore, the cluster decomposition of $\hat{F}^\dagger\hat{F}$ becomes

$$\hat{F}^\dagger\hat{F} = \left(1+\sum_{i<j} X^{(2)}(ij) + \sum_{i<j<k} X^{(3)}(ijk) + \cdots\right). \tag{89}$$

The cluster diagrams have only solid points, and their order of magnitude is Ω^{N_u}.

Step 4 : express the pair function as a sum of cluster diagrams. The ratio of the numerator with the denominator in eq. (79) can be considered as the ratio of two infinite series of cluster terms, each one associated to a cluster diagram.

Let us consider a generic linked (reducible or irreducible) cluster diagram, L_n, of the numerator (see diagram (a) of Fig. 16). Together with L_n there is a series of unlinked diagrams, in which L_n is the part containing the interacting particles. It is easy to verify that the sum of all these diagrams corresponds to $(Cluster\ term\ of\ L_n) \times\ Denominator$.

The same argument holds for any other linked diagrams. This proves the linked cluster property:

$$\begin{aligned} g(1,2) &= \sum (Linked\ Diagrams\ of\ the\ Numerator) \\ &= f^2(r_{12})\left(g_{MF}(1,2) + \sum_{n>2,\alpha} \overline{X}^{(n)}_{Linked}(12;\alpha)\right), \end{aligned} \tag{90}$$

where $\overline{X}^{(n)}_{Linked}(12)$ represents the sum of all linked cluster diagrams with $n-2$ internal points. An analogous property holds for all the n–body distribution functions, and it is not limited to the Jastrow correlation operator.

It should be mentioned that eq. (90) is valid not only in the thermodynamic limit, but at all orders of $1/N$. The same is not true for Bose liquids. There, the sum of linked diagrams (the so called Mayer diagrams) is equal to the pair distribution function in the thermodynamic limit only. In fact, one can use the Fermi cluster expansion to get the various contributions of order $1/N, (1/N)^2, \cdots$. This is obtained by considering the limiting case of the spin degeneracy d going to the number of particles N, and, correspondingly, $k_F \to 0$.

4.2 Diagram classification schemes

We will now discuss some of the classification schemes used for the linked cluster diagrams.

Expansion in the number of points. This was the first kind of classification for Fermi liquid calculations. A given order m is obtained by truncating the sum (90) at $n = m$. For instance, diagram (a) of Fig. 16, which is the only allowed linked diagram, belongs to the third order approximation.

Very soon it was realized that ρ is not a good expansion parameter, unless one restrict himself to the low–density regime, where, however, exact analytical expressions for the binding energy and the momentum distribution exist. Liquid ^{3}He and even nuclear matter cannot be considered as low–density systems.

The expectation value of the hamiltonian at any finite order of the expansion in the number of internal points is unbound, mainly because the normalization of the wave function is not properly taken into account. Low–order variational calculations have been carried out by putting constraints on the variational parameters of the correlation functions.

However, second order approximation can be used for quick and dirty estimates in nuclear matter, once the best variational wave function is known, particularly at the equilibrium density. The main features of the short–range behavior are already reasonably well described by $f(r_{12})$, and the binding energy of strongly correlated Fermi systems mostly depends on the short–range part of the correlation.

These characteristics are used in modern calculations of nuclear matter to get suitable parameterizations of the correlation functions $f^p(r)$ in eq. (22). In practice, a constrained functional variation of the second order (neglecting all the sum in the r.h.s of eq. (90)) expectation value of the hamiltonian is performed, namely

$$\begin{aligned} &\delta_{f^p(r)}\langle H\rangle_2 = 0, \quad \forall p, \\ &f^{p=1}(r=d_1) = 1; \quad f^{p\neq 1}(r=d_p) = 0, \\ &\frac{df^p(r)}{dr}\,|_{r=d_p} = 0, \end{aligned} \tag{91}$$

where d_p are variational parameters[10]. The solution of these Schödinger like equations provides a fairly accurate description of the short range structure of $f^p(r)$. The variational parameters are then obtained by minimizing the expectation value of the hamiltonian, now calculated with a much better approximation (FHNC or FHNC/SOC, depending on the correlation operator).

[10] Other variational parameters β_p are usually introduced in eq. (91). They multiply the strength of the components of the N–N interaction, making H in (91) a sort of effective interaction. However, the optimal values of these parameters are very close to 1.

The most important parameters are the healing distances of the central and tensor correlations, $d_c = d_1 = d_2 = d_3 = d_4$ and $d_t = d_5 = d_6$ ($d_t \sim 3d_c$).

The intermediate and long range behaviors are totally lost by the second order approximation. The expansion (90) is particularly critical when long–range pseudopotentials are used. They behave as α/r^2 at large interparticle distance r. It will be shown in the following of this section that, in such case, any finite order of eq. (90) diverges.

Power series expansion. Another proposed scheme of expansion is based on the number of CI's in the cluster diagrams. Suppose to multiply $h(i,j)$ by a parameter α. The terms of the expansion are the coefficients of a power series in α, and, for this reason, it has been denoted as Power Series (PS) expansion [121]. In the final calculation, the limit $\alpha = 1$ is taken. The zeroth order coincides with $g_{MF}(1,2)$. The first order includes the second order of eq. (79) plus ten three–point diagrams and nine four–point diagrams. In general, a given order of the PS expansion mixes up several order of the previous expansion.

An interesting feature of the PS expansion is that, at any order, all possible sum rules of the pair function are exactly satisfied. One of these is, for instance,the normalization sum rule

$$S(q=0) = 1 + \rho \int d\mathbf{r}_{12}(g(r_{12}) - 1) = 0. \tag{92}$$

This can be shown by the following simple argument. Let us multiply $h(i,j)$ by a constant factor α. We obtain, in this way, a new correlated wave function, for which the sum rules still hold. Since the zeroth order coincides with the uncorrelated wave function, and therefore it satisfies the sum rules, we are left with a series $\sum_{n=1}^{\infty}(PS)_n\alpha^n$ which is equal to zero for any value of α within the convergence radius. It follows that each coefficient $(PS)_n = 0$. This also implies that the PS expansion has no pathologies with long–range correlation functions.

In spite of this important property, which, by the way, is very useful to study cancelation processes among the cluster terms and suggests possible approximation schemes, the PS expansion does not describe properly the short–range behavior of the pair function, and, therefore is not recommended for binding energy calculations. This is because $f^2(r_{12})$ is not a multiplicative factor of each term, as it was in the expansion in the number of points.

Fermi Hyper–Netted Chain scheme. We have learned from the previous sections that, in order to have a good description of both the short and the long range behavior of the pair function, we need a scheme where infinite sets of cluster diagrams are summed up. This is achieved by the Fermi Hyper–Netted Chain (FHNC) summation scheme, which we will describe for the case of a Fermi liquid and a wave function with Jastrow–type correlations (for which FHNC has been originally developed).

We will present here the FHNC scheme in a more general form with respect to that given in previous review papers. *We will not assume that the reducible cluster diagrams cancel out*, which is strictly true only for a normal Fermi fluid described by scalar correlations. This is indeed the case we are considering here for pedagogical reasons, and the above cancelation will manifest itself as a structural property of the corresponding generalized FHNC equations. We hope that the presentation of the FHNC scheme which *sums up reducible and irreducible diagrams*, denoted as RFHNC, will be useful for a better understanding of the more recent applications of the theory.

Linked versus vertex corrected irreducible diagrams. The cluster diagrams contributing to the pair function may be irreducible or reducible. The integral corresponding to a reducible diagram is factorizable in the products of its irreducible parts.

For instance, diagram (a) of Fig. 16 is irreducible. If we add to it an extra CI-line, say h_{34}, we get a reducible diagram (where 3 becomes a reducibility point), and the corresponding cluster term is given by the r.h.s of eq. (87) times $\rho \int d\mathbf{r}_4 h(r_{34})$, namely the cluster term of the added part. Obviously, we can add all the possible linked one–body diagrams (diagrams with one external point) at the point 3.

The net result is that the sum of all the reducible diagrams, having diagram (a) as the irreducible part with the external points, can be represented again by diagram (a), in which, however, the vertex 3 is *renormalized*, in the sense that it is vertex corrected by the number[11] $V(3)$

$$V(3) = 1 + \sum (\ linked \ \ one - body \ \ cluster \ \ terms\ ) . \tag{93}$$

Note that there are no restrictions on the kind of cluster terms in the sum, because only CI lines reach the point 3 in (a), and therefore both CI and exchange lines can have the common point 3 in the vertex correction. If the reducibility point is reached by exchange lines (like for instance one of the solid points of the third diagram on the first row of Fig. 17), then only CI lines could have reached the external point of the vertex correction diagrams. It follows that we have two kind of vertex corrections: V for the reducibility points of type d, with only CI lines (the label d comes from dynamical correlations which is the name originally given to CI), and V_d for the points of type e, with exchange lines.

[11] In the case of the spin–isospin dependent correlation operator of eq. (22), the vertex correction V is not merely a number, but depends on the operatorial components of the CI's at the reducibility point, because there are commutators involved between them. Similarly, in the case of the spin–coherent wave function, the vertex correction depends on the variables ω_3^{left} and ω_3^{right}. For finite systems, like nuclei or helium droplets one can express the pair function as an infinite series of cluster diagrams, very similar to that obtained for Fermi liquids (see [18]). In this case the vertex correction is a function of the coordinate $\mathbf{r}_3$.

Actually, there is a third type of vertex correction, occurring when the reducibility point is an internal point, reached only by exchange lines. This correction cannot be the full V_d, since V_d includes the case of absence of vertex corrections (the factor 1 in eq. (93)). The diagrammatical rule, stating that any solid point must be reached by at least one CI line, forbids this factor, so the vertex correction is $V_d - 1$, in this case. Therefore, taking into account vertex corrections, a diagram like (c) of Fig. 16, without the two superimposed $h(r_{34})$ lines, is allowed, because point 3 is vertex corrected by $V_d - 1$. In conclusion, *VIC diagrams may have internal points reached by exchange lines only.*

The r.h.s. of eq. (93), when multiplied by ρ, corresponds to the cluster expansion of the one–body distribution function, being equal to ρ itself for an infinite system. Therefore, we have the sum rule $V = 1$, serving as an useful check for the accuracy of the vertex corrections calculation.

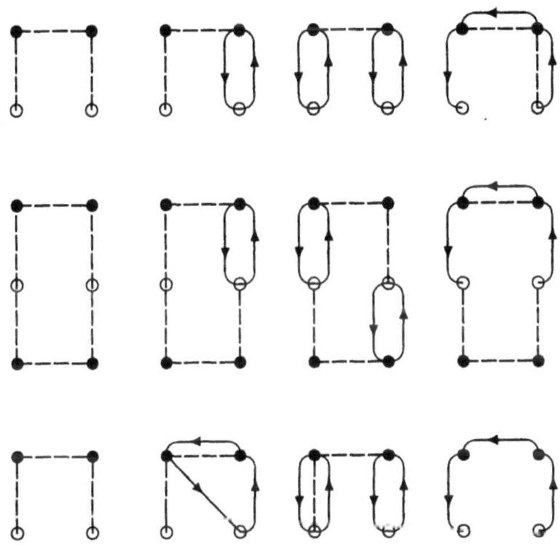

Fig. 17. Classification of FHNC diagrams. Diagrams in the first row are nodal, in the second are composite and in the third are elementary. Diagrams in the first column are of type *dd*; those in the other columns are of types *de*, *ee* and *cc* respectively.

We can repeat the above procedure for the external points of diagram (a), and for all the irreducible diagrams, with the result that *the sum of the linked diagrams may be seen as a sum of vertex corrected irreducible (VIC) diagrams*, namely irreducible diagrams whose points carry the vertex corrections V, V_d or $V_d - 1$.

Classification of the VIC diagrams. The classification follows the Hyper–Netted Chain (HNC) theory, originally developed in statistical physics [122, 123], and later on applied to quantum Bose systems.

Nodal (or Chain) diagrams. These diagrams, denoted as N–diagrams, are characterized by having one or more *nodes*, where a node is an internal point through which any path of correlation lines joining the external points 1 and 2 goes. In other words, they are constituted by one chain of *bonds* joining the two external points. This is the reason why they are often referred to as chain diagrams. For instance, diagram (a) of Fig. 16 and all the diagrams in the first row of Fig. 17 are nodal. Actually, the fourth diagram in the first row (as all the diagrams in that same column) would not be allowed, because the path of exchange lines does not form a closed loop. We consider also such kind of diagrams (those with an open path of exchange lines from 1 to 2) for reasons that will become clear later.

The inclusion of nodal diagrams is important in order to properly consider long–range correlations. This is understood by analyzing a nodal diagram $N^{(n)}$ formed by a chain of n CI lines. For instance, diagram (a) of Fig. 16 is $N^{(2)}$, and the first diagram of Fig. 17 is $N^{(3)}$ [12]. It is easy to verify that the Fourier transform $\tilde{N}^{(n)}(k)$ (where $N(r_{12})$ is not vertex corrected in 1 and 2) is given by

$$\rho V \tilde{N}^{(n)}(k) = (\rho V \tilde{h}(k))^n, \tag{94}$$

and the sum of all the $\tilde{N}^{(n)}$, by

$$\tilde{N}(k) = \sum_{n=2}^{\infty} \tilde{N}^{(n)}(k) = \frac{\rho V \tilde{h}^2(k)}{1 - \rho V \tilde{h}(k)}. \tag{95}$$

Equivalently, in the $\mathbf{r}$–space, $N(r_{12})$ is the solution of the following integral equation

$$N(r_{12}) = \rho V \int d\mathbf{r}_3 h(r_{13})(h(r_{32}) + N(r_{32})). \tag{96}$$

If h is long-ranged, $h(r) \to \alpha/r^2$ for $r \to \infty$, then $\tilde{h}(k) \to -\alpha/(2\pi k)$ for $k \to 0$. It follows that each $\tilde{N}^{(n)}(k)$ is more divergent than $\tilde{h}(k)$ in the long wavelength limit, whereas their sum is well behaved (it diverges as $-\tilde{h}(k)$). This fact tells us, on one side, that the expansion in the number of points

[12] The $1/\rho^2$ factor in the definition of the pair function is formally $1/(V_d(1)\rho(1)V_d(2)\rho(2))$. This means that in infinite Bose system the two external points should not be vertex corrected in the calculation of the pair function. In any case, it is convenient to associate to the VIC diagrams cluster terms having the two external points not vertex corrected.

diverges at any finite order, and, on the other side, that chain summations lead to a proper long range behavior[13].

Not–nodal diagrams. These diagrams, denoted as X–diagrams are subdivided into two classes: *composite* and *elementary* (or *bridge*) diagrams.

The set of *composite* diagrams is constituted by the diagram with a direct CI link plus all the diagrams having at least two chains or a direct link and a chain, which are connected between them at the external points 1 and 2 only. The diagrams in the second row of Fig. 17 are examples of composite diagrams. The corresponding cluster terms can be written, in $\mathbf{r}$–space, as products of N–cluster terms. Each of the composite diagrams of the figure have two chain terms. In the first one the two chains are equal, and, as a consequence, the cluster term must have also a factor $1/s = 1/2!$.

The *elementary diagrams*, denoted as E–diagrams are not–*composite*. They are fully connected diagrams, and some of them are displayed in the last row of Fig. 17. The first one is symmetric under the exchange of the solid points, and, therefore, has again a factor $1/2!$.

Vertex correction diagrams. The linked diagrams contributing to the vertex corrections can be expressed in terms of one–body VIC diagrams through the following equations

$$\begin{aligned} V(1) &= (1 + U_e(1)) \exp(U_d(1)) \\ V_d(1) &= \exp(U_d(1)), \end{aligned} \tag{97}$$

where $U_d(1)$ and $U_e(1)$ are VIC diagrams of type d and e respectively. The exponential in the above equations is due to the fact that any number of d–structures can reach the external point 1. As for the case of the composite diagrams, if these n structures are topologically equal, then the corresponding term has a factor $1/n!$. Therefore, the global contribution is $\sum_n (1/n!) U_d^n(1)$.

4.3 RFHNC summation scheme

The FHNC diagrams can be constructed (and then summed up) by means of an iterative process, in which one uses a limited set of N– and X–diagrams as building block diagrams to form an enlarged set, up to completion of the whole set. For this reason we need to introduce also the diagrams of the type shown in the fourth column of Fig. 17.

Take, for instance, eq.(96). $N(r_{12})$ sums up all the nodal diagrams built with the *bond* $h(r_{ij})$. Suppose we restrict ourselves to diagrams having only

[13] It should also be noticed that ring diagrams in standard perturbation theory are very much related to chain diagrams. The particle–hole bubble in ring diagrams can be splitted up into two contributions by performing the sum on particle state as an unrestricted sum minus a sum on the hole states. The unrestricted sum gives a delta function leading to a chain of two interaction lines. The other term leads to a chain of two interaction lines with an exchange loop in between.

CI lines (the boson–like diagrams). Then, one can consider a new *bond* given by the sum of X–diagrams built up with these chains, namely

$$X(r_{12}) = f^2(r_{12}) \exp[N(r_{12}) + E(r_{12})] - N(r_{12}) - 1. \qquad (98)$$

The presence of the exponential in the above equation has the same explanation as for the vertex corrections of eq. (97).

Now, one can sum up all the N–diagrams built with the new X–bond by solving eq. (96) after the substitution $h \to X$. Inserting back the enlarged set of N-diagrams into eq. (98), one gets a more complete bond. The process is iterated until convergence is reached, within the required numerical accuracy.

The two equations (96) and (98) are the HNC equations and their solution leads to the sum $g_B(r_{12})$ of all the boson–like diagrams

$$g_B(r_{12}) = 1 + X(r_{12}) + N(r_{12}) = f^2(r_{12}) \exp[N(r_{12}) + E(r_{12})]. \qquad (99)$$

There is a bad and a nice feature in the HNC equations. The bad one is that $X(r_{12})$ is defined in terms of the function $E(r_{12})$ which is not evaluable in any known closed form, and therefore we are not able to solve exactly the HNC equations. The simplest approximation is to set $E(r_{12}) = 0$, i.e. to neglect the *elementary* diagrams at all (HNC/0 approximation). One can include the four-body *elementary* structure $E_4(r_{ij})$, shown in Fig. 17, and compute it with $g_B(r_{ij}) - 1$ bonds (HNC/4 approximation), then proceed with the five–body structure and so on. However, in the cases where the *elementary* diagrams give reasonable effects, like in liquid helium, the series $\{HNC/n\}$ does not converge very well.

A more effective procedure (*scaling approximation* (HNC/s)) consists in approximating $E(r_{ij})$ by $\alpha E_4(r_{ij})$, where α is a scaling parameter, determined by requiring that the kinetic energy expectation value calculated with two formally equivalent expressions is the same [42]. This approximation is based on the property that the various $E_n(r_{12})$ roughly scale among themselves. Typical expressions used for the kinetic energy are the *Pandharipande–Bethe* (T_{PB}) and *Jackson–Feenberg* (T_{JF}) expressions[14]. Other global approximation schemes have been proved to be equally reliable, like the so called *interpolating equations* method (HNC/α) [43] or the one used in [6].

The nice feature is that HNC/0 provides a very good description of the long–range part of the pair function, and it is not too bad at short interparticle distance either (although $E(r)$ is a short–range function). For instance, in nuclear matter calculation the elementary diagrams can safely be neglected.

[14] T_{PB} is obtained by simply acting ∇_i^2 on the ket, whereas in T_{JF}, $\Psi^*(\nabla_i^2\Psi)$ is written as $(\Psi^*(\nabla_i^2\Psi) + |\nabla\Psi|^2)/2$. T_{PB} has a two–body operator which has a rather smooth behavior at short interparticle distances, but has also a three-body operator which cannot be neglected. T_{JF} has no three–body term in the Jastrow–Bose case, but the two–body term is more sensitive to the short–range behavior of the pair function. The three–body term arising for Fermi wave functions is in general almost negligible (see [124, 24, 26]) for a more detailed discussion).

In the case of long–range Jastrow correlation, one can see from eq. (99) that the long range part of the pseudopotential is canceled by that of N, and the pair function results to be short–ranged, as it should be.

Other ingredients of the RFHNC scheme are the vertex corrections, whose calculation has to be included in the above iterative process. $U_d(1)$ and $U_e(1)$ can be expressed in terms of the N– and X–diagrams [125, 126]. Suppose we fix our attention on the bosonic diagrams only. One is naively led to think that $U_B(1)$ is given by

$$U_B(1) = \int d\mathbf{r}_2 X(1,2) \exp[U_B(2)]. \tag{100}$$

This is not true, because it includes cluster terms of spurious diagrams. However, one may correct the above expression by *subtracting* out such spuriousities.

One of these is $N^2(r_{12})/2$, which, once integrated over $\mathbf{r}_2$, leads to chain loops already included in $h(r_{12})N(r_{12})$. Moreover, these loops are overcounted in (100) because 2 is now a dummy index. It means that a given diagram is counted as many times as the number of nodes, instead of only once. This reflects the fact that the symmetry factors of the two–body diagrams are generally different from those of the one–body ones. One can verify that such overcounting is exactly corrected by subtracting the expression $\int d\mathbf{r}_2 N(r_{12})X(r_{12})/2$.

In conclusion, the final equation for $U_B(1)$ is given by

$$\begin{aligned} U_B(1) = \int d\mathbf{r}_2 \{ & X(1,2) - \frac{1}{2} N(1,2)(g_B(2,1) - 1) \\ & - E(1,2)\} \exp[U_B(2)] + E(1), \end{aligned} \tag{101}$$

where $E(1)$ corresponds to the sum of the one–body elementary diagrams.

RFHNC equations. In order to apply the RFHNC scheme discussed above, we have to subdivide the set of nodal and not–nodal two–body diagrams and the vertex correction U–diagrams in different classes, according to the type (d or e) of their external points. This is necessary because, in the process of using limited sets of diagrams as building blocks to construct new ones, we cannot join e with e. We also need a label c for the intermediate diagrams having a path of exchange lines from 1 to 2.

The various types of two–body diagrams are labeled with the indices dd, de, ee and cc. They are illustrated in Fig. 17, where the diagrams in the first column are of type dd, in the second de, in the third ee, and in the last cc.

There is a further subdivision to be done for the cc–diagrams. The reason is that points reached only by exchange lines must be vertex corrected by $V_d - 1$ instead of V_d. The cc-diagrams are, therefore, also characterized by the labels hh if both the external points are reached by CI lines, $h\rho(\rho h)$ if point 2(1) is not reached by a CI line and $\rho\rho$ if neither points are reached by CI lines.

The RFHNC equations strictly follow from the RFHNC scheme and the constraints due to the diagrammatical rules for the exchange lines. The chain equations are given by

$$N_{xy}(r_{12}) = \sum_{x'y'} \int d\mathbf{r}_3 X_{xx'}(r_{13}) \zeta_{x'y'} \times \left[X_{y'y}(r_{32}) + N_{y'y}(r_{32})\right], \quad (102)$$

where the *composite* cluster terms are given by

$$X_{dd}(r_{12}) = f^2(r_{12}) \exp[N_{dd}(r_{12}) + E_{dd}(r_{12})] - N(r_{12}) - 1, \quad (103)$$

$$X_{de}(r_{12}) = (g_{dd}(r_{12}) - 1)(N_{de}(r_{12}) + E_{de}(r_{12})) = X_{ed}(r_{21}), \quad (104)$$

$$X_{ee}(r_{12}) = (g_{dd}(r_{12}) - 1)(N_{ee}(r_{12}) + E_{ee}(r_{12}) + (N_{de}(r_{12}) + E_{de}(r_{12}))^2 - d(N_{cc}(r_{12}) + E_{cc}(r_{12}) - \frac{1}{d} l(k_F r_{12}))^2), \quad (105)$$

The indices x and y run over the types d and e of the external points, and $\zeta_{dd} = \rho V$ and $\zeta_{de} = \zeta_{ed} = \rho V_d$, whereas $\zeta_{ee} = 0$.

Similarly, the equations for the cc–diagrams are given by

$$N_{cc}^{xy}(r_{12}) = \sum_{x'y'} \int d\mathbf{r}_3 X_{cc}^{xx'}(r_{13}) \xi_{x'y'} \times \left[X_{cc}^{y'y}(r_{32}) + N_{cc}^{y'y}(r_{32})\right], \quad (106)$$

where the indices x and y run over the labels h and ρ of the external points of the cc–diagrams, $\xi_{hh} = \xi_{h\rho} = \xi_{\rho h} = \rho V_d$ and $\xi_{\rho\rho} = \rho(V_d - 1)$. The X_{cc} functions are given by

$$X_{cc}^{hh}(r_{12}) = (g_{dd}(r_{12}) - 1)(N_{cc}(r_{12}) + E_{cc}(r_{12}) - \frac{1}{d} l(k_F r_{12})) = X_{cc}(r_{12}), \quad (107)$$

$$X_{cc}^{h\rho}(r_{12}) = X_{cc}^{\rho h}(r_{12}) = 0, \quad (108)$$

$$X_{cc}^{\rho\rho}(r_{12}) = -\frac{1}{d} l(k_F r_{12}), \quad (109)$$

where $N_{cc} = \sum_{x,y} N_{cc}^{xy}$ and $E_{cc} = \sum_{x,y} E_{cc}^{xy}$. The RFHNC equations for a finite system are given in [18]).

The equations for the U functions are obtained in a similar way and are given by

$$U_x(1) = U_x = E_x(1) + \int d\mathbf{r}_2 \{\sum_{x'} [(X_{xx'}(r_{12}) - E_{xx'}(r_{12}))\eta_{x'} - \sum_{y'} S_{xx'}(r_{12})(X_{xy'}(r_{12}) + N_{xy'}(r_{12}))\zeta_{x'y'}] + \delta_{xe}[2dS_{cc}(r_{12})(X_{cc}(r_{12}) + N_{cc}(r_{12})) - l(k_F r_{12})(N_{cc}^{\rho h}(r_{12}) + N_{cc}^{\rho\rho}(r_{12}) - \frac{1}{d} l(k_F r_{12}))]\}, \quad (110)$$

with $S_{xx'} = N_{xx'}/2 + E_{xx'}$ (and similarly for S_{cc}), $\eta_d = \rho V$ and $\eta_e = \rho V_d$. It is easy to verify that, for an infinite system and Jastrow correlations, the RFHNC equations coincide with the standard FHNC equations, with the following correspondence between the FHNC and the RFHNC quantities: $Y_{dd} \to Y_{dd}$, $Y_{de} \to V_d Y_{de}$, $Y_{ee} \to V_d^2 Y_{ee}$, where $Y = N, X, E$, and $X_{cc} \to V_d X_{cc}$ and $(N_{cc} - l/2) \to V_d(N_{cc} - l/2)$.

The pair function is given by

$$g(r_{12}) = 1 + (X_{dd}(r_{12}) + N_{dd}(r_{12})) + 2V_d(X_{de}(r_{12}) + N_{de}(r_{12})) + V_d^2(X_{ee}(r_{12}) + N_{ee}(r_{12})). \quad (111)$$

The RFHNC equations can be solved numerically by means of iterative procedures. For dense systems, like liquid helium, the convergence may be difficult to reach, and one may prefer to smooth out the iterative process by standard numerical techniques (see [24, 26] for more details).

Obviously, also for the RFNC equations, one is faced with the problem of the elementary diagrams, which is even more serious than for the case of HNC. Now, there are various classes of $E_{xy}(r)$, and the existence of a scaling property is not at all clear. Nevertheless, the fermion generalization of the *scaling* and *interpolating* approximations have been used in liquid ^{3}He with some success [35, 37].

Another interesting scheme, which has given reasonable results in nuclei [127, 128] and in the Hubbard model [71], follows the PS expansion, and has been denoted as (R)FHNC–n approximation. First order includes elementary diagrams with one CI bond, second order with one and two CI's, and so on.

RFHNC with spin–dependent correlations. The above discussion has been limited to the case of scalar correlations, mainly for pedagogical purposes. Spin–dependent correlations are extremely important in several strongly interacting Fermi systems, particularly in nuclei and nuclear matter, where the interaction is strongly spin–isospin dependent, and *realistic calculations have to take them into account.* RFHNC method has been and is largely used in such calculations.

The technique for summing linked cluster diagrams containing spin-dependent CI's is technically difficult because of the operatorial structure of the correlations and of their non commutativity, which makes a full RFHNC summation prohibitive. A largely used approximation is the so called Single Operator Chain (FHNC/SOC) scheme. A detailed description of this scheme can be found in [22], and we shall limit ourselves to briefly summarize its main features.

1. The underlying idea is that the most important effects of the non–scalar components of the CI occur on the interacting particles. For this reason, the hamiltonian operator is always dressed with the full correlation operator $\hat{F}^\dagger(1,2)\hat{F}(1,2)$.

2. On the other side, for the passive particles, non scalar contributions are much smaller than those coming from scalar correlations, and, therefore, only the leading terms are kept. Such leading terms are the chain diagrams built up with *operatorial bonds*, namely bonds made up of composite diagrams $X^{(n)}(i,j)$ having any number of scalar correlations and one operatorial correlation ($h^{(n)}(r_{ij})\hat{O}^{(n)}(i,j)$ or an exchange operator $l(k_F r_{ij})P_{\sigma\tau}(i,j)$ coming from the Slater Determinant) as $i-j$ direct link. Similarly, the vertex corrections are approximated with single operator rings (SOR)[15].
3. Up to five operatorial lines may reach the external points and they are treated exactly. This leads to a messy, but treatable, algebra.
4. The inclusion of single operator chains implies that the various *nodal* and *composite* functions have as many components as $\hat{F}(1,2)$. Consequently, in nuclear matter calculations, the number of coupled integral equations increases of about a factor 6 with respect to the Jastrow case.
5. Higher order corrections [130, 131, 37] to the RFHNC/SOC have been found to be small both in nuclear matter and in liquid ^{3}He. However, along this scheme, we do not really know, at present, how to perform a next order calculation.

The spin–coherent wave function seems to offer the possibility to go beyond RFHNC/SOC approximation. With this wave function the two–body quantities of the *nodal*, *composite* and *elementary* type depend on the interparticle distance r_{ij}, but also on the angular variable ω_i^L,ω_j^L, ω_i^R, ω_j^R. Such dependence is due to the fact that $|\Psi_{CS}\rangle$ involves integrals of $\{\omega_i^R\}$ and, similarly, $\langle\Psi_{CS}|$ involves integrals in $\{\omega_i^L\}$. Thus, in a given matrix element, there are two angles to be integrated for each particle. In particular, the exchange function $l(k_F r_{ij})$ acquires an additional ω–dependent factor [38].

Numerical evaluation of RFHNC quantities can be made easier by using a discrete set of points to replace the $\{\omega_i^R\}$ integrations. This can also be viewed as using discrete coherent states where the number of functions to be evaluated goes as n^4, where n is the number of $\{\omega_i\}$ points. This number, however, can be greatly reduced by using symmetries, and actual numerical calculations are totally feasible.

5 One–body Green function

The one body Green's function is an essential quantity in many-body physics [132], [133]). We recall that its definition is

[15] The derivation of vertex corrections given in [22] is different from that discussed here in connection with the RFHNC scheme, and it is based on a low order cluster decomposition. However, by taking the SOC approximation of the RFHNC equations and recalling that $\rho\int drl^2(k_F r) = \rho\int drl(k_F r) = 1$, the leading orders of the expressions given there can be recovered [129]).

$$\begin{aligned} G(k,E) &= \langle\bar{0}|a^\dagger_{\mathbf{k}}\frac{1}{H-E_\circ-E-\imath\eta}a_{\mathbf{k}}|\bar{0}\rangle \\ &-\langle\bar{0}|a_{\mathbf{k}}\frac{1}{H-E_\circ+E-\imath\eta}a^\dagger_{\mathbf{k}}|\bar{0}\rangle \\ &= G_h(k,E)+G_p(k,E), \end{aligned} \tag{112}$$

where $|\bar{0}\rangle$ is the true nuclear ground state with energy $E_\circ$. $G_{p(h)}$ is the particle (hole) Green's function, describing the propagation of a particle (hole) state and defined for $\epsilon = -E > (<) e_F$, where e_F is the Fermi energy.

The imaginary parts of the Green's function define the hole and particle spectral functions, $P_h(k,E)$ and $P_p(k,E)$, as

$$\begin{aligned} P_h(k,E) &= \sum_N \left|\langle\bar{0}|a^\dagger_{\mathbf{k}}|\bar{N}(A-1)\rangle\right|^2 \\ &\times\delta(E-E_N(A-1)+E_\circ(A)), \end{aligned} \tag{113}$$

$$\begin{aligned} P_p(k,E) &= \sum_N |\langle\bar{0}|a_{\mathbf{k}}|\bar{N}(A+1)\rangle|^2 \\ &\times\delta(E+E_N(A+1)-E_\circ(A)), \end{aligned} \tag{114}$$

where $|\bar{N}(A\pm 1)\rangle$ are intermediates states of the $(A\pm 1)$-particle systems with energies $E_N(A\pm 1)$. The spectral functions give the joint probability of leaving the nuclear system with an excitation energy E, when removing (P_h) or adding (P_p) a nucleon with momentum $\mathbf{k}$.

Semi-inclusive electron scattering experiments on nuclei, specifically $(e,e'p)$ reactions, provide a quasi direct measure of the spectral function and have clearly shown the effect of the correlations in the form of large deviations from the predictions of simple independent particle models [135, 134, 137].

In [115, 136] the authors have computed the nuclear matter Green's function with the Urbana v_{14}+DDI interaction and using orthogonal CBF theory. We will present here some of the results obtained in those papers.

Variational spectral functions. The *variational* spectral functions, $P^{(o)}_{h(p)}(k)$, are obtained by keeping only the lowest order terms in the perturbative expansion for the full $P_{h(p)}(k)$ (see diagrams (a) and (b) of Fig. 15). It is equivalent to approximate $|\bar{0}\rangle$ and $|\bar{N}\rangle$ with the correlated $|0\rangle$ and $|N\rangle$ CBF states in eqs. (113,114). 1 hole (1 particle) and 2 hole–1 particle (2 particle–1 hole) intermediate states have been used for the variational hole (particle) spectral function.

The term associated with 1 hole (1 particle) state is the dominant onc and it is given by:

$$P^{(o)}_{1h(p)}(k,E) = |\Phi^{h(p)}_{\mathbf{k}}|^2\delta(e^v(k)+E)\Theta[\pm(k-k_F)], \tag{115}$$

where the upper (lower) sign in the argument of the Θ–function applies to the hole (particle) spectral function, $e^v(k) = \langle k|H|h\rangle - E^v_o$ is the variational single particle energy and the overlap matrix element $\Phi^{h(p)}_{\mathbf{k}}$ is given by:

$$\Phi^{h(p)}_{\mathbf{k}} = \langle 0|a^\dagger_{\mathbf{k}}(a_{\mathbf{k}})|h(p) = k\rangle. \tag{116}$$

This intermediate state is the only non vanishing one in any uncorrelated theory, where $\Phi_{unc} = 1$ and $P^{(o)}_{unc,h(p)}(k,E) = P^{(o)}_{unc,1h(p)}(k,E) = \delta(k^2/2m + E)\Theta[\pm(k - k_F)]$. In CBF other intermediate states contribute to the zeroth order orthogonal perturbation theory, as correlations are already built in the basis functions. Hence, part of the strength is moved to higher excitation energies and the overlap matrix element $\Phi^{h(p)}_{\mathbf{k}}$ is appreciably quenched.

The contribution from 2h–1p (2p–1h) states is spread out in energy and can be interpreted as a background contribution to be added to the quasi–particle part of the spectral function, coming from the 1h (1p) state.

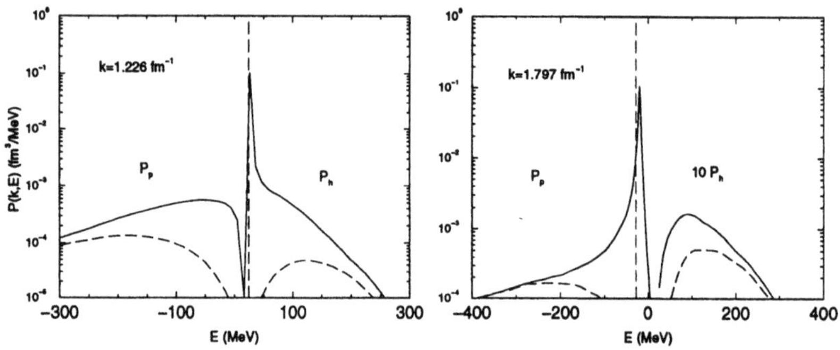

Fig. 18. Spectral functions of nuclear matter below and above $k_F = 1.33 \text{ fm}^{-1}$. From [136]. The dashed lines are the variational spectral functions; the continuous lines are the perturbative results.

The nuclear matter variational spectral functions at two momenta, below and above the Fermi momentum, are shown in Fig.18 as dashed curves. They clearly exhibit the two features previously discussed: the delta peak located at $E = -e^v(k)$, whose strength $Z^v(k)$ is quenched respect to unity, as an effect of the correlations, and the background, due to 2h-1p (2p-1h) intermediate states and extending over a large energy range. $Z^v(k)$ is the variational estimate of the quasiparticle pole, $Z(k)$, given by

$$Z(k < k_F) = \left|\langle\bar{0}|a^\dagger_{\mathbf{k}}|\bar{\mathbf{k}}\rangle\right|^2, \tag{117}$$

and

$$Z(k > k_F) = \left|\langle \bar{0}|a_{\mathbf{k}}|\bar{\mathbf{k}}\rangle\right|^2, \tag{118}$$

where $|\bar{\mathbf{k}}\rangle$ is a hole or a particle state, respectively. For the momenta shown in the figure, the values are $Z^v(k = 1.226\text{fm}^{-1}) = 0.88$ and $Z^v(k = 1.797\text{fm}^{-1}) = 0.90$.

Perturbative corrections to the spectral functions. Perturbative corrections come on account of m particle–n hole orthogonal correlated states admixture in $|\bar{0}, \bar{N}\rangle$. In [115, 136]), 2p–2h admixture in the ground state, $|\bar{0}\rangle$, and 2h(2p)–1p(1h) in the one-hole(particle) state $|\bar{\mathbf{k}}\rangle$, were considered.

The corrections are classified as $\delta P^{GR}_{h(p)}(k, E)$, coming from ground state admixture, and $\delta P^{INT}_{h(p)}(k, E)$, due to corrections to intermediate states. Therefore, the total spectral functions is written as

$$P_{h(p)}(k, E) = P^{(o)}_{h(p)}(k, E) + \delta P^{GR}_{h(p)}(k, E) + \delta P^{INT}_{h(p)}(k, E). \tag{119}$$

2p-2h states contribute for all k-values (both below and above k_F). The correction coming from them can be expressed in terms of the imaginary part of the *correlation* contribution to the self energy $\Sigma^{CO}(\mathbf{k}, E)$, given in eqs.(47, 46).

The self energy has been computed using orthogonal correlated states at low orders in the cluster expansion, consisting in retaining only two-body and three-body separable cluster contributions [104].

The main effects of $\delta P^{GR}_{h(p)}(k, E)$ on the spectral functions are: i) a quenching of the strength of the quasi-particle peak and ii) a shifting of its position of a quantity $\delta e^{CO}_{h(p)}(k)$ given by

$$\delta e^{CO}_h(k) = \frac{1}{\pi} \int_{e^v(k_F)}^{\infty} \frac{Im\Sigma^{CO}(\mathbf{k}, E)}{e^v(k) - E} dE, \tag{120}$$

for the hole part, and

$$\delta e^{CO}_p(k) = \frac{1}{\pi} \int_{-\infty}^{e^v(k_F)} \frac{Im\Sigma^{CO}(\mathbf{k}, E)}{e^v(k) - E} dE, \tag{121}$$

for the particle part. The shifts are the corrections to the real part of the optical potential coming from the correlation part of the self energy.

Intermediate state admixture dramatically affects the shape itself of the spectral functions. The leading contribution comes from the coupling of the 1h (or 1p) CBF states with the 2h–1p (or 2p–1h) ones. The main consequence is that the delta shaped quasiparticle peak acquires a width that is roughly proportional to the imaginary part of the *polarization* self energy, $Im\Sigma^{PO}(\mathbf{k}, E = -e^v(k))$ (see eqs.(49, 48)).

The continuous curves in Fig.18 are the total spectral functions. The height of the quasiparticle peak is noticeably reduced by the perturbative corrections, even in the vicinity of k_F, both for the hole and the particle spectral functions; the spreading of the quasiparticle state, due to the admixture with 2h-1p (2p-1h) states, points to the inadequacy of a quasiparticle description away from the close neighborhood of k_F, where the width of the peak is small.

Perturbative corrections to $Z(k)$ originate, as for the spectral functions, from 2h-2p admixture in the ground state and 2h-1p or 2p-1h admixture in the one-hole or one-particle intermediate states. As a consequence, $Z(k)$ can be written as

$$Z(k) = Z^v(k) + \delta Z^{GR}(k) + \delta Z^{INT}(k). \tag{122}$$

The expressions of $\delta Z^{GR,INT}(k)$ are derived and discussed in [136]. Table 4 gives both $Z^v(k)$ and $Z(k)$ at several momenta. The perturbative corrections are sizeable, mainly around k_F, and provide a value of the quasiparticle pole at k_F of $Z(k_F) = 0.70$, with a 20% reduction respect to the variational estimate.

Table 4. Quasiparticle strengths in nuclear matter at $k_F = 1.33 fm^{-1}$. From [136]).

k/k_F	$Z^v(k)$	$Z(k)$
0.57	0.87	0.88
0.67	0.87	0.81
0.87	0.88	0.71
0.97	0.88	0.70
1.00	0.88	0.70
1.10	0.88	0.70
1.24	0.89	0.76
1.35	0.89	0.80
1.47	0.89	0.85
1.59	0.90	0.89
1.72	0.91	0.91

In finite nuclei, $Z(e \sim e_F)$ is affected by the coupling between single particle states and surface vibrations. The effect is an extra depletion at the Fermi surface, which can be estimated by modifying the imaginary part of the self-energy in such a way to reproduce the measured widths of the states around the Fermi surface. The final value of the quasiparticle strength is

$Z(e_F)+\delta Z_{surf}(e_F) \simeq 0.55$ [114], to be compared with the recent reanalysis of the experimental $(e,e'p)$ data in ^{208}Pb [135, 134, 137], yielding a spectroscopic factor of the $3s_{1/2}$ state of 0.65 ± 0.05.

Momentum distributions. Let's come back, now, to the momentum distribution and discuss it mainly in connection with the spectral functions. The hole and particle momentum distributions $n_h(k)$ and $n_p(k)$ are the occupation and not–occupation probabilities of a state carrying momentum k. They are given by

$$n_{h(p)}(k) = \frac{\langle\bar{0}|a^\dagger_{\mathbf{k}}a_{\mathbf{k}}(a_{\mathbf{k}}a^\dagger_{\mathbf{k}})|\bar{0}\rangle}{\langle\bar{0}|\bar{0}\rangle}, \tag{123}$$

are related to the spectral functions by the sum rules

$$n_h(k) = \int_{-e_F}^{\infty} P_h(k,E)dE, \tag{124}$$

$$n_p(k) = \int_{-\infty}^{-e_F} P_p(k,E)dE, \tag{125}$$

and satisfy the condition:

$$n_h(k) + n_p(k) = 1, \tag{126}$$

for each value of k. The uncorrelated momentum distributions are step functions,

$$n^{UNC}_{h(p)}(k) = \Theta[\pm(k-k_F)], \tag{127}$$

reflecting the property that states having $k < k_F$ are totally occupied, and those with $k > k_F$ are totally empty. The correlations deplete (enhance) the hole (particle) momentum distribution below k_F moving strength to states with $k > (<) k_F$.

The *variational* momentum distributions

$$n^{var}_{h(p)}(k) = \frac{\langle 0|a^\dagger_{\mathbf{k}}a_{\mathbf{k}}(a_{\mathbf{k}}a^\dagger_{\mathbf{k}})|0\rangle}{\langle 0|0\rangle}, \tag{128}$$

are related to the variational spectral functions by

$$n^{var}_{h(p)}(k) = \int P^{(o)}_{h(p)}(k,E)dE, \tag{129}$$

and they exactly satisfy the relation

$$n^{var}_h(k) + n^{var}_p(k) = 1. \tag{130}$$

$n^{var}_{h(p)}(k)$ are given by

$$n^{var}_{h(p)}(k) = |\Phi^{h(p)}_{\mathbf{k}}|^2 \Theta[\pm(k - k_F)] + n_{h(p),c}(k), \tag{131}$$

where $n_{h(p),c}(k)$ are continuous functions given by the integral over the energy of the background part of $P^{(o)}_h$. $|\Phi^h_{k=k_F}|^2$ is the discontinuity of the hole momentum distributions at the Fermi surface and represents the variational estimates of $Z(k = k_F)$.

The perturbative corrections to $n^{var}_{h(p)}(k)$ are written as

$$\delta n_{h(p)}(k) = \int \delta P^{GR}_{h(p)}(k, E) dE, \tag{132}$$

whereas it can be shown that

$$\int \delta P^{INT}_{h(p)}(k, E) dE = 0. \tag{133}$$

As a consequence, $\delta P^{INT}_{h(p)}$ affects the shape of the spectral functions but leaves unchanged their sum rules. Moreover, $\delta n_h(k) = -\delta n_p(k)$, so leaving unaffected condition (130).

The momentum distribution $n_h(k)$, in nuclear matter, has been already shown in Fig. 14. The occupation probabilities are given in table 5. The energy integrals of eqs. (124,125) are carried out up to very large E-values ($|E_{max(min)}| \sim 1$ GeV), so ensuring convergence. The table gives also $n^{var}_{h(p)}(k)$. Condition (130) is violated up to a maximum of 10% because not all the terms entering the cluster expansions of $\langle 0|a^\dagger_{\mathbf{k}} a_{\mathbf{k}} (a_{\mathbf{k}} a^\dagger_{\mathbf{k}})|0\rangle$ have been summed (the FHNC/SOC approximation, as given in [22], has been adopted).

Spectral functions and density response. The study of the inclusive, (e, e'), cross section in nuclei requires the knowledge of the longitudinal and transverse response functions $S_{L,T}(q, \omega)$. At intermediate and high momentum transfers, q, they are heavily influenced by the short range correlations and, therefore, nuclear matter is generally considered to be a reliable model. A short description of the CBF theory for $S_{L,T}(q, \omega)$ is given elsewhere in this book [18]. The simpler density response, $S(q, \omega)$, defined in section 3 as structure function (see eq.(70)), is sometimes used to approximate the previous ones and the spectral functions can be used for its evaluation in the ladder approximation for the two-body Green's function [132].

$S(q, \omega)$ can be written in terms of P_h and P_p and of the particle-hole effective interaction. At the zeroth order, one gets

$$\begin{aligned} S(q, \omega) \simeq S_{LA/0}(q, \omega) &= \int d\mathbf{k} dE dE' P_h(k, E) P_p(|\mathbf{k} + \mathbf{q}|, E') \\ &\times \delta(\omega - E - E'). \end{aligned} \tag{134}$$

Table 5. Occupation probabilities in nuclear matter at $k_F = 1.33\text{fm}^{-1}$. From [136].

k/k_F	$n_h^{var}(k)$	$n_p^{var}(k)$	$n_h(k)$	$n_p(k)$
0.57	0.876	0.020	0.832	0.064
0.67	0.877	0.030	0.825	0.082
0.87	0.882	0.037	0.809	0.110
0.97	0.885	0.041	0.786	0.140
1.10	0.006	0.926	0.042	0.890
1.24	0.007	0.936	0.027	0.916
1.35	0.007	0.946	0.020	0.933
1.47	0.006	0.951	0.016	0.942
1.59	0.006	0.967	0.012	0.961
1.72	0.007	0.967	0.010	0.964

$P_p(k,E)$ gives a measure of the Final State Interactions (FSI) of the struck nucleon traveling through the nuclear medium. In fact, if $P_p(k,E) = \delta(e(k) + E)\Theta(k_F - k)$, the Plane Wave Impulse Approximation (PWIA) to the response is recovered. The validity of these approximations has been object of investigation in literature [139]. Orthogonal CBF theory allows for a consistent calculation of the full density response of eq.(70) [138, 118] as well as of the spectral functions. Therefore, it is possible, within CBF, a systematic study of the validity of eq.(134).

It is worth to notice that the CBF theory of the response, being a non relativistic one, is unreliable in the few GeV region, where the struck nucleon is fully relativistic. In addition, transitions of the nucleons to excited hadronic states become important at these large momentum transfers. In PWIA the struck nucleon is considered to freely move after being hit by the probe and, so, it can be treated relativistically. S_{PWIA} has been computed within CBF in nuclear matter at $q \sim 2\text{GeV}/c$ in [140]. The results have been compared with empirical estimates obtained by extrapolating from data on laboratory nuclei and, if inelasticity and relativity are properly introduced, they accurately describe the quasielastic peak. However, the response at low ω-values, where FSI effects are expected to be important, is grossly underestimated.

The zeroth order ladder approximation of eq.(134) is expected to provide a good description of FSI, even if the use of the non relativistic particle spectral function limits its area of applicability to relatively low q values. At large ω values, FSI do not play any significant role and eq.(134) includes a larger class of intermediate states than the CBF response of [138]. In fact, the perturbative expansion of the polarization propagator was stopped at the

1p-1h level, whereas the expansion of the spectral functions includes up to 2p-1h and 2h-1p correlated intermediate states.

Figure 19 shows the response at two momenta. The full curve is the CBF response $S(q,\omega)$, the dashed curve gives $S_{LA/0}(q,\omega)$ and the dash–dotted curve $S_{PWIA}(q,\omega)$. PWIA overestimates the response at the quasielastic peak but, as q increases, it becomes closer to the other calculations. The inclusion of the FSI, by means of the particle spectral function in $S_{LA/0}$, brings the computed response in better agreement with the full calculation, particularly in the low energy tail of the peak. In the large energy region, as expected, $S_{LA/0}(q,\omega)$ is consistently larger than $S(q,\omega)$.

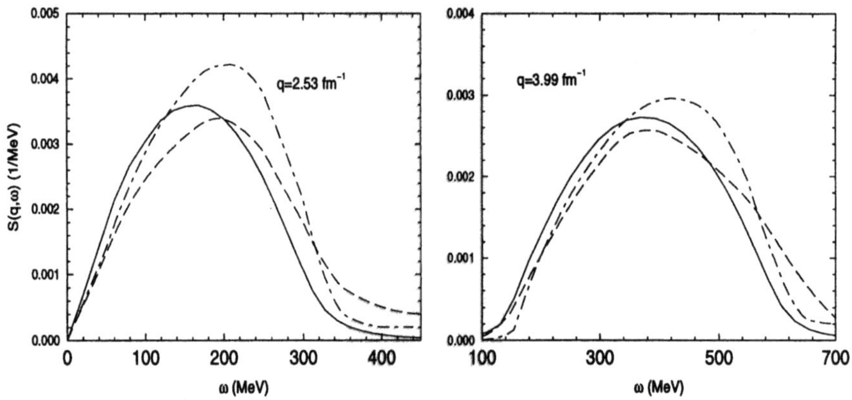

Fig. 19. Density responses of nuclear matter at $k_F = 1.33\ fm^{-1}$ in different approximations. From [136].

At high momentum transfer q, one can treat FSI by using Glauber theory. Such theory, however, needs to be corrected to include N–N correlations. A correlated Glauber theory has been developed in [140]. The calculated response functions are in excellent agreement with experimental data on inclusive e–scattering off nuclei in the GeV region. It is beyond the scope of these lecture notes to give details of these important developments, and the eager reader is referred to [141].

Acknowledgments

We would like to thank Omar Benhar, Orion Ciftja, Saverio Moroni, Vijay Pandharipande, Kevin Schmidt, Ingo Sick and Erio Tosatti for helpful discussions and critical advice. We also acknowledge financial support from the Istituto Nazionale di Fisica Nucleare (Sezione di Trieste), the Istituto Nazionale di Fisica della Materia and the European Community contract, No CHR–XCT94–0456.

References

1. D.D. Osheroff, R.C. Richardson and D.M. Lee, Phys. Rev. Lett. **28**, 885 (1972)
2. S. Fantoni and S. Rosati, Lett. Nuovo Cimento **10**, 545 (1974)
3. E. Krotscheck and M.L. Ristig, Phys. Lett. **A48**, 17 (1971)
4. W.L. McMillan, Phys. Rev. A **138**, 442 (1965)
5. D.M. Ceperley and M.H. Kalos, *Monte Carlo Methods in Statistical Physics* edited by K. Binder, 1979 (Springer–Verlag, Berlin)
6. E. Krotscheck, Phys. Rev. B **33**, 3158 (1986)
7. E. Krotscheck, contribution to this volume
8. S.A. Vitiello and K.E. Schmidt, Phys. Rev. B **46**, 5442 (1992)
9. S. Moroni, S. Fantoni and G. Senatore, Phys. Rev. B **52**, 13547 (1995)
10. J.W. Clark, *Progress in Particle and Nuclear Physics* **2**, edited by D.H. Wilkinson, 1979(Pergamon, Oxford) 89
11. E. Krotscheck and J.W. Clark, Nucl. Phys. **A333**, 77 (1980)
12. S. Fantoni and V.R. Pandharipande, Phys. Rev. C **37**, 1697 (1988)
13. R.F. Bishop, contribution to this volume
14. K.E. Schmidt and M.H. Kalos, *Monte Carlo Methods in Statistical Physics II, Topics in Current Physics*, edited by K. Binder, 1984 (Springer–Verlag, Berlin)
15. K.E. Schmidt and D.M. Ceperley, *Monte Carlo Methods III*, edited by K. Binder, 1991 (Springer–Verlag, Berlin)
16. D.M. Ceperley, Rev. Mod. Phys. **67**, 279 (1995)
17. S. Fantoni and S. Moroni, *Many Body Theory of Correlated Fermion Systems*, edited by J.M. Arias, M. Gallardo and M. Lozano, 1998 (World Scientific), in press
18. A. Fabrocini, contribution to this volume
19. J.G. Zabolitzky, Phys. Rev. B **22**, 2353 (1980)
20. E. Krotscheck, Ann. Phys. (NY) **155**, 1 (1984)
21. M.L. Ristig, J.W. Kim and R. Mehlman, *Condensed Matter Theories, Vol. 11*, edited by E. Ludeña et al, 1995 (Nova Science Publishers, New York)
22. V.R. Pandharipande and R.B. Wiringa, Rev. Mod. Phys. **51**, 821 (1979)
23. S. Rosati, in *International School of Physics Enrico Fermi, Course LXXIX*, edited by A. Molinari, 1981 (North–Holland, Amsterdam), 73
24. J.G. Zabolitzky, in *Advances in Nuclear Physics, Vol. 12*, edited by J.W. Negele and E. Vogt, 1981 (Plenum, New York,London) 1
25. J.P. Blaizot and G. Ripka, *Quantum Theory of Finite Systems*, 1986 (The MIT Press, Cambridge)
26. A. Fabrocini and S. Fantoni, *First International Course on Condensed Matter, ACIF series, Vol. 8* edited by D. Prosperi, S. Rosati and S. Violini, 1987 (World Scientific, Singapore), 87
27. V.R. Pandharipande, I. Sick and P.K.A. de Witt Huberts, Rev. Mod. Phys. **69**, 981 (1997)
28. E. Feenberg, *Theory of Quantum Liquids*, edited by K. Binder, 1969 (Springer, New York)

29. S. Moroni, G. Senatore and S. Fantoni, Phys. Rev. B. **55**, 1040 (1997)
30. S. Fantoni and V.R. Pandharipande, Nucl. Phys. **A427**, 473 (1984)
31. O. Benhar, A. Fabrocini and S. Fantoni, *4th Workshop on Perspectives in Nuclear Physics at Intermediate Energies*, edited by S. Boffi, C. Ciofi degli Atti and M. Giannini, 1989 (World Scientific, Singapore), 333
32. R.A. Aziz et al, J. Chem. Phys. **70**, 4330 (1979)
33. M.H. Kalos et al, Phys. Rev. B **24**, 115 (1981)
34. K.E. Schmidt et al, Phys. Rev. Lett. **47**, 807 (1981)
35. E. Manousakis et al, Phys. Rev. B **28**, 3770 (1983)
36. R.P. Feynman and M. Cohen, Phys. Rev. **102** 1189 (1956)
37. M. Viviani et al, Phys. Rev. B **38**, 4523 (1988)
38. J.W. Lawson et al, Phys. Rev. Lett. **78**, 1846 (1997)
39. O. Ciftja, PhD thesis, International School for Advanced Studies, Trieste (1997)
40. F. Pederiva et al, Phys. Rev. B **53**, 15129 (1996)
41. K.E. Schmidt, private communication
42. Q.N. Usmani, S. Fantoni and V.R. Pandharipande, Phys. Rev. B **26**, 6123 (1983)
43. A. Fabrocini and S. Rosati, Nuovo Cimento D **1**, 165; 567 (1982)
44. S.A. Vitiello et al, Phys. Rev. Lett. **60**, 1970 (1988)
45. L. Reatto and G.L. Masserini, Phys. Rev. B **38**, 4516 (1988)
46. F. Pederiva et al, Phys. Rev. Lett. **72**, 2589 (1994)
47. S. Moroni, L. Reatto and S. Fantoni, Czech. J. Phys. S1 **46** 281 (1996)
48. D.C. Tsui, H.L. Stormer and A.C. Gossard, Phys. Rev. Lett. **48**, 1559 (1982)
49. P.W. Anderson, Science **235**, 1196 (1987)
50. R.B. Laughlin, *The Quantum Hall Effect*, edited by R. Prange and S. Girvin, 1987 (Springer–Verlag, New York)
51. T. Chakraborty and P. Pietaläinen, *The Fractional Quantum Hall Effect*, 1988 (Springer–Verlag, New York)
52. R.B. Laughlin, Phys. Rev. Lett. **50**, 1395 (1983)
53. B.I. Halperin, Helv. Phys. Acta **56**, 75 (1983)
54. J.M. Caillol et al, J. Stat. Phys. **28**, 325 (1982)
55. J.P. Hansen and D. Levesque, J. Phys. C **14**, L603 (1981)
56. O. Ciftja and S. Fantoni, Europhys. Lett. **36**, 663 (1996)
57. J.K. Jain, Phys. Rev. Lett. **63**, 199 (1989)
58. G. Dev and J.K. Jain, Phys. Rev. Lett. **69**, 2483 (1992)
59. X.G. Wu, G. Dev and J.K. Jain, Phys. Rev. Lett. **71**, 153 (1993)
60. O. Ciftja and S. Fantoni, Phys. Rev. B **56**, 6712 (1997)
61. J.K. Jain and R.K. Kamilla, Phys. Rev. B **55**, R4895 (1997)
62. O. Ciftja et al, J. Low Temp. Phys. **108**, 357 (1997)
63. J.G. Bednorz and K.A. Müller, Z. Phys. B **64**, 188 (1986)
64. C.W. Chu et al, Phys. Rev. Lett. **58**, 405 (1987)
65. D. Vollhardt, Rev. Mod. Phys. **99**, 1196 (1984)
66. E. Manousakis, Rev. Mod. Phys. **63**, 1 (1991)
67. E.H. Lieb and F.Y. Wu, Phys. Rev. Lett. **20**, 1445 (1968)
68. S. Fantoni et al, Physica C **153–155**, 1255 (1988)
69. W. Metzner and D. Vollhardt, Phys. Rev. Lett. **59**, 121 (1987)
70. X.Q. Wang et al, Phys. Rev. B **41**, 11479 (1990)
71. X.Q. Wang et al, Phys. Rev. B **46**, 8894 (1992)

72. P. Fazekas and K. Penc, Int. J. Mod. Phys. B **1**, 1021 (1988)
73. H. Yokoyama and H. Shiba, J. Phys. Soc. Jap. **56**, 1490; 3570; 3582 (1987)
74. J. Solyom, Adv. Phys. **28**, 201 (1979)
75. F.C. Zhang et al, Supercond. Sci. Technol. **1**, 36 (1988)
76. G. Kotliar and J. Liu, Phys. Rev. B **38**, 5142 (1988)
77. J.R. Schrieffer, X.G. Wen and S.C. Zhang, Phys. Rev. B **39**, 11663 (1989)
78. R.V. Reid Jr., Ann. Phys. (N.Y.) **50**, 411 (1968)
79. M. Lacombe et al, Phys. Rev. C **21**, 861 (1980)
80. I.E. Lagaris and V.R. Pandharipande, Nucl. Phys. **A359**, 331 (1981)
81. R.B. Wiringa, V.G.J. Stocks and R. Schiavilla, Phys. Rev. C **51**, 38 (1995)
82. J.L. Forest et al, Phys. Rev. C **54**, 646 (1996)
83. A. Akmal and V.R. Pandharipande, Phys. Rev. C **56**, 2261 (1997)
84. R. Schiavilla, V.R. Pandharipande and R.B. Wiringa, Nucl. Phys. **A449**, 219 (1986)
85. J. Fujita and H. Miyazawa, Prog. Theor. Phys. **17**, 360 (1957)
86. B.S. Pudliner, Ph.D. Thesis, University of Illinois at Urbana–Champaign (1996)
87. B.S. Pudliner et al, Phys. Rev. Lett. **74**, 4396 (1995)
88. A. Kievsky, S. Rosati and M. Viviani Nucl. Phys. **A57**, 511 (1994)
89. M. Viviani, R.Schiavilla and A. Kievsky, Phys. Rev. C **54**, 534 (1996)
90. E. Fradkin, *Field Theories of Condensed Matter Systems*, 1991 (Addison–Wesley, Reading MA)
91. P.M.Morse and H. Feshbach, *Methods of Theoretical Physics*, 1953 (McGraw–Hill, New York), Chap. 9, Part II
92. S. Fantoni, Phys. Rev. B **29**, 2544 (1984)
93. C.W. Woo, Phys. Rev. **151**, 138 (1966)
94. J.W. Clark, *Lecture Notes in Physics*, **138**, edited by R. Guardiola and J. Ros, 1981 (Springer, New York), p. 184
95. E. Krotscheck, R.A. Smith and J.W. Clark, *Lecture Notes in Physics*, **142**, 1981 (Springer, New York), 270
96. E. Krotscheck, R.A. Smith and A.D. Jackson, Phys. Lett. B **104**, 421 (1981)
97. A.D. Jackson et al, Nucl. Phys. **A386**, 125 (1982)
98. E. Manousakis and V.R. Pandharipande, Phys. Rev. B **33**, 150 (1986):
99. E. Krotscheck, R.A. Smith, J.W. Clark, and R.M. Panoff, Phys. Rev. B **24**, 6383 (1981)
100. S. Fantoni, V.R. Pandharipande and K.E. Schmidt, Phys. Rev. Lett. **48**, 878 (1982)
101. B.L. Friman and E. Krotscheck, Phys. Rev. Lett. **49**, 1705 (1982)
102. R.B. Wiringa, R.A. Smith and T.L. Ainsworth, Phys. Rev. C **29**, 1207 (1984)
103. I.E. Lagaris and V.R. Pandharipande, Nucl. Phys. **A359**, 349 (1981)
104. S. Fantoni, B.L. Friman and V.R. Pandharipande, Nucl. Phys. **A399**, 51 (1983)
105. A. Fabrocini and S. Fantoni, Phys. Lett. B **298**, 263 (1993)
106. A.A. Abrikosov, L.P. Gorkov and I.E. Dzyaloshinski, *Methods of Quantum Field Theory in Statistical Physics*, 1963 (Dover, New Jork)
107. M. Baldo et al, Nucl. Phys. **A545**, 741 (1992)
108. M. Bauer et al, J. Phys. G **8**, 525 (1982)
109. E. Krotscheck, Phys. Rev. A **26**, 3536 (1982)

110. S. Fantoni, B.L. Friman and V.R. Pandharipande, Phys. Lett. B **104**, 89 (1981)
111. J.W. Negele and K. Yazaki, Phys. Rev. Lett. **47**, 71 (1981)
112. A.B. Migdal, JETP (Sov. Phys.) **5**, 333 (1957)
113. J.M. Luttinger, Phys. Rev. **119**, 1153 (1960)
114. O. Benhar, A. Fabrocini and S. Fantoni, Phys. Rev. C **41**, R24 (1990)
115. O. Benhar, A. Fabrocini and S. Fantoni, Nucl. Phys. **A505**, 267 (1989)
116. P.O. Löwdin, J. Chem. Phys. **18**, 365 (1950)
117. H. Brandow and H. Baird, Rev. Mod. Phys. **39**, 771 (1967)
118. A. Fabrocini and S. Fantoni, Nucl. Phys. **A503**,375 (1989)
119. A. Fabrocini et al, Phys. Rev. B **33**, 6057 (1986)
120. A. Fabrocini et al, Phys. Rev. B **54**, 1035 (1996)
121. S. Fantoni and S. Rosati, Nuovo Cimento, A **20**, 179 (1974)
122. T. Morita, Progr. Theor. Phys. **20**, 920 (1958)
123. J.M. Van Leeuwen, J. Groeneveld and J. De Boer, Physica, **25**, 792 (1959)
124. S. Fantoni and S. Rosati, Phys. Lett. B **84**, 23 (1979)
125. T. Morita and K. Hiroike, Progr. Theor. Phys. **23**, 1003 (1960)
126. S. Fantoni, Nuovo Cimento, A **44**, 191 (1978)
127. G. Có et al, Nucl. Phys. **A549**, 439 (1992)
128. F. Arias de Saavedra et al, Nucl. Phys. **A605**, 359 (1996)
129. A. Fabrocini et al, Phys. Rev. C (1998) in press
130. R.B. Wiringa, V. Fiks and A. Fabrocini, Phys. Rev. C **38**, 1010 (1988)
131. I.E. Lagaris, Anales de Fisica, **81**, 39 (1985)
132. G.E. Brown, *Many Body Physics*, 1972 (North-Holland, Amsterdam)
133. A.L. Fetter and J.D. Walecka, *Quantum Theory of Many-Particle Systems*, 1971 (McGraw-Hill, NY)
134. P.K.A. DeWitt Huberts, J. Phys. G **16**, 507 (1990)
135. S. Frullani and J. Mougey, Adv. Nucl. Phys. **14**, 1 (1984)
136. O. Benhar, A. Fabrocini and S. Fantoni, Nucl. Phys. **A550**, 201 (1992)
137. B. Frois and I. Sick, editors *Modern Topics in Electron Scattering*, (1991) (World Scientific, Singapore)
138. S. Fantoni and V.R. Pandharipande, Nucl. Phys. **A473**, 234 (1987)
139. A.S. Rinat and W.H. Dickhoff, Phys. Rev. B **42**, 1004 (1990)
140. O. Benhar et al, Phys. Rev. C **44**, 2328 (1991)
141. S. Fantoni and I. Sick, *Electron–Nucleus Scattering*, edited by O. Benhar, A. Fabrocini and R. Schiavilla, 1994 (World Scientific, Singapore), 88–120

Inhomogeneous Quantum Liquids: Statics, Dynamics, and Thermodynamics

Eckhard Krotscheck

Institut für Theoretische Physik,
Johannes Kepler Universität, A-4040 Linz, Austria

Abstract. We review the variational theory for inhomogeneous Bose and Fermi fluids. Euler–Lagrange equations are derived for the one- and two-body correlations and systematic approximation methods are physically as well as formally motivated. For bosons, it is shown that the optimized variational method corresponds to a self-consistent approximate summation of ladder and ring diagrams. The connection between time-dependent Hartree theory and the equations of motion for strongly correlated Bose liquids is highlighted.

The finite-temperature extension of the boson theory is discussed. Drawing connections to the random phase approximation and linear response theory at finite temperatures, it is shown that present implementations of a finite-temperature version of variational theories are equivalent to a relatively simple quasiparticle description of the thermodynamics.

Building on the optimized ground states, we next develop the equations of motion method for the study of excitations and linear response in inhomogeneous Bose systems. For weakly interacting systems, the theory is equivalent to the Feynman theory of collective excitations in quantum liquids, or to the random phase approximation. When the interactions are strong, the theory amounts to a theory of effective interactions in which the strongly interacting system is mapped onto a weakly interacting one.

Finally, we turn to a systematic study of Fermi systems. It is shown how the choice of an optimal single-particle basis facilitates the formulation of Euler equations and improves the convergence of cluster expansion and integral equation methods. To interpret the Fermi theory, we develop a version of time-dependent Hartree Fock theory for strongly interacting systems. Unlike time-dependent Hartree theory, new effects, in particular the particle-hole continuum, enter the description of the excited states. We highlight the close relationship between diagrams of the FHNC variational theory and the random phase approximation, and how the time-dependent theory can serve to improve upon the FHNC description of the ground state.

1 Introduction

This lecture series is about the physics of inhomogeneous quantum liquids. The task of dealing with inhomogeneity in a microscopic system might at the first glance appear to be a purely technical one: Once a problem has been solved in a homogeneous geometry, the inhomogeneous and anisotropic case could mean just adding more coordinates, and perhaps spending more computer time.

However, the problem is not as simple:

First, inhomogeneous systems display vastly more complex behavior than homogeneous ones; this series of lectures can only scratch the surface of the variety of features and observations that are a direct consequence of the breaking of translational invariance. The simplest family of systems to study are quantum liquid surfaces and clusters. Then one may look at inhomogeneous electronic systems like metallic surfaces, quasi-two-dimensional conductor or semi-conductor layers or metallic clusters. One can study quantum crystals or electrons in a periodic lattice. Finally, once reliable quantitative methods for nuclear forces have been developed, one might be able to attack the problem of the "classical nucleus". Being able to deal with non-uniform quantum many-particle systems in a systematic and quantitative manner opens a therefore wide range of possibilities.

Second, much of physics is indeed "just" a technical problem. But physics is not only about writing down fundamental equations, but also explaining nature in terms of these fundamental equations. For that purpose, one must be able to solve these equations with today's resources. A part of theoretical physics is indeed the development of methods, and the main purpose of these lectures is indeed about *many-body theories.* However, we are not done with developing theories, it is just as important to apply these theories to describe nature, in other words to do *many-body physics.* The study of a physically rich case within a method will also always tell us about two things: about the nature of physical phenomena, and about our theoretical methods; it can highlight the strengths just as much as it can expose its weaknesses. Asking for a wider applicability of a given physical method will also put more stringent requirements on that method concerning both its consistency and its efficiency.

2 Basic Variational Theory

Among the available theoretical methods, variational approaches have today surfaced as the methods of choice for systems that are dominated by strong, short-range interactions. The power of variational methods has first been fully recognized and — within the computational possibilities available at that time — exploited by Feenberg and his students [1]; many of the techniques used today go back to this work. The essentials of the method have been introduced in Fantoni's lectures; let us therefore look at the technical problem directly from the variational point of view. Inhomogeneity in a many-body system may be caused by a Hamiltonian that contains a symmetry-breaking external potential $U_{\rm ext}(\mathbf{r})$, or by spontaneous symmetry breaking like due to the formation of droplets at low enough average particle densities. To cover all cases, let us assume a system of identical particles in an external field

$U_{\text{ext}}(\mathbf{r})$, interacting via the two-body interaction $v(|\mathbf{r}_i - \mathbf{r}_j|)$,

$$H = -\frac{\hbar^2}{2m}\sum_{i=1}^{N}\nabla_i^2 + \sum_{i=1}^{N} U_{\text{ext}}(\mathbf{r}_i) + \sum_{i<j}^{N} v(|\mathbf{r}_i - \mathbf{r}_j|) \ . \tag{2.1}$$

To incorporate correlation effects, one makes an "intelligent guess" for the wave function of the *Jastrow-Feenberg* form,

$$\Psi_0(1,\ldots,N) = F(\mathbf{r}_1,\ldots,\mathbf{r}_N)\Phi_0(1,\ldots,N) \ , \tag{2.2}$$

$$F(\mathbf{r}_1,\ldots,\mathbf{r}_N) = \exp\frac{1}{2}\left[\sum_{i}^{N} u_1(\mathbf{r}_i) + \sum_{i<j}^{N} u_2(\mathbf{r}_i,\mathbf{r}_j) + \ldots\right] \ ,$$

where the $u_n(\mathbf{r}_1,\ldots,\mathbf{r}_n)$ are n-body *correlation functions,* and $\Phi_0(1,\ldots,N)$ is the wave function of a model ground state. The decomposition of the Feenberg function into one-, two-, ..., n-body correlations is made unique by requiring that each of the $u_n(\mathbf{r}_1,\ldots\mathbf{r}_n)$ satisfies the *cluster property*, i.e. it vanishes whenever one or more of the particles is moved far away from the others. These functions and their interpretation have been introduced in Fantoni's lectures. We shall restrict ourselves here to the case $n = 2$ which is sufficient to derive the correct structure of the theory to be developed. Higher-order correlations improve the quantitative agreement with experiments and can in that sense provide a better understanding of the physics of a many-body system. Especially triplet correlations introduce some new physics into the study of the density-dependence of the roton minimum in ^{4}He, and are relevant in multi-particle scattering processes; they are therefore routinely included in applications. Not much can otherwise be learned from these complications; the modification of the equations that arise from the inclusion of triplet correlations and "elementary diagrams" will be indicated where appropriate. The situation is different for Fermi systems, where the naïve, topological definition of "elementary" diagrams leads to classifications of cluster contributions that violate the Pauli principle [2] and are not sympathetic with the optimization problem [3].

It is indicative for the methodological progress of the past years that perturbative techniques, which are based on Green's functions and Feynman diagrams and have dominated the many-body problem for many years, are hardly mentioned at this school. The physical reason is that one can prove [4] that the *minimum set*[1] of Feynman diagrams needed to get the physics of a many-body system right is the class of parquet-diagrams [8]. This set of Feynman diagrams cannot be summed without further approximations; hence

[1] A little over two years ago, one would have had to state that there is no many-body system on earth whose interactions are weak enough that a perturbative treatment is appropriate. Dilute atomic gases that can be trapped by magnetic fields and show Bose–Einstein condensation [5, 6, 7] are an exception to the rule.

the method is simply impractical for the description of ground state properties of terrestrial quantum fluids. Both "semi-analytic" many-body methods discussed here, Coupled Cluster and optimized (Fermi)– Hypernetted–Chain theory, provide exactly what is missing: justifiable and systematic approximation methods for sets of Feynman diagrams that cannot be calculated exactly.

To motivate some of the further course of these lecture notes as well as to preview some of the results of the sections on ground state theory, let us elaborate briefly on the concepts of consistency and practicality outlined above:

- (i) A satisfactory theory for a quantum system should be able to *decide* whether at zero temperature a given number N of interacting particles within a certain volume V is in a homogeneous or in an inhomogeneous state, in particular it should not permit the calculation of systems that are unstable against infinitesimal perturbations.
- (ii) The *physical* consistency requirement spelled out in item (i) was translated into the language of diagrammatic perturbation theory [4] and has led to the conclusion that the "parquet" set of Feynman diagrams is the minimum requirement. Along the same lines it was proven that the variational/integral equation methods to be presented here provide, already in their simplest version, an exact summation of all rings and ladders, as well as an approximate summation of all other diagrams of the parquet set.[2].
- (iii) To achieve the necessary *practicality* it will be shown that variational/integral equation theories for *homogeneous* systems require very small computational effort. They therefore provide indeed the possibility to handle the additional complications caused by symmetry breaking.

The ground state problem of an inhomogeneous quantum many-body system is just one task to be solved. The next problem — more demanding but also closer to experimental reality — is how one probes these systems. For that, we need to look at elementary excitations and scattering processes. A carefully designed ground state theory is essential: calculating excited states on the background of a poor ground-state will inevitably give rise to spurious instabilities. Keeping this in mind, we will develop the method of time-dependent correlation functions as a general tool. This is partly new territory; only the simplest implementations of that method are the well established Feynman or random phase approximation (RPA). The theory is quite well developed for Bose systems, but still offers challenges for excitations in interacting Fermi systems.

Since the scene has been set in Fantoni's lectures, let us jump right into the middle of the action and outline the recurring issues of this lecture series.

[2] The statement will need some qualification for fermions, we shall elaborate on this point in Section 4

With the wave function (2.2) given, we define as usual the n-body densities

$$\rho_n(\mathbf{r}_1,\ldots,\mathbf{r}_n) = \frac{N!}{(N-n)!}\frac{\int d^3r_{n+1}\ldots d^3r_N\,|\Psi_0(\mathbf{r}_1,\ldots,\mathbf{r}_N)|^2}{\int d^3r_1\ldots d^3r_N\,|\Psi_0(\mathbf{r}_1,\ldots,\mathbf{r}_N)|^2}\,, \tag{2.3}$$

as well as the two-body distribution function

$$g(\mathbf{r}_1,\mathbf{r}_2) = \frac{\rho_2(\mathbf{r}_1,\mathbf{r}_2)}{\rho_1(\mathbf{r}_1)\rho_1(\mathbf{r}_2)}\ . \tag{2.4}$$

The primary quantity of interest is then the energy expectation value of the Hamiltonian (2.1) for the wave function (2.2), which is generally a functional of the model wave function $\Phi_0(1,\ldots,N)$ and the correlations $u_n(\mathbf{r}_1,\ldots\mathbf{r}_n)$,

$$E_{\mathrm{var}}[u_1,u_2,\ldots] = \frac{\int d^3r_1\ldots d^3r_N\Psi_0^*(\mathbf{r}_1,\ldots,\mathbf{r}_N)H\Psi_0(\mathbf{r}_1,\ldots,\mathbf{r}_N)}{\int d^3r_1\ldots d^3r_N\,|\Psi_0(\mathbf{r}_1,\ldots,\mathbf{r}_N)|^2}\ . \tag{2.5}$$

Once we have the energy as a functional of the correlations $u_n(\mathbf{r}_1,\ldots\mathbf{r}_n)$ (or any other set of functions that is has a unique, one-to-one relationship to these correlation functions) we should optimize the correlations by minimizing the energy within the allowed function space,

$$\frac{\delta E_{\mathrm{var}}[u_1,u_2,\ldots]}{\delta u_n(\mathbf{r}_1,\ldots\mathbf{r}_n)} = 0\ . \tag{2.6}$$

The optimization (2.6) is, as we shall see, the key to both the efficiency and consistency of the theory. Moreover, the *requirement* that the theories to be developed should be optimizable will be a strong selective criterion for acceptable approximations. Finally, only the optimization will provide a vehicle to uniquely determine the ingredients of our theory and thereby permit the comparison with other theories that might use different auxiliary quantities but should eventually arrive at the same answer for physical observables.

2.1 Hartree and Hartree–Fock Theory

To motivate some of our notations and procedures used later, let us briefly discuss weakly interacting systems.

In a weakly interacting system, perturbation theory is assumed to be valid[3]. Perturbation theory starts with an "unperturbed" Hamiltonian

$$H_0 = \sum_{i=1}^{N}\left[-\frac{\hbar^2}{2m}\nabla_i^2 + U_{\mathrm{ext}}(\mathbf{r}_i)\right] = \sum_{i=1}^{N} h_0(i) \tag{2.7}$$

[3] This statement should be taken with a grain of salt because one can demonstrate [9] that the energy per particle has a branchcut at zero interaction strength. In other words it is a non-analytic function of the interaction strength around zero and can, hence, *never* be expanded in a power series in the interaction strength.

and considers the interaction as a perturbation. The unperturbed wave function of the ground state is the (anti-)symmetrized product of the single-particle wave functions $\phi_i(\mathbf{r})$ determined by the one-body equation

$$h_0\phi_i(\mathbf{r}) = \epsilon_i\phi_i(\mathbf{r}) \ . \tag{2.8}$$

The ground state of bosons is simply the product of the lowest eigenfunctions of h_0,

$$\Phi_0^{\mathrm{H}}(\mathbf{r}_1,\ldots,\mathbf{r}_N) = \prod_i \phi_0(\mathbf{r}_i) \ . \tag{2.9}$$

This Hartree wave function is readily identified with a simple version of our variational wave function (2.2) obtained by setting $u_n(\mathbf{r}_1,\ldots,\mathbf{r}_n) = 0$ for $n \geq 2$ and identifying $\phi_0(\mathbf{r}) = \exp\frac{1}{2}u_1(\mathbf{r})$. The ground state density $\rho_1(\mathbf{r})$ corresponding to a wave function (2.9) is

$$\rho_1^{\mathrm{H}}(\mathbf{r}) = N\,|\phi_0(\mathbf{r})|^2 \ , \tag{2.10}$$

and the energy is, in first order perturbation theory[4]

$$\begin{aligned} E_{\mathrm{H}} &= N\epsilon_0 + \frac{N(N-1)}{2}\int d^3r d^3r' |\phi_0(\mathbf{r})|^2|\phi_0(\mathbf{r}')|^2 v(|\mathbf{r}-\mathbf{r}'|) \\ &= \frac{\hbar^2}{2m}\int d^3r \left|\nabla\sqrt{\rho_1(\mathbf{r})}\right|^2 + \int d^3r \rho_1(\mathbf{r})U_{\mathrm{ext}}(\mathbf{r}) \\ &\quad + \frac{1}{2}\int d^3r d^3r' \rho_1(\mathbf{r})\rho_1(\mathbf{r}')v(|\mathbf{r}-\mathbf{r}'|) \ . \end{aligned} \tag{2.11}$$

One can do better than first order perturbation theory: The energy is the expectation value of the Hamiltonian (2.1) with respect to the "Hartree" wave function (2.9). As such, it is an upper bound for the ground state energy, no matter which one-body wave function is used. We can therefore construct a better ground state by minimizing the energy (2.11) with respect to the single-particle functions $\phi_0(\mathbf{r})$ or, which is equivalent in this case, with respect to the density $\rho_1(\mathbf{r})$. Keeping a fixed particle number

$$N = \int d^3r \rho_1(\mathbf{r}) \tag{2.12}$$

as constraint, we find easily

$$\frac{1}{2}\frac{\delta(E_{\mathrm{H}} - \mu N)}{\delta\sqrt{\rho_1(\mathbf{r})}} = \left[-\frac{\hbar^2}{2m}\nabla^2 + U_{\mathrm{ext}}(\mathbf{r}) + V_{\mathrm{H}}(\mathbf{r}) - \mu\right]\sqrt{\rho_1(\mathbf{r})} = 0 \tag{2.13}$$

where

$$V_{\mathrm{H}}(\mathbf{r}) = \int d^3r' \rho_1(\mathbf{r}')v(|\mathbf{r}-\mathbf{r}'|) \tag{2.14}$$

is a self-consistent "Hartree" potential.

[4] We always assume large particle numbers N.

The fermion case is similar, but the energy does not allow for such a straightforward formulation in terms of the one-body density. The wave function is the Slater determinant of single-particle orbitals,

$$\Phi_0^{\mathrm{HF}} = \frac{1}{\sqrt{N!}} \det\{\phi_j(\mathbf{r}_i)\chi(i)\} \tag{2.15}$$

where the $\chi(i)$ are spin- (and possibly isospin) eigenfunctions. For the calculation of the energy we also need the off-diagonal one-body density matrix corresponding to this state

$$\rho_1^{\mathrm{F}}(\mathbf{r},\mathbf{r}') = \sum_i n(i)\phi_i^*(\mathbf{r})\phi_i(\mathbf{r}') \ , \tag{2.16}$$

where the $n(i)$ are the fermion occupation numbers,

$$n(i) = \begin{cases} 1 & \text{for} \quad \epsilon_i < E_{\mathrm{F}}, \\ 0 & \text{for} \quad \epsilon_i > E_{\mathrm{F}}, \end{cases} \tag{2.17}$$

and E_{F} is the Fermi energy. The physical one-body density is the diagonal value of the density matrix,

$$\rho_1^{\mathrm{HF}}(\mathbf{r}) = \rho_1^{\mathrm{F}}(\mathbf{r},\mathbf{r}) = \sum_i n(i)\left|\phi_i(\mathbf{r})\right|^2 \ . \tag{2.18}$$

The total energy is then

$$\begin{aligned} E_{\mathrm{HF}} = & \frac{\hbar^2}{2m}\sum_i n(i)\int d^3r\left|\nabla\phi_i(\mathbf{r})\right|^2 + \int d^3r U_{\mathrm{ext}}(\mathbf{r})\rho_1(\mathbf{r}) \\ & +\frac{1}{2}\int d^3r d^3r' v(|\mathbf{r}-\mathbf{r}'|)\left[\rho_1(\mathbf{r})\rho_1(\mathbf{r}') - \frac{1}{\nu}\left|\rho_1^{\mathrm{F}}(\mathbf{r},\mathbf{r}')\right|^2\right] \end{aligned} \tag{2.19}$$

where ν is the spin-degeneracy of the single particle orbitals. This expression is now used to optimize the (occupied states) by minimizing the energy

$$\frac{\delta E_{\mathrm{HF}}}{\delta\phi_i(\mathbf{r})} = 0 \ , \tag{2.20}$$

which leads to the famous Hartree–Fock equation

$$\left[-\frac{\hbar^2}{2m}\nabla^2 + U_{\mathrm{ext}}(\mathbf{r}) + V_{\mathrm{H}}(\mathbf{r})\right]\phi_i(\mathbf{r}) - \int d^3r' v_{\mathrm{F}}(\mathbf{r},\mathbf{r}')\phi_i(\mathbf{r}') = \epsilon_i\phi_i(\mathbf{r}) \ , \tag{2.21}$$

where the $V_{\mathrm{H}}(\mathbf{r})$ is given in (2.14), and the antisymmetry contributes the non-local "Fock term"

$$v_{\mathrm{F}}(\mathbf{r},\mathbf{r}') = \frac{1}{\nu}\rho_1^{\mathrm{F}}(\mathbf{r},\mathbf{r}')v(|\mathbf{r}-\mathbf{r}'|) \ . \tag{2.22}$$

The most obvious problem of both the Hartree and the Hartree–Fock theory is that these approximations are inapplicable for any interaction that has a strong, short-range component. Further scrutiny of the "Hartree theory" reveals more physical problems, with the consequence that theories based on a single-particle picture are mostly usable only in conjunction with phenomenological, effective interactions. From the *microscopic* point of view, these theories are nevertheless quite useful because, as we shall see, microscopic theories can be interpreted, to a large extent, as methods providing meaningful effective interactions for the Hartree or Hartree–Fock theory.

2.2 Density Functional Theory

The most popular theory for the study of electrons in metals and metal surfaces is the density-functional theory [10]. Much excellent literature on density functional theory of electronic systems exists and it is not the purpose of this lecture series to add to that list; the reader is directed to, for example, the review article by Kohn and Vashishta [11] for details.

Central to density functional theory is the Kohn–Hohenberg theorem [10, 12] stating that the ground-state energy of an inhomogeneous many-body system is a unique function of its density $E = E[\rho_1(\mathbf{r})]$, and that the ground state density is the one that minimizes the total energy. This energy functional $E[\rho_1(\mathbf{r})]$ is, of course, generally not known. Of great practical value is the local density approximation (LDA) which assumes that the correlation energy (we will later be more precise with the definition of this correlation energy) can be written as

$$E_c[\rho_1(\mathbf{r})] = \int d^3r \epsilon_c(\rho_1(\mathbf{r})) \ , \tag{2.23}$$

where the correlation energy density $\epsilon_c(\rho_1(\mathbf{r}))$ is assumed to be that of the homogeneous system at the density $\rho_1(\mathbf{r})$. For fermions, the total density is then written in the form (2.18) in terms of single particle orbitals which are, in turn, obtained from the variational principle

$$\frac{\delta E}{\delta \phi_i}(\mathbf{r}) = 0 \ . \tag{2.24}$$

Again, a legitimate interpretation of microscopic theories of the kind presented here is that they provide information on the energy functional. Unfortunately, functionals that are derivable from microscopic theories are more complicated than one would like them to be for "black-box" applications.

2.3 Time-Dependent Hartree Theory

Even in a weakly interacting system, the ground state calculation is only the beginning of the story. To motivate some of the more advanced methods to be discussed later, and also to highlight the difference between the "weakly interacting" and "strongly interacting" theories, let us discuss how to deal with excitations. We formulate the theory as a linear response theory, where the system is exposed to a weak, time-dependent external potential $\delta U_{\rm ext}(\mathbf{r};t)$. One must stay within the single-particle picture for a simple reason: Since the energy has been minimized within a restricted space of wave functions only, every attempt to go beyond this restricted space has the potential of leading to spurious unstable excitations, simply telling us that the Hartree(–Fock) wave function does not minimize the energy within a larger function space.

Let us start with a weakly interacting Bose system; the case of a Fermi system will be treated later. The perturbed Hamiltonian is

$$H(t) = -\frac{\hbar^2}{2m}\sum_{i=1}^{N}\nabla_i^2 + \sum_{i=1}^{N} U_{\rm ext}(\mathbf{r}_i) + \sum_{i<j}^{N} v(|\mathbf{r}_i - \mathbf{r}_j|) + \sum_{i=1}^{N}\delta U_{\rm ext}(\mathbf{r}_i;t) \ . \quad (2.25)$$

We postulate that – for bosons – the wave function is a product of single particle functions (2.9) satisfying the Hartree equation (2.13). To describe excited states within this framework, we can make these one-body orbitals time-dependent and write the wave function of the perturbed state as

$$\begin{aligned} \left|\Psi_{\rm X}^{\rm H}(t)\right\rangle &= \exp\left[-\frac{\rm i}{\hbar}E_H t\right]\left|\Psi_0^{\rm H}(t)\right\rangle \ , \\ \left|\Psi_0^{\rm H}(t)\right\rangle &= \frac{\exp\frac{1}{2}\delta U(t)\left|\Psi_0^{\rm H}\right\rangle}{\left\langle\Psi_0^{\rm H} \mid \exp\Re e\,\delta U(t) \mid \Psi_0^{\rm H}\right\rangle^{1/2}} \ , \\ \delta U(t) &= \sum_i \delta u_1(\mathbf{r}_i;t) \ . \end{aligned} \quad (2.26)$$

In analogy to defining the model ground state by a variational principle, we determine the *perturbed* system by a stationarity principle [13, 14]. Define an action

$$\begin{aligned} \mathcal{S}\left[\delta u_1, \delta u_1^*\right] &= \int dt \mathcal{L}(t) \ , \\ \mathcal{L}(t) &= \left\langle \Psi_X^{\rm H}(t) \mid H(t) - {\rm i}\hbar\frac{\partial}{\partial t} \mid \Psi_X^{\rm H}(t)\right\rangle \\ &= \left\langle \Psi_0^{\rm H}(t) \mid H(t) - E_{\rm H} - {\rm i}\hbar\frac{\partial}{\partial t} \mid \Psi_0^{\rm H}(t)\right\rangle \end{aligned} \quad (2.27)$$

and require that the action integral be stationary under the variations $\delta u_1(\mathbf{r}_i;t)$,

$$\delta\mathcal{S}\left[\delta u_1(\mathbf{r};t), \delta u_1^*(\mathbf{r};t)\right] = 0 \ . \quad (2.28)$$

It is for this case very easy to work out the action integral $\mathcal{S}$, we only need to allow for time-dependent densities and calculate the time-derivative term to second order in $\delta\dot{u}_1(\mathbf{r};t)$:

$${\rm i}\hbar\frac{\partial}{\partial t}\left|\Psi_0^{\rm H}(t)\right\rangle = \frac{{\rm i}\hbar}{2}\delta\dot{U}(t)\left|\Psi_0^{\rm H}(t)\right\rangle - \frac{{\rm i}\hbar}{2}\left|\Psi_0^{\rm H}(t)\right\rangle\left\langle\Psi_0^{\rm H}(t) \mid \Re e\,\delta\dot{U}(t) \mid \Psi_0^{\rm H}(t)\right\rangle$$

and

$$\mathrm{i}\hbar\left\langle \Psi_0^{\mathrm{H}}(t)\left|\frac{\partial}{\partial t}\right|\Psi_0^{\mathrm{H}}(t)\right\rangle = \frac{\mathrm{i}\hbar}{4}\int d^3r\rho_1(\mathbf{r};t)\left[\delta\dot{u}_1(\mathbf{r};t)-\delta\dot{u}_1^*(\mathbf{r};t)\right] \ . \tag{2.29}$$

For the remaining parts of the Lagrangian, we simply replace the ground state density by the time-dependent density and add the perturbing external potential. Finally, we must expand the action integral to second order in the fluctuations. There are no first order terms since the energy is stationary by construction of the unperturbed wave function. The calculation is straightforward and leads to

$$\begin{aligned}\mathcal{L}(t) &= \frac{\hbar^2}{8m}\int d^3r\rho_1(\mathbf{r})\left|\nabla\delta u_1(\mathbf{r};t)\right|^2\\ &+\frac{1}{2}\int d^3rd^3r'\rho_1(\mathbf{r})\rho_1(\mathbf{r}')v(|\mathbf{r}-\mathbf{r}'|)\Re e\,\delta u_1(\mathbf{r};t)\Re e\,\delta u_1(\mathbf{r}';t)\\ &+\int d^3r\rho_1(\mathbf{r})\Re e\,\delta u_1(\mathbf{r};t)\delta U_{\mathrm{ext}}(\mathbf{r};t)-\frac{\mathrm{i}\hbar}{4}\int d^3r\rho_1(\mathbf{r})\delta u_1^*(\mathbf{r};t)\delta\dot{u}_1(\mathbf{r};t) \ .\end{aligned} \tag{2.30}$$

The corresponding equations of motion are obtained by variation with respect to $\delta u_1(\mathbf{r};t)$ and $\delta u_1^*(\mathbf{r};t)$, they are

$$\begin{aligned}\frac{4\delta S}{\delta u_1^*(\mathbf{r})} &= -\frac{\hbar^2}{2m}\nabla\rho_1(\mathbf{r})\nabla\delta u_1(\mathbf{r};t)+2\int d^3r'\rho_1(\mathbf{r})v(|\mathbf{r}-\mathbf{r}'|)\rho_1(\mathbf{r}')\Re e\,\delta u_1(\mathbf{r}';t)\\ &\quad+\rho_1(\mathbf{r})\left[2\delta U_{\mathrm{ext}}(\mathbf{r};t)-\mathrm{i}\hbar\delta\dot{u}_1(\mathbf{r};t)\right]=0 \ ,\end{aligned} \tag{2.31}$$

$$\begin{aligned}\frac{4\delta S}{\delta u_1(\mathbf{r})} &= -\frac{\hbar^2}{2m}\nabla\rho_1(\mathbf{r})\nabla\delta u_1^*(\mathbf{r};t)+2\int d^3r'\rho_1(\mathbf{r})v(|\mathbf{r}-\mathbf{r}'|)\rho_1(\mathbf{r}')\Re e\,\delta u_1(\mathbf{r}';t)\\ &\quad+\rho_1(\mathbf{r})\left[2\delta U_{\mathrm{ext}}(\mathbf{r};t)+\mathrm{i}\hbar\delta\dot{u}_1^*(\mathbf{r};t)\right]=0 \ .\end{aligned} \tag{2.32}$$

In view of further developments, it is useful to introduce a few abbreviations. A slightly more compact formulation can be reached by introducing

$$H_1(\mathbf{r}) = -\frac{\hbar^2}{2m}\frac{1}{\sqrt{\rho_1(\mathbf{r})}}\nabla\cdot\left[\rho_1(\mathbf{r})\nabla\frac{1}{\sqrt{\rho_1(\mathbf{r})}}\right] = \frac{\hbar^2}{2m}\left[-\nabla^2+\frac{\nabla^2\sqrt{\rho_1(\mathbf{r})}}{\sqrt{\rho_1(\mathbf{r})}}\right] \ . \tag{2.33}$$

This is the one body operator whose ground state is the $\sqrt{\rho_1(\mathbf{r})}$ and whose ground-state energy is zero. Evidently, $H_1(\mathbf{r})$ is positive-semidefinite.

Also, introduce the "tilde notation"

$$\tilde{A}(\mathbf{r}_1,\mathbf{r}_2) \equiv \sqrt{\rho_1(\mathbf{r}_1)}A(\mathbf{r}_1,\mathbf{r}_2)\sqrt{\rho_1(\mathbf{r}_2)} \tag{2.34}$$

and write convolution integrals in the abbreviated form

$$[A*B](\mathbf{r}_1,\mathbf{r}_2) \equiv \int d^3r_3\rho_1(\mathbf{r}_3)A(\mathbf{r}_1,\mathbf{r}_3)B(\mathbf{r}_3,\mathbf{r}_2) \ , \tag{2.35}$$

$$[\tilde{A} * \tilde{B}](\mathbf{r}_1, \mathbf{r}_2) \equiv \int d^3r_3 \tilde{A}(\mathbf{r}_1, \mathbf{r}_3)\tilde{B}(\mathbf{r}_3, \mathbf{r}_2) \ . \tag{2.36}$$

Functions of one variable are interpreted as diagonal, i.e. $u_1(\mathbf{r}) \equiv u_1(\mathbf{r})\delta(\mathbf{r} - \mathbf{r}')$; the "tilde" notation and the convolution product are expressed in an obvious manner, e.g.

$$\tilde{u}_1(\mathbf{r}) \equiv \sqrt{\rho_1(\mathbf{r})}u_1(\mathbf{r}) \tag{2.37}$$

and

$$[A * u_1](\mathbf{r}) \equiv \int d^3r' A(\mathbf{r}, \mathbf{r}')\rho_1(\mathbf{r}')u_1(\mathbf{r}') \tag{2.38}$$

or

$$[\tilde{A} * \tilde{u}_1](\mathbf{r}) \equiv \int d^3r' \tilde{A}(\mathbf{r}, \mathbf{r}')\tilde{u}_1(\mathbf{r}') \ . \tag{2.39}$$

The unit operator and the inverse will generally be interpreted in the sense of the convolution product (2.36). Then we can write the equations of motion (2.31), (2.32) in a "supermatrix" form

$$\begin{pmatrix} H_1 + \tilde{v} & \tilde{v} \\ \tilde{v} & H_1 + \tilde{v} \end{pmatrix} \begin{pmatrix} \delta\tilde{u}_1 \\ \delta\tilde{u}_1^* \end{pmatrix} + 2 \begin{pmatrix} \tilde{U}_{\text{ext}} \\ \tilde{U}_{\text{ext}} \end{pmatrix} = \mathrm{i}\hbar \frac{\partial}{\partial t} \begin{pmatrix} \delta\tilde{u}_1 \\ -\delta\tilde{u}_1^* \end{pmatrix} \ , \tag{2.40}$$

where the "matrix product" is defined in the sense of the convolution product. The "supermatrix" form of the time-dependent Hartree equation is, at this point, of course a luxury, it has been introduced at this early stage in order to make later results for both interacting bosons and fermions recognizable.

We are interested in the real density fluctuation

$$\delta\rho_1(\mathbf{r}; t) = \rho_1(\mathbf{r})\Re e\, \delta u_1(\mathbf{r}; t) \tag{2.41}$$

or

$$\delta\sqrt{\rho_1(\mathbf{r}; t)} = \frac{1}{2}\sqrt{\rho_1(\mathbf{r})}\Re e\, \delta u_1(\mathbf{r}; t) = \frac{1}{2}\Re e\, \delta\tilde{u}_1(\mathbf{r}; t) \tag{2.42}$$

induced by the external potential $\delta U(\mathbf{r}; t)$. By adding and subtracting the equations of motion and assuming harmonic time dependence, we find finally

$$\begin{aligned} \left[H_1^2(\mathbf{r}) - \hbar^2\omega^2\right] \delta\sqrt{\rho_1(\mathbf{r}; \omega)} + 2H_1(\mathbf{r}) \int d^3r' \tilde{v}(\mathbf{r}, \mathbf{r}')\delta\sqrt{\rho_1(\mathbf{r}'; \omega)} \\ = -H_1(\mathbf{r})\delta U_{\text{ext}}(\mathbf{r}; \omega)\sqrt{\rho_1(\mathbf{r})} \ . \end{aligned} \tag{2.43}$$

With this, we have accomplished two things: We have identified the density-density response function which we write in a formal operator form as

$$\chi(\mathbf{r}, \mathbf{r}'; \omega) = 2\sqrt{\rho_1(\mathbf{r})} \left[\hbar^2\omega^2 - H_1^2 - 2H_1\tilde{v} + \mathrm{i}\eta\right]^{-1} (\mathbf{r}, \mathbf{r}')H_1(\mathbf{r}')\sqrt{\rho_1(\mathbf{r}')} \ . \tag{2.44}$$

(A small imaginary part in the energy denominator ensures causality). We also can identify the normal modes of the system as the zeroes of the operator $H_1^2 + 2H_1\tilde{V} - \hbar^2\omega^2$, in other words the normal mode frequencies are the eigenvalues of

$$\int d^3r' H_1(\mathbf{r}) \left[H_1(\mathbf{r})\delta(\mathbf{r}-\mathbf{r}') + 2\tilde{v}(\mathbf{r},\mathbf{r}')\right] \phi_\epsilon(\mathbf{r}') = \epsilon^2 \phi_\epsilon(\mathbf{r}) \ . \tag{2.45}$$

This is a generalized eigenvalue problem, familiar from the theory of small oscillations in classical mechanics. A condition for the existence of real eigenvalues is that at least one of the operators, $H_1(\mathbf{r})$ or $H_1(\mathbf{r})\delta(\mathbf{r}-\mathbf{r}')+2\tilde{v}(\mathbf{r},\mathbf{r}')$ is positive (semi-)definite. Both operators must be positive (semi-)definite to have non-negative eigenvalues ϵ^2. Whereas $H_1(\mathbf{r})$ is positive semi-definite by construction, the requirement of positivity of the second operator is a non-trivial condition on the interaction. We will encounter the above equation again when we study strongly interaction systems, all that will change is that the bare interaction $v(|\mathbf{r}-\mathbf{r}'|)$ is replaced by an effective interaction.

The result (2.44) may be cast into an more familiar form by defining a "non-interacting" response function

$$\chi_0(\mathbf{r},\mathbf{r}';\omega) = 2\sqrt{\rho_1(\mathbf{r})}\left[\left[\hbar^2\omega^2 - H_1^2 + i\eta\right]^{-1} * H_1\right](\mathbf{r},\mathbf{r}')\sqrt{\rho_1(\mathbf{r}')}. \tag{2.46}$$

and noting that (2.44) may also be written in the form

$$\chi^{-1}(\mathbf{r},\mathbf{r}';\omega) = \chi_0^{-1}(\mathbf{r},\mathbf{r}';\omega) - v(|\mathbf{r}-\mathbf{r}'|) \ . \tag{2.47}$$

Up to this point, the formulation of the theory might look complicated, to simplify the results let us therefore take the limit of the translationally invariant system. In this case, we can go to momentum space, the one body operator $H_1(\mathbf{r})$ turns into the kinetic energy

$$H_1(\mathbf{r}) \quad \rightarrow \quad t(q) = \frac{\hbar^2 q^2}{2m} \ , \tag{2.48}$$

and we obtain

$$\chi(q;\omega) = \frac{2\rho t(q)}{\hbar^2\omega^2 - t^2(q) - 2t(q)\tilde{v}(q) + i\eta} \ . \tag{2.49}$$

The response function (2.49) can also be written in a familiar RPA form by introducing the response function of the non-interacting system

$$\chi_0(q,\omega) = \frac{\rho}{\hbar\omega - t(q) + i\eta} - \frac{\rho}{\hbar\omega + t(q) + i\eta} = \frac{2\rho t(q)}{\hbar^2\omega^2 - t^2(q) + i\eta} \ . \tag{2.50}$$

The full response function is then obtained in the usual manner

$$\chi(q,\omega) = \frac{\chi_0(q,\omega)}{1 - \chi_0(q,\omega)v(q)} \ , \tag{2.51}$$

which is identical to (2.49). The excitation energies are the poles of the response function

$$\epsilon(q) = \sqrt{t^2(q) + 2t(q)\tilde{v}(q)} \ . \tag{2.52}$$

From the response function, we also get the static structure function

$$S(q) = -\int \frac{d(\hbar\omega)}{\rho\pi} \Im m\, \chi(q;\omega) = \frac{t(q)}{\epsilon(q)} = \frac{t(q)}{\sqrt{t^2(q) + 2t(q)\tilde{v}(q)}} , \tag{2.53}$$

which is the famous Feynman relation [15].

We have shown here how excitations are treated within time-dependent Hartree theory. But we see that the Hartree theory is in a sense inconsistent since we could now calculate a new, non-trivial pair distribution function and a new energy, using coupling constant integration. This energy would not be equal to the Hartree energy.

2.4 Pair Correlations and the Optimization Problem

We have so far considered only model systems, and the techniques developed were useful perhaps in connection with phenomenological effective interactions and to get some qualitative features, but inadequate for a realistic, strongly interacting many-body system. We will start now with the theory for the simplest *strongly* correlated many body system, namely a uniform liquid of bosons, interacting through a central potential, at zero-temperature. To be definite, we may think of ^{4}He. Several important things can be learned from carefully analyzing the bulk liquid; we shall therefore review this problem for the purpose of having a back-reference to simpler equations when we look at the more complicated inhomogeneous equations. Also, much of the motivation for why one formulates the inhomogeneous problem in the way presented here is drawn from the experience with the homogeneous geometry.

Start from the wave function (2.2) with only $u_2(\mathbf{r}_i, \mathbf{r}_j) \neq 0$, and, as appropriate for bosons, $\Phi_0(1, \ldots, N) = 1$. If one decides to use Monte Carlo techniques, the game is essentially over and all one can do is to evaluate the expectation values of interest as accurately as possible or affordable. Of course, exact analytic features of the mathematical problem of evaluating and optimizing the ground state prevail independently of how the energies are calculated at the end, hence analytic and semi-analytic results have their feedback on simulation techniques.

When integral equations are used, it is best to perform the functional optimization (2.6) of the correlations by minimizing the energy expectation value (2.5). Care must be exercised in the choice of an approximation method because there is no guarantee that the exact solution of an approximate variational problem will also be an approximate solution of the exact variational problem. The safest way to proceed therefore is to first formulate the exact variational problem, and then see which approximations are legitimate.

Let us start by calculating the expectation value for the Hamiltonian (2.1) with $U_{\rm ext}(\mathbf{r}) = 0$ for our special case

$$E_{\rm var}[u_2] = \frac{\int d^3r_1 \ldots d^3r_N \Psi_0^*(\mathbf{r}_1, \ldots, \mathbf{r}_N) H \Psi_0(\mathbf{r}_1, \ldots, \mathbf{r}_N)}{\int d^3r_1 \ldots d^3r_N \left|\Psi_0(\mathbf{r}_1, \ldots, \mathbf{r}_N)\right|^2} . \tag{2.54}$$

To simplify the kinetic energy, use the Jackson–Feenberg identity

$$F\nabla^2 F = \frac{1}{2}(F^2\nabla^2 + \nabla^2 F^2) + \frac{1}{2}F^2[\boldsymbol{\nabla},[\boldsymbol{\nabla},\ln F]] - \frac{1}{4}[\boldsymbol{\nabla},[\boldsymbol{\nabla},F^2]] \quad (2.55)$$

to rewrite the energy as

$$\begin{aligned} E_{\mathrm{var}}[u_2] &= \frac{\int d^3r_1 \dots d^3r_N \, |\Psi_0(\mathbf{r}_1,\dots,\mathbf{r}_N)|^2 \sum_{i<j} v_{\mathrm{JF}}(|\mathbf{r}_i - \mathbf{r}_j|)}{\int d^3r_1 \dots d^3r_N \, |\Psi_0(\mathbf{r}_1,\dots,\mathbf{r}_N)|^2} \\ &= N\frac{\rho}{2}\int d^3r \left[v(r) - \frac{\hbar^2}{4m}\nabla^2 u_2(r)\right] g(r) \equiv N\frac{\rho}{2}\int d^3r v_{\mathrm{JF}}(r) g(r) \quad (2.56) \end{aligned}$$

where $v_{\mathrm{JF}}(r)$ is the "Jackson–Feenberg effective interaction". To arrive at workable formulas for the optimization problem is to use now the HNC equations

$$g(r) = \mathrm{e}^{[u_2(r)+N(r)+E(r)]} \;, \quad (2.57)$$

$$\tilde{N}(k) = \frac{(S(k)-1)^2}{S(k)} \;, \quad (2.58)$$

where

$$S(k) = 1 + \rho \int d^3r [g(r) - 1] \mathrm{e}^{i\mathbf{k}\cdot\mathbf{r}} \quad (2.59)$$

is the static structure function (see above), and we have defined the "dimensionless" Fourier transform of an arbitrary function $f(r)$ as

$$\tilde{f}(k) = \rho \int d^3r f(r) \mathrm{e}^{i\mathbf{k}\cdot\mathbf{r}} \;. \quad (2.60)$$

$N(r)$ is the set of "nodal", and $E(r)$ the set of "elementary" diagrams; further auxiliary quantities that we will need are

$$h(r) = g(r) - 1 \quad (2.61)$$

and the "non-nodal" function

$$X(r) = h(r) - N(r) \quad (2.62)$$

as well as the relationships following from the above definitions

$$\begin{aligned} S(k) &= \frac{1}{1-\tilde{X}(k)} = 1 + \tilde{h}(k) \\ \tilde{N}(k) &= \frac{\tilde{X}^2(k)}{1-\tilde{X}(k)} = \frac{\tilde{h}^2(k)}{1+\tilde{h}(k)} \;. \end{aligned} \quad (2.63)$$

The HNC equation (2.57) is used to eliminate the pair correlation functions from the energy (2.56),

$$\begin{aligned} g(r)\nabla^2 u_2(r) &= g(r)\nabla^2 \left[\ln g(r) - N(r) - E(r)\right] \\ &= -4\left|\nabla\sqrt{g(r)}\right|^2 + \nabla^2 g(r) - g(r)\nabla^2\left[N(r) + E(r)\right] \;. \end{aligned} \quad (2.64)$$

This leads to an expression for the energy per particle

$$\frac{E_{\rm var}[g]}{N} = \frac{E_{\rm R}}{N} + \frac{E_{\rm Q}}{N} + \frac{E_{\rm e}}{N} , \tag{2.65}$$

$$\frac{E_{\rm R}}{N} = \frac{\rho}{2}\int d^3r \left[g(r)v(r) + \frac{\hbar^2}{m}\left|\nabla\sqrt{g(r)}\right|^2 \right] , \tag{2.66}$$

$$\frac{E_{\rm Q}}{N} = -\frac{1}{4}\int \frac{d^3k}{(2\pi)^3\rho} t(k)\tilde{N}(k)\tilde{h}(k) , \tag{2.67}$$

$$\frac{E_{\rm e}}{N} = -\frac{1}{4}\int \frac{d^3k}{(2\pi)^3\rho} t(k)(S(k)-1)\tilde{E}(k) , \tag{2.68}$$

which is formulated entirely in terms of the pair distribution function $g(r)$. The observation that (2.66)-(2.68) contain no reference to $u_2(r)$ is particularly important and, in fact, (2.66)-(2.68) they can be derived in various ways without ever mentioning a Jastrow–Feenberg wave function [8].

We may now consider $g(r)$ as the unknown function in our variational problem. It is straightforward to derive the Euler equation

$$\frac{\delta E[g]}{\delta g(r)} = 0 \tag{2.69}$$

for the $g(r)$ that minimizes the energy (2.65). The resulting Euler equation can be written in two useful algebraically different forms which highlight different aspects of the problem. One form focuses on the momentum-space structure, it is

$$S(k) = \frac{1}{\sqrt{1 + \frac{2}{t(k)}\tilde{V}_{\rm p-h}(k)}} \tag{2.70}$$

with

$$V_{\rm p-h}(r) = g(r)\left[v(r) + \Delta V_{\rm e}(r)\right] + \frac{\hbar^2}{m}|\nabla\sqrt{g(r)}|^2 + [g(r)-1]w_I(r) \tag{2.71}$$

and the "induced interaction"

$$\tilde{w}_{\rm I}(k) = -\frac{1}{2}t(k)[2S(k)+1][1-S^{-1}(k)]^2 . \tag{2.72}$$

An alternative form emphasizes the coordinate space features of $g(r)$:

$$\frac{\hbar^2}{m}\nabla^2\sqrt{g(r)} = \left[v(r) + \Delta V_{\rm e}(r) + w_{\rm I}(r)\right]\sqrt{g(r)} . \tag{2.73}$$

The correction term

$$\Delta V_{\rm e}(r) = \frac{\hbar^2}{4m}\left[\nabla^2 E(r) + \rho\int d^3r' \frac{\delta E(r')}{\delta g(r)}\nabla^2 g(r')\right] \tag{2.74}$$

is due to elementary diagrams and multiparticle correlations. This term is the only one that contains approximations. The structure of the equations derived here is independent of the specific approximation chosen for these quantities which identifies the HNC hierarchy as a legitimate approximation scheme as outlined above.

Equation (2.73) has the form of a two-particle s-wave Schrödinger equation, the interaction potential being given by the quantity in square brackets. Since the potential is dominant in the core region, the short-range correlations are seen to be insensitive to the specific form of the elementary-diagram contributions or the induced interaction $w_{\rm I}(r)$, who contribute a "mean field". With the formulation (2.73), we have taken care of the short-range correlations caused by a strong interaction.

Equation (2.70) shows another aspect: This equation is identical to our result (2.53) derived above from time-dependent Hartree theory and the linear response theory of a weakly interacting system with the interaction $\tilde{V}_{\rm p-h}(k)$. Hence, we have mapped the problem of a strongly interacting Bose liquid with a bare interaction to that of a weakly interacting system with an effective interaction. We also have given prescriptions for calculating effective interactions and have shown the usefulness of our approach in coordinate and momentum space.

The momentum-space formulation suggests an iterative procedure for solution of the equations of the following form: Starting from a reasonable first guess for $g(r)$ and the corresponding $S(k)$,

- (1) calculate $V_{\rm e}(r)$ from (2.74), $w_I(r)$ from (2.72) and $V_{\rm p-h}(r)$ from (2.71) for a given n-th order approximation $g_n(r)$,
- (2) Fourier transform and calculate a new $S(k)$ from (2.70),
- (3) construct from that $S(k)$ a new "raw" $g^{(0)}_{n+1}(r)$.
- (4) regularize the short-range part by $g_{n+1}(r) = g_n(r)\exp[g^{(0)}_{n+1}(r) - g_n(r)]$ and repeat the procedure until convergence is reached.

This iteration path, which is known as the "paired-phonon-analysis" (PPA) after its first derivation [1], normally converges quite rapidly, provided one starts with a reasonable initial guess. Counting the number of operations for each iteration makes it clear that one iteration of the PPA optimization procedure is hardly more time consuming than one iteration of the HNC equations (2.57), (2.58) for a given, fixed pair correlation function $u_2(r)$.

It is, at this juncture, not expected of the reader to appreciate the choice of the name "particle-hole interaction" for the effective potential $V_{\rm p-h}(r)$. This will become clear later when we study excited states in a Fermi system.

Let us return to the physical content of the Euler equations. To have a acceptable solution, the term under the square-root in (2.70) should be positive, in particular $\tilde{V}_{\rm p-h}(0^+) > 0$. From our analysis of the excitations of a weakly interacting system (it is plausible that the same types of arguments are also valid for strongly interacting systems, this will be shown later) we find from (2.52) that

$$\epsilon(q) = \frac{t(q)}{S(q)} \sim \frac{\hbar q}{m}\sqrt{m\tilde{V}_{\mathrm{p-h}}(0^+)} \quad \text{as} \quad q \to 0^+ \ . \tag{2.75}$$

Such excitations in a liquid with a linear dispersion law are sound waves with

$$\epsilon(q) = \hbar c q \quad \text{as} \quad q \to 0^+ \ , \tag{2.76}$$

where c is the *hydrodynamic speed of sound* defined through

$$mc^2 = \frac{\mathrm{d}}{\mathrm{d}\rho}\rho^2 \frac{\mathrm{d}}{\mathrm{d}\rho}\frac{E}{N} \ , \tag{2.77}$$

hence one should be able to identify $\tilde{V}_{\mathrm{p-h}}(0^+)$ with the *incompressibility* mc^2. Such an identification is not *a priori* obvious; it takes a different, diagrammatic derivation of our equations to clarify this relationship. We go back to the energy expression (2.56) and carry out the variation with respect to $u_2(r)$:

$$\frac{\hbar^2}{4m}\nabla^2 g(r) = \int d^3r' v_{\mathrm{JF}}(r')\frac{\delta g(r')}{\delta u_2(r)} \equiv g'(r) \ . \tag{2.78}$$

The function $g'(r)$ is to be calculated from the energy expectation value (2.54) while holding the Jackson–Feenberg interaction $v_{\mathrm{JF}}(r)$ constant. Going back to the definition of the pair distribution function, we find for the contribution to the functional derivative from the implicit dependence,

$$\left.\frac{\delta E_{\mathrm{var}}}{\delta u_2(\mathbf{r},\mathbf{r}')}\right|_{v_{\mathrm{JF}}} = \frac{\delta}{\delta u_2(\mathbf{r},\mathbf{r}')}\frac{\int d^3r_1 \dots d^3r_N |\Psi_0(\mathbf{r}_1,\dots,\mathbf{r}_N)|^2 \sum_{i<j} v_{\mathrm{JF}}(\mathbf{r}_i,\mathbf{r}_j)}{\int d^3r_1 \dots d^3r_N |\Psi_0(\mathbf{r}_1,\dots,\mathbf{r}_N)|^2}$$

$$= \frac{N!}{(N-2)!}\frac{\int d^3r_3 \dots d^3r_N |\Psi_0(\mathbf{r},\mathbf{r}',\dots,\mathbf{r}_N)|^2 \left[\sum_{i<j} v_{\mathrm{JF}}(\mathbf{r}_i,\mathbf{r}_j) - E_{\mathrm{var}}\right]}{\int d^3r_1 \dots d^3r_N |\Psi_0(\mathbf{r}_1,\dots,\mathbf{r}_N)|^2} \ . \tag{2.79}$$

Another way to calculate the same quantity is to introduce a *generalized correlation function*

$$u_2(\mathbf{r}_i,\mathbf{r}_j;\beta) \equiv u_2(\mathbf{r}_i,\mathbf{r}_j) + \beta v_{\mathrm{JF}}(\mathbf{r}_i,\mathbf{r}_j) \ , \tag{2.80}$$

define in analogy to (2.4) a generalized pair-distribution function $g(r;\beta)$ and verify

$$\left.\frac{\delta E_{\mathrm{var}}[u_2]}{\delta u_2(r)}\right|_{v_{\mathrm{JF}}} = \left.\frac{\partial}{\partial\beta}\rho_2(r;\beta)\right|_{\beta=0} = \rho^2\left.\frac{\partial}{\partial\beta}g(r;\beta)\right|_{\beta=0} \ . \tag{2.81}$$

The last line justifies the use of the prime notation $g'(r)$. Thus, we may identify

$$g'(r) = \left.\frac{\partial}{\partial\beta}g(r;\beta)\right|_{\beta=0} \ . \tag{2.82}$$

The usefulness of the generating functional formalism is now evident. It is not necessary to carry out an independent cluster-expansion and subsequent resummation for $g'(r)$. Rather, one only needs to reinterpret the $u_2(r)$ in (2.57) as $u_2(r) + \beta v_{\rm JF}(r)$ and perform the β derivative directly on the HNC equations. Moreover, the representation of the pair distribution function by a cluster expansion defines the diagrammatic rules for the construction of $g'(r)$: The diagrammatic representation of the cluster expansion for $g'(r)$ is generated from the expansion for $g(r)$ by replacing, in turn, each $i-j$ correlation line in each diagram by a modified bond representing the function $e^{u_2(r_{ij})}v_{JF}(r_{ij})$.

Implementing our β-derivative algorithm for the calculation of $g'(r)$, the HNC equations (2.57) and (2.58) yields an associated set of linear equations, the so-called HNC′ equations:

$$\begin{aligned} X'(r) &= g(r)\left[v_{\rm JF}(r) + N'(r) + E'(r)\right] - N'(r), \\ \tilde{N}'(k) &= \left[S^2(k) - 1\right]\tilde{X}'(k) \ . \end{aligned} \tag{2.83}$$

In terms of these quantities, we may construct $g'(r)$ and its Fourier transform $S'(k)$ as

$$\begin{aligned} g'(r) &= g(r)\left[v_{\rm JF}(r) + N'(r) + E'(r)\right] \ , \\ S'(k) &= \tilde{g}'(k) = S^2(k)\tilde{X}'(k) \ . \end{aligned} \tag{2.84}$$

Inserting these relationships in the Euler equation (2.78) and comparing with (2.70), we obtain the sought-for diagrammatic identification of the effective interaction

$$\tilde{V}_{ph}(k) = \tilde{X}'(k) - \frac{1}{2}t(k)\tilde{X}(k) \ . \tag{2.85}$$

Of course, one can now again eliminate $\nabla^2 u_2(r)$ from $X'(r)$ using (2.64) and arrives at the previous formula (2.71). We are here interested in the long-wavelength limit of $\tilde{V}_{ph}(k)$. Since $\tilde{X}(k) = 1 - S^{-1}(k) \sim k^{-1}$ for $k \to 0^+$, we only need to calculate the long wavelength limit of $\tilde{X}'(k)$. Recall that $X'(r)$ is by definition the sum of all non-nodal 1–2 diagrams which may be constructed from an arbitrary number of correlation factors $f^2(r_{ij}) - 1$ and precisely one screened interaction $e^{u_2(r_{ij})}v_{\rm JF}(r_{ij})$. These may be constructed from a diagrammatic expansion of the energy expectation value by "opening", in turn, each possible pair of internal points. For an n-body diagram, this may be done in $n(n-1)/2$ ways. Fourier transforming and taking the limit $k \to 0^+$ means, in turn, "closing" the external points, which leaves us with $n(n-1)$ times the original energy diagram; note that the energy carries an overall factor 1/2 on all diagrams, cf. (2.56). Accordingly, an energy expansion

$$\frac{E_{\rm var}}{N} = \sum_n \frac{(\Delta E)_n}{N} = \frac{1}{2}\sum_n \rho^{n-1} e_n \tag{2.86}$$

leads to

$$\tilde{X}'(0^+) = \frac{1}{2}\sum_n n(n-1)\rho^{n-1}e_n = \frac{\partial}{\partial\rho}\rho^2\frac{\partial}{\partial\rho}\frac{E}{N} \equiv mc_*^2 \ , \tag{2.87}$$

which defines an approximation c_* for the speed of sound. In arriving at this result, we note that an n-body diagram contributing to E/N carries an explicit density factor ρ^{n-1}. The identification of the second member of (2.87) with the first requires that the partial derivatives only involve the explicit density dependence of the energy diagrams, the implicit density dependence of the correlations $u_2(r)$ being disregarded.

From this derivation we see that the slope of the excitation spectrum as determined by the particle-hole interaction of optimized Jastrow theory deviates from that given by the Feynman formula (2.77) by an amount determined by the *density dependence* of $u_2(r)$. Assuming that $u_2(r)$ is optimized for all densities, we may derive the inequality

$$\begin{aligned} mc^2 &= mc_*^2 - \frac{1}{V}\int d^3r\, d^3r' \frac{\delta^2 E_{\rm var}}{\delta u_2(\mathbf{r})\delta u_2(\mathbf{r}')}\frac{\mathrm{d}u_2(\mathbf{r})}{\mathrm{d}\rho}\frac{\mathrm{d}u_2(\mathbf{r}')}{\mathrm{d}\rho} \\ &\leq mc_*^2 \ , \end{aligned} \tag{2.88}$$

where inequality follows from the assumption that the variational problem is stable. Thus, c_* determined from the particle-hole interaction is larger than the speed of sound c determined from the density dependence of the energy expectation value based on the full thermodynamic relation $\partial P/\partial\rho = mc^2$, under the assumptions that (i) the Jastrow state is stable and (ii) the optimal pair correlation function is density dependent. We conclude this section by remarking that this analysis has assumed that all elementary diagrams have been retained; at any level of HNC approximation one has further inconsistencies that cannot be rigorously estimated.

As a remark at the closure of this section, it is worth trying the "view from the top". It appears that the variational theory has arrived at two very plausible and generic equations. One is the Bogoljubov (or RPA) equation (2.70) and the other is the coordinate-space equation (2.73). We can write this equation in the form

$$\frac{\hbar^2}{m}\nabla^2\sqrt{g(r)} = V_{\rm p-p}(r)\sqrt{g(r)} \ . \tag{2.89}$$

which we identify with the coordinate-space equation as the boson Bethe-Goldstone equation. We call $V_{\rm p-p}(r)$ the "particle-particle interaction" about which we know only that it should be equal to the bare interaction at short distances. We now have two equations, (2.70) and (2.89), for the pair distribution function $g(r)$ or its Fourier transform $S(k)$; let us forget where they came from. The two effective interactions, $V_{\rm p-p}(r)$ and $V_{\rm p-h}(r)$ are not independent and the question arises what the relationship between these two interactions is. A number of simple algebraic manipulations lead to the statement that both of these effective interactions can be expressed in terms of one unknown common vertex function $\Delta V_e(r)$, where $V_{p-h}(r)$ is given by (2.71) and $V_{p-p}(r)$ is the term in $v(r) + \Delta V_e(r) + w_I(r)$ in (2.73). Of course, the simple equivalence requirement between a distribution function calculated by a Bogoljubov or a Bethe-Goldstone equation does not provide a method for obtaining the $\Delta V_e(r)$, but it shows the very weak assumptions under which the HNC/EL equations can be derived without ever mentioning a Jastrow–Feenberg wave function.

2.5 Applications of the Ground State Theory. Homogeneous ^{4}He in Two and Three Dimensions

Once we have arrived at what may be thought of a "generic" many-body theory it is time to confront it with experiments. Three-dimensional (3D), homogeneous ^{4}He has been studied thoroughly both theoretically and experimentally [16]; for homogeneous, two-dimensional (2D) ^{4}He we must rely on the Monte Carlo data [17, 18]. The purpose of studying the bulk liquids in these cases also serves as an essential test for what it takes to get correct results in the symmetry broken geometry.

To be quantitative, one must include elementary diagrams and triplet correlations. These improvements are tedious and shall not be derived here; the reader is referred to the original literature for details [19, 20]. To save some of the most laborious and unrewarding aspects, one takes a shortcut, calculates only a few elementary diagrams and then scales these by a single parameter such that the experimental bulk 3D binding energy is reproduced.

The usual lower limit of the density for the 3D liquid is the saturation density of 0.0218 Å^{-3}, but one can in principle lower the density into a regime of negative pressure down to the spinodal point at approximately 0.016 Å^{-3} where the compressibility vanishes. The freezing density of the 3D liquid at zero temperature and 25 atm pressure is about 0.026 Å^{-3}. There, the pair distribution function develops long-range oscillations reflecting the long-range order of the solid phase. In 2D, a liquid state is found in the density range between the spinodal point around 0.031 Å^{-2} and the freezing density 0.07 Å^{-2}. The liquid densities of the 3D and 2D systems barely overlap: The interparticle spacing of the 2D upper limit 0.07 Å^{-2} corresponds to the 3D density of 0.0185 Å^{-3}.

The results for the energy per particle as a function of density are shown, in 3D and 2D, in Figs. 1. The agreement with the experiments and the Monte Carlo results is quite satisfactory within the entire density range. In Fig. 1 we have also included a cubic polynomial fit to the experimental (in 3D) or Monte Carlo (in 2D) equation of state. This polynomial fit is used to calculate the pressure,

$$\frac{P}{\rho} = \rho \frac{d(E/N)}{d\rho} . \tag{2.90}$$

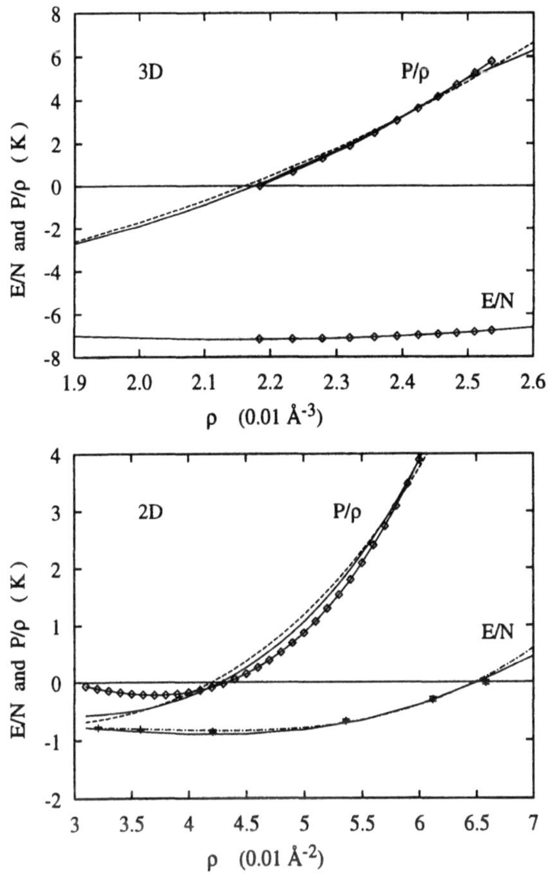

Fig. 1. The energy per particle and pressure/density as a function of density for 2D and 3D liquid ^{4}He. The 3D results are compared with experiments [16] (lines with diamonds) and the 2D case with the Monte Carlo results [17] (dots with error bars). The dash-dotted line is a cubic fit to the Monte Carlo results and the pressure (line with diamonds) is calculated from that fit. The solid lines show the calculated E/N and p/ρ and the dashed lines for the pressure are derived from the cubic fit to the energy.

Agreement between experimental (in 3D) or Monte Carlo (in 2D) structure and distribution functions is equally satisfactory [20].

2.6 Variational Theory at Finite Temperature

Having developed a quantitatively successful theory for strongly interacting quantum fluids at zero temperature, it is natural to seek an extension to finite temperatures. To design a strategy, it is worth recalling *why* the zero temperature theory has been so successful: The reason for that is that one has introduced a sufficiently manageable wave function that allowed for an efficient calculation of a large number of diagrams.

For the development of the finite-temperature theory, we follow largely the path charted by Campbell and collaborators [21, 22]. At finite temperatures, the density matrix $\rho(\mathbf{r}_1,\ldots\mathbf{r}_N;\mathbf{r}'_1,\ldots\mathbf{r}'_N)$ takes the place of the wave function. The relevant minimum principle for a trial density matrix in this formulation is the Gibbs-Delbrück-Molière principle for the Helmholtz free energy F, i.e.,

$$F \le F_{\rm var} \equiv tr(H\rho_{\rm var}) + T\, tr(\rho_{\rm var}\ln\rho_{\rm var}) \,, \tag{2.91}$$

where the subscript "var" signifies *variational* or *trial*, and the two terms are the trial internal energy and entropy term $(-TS)$, respectively. Here, T is the temperature, and we set the Boltzmann constant $k_{\rm B}$ to unity throughout. A variation over all possible choices for $\rho_{\rm var}$ subject to the constraints that it satisfies the appropriate statistics (boson in the present case) and is a proper density matrix gives the exact result $\rho_{\rm var} = Z^{-1}\exp(-\beta H)$, where Z is the partition function and $\beta = 1/T$.

In the absence of a practical representation of this exact solution, $\rho_{\rm var}$ may be varied over a restricted space of trial operators. Moving the variational theory to finite temperatures is accomplished by writing the trial density matrix in a form similar to the ground state wave function by first expressing $\rho_{\rm var}$ in coordinate representation:

$$\begin{aligned}&\rho_{\rm var}(\mathbf{r}_1,\ldots,\mathbf{r}_N;\mathbf{r}'_1,\ldots,\mathbf{r}'_N) =\\ &\Psi_{\rm var}(\mathbf{r}_1,\ldots,\mathbf{r}_N)Q_{\rm var}(\mathbf{r}_1,\ldots,\mathbf{r}_N;\mathbf{r}'_1,\ldots,\mathbf{r}'_N)\Psi^*{}_{\rm var}(\mathbf{r}'_1,\ldots,\mathbf{r}'_N)\end{aligned} \tag{2.92}$$

where the function $Q_{\rm var}$ is the *incoherence function* which links the primed and unprimed coordinates. At zero temperature, $Q_{\rm var}$ is a constant and $\Psi_{\rm var}$ is the exact ground state wavefunction. At non-zero temperatures $Q_{\rm var}$ is not a constant and $\Psi_{\rm var}$ picks up a temperature dependence. In our case, $\Psi_{\rm var}$ is symmetric under interchange of the particle coordinates, and $Q_{\rm var}$ is symmetric under interchange of primed and unprimed coordinates *separately*. $\Psi_{\rm var}$ and $Q_{\rm var}$ are real in the absence of flow.

The next step is to use manageable choices for $\Psi_{\rm var}$ and $Q_{\rm var}$, and then to derive and solve Euler-Lagrange equations for these functions by minimizing the trial free energy. The $\Psi_{\rm var}$ should be identified with our trial wave function (2.2). The incoherence factor may also be written as a Feenberg function[5], with the provison that $\gamma(\mathbf{r}_i,\mathbf{r}'_j)$ vanishes in the limit that $\left|\mathbf{r}_i - \mathbf{r}'_j\right| \to \infty$

$$Q_{\rm var}(\mathbf{r}_1,\ldots,\mathbf{r}_N;\mathbf{r}'_1,\ldots,\mathbf{r}'_N) = \exp\left[\sum_{i,j}\gamma(\mathbf{r}_i,\mathbf{r}'_j)\right] \,. \tag{2.93}$$

[5] Consistent with the rest of this work, we restrict ourselves to pair correlations in the incoherence function although triplet correlations have been successfully applied [23].

Calculating the variational internal energy is straightforward because the density matrix collapses, in the diagonal limit, to the square of a Jastrow-Feenberg wave function for a correlation function $u_2(r_{ij})+2\gamma(r_{ij})$. Only one additional kinetic energy term survives due to the incoherence function, and the variational internal energy of the homogeneous liquid is

$$\frac{U_{\rm var}}{N} = \frac{\rho}{2}\int d^3r g(r)\left[v(r) - \frac{\hbar^2}{4m}\nabla^2(u_2(r)+2\gamma(r))\right] + \int \frac{d^3k}{(2\pi)^3\rho} t(k)\tilde{\gamma}(k) \ , \tag{2.94}$$

where the pair distribution function $g(r)$ has been calculated for a Jastrow-Feenberg wave function with correlation factor $u_2(r)+2\gamma(r)$. We can redefine $u_2(r)$ and realize that the *only* new term is the last term in (2.94).

The much harder problem is the calculation of the entropy

$$S_{\rm var} = T\,{\rm Tr}\left(\rho_{\rm var}\ln\rho_{\rm var}\right) \ . \tag{2.95}$$

Present techniques to calculate the entropy [22, 24] employ the so-called "replica trick": One calculates for all values of n the traces

$$W(n) = {\rm Tr}\left(\rho_{\rm var}^n\right) \tag{2.96}$$

and then tries to identify the $W(n)$ with an *analytic function* $W(\alpha)$ that reproduces the above traces for integer n. One can then obtain

$${\rm Tr}\left(\rho_{\rm var}\ln\rho_{\rm var}\right) = \left.\frac{\partial}{\partial\alpha}W(\alpha)\right|_{\alpha=0} \ . \tag{2.97}$$

The difficulty remains to calculate the $W(n)$, and up to now, only rather simple approximations have been derived that permit the use of the replica trick. The approximations amount to – as we shall see – a quasiparticle approximation of the entropy which can be alternatively be derived by assuming that the thermodynamics is determined by populating Feynman phonons with a boson occupation number. The result of the approximation appears much more plausible than the individual steps leading to it, and the final answer for the entropy is

$$\frac{TS}{N} = \frac{1}{\beta N}\sum_{\mathbf{k}}\left[(1+n(k))\ln(1+n(k)) - n(k)\ln n(k)\right] \ , \tag{2.98}$$

where the $n(k)$ is the positive root of the equation $n(k)(n(k)+1) = \tilde{\gamma}(k)S(k)$. The identification of $n(k)$ with an occupation number is at this point a matter of observation and formal analogy, but not a matter of physical insight.

Once the free energy functional is determined, the remaining manipulation are straightforward: Convenient free functions are the pair distribution function $g(r)$ and the occupation number $n(k)$. The free energy functional is then

$$\frac{F_{\rm var}}{N} = \frac{E_{\rm var}[g(r;T)]}{N} + \int \frac{d^3k}{(2\pi)^3\rho}\frac{t(k)}{S(k)}n(k)(1+n(k)) - \frac{TS}{N}\,, \tag{2.99}$$

where the first term, $E_{\rm var}[g(r;T)]$ is formally identical to (2.65), but depends implicitly on the temperature through the temperature dependence of $g(r;T)$. The free energy is now minimized with respect to these two functions which may be considered independent.

We first look at the Euler equation for the pair distribution function $g(r)$. Manipulations paralleling those for the derivation of the zero-temperature Euler equation lead to the finite-temperature generalization of (2.70), which is

$$S(k) = \frac{1+2n(k)}{\sqrt{1+\frac{2}{t(k)}\tilde{V}_{\rm p-h}(k)}} = (1+2n(k))\frac{t(k)}{\epsilon(k)}\,, \tag{2.100}$$

$$\epsilon(k) = \sqrt{t^2(k) + 2t(k)\tilde{V}_{\rm p-h}(k)}\,, \tag{2.101}$$

where $V_{\rm p-h}(r)$ is given by (2.71) with a modified "induced interaction" $w_{\rm I}(r)$

$$\tilde{w}_{\rm I}(k) = -\frac{t(k)}{2S^2(k)}[2S(k)+1][S(k)-1]^2 + \tilde{v}^*(k) \tag{2.102}$$

and

$$\tilde{v}^*(k) = -2\frac{t(k)}{S^2(k)}n(k)(1+n(k))\,. \tag{2.103}$$

Carrying out the variation of the free energy with respect to the new variable $n(k)$ leads to a familiar looking result:

$$(2\pi)^3\rho\frac{\delta F_{\rm var}}{\delta n(k)} = 0 = \frac{t(k)}{\epsilon(k)} - T\ln\left[\frac{1+n(k)}{n(k)}\right]\,, \tag{2.104}$$

with the solution

$$n(k) = \frac{1}{\exp(\beta\epsilon(k))-1}\,. \tag{2.105}$$

It is now apparent that, if $\tilde{V}_{\rm p-h}(k)$ were temperature independent, the $\epsilon(k)$ would indeed be the Feynman excitation energies of the ground state. In that sense, the identification of $n(k)$ with a boson occupation number is quite plausible.

A number of things should be verified to fully understand the content and the structure of the theory. One important test is that one should recover the "quasiparticle approximation" in the low-temperature limit. The internal energy should, for low temperatures, behave as

$$\frac{\Delta U_{\rm var}(T)}{N} \equiv \frac{U_{\rm var}(T)}{N} - \frac{U_{\rm var}(0)}{N} = \int \frac{d^3k}{(2\pi)^3\rho}\epsilon(k)n(k)\,, \tag{2.106}$$

where the $n(k)$ are the boson occupation numbers (2.105). To derive the low temperature limit one must retain the temperature-dependence of the distribution and structure functions. Since we are looking at low-temperature expansions, we use the zero-temperature Euler equation as reference; the temperature will be indicated by a subscript. We may assume that the temperature dependence of the particle-hole interaction is of higher order in the temperature because the low-lying excitations are dominated by the finite value $\tilde{V}_{\rm p-h}(0^+)$. Using the Euler equation for both finite and zero temperature we get $S_T(k) = (1 + 2n(k))S_0(k)$. To rewrite everything in momentum space, we use the coordinate-space Euler equation (2.73) and calculate

$$\begin{aligned}
\frac{\Delta E_{\rm R}(T)}{N} &= \frac{E_{\rm R}(T) - E_{\rm R}(0)}{N} \qquad (2.107)\\
&= \frac{\rho}{2}\int d^3r\sqrt{g_T(r)}\left[v(r) - \frac{\hbar^2}{m}\nabla^2\right]\sqrt{g_T(r)}\\
&\quad -\frac{\rho}{2}\int d^3r\sqrt{g_0(r)}\left[v(r) - \frac{\hbar^2}{m}\nabla^2\right]\sqrt{g_0(r)}\\
&= \frac{1}{2}\int\frac{d^3k}{(2\pi)^3\rho}\left[-(S_T(k)-1)(\tilde{w}_T(k) + \tilde{v}^*(k)) + (S_0(k)-1)\tilde{w}_0(k)\right] .
\end{aligned}$$

We now can collect the $E_{\rm Q}$ terms and find

$$\begin{aligned}
\frac{\Delta U_{\rm var}(T)}{N} &= \int\frac{d^3k}{(2\pi)^3\rho}\frac{\hbar^2k^2}{2mS_0(k)}n(k)\left[1 - S_0^2(k) + (1+n(k))S_0^3(k)\right]\\
&= \int\frac{d^3k}{(2\pi)^3\rho}\epsilon(k)n(k)\times\left[1 + O(T^2)\right] . \qquad (2.108)
\end{aligned}$$

To prove the $T \to 0$ limit in the last line, one assumes a phonon spectrum $\omega(k) = ck$, the Bose distribution $n(k) = 1/\left[\exp\beta\hbar\omega(k) - 1\right]$, and introduce the dimensionless coordinate $x = \beta\hbar ck$. Then it is seen that every k-factor that appears *not* in the momentum distribution contributes a T factor to the energy correction. In other words, the second term in (2.108) is smaller by a factor T^2 and the third one by T^3. In other words, the "textbook" result does not come out rigorously, but in the $T \to 0$ limit. This is of course what one would expect. Technically, our derivation highlights the importance of optimization: It is hard to see how one could get the low-temperature properties of a Bose liquid right by playing with parametrized trial functions.

To conclude these considerations, let us again try the "view from the top". As for the zero-temperature case, one can interpret the theory in terms of an RPA equation and a Bethe-Goldstone equation. The response function $\chi(k,\omega)$ is identical to (2.49) with the bare potential replaced by $\tilde{V}_{\rm p-h}(q)$, and the connection between the dynamic susceptibility and the dynamic structure function is at finite temperature

$$S(k,\omega) = -\frac{1}{\pi}\frac{1}{1-\exp(-\hbar\omega/T)}\Im m\,\chi(k,\omega) , \qquad (2.109)$$

and the static structure function can be calculated by frequency integration

$$S(k) = \int_{-\infty}^{\infty} d(\hbar\omega) S(k,\omega) = \frac{t(k)}{\epsilon(k)} \coth \frac{1}{2}\beta\epsilon(k) = \frac{t(k)}{\epsilon(k)}(1 + 2n(k))$$

which is evidently the same as (2.100). Thus, we can understand the Euler equations of the theory in the same manner as those of the zero-temperature theory, with the only modification that the random-phase approximation for $S(k)$ is calculated at finite temperature.

These arguments do not yet provide the free energy $F_{\rm var}$ or the internal energy $U_{\rm var}$, which is usually calculated by coupling constant integration. One can avoid solving the RPA equations for all coupling strengths if the equations of motion can be obtained from a variational principle. To demonstrate this, consider $F_{\rm var}[g,n]$ as a functional of the two independent functions $g(r)$ and $n(k)$; these functions are determined by the two extremum conditions

$$\frac{\delta F_{\rm var}}{\delta g}(r) = 0 \quad \text{and} \quad \frac{\delta F_{\rm var}}{\delta n}(k) = 0 \; . \tag{2.110}$$

Consider now a generalized $F_{\rm var}[g,n](\lambda) = F_{\rm var}[g,n,\lambda v(r)]$. Then

$$\begin{aligned} \frac{\mathrm{d}F_{\rm var}}{\mathrm{d}\lambda} &= \frac{\partial F_{\rm var}}{\partial \lambda} + \int d^3r \frac{\delta F}{\delta g}(r)\frac{dg(r)}{d\lambda} + \int d^3r \frac{\delta F}{\delta n}(k)\frac{dn(k)}{d\lambda} \\ &= \frac{\partial F_{\rm var}}{\partial \lambda} = N\frac{\rho}{2}\int d^3r v(r) g(r) \end{aligned} \tag{2.111}$$

due to the two optimization conditions. The coupling constant integration can be done in two ways: One is to keep the $n(k)$ a fixed, unknown function and to retain only the dependence of the pair correlations on the coupling strength. Then, the result of the integration is the internal energy. The second way is to assume some relationship between $g(r)$, $n(k)$, and $\lambda v(r)$. The coupling-constant integration then leads to the full free energy.

The virtue of the direct variational approach is that the free energy and the internal energy functional are known, and the coupling constant integration never needs to be carried out. But unfortunately, this is about it. Clearly, the level of the theory presented here can be derived from very basic principles without introducing a variational density matrix. It also shows where potential improvements must be sought: These lie in a better evaluation of the entropy, utilizing the technical simplifications provided by the local ansatz for the density matrix (2.92).

3 Inhomogeneous Bose Liquids

3.1 Cluster Expansions and the HNC Equations for Inhomogeneous Bose Systems

After this introductory survey, we are ready to generalize the variational theory to an inhomogeneous geometry. The essential new object that must be dealt with is the one-body function $u_1(\mathbf{r})$; we still set the model wave function $\Phi_0(1,\dots,N)=1$. To derive the HNC/EL theory for inhomogeneous systems, we must rederive our cluster expansion and HNC summation methods for the non-uniform geometry. The definition of the Feenberg function (2.2) has already provided for the possibility of symmetry breaking which we now explore. The derivation given in this section will be somewhat heuristic "by inspection", we refer to Ref. [25] for a rigorous, but somewhat lengthy diagrammatic proof.

The central quantity of interest is the normalization integral

$$I=\int d^3r_1\dots d^3r_N \exp\left[\sum_i u_1(\mathbf{r}_i)+\sum_{i<jN} u_2(\mathbf{r}_i,\mathbf{r}_j)\right] \tag{3.1}$$

from which many quantities of interest can be derived by variation, for example the one-body density

$$\rho_1(\mathbf{r})=\frac{\delta \ln I}{\delta u_1(\mathbf{r})}\ , \tag{3.2}$$

and the two-body density

$$\rho_2(\mathbf{r},\mathbf{r}')=2\frac{\delta \ln I}{\delta u_2(\mathbf{r},\mathbf{r}')}\ . \tag{3.3}$$

The expansion of the normalization integral I in terms of pair correlation functions $h(\mathbf{r}_i,\mathbf{r}_j)$ is carried out in exactly the same manner as in the bulk, but we must of course keep the one-body function $u_1(\mathbf{r})$,

$$\begin{aligned} I &= \int d^3r_1\dots d^3r_N \prod_{1\le i\le N} e^{u_1(\mathbf{r}_i)}\Big\{1+\sum_{1\le i\le N} h(\mathbf{r}_i,\mathbf{r}_j) \\ &\quad +\frac{1}{2}\sum_{\substack{i<j,k<l\\(i,j)\neq(k,l)}} h(\mathbf{r}_i,\mathbf{r}_j)h(\mathbf{r}_k,\mathbf{r}_l)+\dots\Big\} \\ &\equiv \Omega^N\Big\{1+\frac{N(N-1)}{2\Omega^2}\int d^3r_1d^3r_2 e^{u_1(\mathbf{r}_1)+u_1(\mathbf{r}_2)}h(\mathbf{r}_1,\mathbf{r}_2)+\dots\Big\} \end{aligned} \tag{3.4}$$

where we have defined

$$\Omega\equiv\int d^3r e^{u_1(\mathbf{r})}\ . \tag{3.5}$$

In the limit of the infinite system, Ω is the normalization volume. In a *finite* system like a droplet, Ω will be approximately proportional to the size of the system, but more definite statements can not be made.

However, we can, as long as we work with an unspecified number of particles, neither neglect terms of relative order Ω^{-1}, nor prove the the cancellation of reducible diagrams. We will therefore assume that Ω is of the order of the particle number N, and that the N is large such that we need to retain only the leading order terms in the particle number. The contributions remaining finite in this limit are shown, together with their associated weight factors, in Table 1.

Table 1. The first few cluster contributions to the generating functional G and their relative weight factors in the limit of large particle numbers. The first column gives the order of the diagram, the second column the integrand of the cluster integral, the third column the diagrammatic representation, and the last column the topological weight factor to leading order in N.

Order	Type	Diagram	Weight
1	$h(1,2)$		$\frac{N^2}{2\Omega^2}$
2	$h(1,2)h(2,3)$		$\frac{N^3}{2\Omega^3}$
2	$h(1,2)h(3,4)$	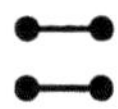	$-\frac{N^3}{2\Omega^4}$
3	$h(1,2)h(2,3)h(3,1)$		$\frac{N^3}{3!\Omega^3}$
3	$h(1,2)h(2,3)h(3,4)$		$\frac{N^4}{2\Omega^4}$
3	$h(1,2)h(1,3)h(1,4)$		$\frac{N^4}{3!\Omega^4}$
3	$h(1,2)h(2,3)h(4,5)$	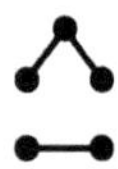	$-\frac{3N^4}{2\Omega^5}$
3	$h(1,2)h(3,4)h(5,6)$	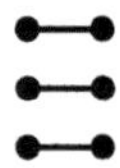	$\frac{5N^4}{3!\Omega^6}$

With this definition of the volume factor Ω, and the additional (preliminary) convention that, in the usual diagrammatic expansions, the filled circle implies integration over the configuration space of the particle, *weighted by a factor* $e^{u_1(\mathbf{r})}$, the diagrammatic expansion of I/Ω^N is, for the non-uniform

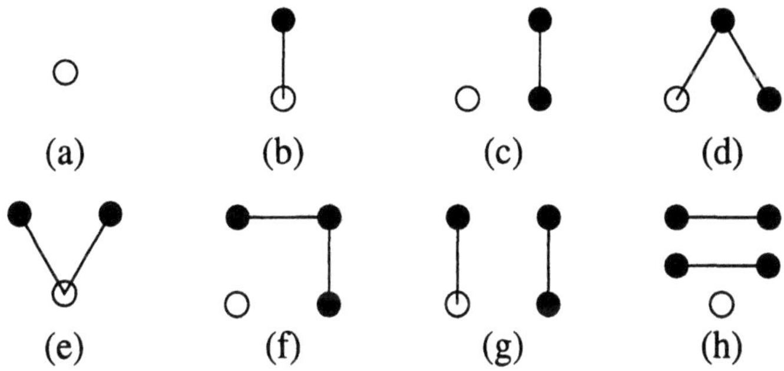

Fig. 2. Some diagrammatic contributions to the one-body density. The diagrams shown are generated by variation of the diagrams given in Table 1. Diagram (a) originates from the first term in the expansion 3.7, the others from the first three diagrams given in Table 1. The weight factors of diagrams (b)-(h) are N/Ω, $-N/\Omega^2$, N^2/Ω^2, $N^2/2\Omega^2$, $-3N^2/2\Omega^3$, $-2N^2/\Omega^3$, and $2N^2/\Omega^4$, respectively.

system, basically the same as the one for the uniform system. Similarly, the diagrammatic expansion of the generating function

$$G = \ln I \tag{3.6}$$

is formally the same *except* that we also have a leading order, one-body term $(\Delta G)_1 = N \ln \Omega$, i.e. the expansion is

$$G = \sum_{n=1}^{\infty} (\Delta G)_n = N \ln \Omega + \sum_{n=2}^{\infty} (\Delta G)_n. \tag{3.7}$$

The new feature of the expansion is that it contains *reducible* and *unlinked* diagrams; these have already been mentioned in Fantoni's lectures. The table shows explicitly the cancellation between reducible diagrams of leading order in N, and unlinked diagrams of lower order in N. Note that, in the limit of an infinite, translationally invariant system, each unlinked portion contributes a volume factor. Hence, the unlinked portions of the expansion of the generating functional must not be omitted.

The one-body density is obtained from the expansion (3.7) for G through (3.2); note that we must also carry out the variation of the normalization factors Ω with respect to $u_1(\mathbf{r})$. The result of the variation is again most conveniently derived and expressed diagrammatically; the variation with respect to the one-body function corresponds to replacing, in turn, each filled circle by an open circle, which represents a function $N e^{u_1(\mathbf{r})}/\Omega$. The result of carrying out these operations on the diagrams of table 1 is shown in Fig. 2. These diagrams are all contributions to the one body density containing up to two correlation functions $h(\mathbf{r}_i, \mathbf{r}_j)$.

As one goes to higher order, the diagrammatic manipulations become increasingly more tedious, the representation given in Fig. 2 should, for the time being, serve our purpose. Similarly, we obtain the pair density through the variation (3.3),

$$\rho_2(\mathbf{r},\mathbf{r}') = 2\frac{\delta G}{\delta u_2(\mathbf{r},\mathbf{r}')}$$
$$= \left(\frac{N}{\Omega}\right)^2 e^{u_1(\mathbf{r})+u_1(\mathbf{r}')+u_2(\mathbf{r},\mathbf{r}')}\left[1+\sum_{n=1}^{\infty}(\Delta C)_n(\mathbf{r},\mathbf{r}')\right] . \qquad (3.8)$$

The diagrammatic representation of all diagrams $(\Delta C)_n(\mathbf{r},\mathbf{r}')$ generated from those shown in Table 1 is shown in Fig. 3.

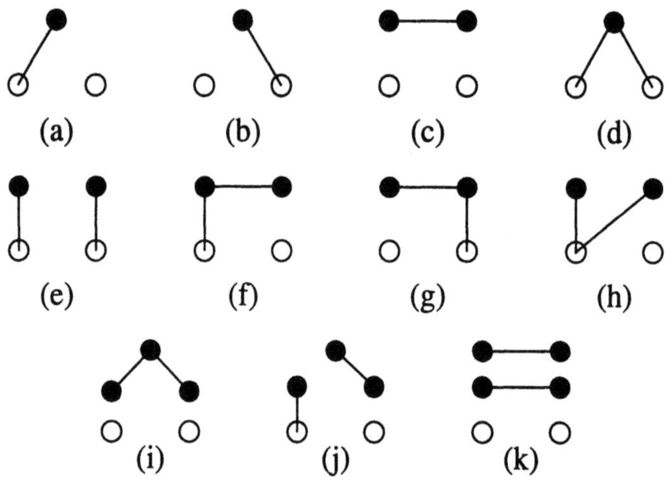

Fig. 3. Some diagrammatic contributions to the two-body density. The diagrams originate from the variation of the diagrams shown in Table 1 with respect to the pair correlation function. The weight factors of diagrams are N/Ω, (diagram (a) and (b), and (d)), $-2N/\Omega^2$ (diagram (c)), N^2/Ω^2 (diagrams (e), (f), and (g)), N^2/Ω^2, (diagram (h)) $-3N^2/\Omega^3$ (diagram (i)), $-6N^2/\Omega^3$ (diagram (j)), and $5N^2/\Omega^4$ (diagram (k)), respectively. Diagrams (h) and (j) are understood to be symmetrized with respect to interchange of the external points.

The resulting expansion appears to become quite complicated. However, we can, by simple rearrangement of the diagrams shown in Fig. 3 show that the pair density can be written in the form

$$\rho_2(\mathbf{r},\mathbf{r}') = \rho_1(\mathbf{r})\rho_1(\mathbf{r}')g(\mathbf{r},\mathbf{r}') , \qquad (3.9)$$

where $g(\mathbf{r},\mathbf{r}')$ has the same diagrammatic representation as the pair distribution function for the uniform system, *provided the filled circle is reinterpreted as* $\int d^3r_i\rho_1(\mathbf{r}_i)$. The diagrams shown in Fig. 3 are the expansion, to second order in the correlation line, of the function $C(\mathbf{r},\mathbf{r}')$ as defined in Eq. (3.8).

At the level presented here, the derivation has been carried out only to a rather low order, and the reader may appreciate the tedium of going much higher. We have chosen here, as earlier, a "pedestrian" route. The derivation presented here allows, in principle, also the calculation of correction terms of order $1/N$.

Once the general construction rule for a diagrammatic expansion of the pair distribution function has been established, it is straightforward to derive HNC equations for the summation of infinite classes of diagrams. The only change is that the bulk density factor appearing in the internal points of the diagrams in the homogeneous geometry gets replaced by a factor $\rho_1(\mathbf{r})$ in the inhomogeneous system. Computational convenience is lost since one cannot calculate convolution products in momentum space. The summation of chain diagrams $N(\mathbf{r}_1, \mathbf{r}_2)$ from non-nodal diagrams $X(\mathbf{r}_1, \mathbf{r}_2)$ assumes the form

$$N(\mathbf{r}_1, \mathbf{r}_2) = \int d^3 r_3 \rho_1(\mathbf{r}_3)[g(\mathbf{r}_1, \mathbf{r}_3) - 1]X(\mathbf{r}_3, \mathbf{r}_2) \ , \tag{3.10}$$

where

$$X(\mathbf{r}_1, \mathbf{r}_2) \equiv g(\mathbf{r}_1, \mathbf{r}_2) - 1 - N(\mathbf{r}_1, \mathbf{r}_2) \tag{3.11}$$

and the parallel connections yield

$$g(\mathbf{r}_1, \mathbf{r}_2) = \exp\left[u_2(\mathbf{r}_1, \mathbf{r}_2) + N(\mathbf{r}_1, \mathbf{r}_2) + E(\mathbf{r}_1, \mathbf{r}_2)\right] \ . \tag{3.12}$$

The static structure function

$$S(\mathbf{r}, \mathbf{r}') = \delta(\mathbf{r}, \mathbf{r}') + \tilde{h}(\mathbf{r}, \mathbf{r}') \tag{3.13}$$

and the nodal function are related through

$$S(\mathbf{r}, \mathbf{r}') = \left[\mathbb{1} - \tilde{X}\right]^{-1}(\mathbf{r}, \mathbf{r}') \ . \tag{3.14}$$

$E(\mathbf{r}_1, \mathbf{r}_2)$ is again the infinite series of "elementary" diagrams which can be expressed as multi-dimensional integrals involving $\rho_1(\mathbf{r})$ and $g(\mathbf{r}_1, \mathbf{r}_2)$. From the equations it is obvious that the assumption of a simple (for example isotropic) form of $u_2(\mathbf{r}_1, \mathbf{r}_2)$ provides no technical advantage.

3.2 Variational Energy Expectation Value and Euler Equations

Provided we neglect three-body correlations, the knowledge of the one- and the two-body functions $u_1(\mathbf{r})$, $\rho_1(\mathbf{r})$, $u_2(\mathbf{r}_1, \mathbf{r}_2)$, and $\rho_2(\mathbf{r}_1, \mathbf{r}_2)$ is sufficient to calculate the energy expectation value

$$E[u_i] = \frac{1}{I} \int d^3 r_1 \ldots d^3 r_N \Psi_0(\mathbf{r}_1, \ldots, \mathbf{r}_N) H \Psi_0(\mathbf{r}_1, \ldots, \mathbf{r}_N) \ . \tag{3.15}$$

As usual, the expectation value of the kinetic energy operator is transformed using the Jackson–Feenberg identity (2.55) and expressed in terms of the one- and the two-body densities:

$$\langle T\rangle = -\int d^3r_1\rho_1(\mathbf{r}_1)\frac{\hbar^2}{8m}\nabla_1^2 u_1(\mathbf{r}_1) \\ -\frac{1}{2}\int d^3r_1 d^3r_2\rho_2(\mathbf{r}_1,\mathbf{r}_2)\frac{\hbar^2}{8m}[\nabla_1^2+\nabla_2^2]u_2(\mathbf{r}_1,\mathbf{r}_2)\ . \tag{3.16}$$

The energy expectation value is still a functional of the four unknown functions, $u_1(\mathbf{r})$ and $u_2(\mathbf{r}_1,\mathbf{r}_2)$ as well as $\rho_1(\mathbf{r}_1)$ and $\rho_2(\mathbf{r}_1,\mathbf{r}_2)$. To eliminate one one-body and one two-body function, we use the Born–Green–Yvon (BGY) equation for the one-body function and the hypernetted chain (HNC) equation for the two-body functions to relate the wave function to the physical distribution functions $\rho_1(\mathbf{r})$ and $g(\mathbf{r},\mathbf{r}')$. The BGY equation is obtained by calculating the gradient of the density $\rho_1(\mathbf{r}_1)$ from its definition (2.3), using the wave function (2.2).

$$\boldsymbol{\nabla}_1\rho_1(\mathbf{r}_1) = \rho_1(\mathbf{r}_1)\boldsymbol{\nabla}_1 u_1(\mathbf{r}_1) + \rho_1(\mathbf{r}_1)\int d^3r_2\rho_1(\mathbf{r}_2)g(\mathbf{r}_1,\mathbf{r}_2)\boldsymbol{\nabla}_1 u_2(\mathbf{r}_1,\mathbf{r}_2)\ . \tag{3.17}$$

Equation (3.17) is exact for the given wave function; a third term is added to the right hand side if triplet correlations are included. We use (3.17) to eliminate the explicit appearance of the one-body factor $u_1(\mathbf{r})$ in the energy expectation value and consider the physically observable $\rho_1(\mathbf{r})$ as the independent one-body function. This elimination of a quantity that we do not need to know, namely $u_1(\mathbf{r})$, in favour of a quantity that we would like to know, namely $\rho_1(\mathbf{r})$, is not only convenient but also important for a number of reasons. It is *convenient* that we do not need to calculate the point-reducible diagrams shown in Fig. 2. It is also *convenient* that the only remaining approximate relationship is the choice of the elementary diagrams $E(\mathbf{r}_1,\mathbf{r}_2)$. Third, one can, in general, one not guarantee that the solution of the two-body Euler–Lagrange equation in an inhomogeneous geometry satisfies the "cluster property" $u_2(\mathbf{r}_1,\mathbf{r}_2) \to 0$ for $|\mathbf{r}_1-\mathbf{r}_2| \to \infty$ for *any* given $u_1(\mathbf{r})$, one might have chosen a non-optimal one-body function, and the two-body function would have to compensate for that. This would lead to a breakdown of the HNC resummation procedure. Thus, the technical *importance* of reformulating the variational problem in terms of the physics observables $\rho_1(\mathbf{r})$ and $g(\mathbf{r},\mathbf{r}')$ id that it removes this difficulty.

The energy expectation value is now in a form that is useful for our further manipulations:

$$E = \frac{\hbar^2}{2m}\int d^3r\left|\nabla\sqrt{\rho_1(\mathbf{r})}\right|^2 + \int d^3r\rho_1(\mathbf{r})U_{\rm ext}(\mathbf{r}) \\ +\frac{1}{2}\int d^3r_1 d^3r_2\rho_1(\mathbf{r}_1)\rho_1(\mathbf{r}_2)g(\mathbf{r}_1,\mathbf{r}_2)v_{JF}(\mathbf{r}_1,\mathbf{r}_2) \\ = T + E_{\rm ext} + E_{\rm c}\ . \tag{3.18}$$

Here, $v_{\rm JF}(\mathbf{r}_1,\mathbf{r}_2)$ is the generalization of the Jackson–Feenberg interaction introduced in (2.56) to an inhomogeneous geometry,

$$v_{\rm JF}(\mathbf{r}_1,\mathbf{r}_2) \equiv v(|\mathbf{r}_1-\mathbf{r}_2|)$$
$$-\frac{\hbar^2}{8m}\left[\frac{1}{\rho_1(\mathbf{r}_1)}\boldsymbol{\nabla}_1\rho_1(\mathbf{r}_1)\cdot\boldsymbol{\nabla}_1+\frac{1}{\rho_1(\mathbf{r}_2)}\boldsymbol{\nabla}_2\rho_1(\mathbf{r}_2)\cdot\boldsymbol{\nabla}_2\right]u_2(\mathbf{r}_1,\mathbf{r}_2) \quad (3.19)$$

Two new, but familiar terms appear in (3.18). The first one,

$$T=\frac{\hbar^2}{2m}\int d^3r\left|\nabla\sqrt{\rho_1(\mathbf{r})}\right|^2 \quad (3.20)$$

is the kinetic energy of a non-interacting, inhomogeneous Bose system whose single-particle wave function is $\sqrt{\rho_1(\mathbf{r})}$. The second term,

$$E_{\rm ext}=\int d^3r\rho_1(\mathbf{r})U_{\rm ext}(\mathbf{r}) \quad (3.21)$$

is the energy of the system in the external field $U_{\rm ext}(\mathbf{r})$. Both of these terms have also been obtained in the Hartree theory. The interactions of the particles give rise to the third term, $E_{\rm c}$, which we shall call the "correlation energy". This term contains all the physics of the interactions.

As above, we can now use either the "prime-equation" technique or the direct variation with respect to $g(\mathbf{r}_1,\mathbf{r}_2)$ to find optimal correlation- or distribution functions. Since we will have to use the "prime-equations" for the fermion problem later, we use here the more straightforward way of direct variation. We use again the HNC relationship (3.12) between $u_2(\mathbf{r}_1,\mathbf{r}_2)$ and $g(\mathbf{r}_1,\mathbf{r}_2)$ to eliminate $u_2(\mathbf{r}_1,\mathbf{r}_2)$ from the energy. In analogy to (2.66), (2.67) and (2.68) one finds

$$E_{\rm c}=E_{\rm R}+E_{\rm Q}+E_{\rm e}\ , \quad (3.22)$$

$$E_{\rm R}=\frac{1}{2}\int d^3r_1d^3r_2\rho_1(\mathbf{r}_1)\rho_1(\mathbf{r}_2)\Bigg[g(\mathbf{r}_1,\mathbf{r}_2)v(|\mathbf{r}_1-\mathbf{r}_2|)$$
$$+\frac{\hbar^2}{2m}\left[\left|\boldsymbol{\nabla}_1\sqrt{g(\mathbf{r}_1,\mathbf{r}_2)}\right|^2+\left|\boldsymbol{\nabla}_2\sqrt{g(\mathbf{r}_1,\mathbf{r}_2)}\right|^2\right]\Bigg]\ , \quad (3.23)$$

$$E_{\rm Q}=-\frac{1}{4}\int d^3r_1d^3r_2\tilde{N}(\mathbf{r}_1,\mathbf{r}_2)H_1(\mathbf{r}_1)\tilde{h}(\mathbf{r}_1,\mathbf{r}_2)\ , \quad (3.24)$$

$$E_{\rm e}=-\frac{1}{4}\int d^3r_1d^3r_2\tilde{E}(\mathbf{r}_1,\mathbf{r}_2)H_1(\mathbf{r}_1)\tilde{h}(\mathbf{r}_1,\mathbf{r}_2)\ . \quad (3.25)$$

The functions $\rho_1(\mathbf{r})$ and $g(\mathbf{r},\mathbf{r}')$ are determined by minimization of the total energy. Out of these two optimization conditions, the one for the pair correlations is the more interesting one. Since $\rho_1(\mathbf{r})$ is the second independent function, we can do the variation at fixed density and need to consider only the "correlation energy"

$$\frac{\delta E_c[\rho_1, g]}{\delta g(\mathbf{r}_1, \mathbf{r}_2)} = 0. \tag{3.26}$$

The energy expression (3.23), (3.24) (ignoring, for the time being, the term (3.25)) allows, similar to the homogeneous case, the straightforward derivation of a variety of representations of the Euler–Lagrange (3.26). In line with the practicality arguments given above, we are looking for a generalization of (2.70). Skipping the algebraic details, we assert that one arrives at the Euler–Lagrange equation in the form

$$\left[\tilde{X} * H_1\tilde{X}\right](\mathbf{r}_1, \mathbf{r}_2) - [H_1(\mathbf{r}_1) + H_1(\mathbf{r}_2)]\,\tilde{X}(\mathbf{r}_1, \mathbf{r}_2) = 2\tilde{V}_{\rm p-h}(\mathbf{r}_1, \mathbf{r}_2) \tag{3.27}$$

or, using the relationship (3.14)

$$\left[S^{-1} * H_1 * S^{-1} - H_1\right](\mathbf{r}_1, \mathbf{r}_2) = 2\tilde{V}_{\rm p-h}(\mathbf{r}_1, \mathbf{r}_2)\ . \tag{3.28}$$

To see that (3.27) is indeed the generalization of (2.70) to the inhomogeneous case, assume that the density is constant and that both $X(\mathbf{r}_1, \mathbf{r}_2)$ and $V_{\rm p-h}(\mathbf{r}_1, \mathbf{r}_2)$ depend only on $r = |\mathbf{r}_1 - \mathbf{r}_2|$. Then we can rewrite (3.28) in momentum space

$$t(k)\left[\frac{1}{S^2(k)} - 1\right] = 2\tilde{V}_{\rm p-h}(k) \tag{3.29}$$

which is equivalent to (2.70).

For completeness, we also spell out the particle-hole interaction

$$\begin{aligned} V_{\rm p-h}(\mathbf{r}_1, \mathbf{r}_2) &= g(\mathbf{r}_1, \mathbf{r}_2)\left[v(|\mathbf{r}_1 - \mathbf{r}_2|) + \Delta V_e \mathbf{r}_1, \mathbf{r}_2)\right] \\ &+ \frac{\hbar^2}{2m}\left[|\nabla_1\sqrt{g(\mathbf{r}_1, \mathbf{r}_2)}|^2 + |\nabla_2\sqrt{g(\mathbf{r}_1, \mathbf{r}_2)}|^2\right] + h(\mathbf{r}_1, \mathbf{r}_2) w_I(\mathbf{r}_1, \mathbf{r}_2) \end{aligned} \tag{3.30}$$

with the induced interaction

$$\tilde{w}_I(\mathbf{r}_1, \mathbf{r}_2) = -\frac{1}{2}\left[[H_1(\mathbf{r}_1) + H_1(\mathbf{r}_2)]\tilde{N}(\mathbf{r}_1, \mathbf{r}_2) + [\tilde{X} * H_1 * \tilde{X}](\mathbf{r}_1, \mathbf{r}_2)\right]\ . \tag{3.31}$$

$\Delta V_e \mathbf{r}_1, \mathbf{r}_2)$ is again the correction from elementary diagram and triplet correlations.

Finally, we must also derive an Euler equation for the density. This equation is relatively un-interesting and produces no particular challenge for its numerical solution, we spell it out here for completeness. The density variation must be subjected to the constraint of given particle number

$$N = \int d^3r \rho_1(\mathbf{r})\ , \tag{3.32}$$

i.e. we carry out the variation

$$\frac{1}{2}\frac{\delta(E-\mu N)}{\delta\sqrt{\rho_1(\mathbf{r})}} = \left[-\frac{\hbar^2}{2m}\nabla^2 + U_{\rm ext}(\mathbf{r}) - \mu\right]\sqrt{\rho_1(\mathbf{r})} + \frac{1}{2}\frac{\delta E_c}{\delta\sqrt{\rho_1(\mathbf{r})}} = 0 \ ,$$
$$\left[-\frac{\hbar^2}{2m}\nabla^2 + U_{\rm ext}(\mathbf{r}) + V_{\rm H}(\mathbf{r})\right]\sqrt{\rho_1(\mathbf{r})} = \mu\sqrt{\rho_1(\mathbf{r})} \ , \tag{3.33}$$

where

$$V_{\rm H}(\mathbf{r}) = \frac{\delta E_c}{\delta\rho_1(\mathbf{r})} \tag{3.34}$$

is a generalized Hartree potential, which depends implicitly on the density and the pair distribution function. Its specific form has been derived in Ref. [26]; the form presented here has turned out to have good convergence properties in numerical optimizations.

$$V_{\rm H}(\mathbf{r}) = V_{\rm H}^{(1)}(\mathbf{r}) + V_{\rm H}^{(2)}(\mathbf{r})$$
$$V_{\rm H}^{(1)}(\mathbf{r}) = \int d^3r_2\rho_1(\mathbf{r}_2)\left[V_{\rm p-h}(\mathbf{r},\mathbf{r}_2) - \frac{1}{2}h(\mathbf{r},\mathbf{r}_2)w_{\rm I}(\mathbf{r},\mathbf{r}_2)\right] \tag{3.35}$$
$$V_{\rm H}^{(2)}(\mathbf{r}) = -\frac{\hbar^2}{16m\rho_1(\mathbf{r})}\nabla\cdot\left[\rho_1(\mathbf{r})\nabla\int d^3r_2\rho_1(\mathbf{r}_2)h(\mathbf{r},\mathbf{r}_2)N(\mathbf{r},\mathbf{r}_2)\right] \ .$$

One might interpret $V_{\rm H}^{(1)}(\mathbf{r})$ as a "volume term" and $V_{\rm H}^{(2)}(\mathbf{r})$ as a "curvature term" in the surface, note that $V_{\rm H}^{(1)}(\mathbf{r})$ survives in the bulk liquid whereas $V_{\rm H}^{(2)}(\mathbf{r})$ vanishes. There are also corrections from elementary diagrams; these terms are not spelled out here for brevity.

3.3 Numerical Optimization

The key equation of our theory is the Euler equation (3.27) for the inhomogeneous system. In the bulk case, the equation is trivially solved by (2.70), but in the inhomogeneous case, (3.27) is a second-order, non-linear partial integro-differential equation in a large number of variables. The theory is useless unless one can construct an efficient algorithm for the solution of this equation. This algorithm is very closely related to the linear response theory of the non-uniform system and we therefore discuss these things here together.

Consider our time-dependent Hartree theory. In (2.45) we have derived how both the energies and the densities of the excitations of a weakly interacting Bose gas are obtained. Replacing the weak, bare interaction by our effective particle-hole interaction, the equation is

$$\int d^3r' H_1(\mathbf{r})\left[H_1(\mathbf{r})\delta(\mathbf{r}-\mathbf{r}') + 2\tilde{V}_{\rm p-h}(\mathbf{r},\mathbf{r}')\right]\phi_n(\mathbf{r}') = \epsilon_n^2\phi_n(\mathbf{r}) \ . \tag{3.36}$$

The eigenvectors of the adjoint equation

$$\int d^3r' \left[H_1(\mathbf{r})\delta(\mathbf{r}-\mathbf{r}') + 2\tilde{V}_{\mathrm{p-h}}(\mathbf{r},\mathbf{r}')\right] H_1(\mathbf{r}')\psi_n(\mathbf{r}') = \epsilon_n^2 \psi_n(\mathbf{r}) \tag{3.37}$$

are related to those of (3.36) by

$$\phi_n(\mathbf{r}) = \frac{1}{\epsilon_n} H_1(\mathbf{r})\psi_n(\mathbf{r}) \ , \tag{3.38}$$

where the normalization factor is a matter of convenience. The states $\psi_n(\mathbf{r})$ and $\phi_n(\mathbf{r})$ then satisfy the completeness and closure relations

$$(\psi_n \mid H_1 \mid \psi_m) = \epsilon_n \delta_{m,n} \tag{3.39}$$

$$\sum_m \psi_n(\mathbf{r})\phi_n(\mathbf{r}) = \delta(\mathbf{r}-\mathbf{r}') \tag{3.40}$$

It is then straightforward to verify that the solutions $\tilde{X}(\mathbf{r},\mathbf{r}')$ of (3.27) and $S(\mathbf{r},\mathbf{r}')$ of (3.28) can be constructed by the normal-mode decomposition

$$\tilde{X}(\mathbf{r},\mathbf{r}') = \delta(\mathbf{r}-\mathbf{r}') - \sum_n \psi_n(\mathbf{r})\psi_n(\mathbf{r}) \tag{3.41}$$

$$S(\mathbf{r},\mathbf{r}') = \sum_n \phi_n(\mathbf{r})\phi_n(\mathbf{r}) \ . \tag{3.42}$$

Given $S(\mathbf{r},\mathbf{r}')$ we can proceed through the normal iterations described in the discussion of the homogeneous case. The remaining operations are just matrix products and do not require any specific numerical techniques.

3.4 Excitations and the Static Structure Factor

It appears at that we have, along with the numerical optimization, also calculated the excitation energies of the system. However, this is still only an observation based on the similarity of equations. To conclude the formal parts of our analysis we now show that the eigenvalues ϵ_n of (3.37) are indeed the excitation energies of the system, and that the static structure function (3.42) is the generalization of the fluctuation-dissipation theorem (2.53) to the non-uniform system.

We proceed in analogy to section 2.3 and subject the system to a weak, time-dependent external potential $\delta U_{\mathrm{ext}}(\mathbf{r};t)$. Generalizing the time-dependent Hartree theory to correlated wave functions, we assume[6] that the system responds by a time-dependent change of the one-body part of the wave function $u_1(\mathbf{r})$:

$$u_1(\mathbf{r};t) = u_1(\mathbf{r}) + \delta u_1(\mathbf{r};t) \ . \tag{3.43}$$

[6] To be completely consistent, one should also assume at least time-dependent pair correlations. This leads indeed to significant quantitative improvemtns of the description of physical phenomena, and to new effects like phonon-damping and phonon-splitting but is beyond the scope of these lectures.

Going beyond the ansatz (2.27), we build the excited states on the correlated ground state $|\psi_0\rangle$ instead of the Hartree approximation $|\psi_0^{\rm H}\rangle$

$$
\begin{aligned}
|\Psi_{\rm X}(t)\rangle &= \exp\left[-\frac{\rm i}{\hbar}E_{\rm var}t\right]|\Psi_0(t)\rangle \ , \\
|\Psi_0(t)\rangle &= \frac{\exp\frac{1}{2}\left[\delta U(t)\right]|\Psi_0\rangle}{\langle\Psi_0 \mid \exp\left[\Re e\delta U(t)\right] \mid \Psi_0\rangle^{1/2}} \ .
\end{aligned}
\tag{3.44}
$$

The wave function of the perturbed system is again determined by the action principle (2.27). Using our assumption that only the one-body component $u_1(\mathbf{r},t)$ is time dependent, we find

$$
\begin{aligned}
\mathcal{L} = \ & \langle\Psi_0(t)|H - E_{\rm var}|\Psi_0(t)\rangle + \\
& + \int d^3r\rho_1(\mathbf{r},t)\left[U_{\rm ext}(\mathbf{r},t) - \frac{i\hbar}{4}(\dot{u}_1(\mathbf{r},t) - \dot{u}_1^*(\mathbf{r},t))\right].
\end{aligned}
\tag{3.45}
$$

The manipulations have up to now been identical to time-dependent Hartree theory. To obtain linear equations of motion for $\delta u_1(\mathbf{r},t)$, we expand the Lagrangian (3.45) to second order in $\delta u_1(\mathbf{r},t)$ and $\delta u_1^*(\mathbf{r},t)$. The problem is simplified considerably since the static correlation functions, $u_1(\mathbf{r})$ and $u_2(\mathbf{r}_i,\mathbf{r}_j)$ have been obtained by optimization. This is a *necessary* condition; the further derivations will not change if one optimizes higher-order correlations or assumes that $|\Psi_0\rangle$ is the exact ground state. The optimization of the one-body correlations, has the consequence that no linear terms in $u_1(\mathbf{r},t)$ occur in (3.45). The two-body optimization has the consequence that

$$
\begin{aligned}
& \int d^3r d^3r' \left(\frac{\delta^2 E_{\rm var}}{\delta u_1(\mathbf{r})\delta u_1(\mathbf{r}')}\right)\delta u_1(\mathbf{r};t)\delta u_1(\mathbf{r};t) \\
&= \frac{1}{4}\int d^3r d^3r' \left(\frac{\delta E_{\rm var}}{\delta u_2(\mathbf{r},\mathbf{r}')}\right)\delta u_1(\mathbf{r};t)\delta u_1(\mathbf{r}';t) = 0 \ .
\end{aligned}
\tag{3.46}
$$

Using (3.46), we find

$$
\langle\Psi_0(t)|H - E_{\rm var}|\Psi_0(t)\rangle = \frac{\hbar^2}{8m}\int d^3r\rho_1(r)|\nabla\delta u_1(\mathbf{r};t)|^2 + 0(|\delta u_1(\mathbf{r};t)|^3). \tag{3.47}
$$

The interaction terms that occured in the time-dependent Hartree theory (cf. (2.30)) has disappeared as a consequence of the fact that the wave function contains optimized pair correlations.

By directly expanding the density $\rho_1(\mathbf{r})$ (Eq. (2.3) for $n = 1$) for small changes in $u_1(\mathbf{r})$, we obtain the relationship between a change in the physical density and a change in the one body wave function:

$$
\begin{aligned}
\delta\rho_1(\mathbf{r},t) &= \rho_1(\mathbf{r})\Re e\,\delta u_1(\mathbf{r};t) + \int d^3r'\left[\rho_2(\mathbf{r},\mathbf{r}') - \rho_1(\mathbf{r})\rho_1(\mathbf{r}')\right]\Re e\,\delta u_1(\mathbf{r}';t) \\
&= \sqrt{\rho_1(\mathbf{r})}\left[S * \Re e\,\delta\tilde{u}_1\right](\mathbf{r};t) \ ,
\end{aligned}
\tag{3.48}
$$

and, using the (exact) chain equation (3.10), the inverse

$$\begin{aligned}\Re e\, \delta\tilde{u}_1(\mathbf{r},t) &= \delta\rho_1(\mathbf{r};t)/\sqrt{\rho_1(\mathbf{r})} - \int d^3r' \tilde{X}(\mathbf{r},\mathbf{r}')\delta\rho_1(\mathbf{r}';t)/\sqrt{\rho_1(\mathbf{r})} \\ &= \left[\left[\mathbb{1} - \tilde{X}\right] * \delta\rho_1/\sqrt{\rho_1(\mathbf{r})}\right](\mathbf{r};t) \ . \end{aligned} \tag{3.49}$$

In passing, we note that (3.47)-(3.49) lead immediately, for adiabatic changes in the physical one-body density, to

$$V_{\text{p-h}}(\mathbf{r},\mathbf{r}') = \frac{\delta^2 E_{\text{var}}}{\delta\rho_1(\mathbf{r})\delta\rho_1(\mathbf{r}')} \ , \tag{3.50}$$

where the variation is carried out for fixed correlation functions.

We are now ready to derive the remaining parts of the action integral. Inserting (3.48) in the second part of (3.45) and omitting zeroth-order terms gives

$$\begin{aligned}&\int d^3r\rho_1(\mathbf{r};t)\left[\delta U_{\text{ext}}(\mathbf{r};t) - \frac{i\hbar}{2}\Im m\, \dot{u}_1(\mathbf{r};t)\right] \\ &= \int d^3r \Re e\, \delta\tilde{u}_1(\mathbf{r},t)\left[S * \left[\delta\tilde{U}_{ext} - \frac{i\hbar}{2}\Im m\, \delta\dot{\tilde{u}}_1\right]\right](\mathbf{r};t) \ . \end{aligned} \tag{3.51}$$

Evaluating the stationarity conditions leads to the compact equations of motion

$$\begin{aligned} 0 = 4\frac{\delta S}{\delta\tilde{u}_1^*(\mathbf{r})} &= H_1(\mathbf{r})\delta\tilde{u}_1(\mathbf{r};t) + \left[S * (2\delta\tilde{U}_{\text{ext}} - i\hbar\delta\dot{\tilde{u}}_1)\right](\mathbf{r};t) \ , \\ 0 = 4\frac{\delta S}{\delta\tilde{u}_1(\mathbf{r})} &= H_1(\mathbf{r})\delta\tilde{u}_1^*(\mathbf{r};t) + \left[S * (2\delta\tilde{U}_{\text{ext}} + i\hbar\delta\dot{\tilde{u}}_1^*)\right](\mathbf{r};t) \ . \end{aligned} \tag{3.52}$$

In analogy to our "supermatrix" form (2.40) we can write the equations of motion as

$$\begin{pmatrix} H_1 & 0 \\ 0 & H_1 \end{pmatrix}\begin{pmatrix} \delta\tilde{u}_1 \\ \delta\tilde{u}_1^* \end{pmatrix} + 2\begin{pmatrix} \tilde{U}_{\text{ext}} \\ \tilde{U}_{\text{ext}} \end{pmatrix} = i\hbar\frac{\partial}{\partial t}\begin{pmatrix} S & 0 \\ 0 & -S \end{pmatrix}\begin{pmatrix} \delta\tilde{u}_1 \\ \delta\tilde{u}_1^* \end{pmatrix} \ . \tag{3.53}$$

The difference to the time-dependent Hartree equation is that the *interaction part* of the equations of motion consists of kinetic energy only, all the dynamics of the system is in the *correlation part.* The form (2.40) can be derived by the following transformation: Define the *correlation matrix*

$$\mathbf{N} = \begin{pmatrix} \tilde{h} & \tilde{h} \\ \tilde{h} & \tilde{h} \end{pmatrix} \tag{3.54}$$

and a new set of variables

$$\begin{pmatrix} \delta\tilde{v}_1 \\ \delta\tilde{v}_1^* \end{pmatrix} = \left(\mathbb{1} + \frac{1}{2}\mathbf{N}\right)\begin{pmatrix} \delta\tilde{u}_1 \\ \delta\tilde{u}_1^* \end{pmatrix} \ , \tag{3.55}$$

note that $\Re e\, \tilde{v}_1(\mathbf{r};t) = 2\delta\sqrt{\rho_1(\mathbf{r};t)}$. Transforming (3.53) to these new variables and using the Euler equation (3.27) yields a new, but equivalent, form that is *identical* to the weakly interacting result (2.40) except that the bare interaction $v(|\mathbf{r}-\mathbf{r}'|)$ has been replaced by the effective interaction $V_{\text{p-h}}(\mathbf{r},\mathbf{r}')$:

$$\begin{pmatrix} H_1 + \tilde{V}_{\text{p-h}} & \tilde{V}_{\text{p-h}} \\ \tilde{V}_{\text{p-h}} & H_1 + \tilde{V}_{\text{p-h}} \end{pmatrix} \begin{pmatrix} \delta\tilde{v}_1 \\ \delta\tilde{v}_1^* \end{pmatrix} + 2 \begin{pmatrix} \tilde{U}_{\text{ext}} \\ \tilde{U}_{\text{ext}} \end{pmatrix} = i\hbar\frac{\partial}{\partial t} \begin{pmatrix} \delta\tilde{v}_1 \\ -\delta\tilde{v}_1^* \end{pmatrix} . \tag{3.56}$$

Hence, the further manipulations to obtain the density-density response function and to identify the collective modes need not be repeated.

We can also read off the frequencies of the collective modes from (3.53). These are found by the solutions of the generalized eigenvalue problem

$$H_1(\mathbf{r})\psi_n(\mathbf{r}) = \epsilon_n \left[S * \psi_n\right](\mathbf{r}) \ , \tag{3.57}$$

which is immediately recognized as the generalization of the Feynman spectrum to a non-uniform geometry.

The final task is to demonstrate that the solution of the Euler equation (3.42) is the inhomogeneous version of the fluctuation-dissipation theorem. Assume that we have solved the eigenvalue problem (3.37). Using the normalization and closure properties (3.39) and (3.40) we find the normal mode decomposition

$$\left[\hbar^2\omega^2 - H_1^2 - 2H_1\tilde{V}_{\text{p-h}} + i\eta\right](\mathbf{r},\mathbf{r}') = \sum_n \left[\hbar^2\omega^2 - \epsilon_n^2 + i\eta\right]\phi_n(\mathbf{r})\psi_n(\mathbf{r}') \ , \tag{3.58}$$

its inverse

$$\left[\hbar^2\omega^2 - H_1^2 - 2H_1\tilde{V}_{\text{p-h}} + i\eta\right]^{-1}(\mathbf{r},\mathbf{r}') = \sum_n \phi_n(\mathbf{r}')\frac{1}{\hbar^2\omega^2 - \epsilon_n^2 + i\eta}\psi_n(\mathbf{r}') \ , \tag{3.59}$$

and, hence,

$$\chi(\mathbf{r},\mathbf{r}';\omega) = 2\sqrt{\rho_1(\mathbf{r})}\sum_n \psi_n(\mathbf{r}')\frac{\epsilon_n}{\hbar^2\omega^2 - \epsilon_n^2 + i\eta}\phi_n(\mathbf{r}')\sqrt{\rho_1(\mathbf{r}')} \ . \tag{3.60}$$

Equation (3.42) follows now by frequency integration.

We have seen here how the introduction of correlations maps the problem of a strongly interacting system to that of a weakly interacting system. The "supermatrix" form of the equations of motion might appear to be an unnecessary formality; it was introduced in view of the Fermion counterpart of the corelated linear response theory derived in section 4. We also note that (3.55) uses ground state properties only and could be used with input data from a Monte–Carlo calculation.

The present formulation of linear response theory in a highly correlated system has only highlighted relationships to weakly interacting systems which should be familiar to the reader. This does not necessarily mean that the theory quantitatively accurate in, for example, ^{4}He. It is well known that the simple Feynman dispersion relation $\epsilon(q) = t(q)/S(q)$ overestimates the experimental dispersion relation by about a factor of two in the regime where the wavelength becomes comparable to the particle distance. There is no reason to assume that the theory is better in the inhomogeneous case with perhaps the dispersion relation of surface excitations which propagate at low densitites. The plausible generalization of the method of time-dependent correlation functions is to allow for time-dependent pair correlation $\delta u_2(\mathbf{r}, \mathbf{r}'; t)$. This generalization leads to significant improvements over the Feynman dispersion relation and to quantitative agreement with experiments.

4 Fermion Theory

This section generalizes our theory to Fermi systems and develops a general framework in which the variational theory of inhomogeneous Fermi systems can be formulated and applied. The boson theory has already stressed the importance of optimization for the dynamics, thermodynamics, and the relationship to excitations as well as for the practicality of the theory. There is no reason that these aspects should be less important in a Fermi system.

We now must specify the model wave function $|\Phi_0\rangle$. Per definition, $|\Phi_0\rangle$ is the ground state wave function of a suitably chosen model system which reflects the *statistics* and the *symmetries* of the physical system under consideration. In a normal N-body Fermi system, $|\Phi_0\rangle$ is the Slater determinant of the N lowest eigenstates of an effective one-body Hamiltonian. We are facing the new task of choosing these single-particle orbitals or the one-body hamiltonian generating them.

Superficially, the one-body correlation function $u_1(\mathbf{r})$ is redundant. However, it is important to keep this one-body function for the same reason as above: There is no guarantee that the the solution of the two-body Euler equation

$$\frac{E_{\text{var}}}{\delta u_2(\mathbf{r}_1, \mathbf{r}_2)} = 0 \tag{4.1}$$

satisfies the cluster property $u_2(\mathbf{r}_1, \mathbf{r}_2) \to 0$ for $|\mathbf{r}_1 - \mathbf{r}_2| \to \infty$ for any set of single-particle orbitals chosen — perhaps for convenience — to define the model state $|\Phi_0\rangle$. The way to proceed is therefore to construct an "optimal" single-particle basis which minimizes the energy expectation value, and to choose the $u_1(\mathbf{r})$ such that the cluster-property is guaranteed.

A part of our task is to develop a systematic cluster expansion and summation method. In line with the arguments given above we insist to define a hierarchy of approximations that is suitable for optimization. As in earlier work for bulk [2, 27, 28] and inhomogeneous [29, 30, 31] Fermi liquids , special attention must be paid to the proper treatment of the Pauli principle. This calls for consistent truncation schemes in the cluster expansions of one- and two-body quantities. We will describe only the simplest version of the theory; in this version the scheme is applicable to inhomogeneous electrons as in atoms, metal clusters, or metal surfaces. An application to the denser system ^{3}He is considerably more laborious but should not pose fundamental difficulties.

4.1 Single-Particle Orbitals

In the Fermi system under consideration here, the "non-interacting" model state $|\Phi_0\rangle$ (cf. (2.2)) is the determinant of a set of single-particle orbitals $\phi_i(j) \equiv \phi_i(\mathbf{r}_j)\chi(j)$, $(i,j = 1,..,N)$, i.e.

$$|\Phi_0\rangle = \det\left[|\phi_j(i)\chi(i)\rangle\right] \ . \tag{4.2}$$

The $\chi(i)$ are the spin-eigenfunctions. No assumption is made yet on how the single-particle states are chosen.

All physical observables are *in principle* independent of the single-particle basis. In practice, however, one must work with expansions and approximations, and the choice of the single particle basis may well affect the structure of the expansion and the convergence rate of an approximation scheme. Thus, our first objective is to define a procedure for generating an "optimal" single particle basis.

Let us first focus on the one-body term $u_1(\mathbf{r})$. For that, we start with the correlated wave function (2.2) *without* two-body and higher correlations,

$$\left|\Psi_0^{(1)}\right\rangle = \exp\left[\frac{1}{2}\sum_i u_1(\mathbf{r}_i)\right]|\Phi_0\rangle \ , \tag{4.3}$$

and calculate the distribution functions generated by this wave function. We postulate that the system is localized and that the single-particle orbitals are asymptotically exact, i.e. that $u_1(\mathbf{r}) \to 0$ as $r \to \infty$. We may then expand in powers of the function $h_1(\mathbf{r}) \equiv \exp[u_1(\mathbf{r})] - 1$. As usual, it is convenient to derive a cluster expansion for the generating functional

$$G_\mathrm{F} \equiv \ln\left\langle \Phi_0 \left| \exp\left[\sum_i u_1(\mathbf{r}_i)\right]\right| \Phi_0 \right\rangle = \sum_n (\Delta G_\mathrm{F})_n \ . \tag{4.4}$$

The first terms in this expansion, from which we can see the principle, are

$$(\Delta G_\mathrm{F})_1 = \sum_i \langle i|h_1(\mathbf{r})|i\rangle = \int d^3r h_1(\mathbf{r})\rho_1^\mathrm{F}(\mathbf{r},\mathbf{r}) \ , \tag{4.5}$$

$$\begin{aligned}(\Delta G_\mathrm{F})_2 &= \frac{1}{2!}\sum_{i,j}\left[\langle ij|h_1(\mathbf{r}_1)h_1(\mathbf{r}_2)|ij\rangle_a - \langle i|h_1(\mathbf{r})|i\rangle\langle j|h_1(\mathbf{r})|j\rangle\right] \\ &= -\frac{1}{2}\sum_{i,j}\langle ij|h_1(\mathbf{r}_1)h_1(\mathbf{r}_2)|ji\rangle \\ &= -\frac{1}{2\nu}\int d^3r_1 d^3r_2 h_1(\mathbf{r}_1)h_1(\mathbf{r}_2)\rho_1^\mathrm{F}(\mathbf{r}_1,\mathbf{r}_2)\rho_1^\mathrm{F}(\mathbf{r}_2,\mathbf{r}_1) \ . \end{aligned} \tag{4.6}$$

In (4.5)–(4.6) we have abbreviated the single-particle wave functions $\phi_i(\mathbf{r}_j)\chi(j)$ by their labels i; the subscript a indicates antisymmetrization. $\rho_1^{\rm F}(\mathbf{r}_i, \mathbf{r}_j)$ is the one-body density matrix (2.16) of the model system. The sequence of approximants (4.5)–(4.6) to the generating functional $G_{\rm F}$ reveals the general formation law which may be verified, with increasing complexity, at higher orders of the cluster expansion (4.4):

$$G_{\rm F} = \sum_{n=1}^{\infty} \frac{(-\nu)^{1-n}}{n} \int d^3r_1 \dots d^3r_n h_1(\mathbf{r}_1)\dots h_1(\mathbf{r}_n)\rho_1^{\rm F}(\mathbf{r}_1,\mathbf{r}_2)\dots\rho_1^{\rm F}(\mathbf{r}_n,\mathbf{r}_1) \ . \tag{4.7}$$

The next step is to construct the one-, and two-body densities by functional variation.

$$\frac{\delta G_{\rm F}}{\delta u_1(\mathbf{r})} = \rho_1^{\rm F}(u_1|\mathbf{r}) \ , \tag{4.8}$$

$$\frac{\delta^2 G_{\rm F}}{\delta u_1(\mathbf{r}_1)\delta u_1(\mathbf{r}_2)} = \rho_2^{\rm F}(u_1|\mathbf{r}_1,\mathbf{r}_2) - \rho_1^{\rm F}(u_1|\mathbf{r}_1)\rho_1^{\rm F}(u_1|\mathbf{r}_2) \ . \tag{4.9}$$

Our notation is intended to remind of the explicit dependence of the densities on the one-body function $u_1(\mathbf{r})$. The superscript F indicates that the two-body correlations are turned off. The one-body density is

$$\begin{aligned}\rho_1^{\rm F}(u_1|\mathbf{r}) = [1+h_1(\mathbf{r})]\Big\{\rho_1^{\rm F}(\mathbf{r},\mathbf{r}) - \frac{1}{\nu}\left[\rho_1^{\rm F} * h_1 * \rho_1^{\rm F}\right](\mathbf{r},\mathbf{r}) \\ +\frac{1}{\nu^2}\left[\rho_1^{\rm F} * h_1 * \rho_1^{\rm F} * h_1 * \rho_1^{\rm F}\right](\mathbf{r},\mathbf{r}) + - \dots\Big\} \equiv \rho_1^{\rm F}(u_1|\mathbf{r},\mathbf{r}),\end{aligned} \tag{4.10}$$

which suggests the definition of a generalized one-body density matrix

$$\begin{aligned}\rho_1^{\rm F}(u_1|\mathbf{r},\mathbf{r}') \equiv \sqrt{1+h_1(\mathbf{r})}\Big\{\rho_1^{\rm F}(\mathbf{r},\mathbf{r}') - \frac{1}{\nu}[\rho_1^{\rm F} * h_1 * \rho_1^{\rm F}](\mathbf{r},\mathbf{r}') \\ +\frac{1}{\nu^2}[\rho_1^{\rm F} * h_1 * \rho_1^{\rm F} * h_1 * \rho_1^{\rm F}](\mathbf{r},\mathbf{r}') - + \dots\Big\}\sqrt{1+h_1(\mathbf{r}')} \ .\end{aligned} \tag{4.11}$$

(Note that $\rho_1^{\rm F}(u_1 = 0|\mathbf{r},\mathbf{r}') \equiv \rho_1^{\rm F}(\mathbf{r},\mathbf{r}')$).

Higher distribution functions are conveniently expressed in terms of the generalized one-body density matrix $\rho_1^{\rm F}(u_1|\mathbf{r},\mathbf{r}')$. They assume a form which is structurally identical to the form of these functions in the absence of the one-body correlation factor. In particular we have

$$\rho_2^{\rm F}(u_1|\mathbf{r},\mathbf{r}') = \rho_1^{\rm F}(u_1|\mathbf{r},\mathbf{r})\rho_1^{\rm F}(u_1|\mathbf{r}',\mathbf{r}') - \frac{1}{\nu}\rho_1^{\rm F}(u_1|\mathbf{r},\mathbf{r}')\rho_1^{\rm F}(u_1|\mathbf{r}',\mathbf{r}). \tag{4.12}$$

We now note that the generalized density matrix $\rho_1^{\rm F}(u_1|\mathbf{r},\mathbf{r}')$ has all properties of the "non-interacting" system. These properties are

– (i) The generalized density matrix preserves the particle number, i.e.

$$\int d^3r \rho_1^{\rm F}(u_1|\mathbf{r},\mathbf{r}) = \int d^3r \rho_1^{\rm F}(u_1 = 0|\mathbf{r},\mathbf{r}) = N. \tag{4.13}$$

– (ii) The generalized density matrix is a projection operator in the sense of our convolution product definition, i.e.

$$[\rho_1^{\rm F}(u_1) * \rho_1^{\rm F}(u_1)](\mathbf{r},\mathbf{r}') = \nu\rho_1^{\rm F}(u_1|\mathbf{r},\mathbf{r}'). \tag{4.14}$$

– (iii) The generalized density matrix has a spectral representation

$$\rho_1^{\rm F}(u_1|\mathbf{r},\mathbf{r}') = \sum_i n(i)\psi_i^*(u_1|\mathbf{r})\psi_i(u_1|\mathbf{r}'), \tag{4.15}$$

i.e. $\rho_1^{\rm F}(u_1|\mathbf{r},\mathbf{r}')$ is for all $u_1(\mathbf{r})$ the density matrix of a "non-interacting" system described by the Slater determinant of a new set of single-particle orbitals $\psi_i(u_1|\mathbf{r})$.

Properties (i) and (ii) are easily proven by direct evaluation using the definition (4.11) and the fact that $\rho_1^{\rm F}(u_1 = 0|\mathbf{r},\mathbf{r}')$ is a projection operator. Property (iii) is just the spectral representation of a projection operator.

4.2 Cluster Expansions for Inhomogeneous Fermi Systems

The next task in the development of a variational theory for inhomogeneous systems is the derivation of cluster expansion- and summation methods for the interesting physical quantities. We again start with a cluster expansion for the generating functional

$$G = \ln\left\langle \Phi_0 \left| \exp\left[\sum_i u_1(\mathbf{r}_i) + \sum_{i<j} u_2(\mathbf{r}_i,\mathbf{r}_j)\right]\right| \Phi_0 \right\rangle = \sum_n (\Delta G)_n\ . \tag{4.16}$$

As previously, we assume $N \gg 1$. The derivation of such cluster expansions is nowadays a technical exercise; the usual diagrammatic notation (see also Ref. [32]) can be carried over to the inhomogeneous case with only slight modifications. The resulting cluster expansions differ slightly from those of the Bose system where we were able to eliminate *all* point-reducible diagrams. We will not completely succeed in doing that for the fermion case, hence we need to spell out the diagrammatic rules specifically applicable for fermion expansions: Cluster contributions to the generating functional G and related quantities are represented as diagrams using the following graphical elements:

– (i) Open circles ("external" points) represent the coordinates of particles. Filled circles ("internal" points) involve an integration over the coordinate space of that particle and a spin sum. We deviate here from the usual convention which includes a density factor.
– (ii) Dashed lines ("correlation lines") between two circles i and j represent dynamical correlations $h_2(\mathbf{r}_i,\mathbf{r}_j) = \exp[u_2(\mathbf{r}_i,\mathbf{r}_j)] - 1$.

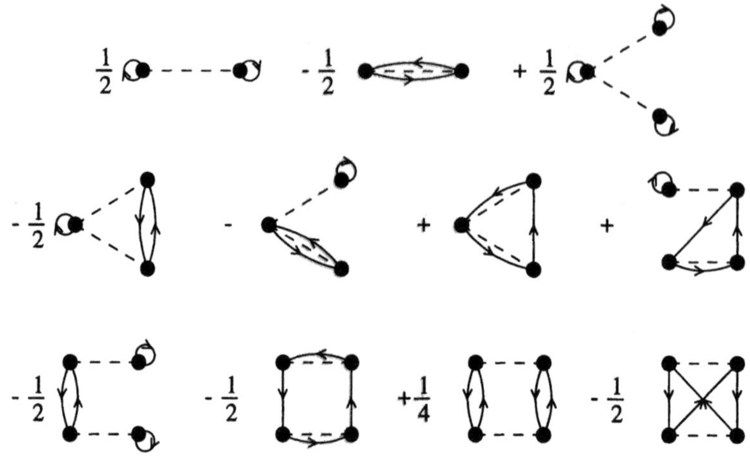

Fig. 4. All diagrams in the diagrammatic representation of the cluster expansion (4.16) are shown which contain one or two correlation lines. Diagrams 3, 5, 7, and 8 are reducible.

- (iii) Solid, oriented lines from point i to point j represent one-body density matrices ("exchange lines") $\rho_1^{\rm F}(u_1|\mathbf{r}_i, \mathbf{r}_j)$.

The expansion of G in terms of correlation functions and exchange functions (one-body density matrices) is represented by the set of all topologically distinct connected diagrams without external points constructed after the following rules:

- (iv) Each n-body diagram has a counting factor $1/n!$.
- (v) Each point is attached by at least one correlation line $h_2(\mathbf{r}_i, \mathbf{r}_j)$. Two different points may be connected by at most one correlation line.
- (vi) Each point is attached by exactly one incoming and one outgoing exchange line.
- (vii) Exchange lines occur always in closed polygons and carry a factor $(-\nu)^{1-n}$, where n is the number of points connected by the exchange loop.

Fig. 4 shows all cluster contributions to G having one or two correlation lines $h_2(\mathbf{r}_i, \mathbf{r}_j)$. The new feature compared to the homogeneous system is that the expansion is reducible (Diagrams 3, 5, 7, and 8 of Fig. 4). Otherwise, the same topological rules apply: In the limit of the bulk system, the reducible diagrams cancel due to momentum conservation.

The second step in the development of a variational theory is the calculation of distribution functions and the classification of (topologically defined) subsets of diagrams which may be summed by integral equation methods.

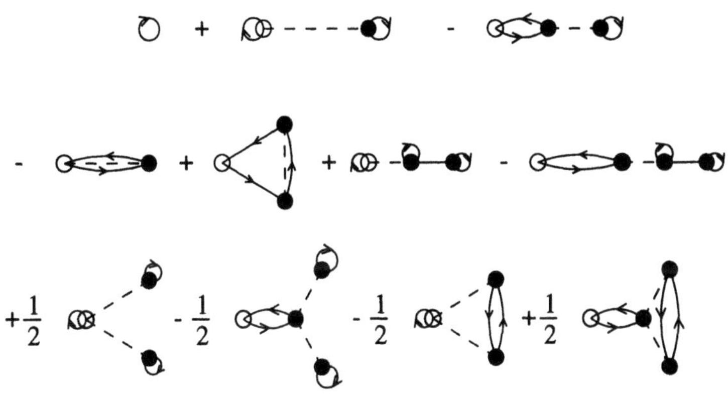

Fig. 5. Graphical representation of the first few terms in the cluster expansion of the one-body density. The first diagram is the Hartree–Fock density, diagrams 2, 3, and 6-11 are included in the definition of the generalized density matrix $\rho_1^{\mathrm{F}}(u_\gamma|\mathbf{r}_i,\mathbf{r}_j)$, cf. (4.19). Note that the exchange line contains the one-body function $u_1(\mathbf{r})$, i.e. it represents the function $\rho_1^{\mathrm{F}}(u_1|\mathbf{r},\mathbf{r}')$, cf. (4.11).

These subsets then serve for a compact representation of physical quantities of interest. To find such classifications we study first the one-body density $\rho_1(\mathbf{r})$ which is obtained by functional variation

$$\rho_1(\mathbf{r}) = \frac{\delta G}{\delta u_1(\mathbf{r})} = \int d^3r_1 d^3r_2 \frac{\delta G}{\delta \rho_1^{\mathrm{F}}(u_1|\mathbf{r}_1,\mathbf{r}_2)} \frac{\delta \rho_1^{\mathrm{F}}(u_1|\mathbf{r}_1,\mathbf{r}_2)}{\delta u_1(\mathbf{r})} \ . \tag{4.17}$$

The second part of (4.17) follows since the function $u_1(\mathbf{r})$ appears in the diagrammatic expansion of G only in $\rho_1^{\mathrm{F}}(u_1|\mathbf{r},\mathbf{r}')$. From (4.11) we can calculate the variation

$$\begin{aligned}\frac{\delta \rho_1^{\mathrm{F}}(u_1|\mathbf{r}_1,\mathbf{r}_2)}{\delta u_1(\mathbf{r})} &= \frac{1}{2}\rho_1^{\mathrm{F}}(u_1|\mathbf{r}_1,\mathbf{r}_2)[\delta(\mathbf{r}_1-\mathbf{r}) + \delta(\mathbf{r}-\mathbf{r}_2)] \\ &\quad - \frac{1}{\nu}\rho_1^{\mathrm{F}}(u_1|\mathbf{r}_1,\mathbf{r})\rho_1^{\mathrm{F}}(u_1|\mathbf{r},\mathbf{r}_2) \ . \end{aligned} \tag{4.18}$$

The graphical representation of the first few diagrams contributing to $\rho_1(\mathbf{r})$ is shown in Fig. 5. Note that the volume integral of all correction terms to the "Hartree–Fock" density $\rho_1^{\mathrm{F}}(\mathbf{r},\mathbf{r})$ (first diagram in Fig. 5) vanishes due to the convolution property (4.14).

One class of contributions to $\rho_1(\mathbf{r})$ includes those which can be written in the form $\rho_1^{\mathrm{F}}(\mathbf{r})\{\exp[\gamma(\mathbf{r})]-1\}$. Diagrams 2, 6, 8, and 10 of Fig. 5 belong to this class. The same subset of diagrams which generates the factor $\{\exp[\gamma(\mathbf{r})]-1\}$ may be connected through an exchange loop to the reference point, examples of such structures are diagrams 3, 7, 9, and 11 of Fig. 5. In fact, we may connect any number of such sets of diagrams by a single exchange path with the reference point $\mathbf{r}$. The same construction may be applied at any internal point and results in a second "dressing" of the exchange function of the same form as discussed in (4.11), but with $u_1(\mathbf{r})$ replaced by $\gamma(\mathbf{r})$. Graphically, $\{\exp[\gamma(\mathbf{r})] - 1\}$ is the set of all diagrams with one reference point $\mathbf{r}$, having no exchange line attached to that point ("d-reducible" diagrams). We are led to the final form of the "dressed" one-body density matrix

$$\rho_1^{\mathrm{F}}(u_1 + \gamma|\mathbf{r},\mathbf{r}') = \sqrt{1 + h_\gamma(\mathbf{r})}\Big\{\rho_1^{\mathrm{F}}(u = 0|\mathbf{r},\mathbf{r}')$$

$$-\frac{1}{\nu}[\rho_1^{\mathrm{F}}(u=0) * h_\gamma * \rho_1^{\mathrm{F}}(u=0)](\mathbf{r},\mathbf{r}') + - \dots\Big\}\sqrt{1 + h_\gamma(\mathbf{r}')}, \quad (4.19)$$

where we have abbreviated

$$h_\gamma(\mathbf{r}) \equiv \exp[u_1(\mathbf{r}) + \gamma(\mathbf{r})] - 1 \equiv \exp[u_\gamma(\mathbf{r})] - 1 \ . \quad (4.20)$$

The analysis of point-reducible diagrams corresponds to the one given in Fantoni's lectures. If we choose to parameterize the correlations and the single particle orbitals, our analysis ends with the formal result (4.19). In the case of bosons, the important simplification was to eliminate the one-body function $u_1(\mathbf{r})$ in favor of the one-body density $\rho_1(\mathbf{r})$. We can do the same elimination here: If we consider the states $\psi_i(u_\gamma|\mathbf{r}')$ as the primary variables, most point-reducible diagrams (specifically, the "d-reducible" diagrams) are implicitly summed in these states. We will therefore from now on use the oriented solid line to represent the fully dressed density matrix $\rho_1^{\mathrm{F}}(u_\gamma|\mathbf{r},\mathbf{r}')$, and omit in our graphical representation the sets of diagrams summed in this function. We can therefore supplement the diagrammatic rules spelled out above by

– (viii) The expansion contains no "d-reducible" diagrams.

In terms of the dressed exchange function $\rho_1^{\mathrm{F}}(u_\gamma|\mathbf{r},\mathbf{r}')$, the graphical expansion of the one-body density becomes now rather compact. The leading term, which integrates to the total particle number, is simply the generalized Hartree–Fock density $\rho_1^{\mathrm{F}}(u_\gamma|\mathbf{r},\mathbf{r})$. The only remaining first order terms are diagrams 4 and 5 of Fig. 5. These two diagrams, and proper combinations of higher order diagrams, integrate pairwise to zero. The rules for the representation of the generating functional G are a bit more complicated; see Ref. [33] for details.

4.3 Energy Expectation Value and Optimal One-Body Correlations

To complete the elimination of point-reducible diagrams, we must show that they can also be eliminated from the variational energy. The essential complication is that one must carry out the derivatives of the kinetic energy *before* the transformation to the new basis $\{\psi_i(u_1|\mathbf{r})\}$ is performed.

It is not yet necessary to make explicit use of a specific method like the FHNC to relate the two-body density to the pair correlation function. In the Bose case, the gradient of the one-body function $u_1(\mathbf{r})$ was eliminated in favor of $\rho_1(\mathbf{r})$ by the Born–Green–Yvon equation (3.17). A similar transformation of the independent one-body function must be made in the case of Fermi statistics. The Fermi generalization of the BGY equation may be derived in the same manner, it is

$$\boldsymbol{\nabla}\rho_1(\mathbf{r}) = \rho_1(\mathbf{r})\boldsymbol{\nabla}u_1(\mathbf{r}) + \int d^3r'\rho_2(\mathbf{r},\mathbf{r}')\boldsymbol{\nabla}_r u_2(\mathbf{r},\mathbf{r}') + (\boldsymbol{\nabla}_\phi + \boldsymbol{\nabla}_{\phi^*})\,\rho_1(\mathbf{r}) \tag{4.21}$$

Here, $\boldsymbol{\nabla}_{\phi^*}$ and $\boldsymbol{\nabla}_\phi$ are operators that differentiate only the single-particle wave functions $\phi_i^*(\mathbf{r})$ and $\phi_i(\mathbf{r})$ in $\rho(\mathbf{r})$.

Using the Jackson–Feenberg identity (2.55) we find the energy (2.5) for a Hamiltonian (2.1) is

$$E_{\text{var}} = T_{\text{F}} + E_{\text{ext}} + E_1^* + E_2^* + T_{\text{JF}}^* \ , \tag{4.22}$$

where E_{ext} is the same as in (3.18), and

$$T_{\text{F}} = \frac{1}{2}\frac{\langle \Phi_0 \mid TF^2 + F^2T \mid \Phi_0\rangle}{\langle \Phi_0 \mid F^2 \mid \Phi_0\rangle} = -\frac{\hbar^2}{4m}\int d^3r \left[\nabla_{\phi^*}^2 + \nabla_\phi^2\right]\rho_1(\mathbf{r}) \ , \tag{4.23}$$

$$E_1^* = -\frac{\hbar^2}{8m}\left\langle \Psi_0 \middle| \sum_i \nabla^2 u_1(\mathbf{r}_i) \middle| \Psi_0 \right\rangle = -\frac{\hbar^2}{8m}\int d^3r \rho_1(\mathbf{r})\nabla^2 u_1(\mathbf{r}) \ , \tag{4.24}$$

$$E_2^* = \left\langle \Psi_0 \middle| \sum_{i<j}\left[v(|\mathbf{r}_i - \mathbf{r}_j|) - \frac{\hbar^2}{8m}(\nabla_i^2 + \nabla_j^2)u_2(\mathbf{r}_i,\mathbf{r}_j)\right] \middle| \Psi_0 \right\rangle \tag{4.25}$$

$$= \frac{1}{2}\int d^3r_1 d^3r_2 \rho_2(\mathbf{r}_1,\mathbf{r}_2)\left[v(|\mathbf{r}_1 - \mathbf{r}_2|) - \frac{\hbar^2}{8m}(\nabla_1^2 + \nabla_2^2)u_2(\mathbf{r}_1,\mathbf{r}_2)\right] \ ,$$

$$T_{JF}^* = \frac{\hbar^2}{8m}\left\langle \Psi_0 \middle| \sum_i \frac{(\nabla_i^2 F^2)}{F^2} \middle| \Psi_0 \right\rangle = \frac{\hbar^2}{8m}\int d^3r \left(\nabla_\phi + \nabla_{\phi^*}\right)^2 \rho_1(\mathbf{r}) \ . \tag{4.26}$$

Next, we must rewrite the derivatives of the bare Hartree–Fock density matrix $\rho_1^F(\mathbf{r},\mathbf{r}')$ in terms of the derivatives of the "dressed" density matrix $\rho_1^F(u_\gamma|\mathbf{r},\mathbf{r}')$. Using the notation ∇_ψ and ∇_{ψ^*} for the derivative operators acting on the eigenstates $\psi_i(u_\gamma|\mathbf{r})$ of the density matrix $\rho_1^F(u_\gamma|\mathbf{r},\mathbf{r}')$ and their complex conjugates, one finds a result identical to (4.23)–(4.26) with all ∇_ϕ operators replaced by ∇_ψ operators, and *the one-body function* $u_1(\mathbf{r})$ *replaced by* $-\gamma(\mathbf{r})$. The $u_1(\mathbf{r})$ does not appear explicitly in the equations; it only appears implicitly in the one-body density matrix.

We can now formulate the optimization condition for the one-body function $u_1(\mathbf{r})$. Since $u_1(\mathbf{r})$ contributes to the energy only through the density matrix $\rho_1^F(u_\gamma|\mathbf{r},\mathbf{r}')$, and we can think of this density matrix to be represented in the form (2.10), we obtain

$$\frac{\delta E}{\delta u_1(\mathbf{r})} = \int d^3r' \left[\sum_i \frac{\delta E}{\delta \psi_i(u_\gamma|\mathbf{r}')} \frac{\delta \psi_i(u_\gamma|\mathbf{r}')}{\delta u_1(\mathbf{r})} + \text{c.c} \right] = 0 \ . \tag{4.27}$$

In other words, if the energy is minimized with respect to the transformed single-particle orbitals $\psi_i(u_\gamma|\mathbf{r})$, i.e. $\delta E/\delta\psi_i(u_\gamma|\mathbf{r}) = 0$, then it is also stationary with respect to $u_1(\mathbf{r})$. If we determine the single-particle orbitals by minimization of the ground-state energy, we are free to choose *any* $u_1(\mathbf{r})$ that is convenient for further manipulations. The choice $u_1(\mathbf{r}) = -\gamma(\mathbf{r})$ eliminates all d-reducible diagrams from the cluster expansion and is therefore the obvious one. In that case we recover the original set of orbitals: $\{\psi_i(u_\gamma = 0|\mathbf{r})\} = \{\phi_i(\mathbf{r})\}$; and the density matrix $\rho_1^F(u_\gamma|\mathbf{r},\mathbf{r}')$ reverts to $\rho_1^F(\mathbf{r},\mathbf{r}')$.

Using (4.21), we can, after some straightforward algebra, write the energy expectation value in the form

$$E_{\text{var}} = T_{\text{F}} + E_{\text{ext}} + E_{\text{c}} = T_{\text{F}} + E_{\text{ext}} + E_2 + T_{\text{JF}} \ , \tag{4.28}$$

$$E_2 = \frac{1}{2}\int d^3r d^3r' \rho_2(\mathbf{r},\mathbf{r}') v_{\text{JF}}(\mathbf{r},\mathbf{r}'), \tag{4.29}$$

$$T_{JF} = \frac{\hbar^2}{8m}\int d^3r \Big\{ \nabla \ln \rho_1(\mathbf{r}) \cdot [\nabla\rho_1(\mathbf{r}) - (\nabla_\phi + \nabla_{\phi^*})\rho_1(\mathbf{r})] + (\nabla_\phi + \nabla_{\phi^*})^2 \rho_1(\mathbf{r}) \Big\} \ , \tag{4.30}$$

where v_{JF} is the Jackson-Feenberg effective interaction defined in (3.19).

With (4.23), (4.29) and (4.30), we have derived a form of the energy expectation value that does not contain the one-body functions $u_1(\mathbf{r})$ or $\gamma(\mathbf{r})$ explicitly. This step furnishes an important simplification of the formalism since none of these one-body functions need to be referenced from here on.

We are now ready to perform the variation with respect to the single-particle orbitals more explicitly, giving the one-body equation

$$H_{\text{CHF}}\phi_i(\mathbf{r}) \equiv \left[-\frac{\hbar^2}{2m}\nabla^2 + U_{\text{ext}}(\mathbf{r}) \right] \phi_i(\mathbf{r}) + \int d^3r' \frac{\delta E_{\text{c}}}{\delta \rho_1^F(\mathbf{r},\mathbf{r}')} \phi_i(\mathbf{r}') = \epsilon_i \phi_i(\mathbf{r}) \ . \tag{4.31}$$

One must isolate the proper one-body terms from the density in order to get the form (4.31), the remaining terms depend on the approximation chosen for the energy functional. In recognition of the fact that the last term on the left hand side of this equation includes correlations, we shall call (4.31) the correlated Hartree–Fock equation.

4.4 Fermi–Hypernetted Chain Equations for Inhomogeneous Systems

Of course, for the calculation of distribution functions through the algorithms (4.17) and

$$\rho_2(\mathbf{r},\mathbf{r}') = \frac{\delta^2 G}{\delta u_1(\mathbf{r})\delta u_1(\mathbf{r}')} + \rho_1(\mathbf{r})\rho_1(\mathbf{r}') \tag{4.32}$$

one must still observe (4.18). Our next task in the development of the variational theory of inhomogeneous Fermi systems is the derivation of integral equations summing, for a given two-body correlation function $u_2(\mathbf{r},\mathbf{r}')$, infinite series of cluster contributions to the two-body density. Care is needed to construct consistent approximation schemes for the one-body and the two-body functions: Consider the sequential relation between the one-body and the two-body density

$$-\rho_1(\mathbf{r}_1) = \int d^3r_2\,[\rho_2(\mathbf{r}_1,\mathbf{r}_2) - \rho_1(\mathbf{r}_1)\rho_1(\mathbf{r}_2)]\,. \tag{4.33}$$

We show in Fig. 6 all diagrams contributing to $\rho_2(\mathbf{r}_1,\mathbf{r}_2) - \rho_1(\mathbf{r}_1)\rho_1(\mathbf{r}_2)$ containing zero or one correlation function. Performing the integral (4.33) we see that the first diagram shown in Fig. 6 (i.e. the one containing no correlation function) integrates to the generalized Hartree–Fock density $\rho_1^{\mathrm{F}}(\mathbf{r},\mathbf{r})$. The remaining diagrams integrate to the contributions to $\rho_1(\mathbf{r})$ containing one correlation function, i.e. to the diagrams 4 and 5 of Fig. 5. We see that

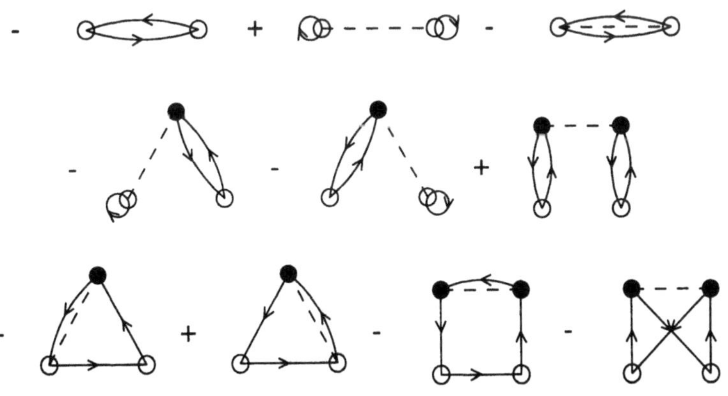

Fig. 6. Diagrammatic representation of all diagrams contributing to the two-body density $\rho_2(\mathbf{r}_1,\mathbf{r}_2)$ containing no more than one correlation line. Note that we assume that the one-body density matrix (solid, oriented line) is the one obtained from the optimized single-particle basis.

consistent approximations must be made for the one- and the two-body densities in order to guarantee the sequential relation (4.33). A possible, but not the only, guiding principle is to classify the cluster contributions according to the number of correlation lines involved. We also note that the present consideration provides just one of several consistency requirement. In a strongly interacting system one will also require that the two-body density is proportional to $1 + h_2(\mathbf{r}, \mathbf{r}')$. Both of the above requirements can be satisfied rigorously only by the summation of an infinite series of three- and four-point diagrams. The proper way to deal with such an inconsistency (other than solving three- and four-point equations) is *not* to ignore one of the two features, but rather use approximations that are appropriate for the important feature of the quantity under consideration. Then, one must accept a slight inconsistency between the two-body density and the static form factor [34] which can be used as a convergence test. The same is true for the Euler equations: One must first derive an exact Euler equation, then choose approximations at a level at which one wants to deal with them, and then calculate the energy expectation value for the optimized functions.

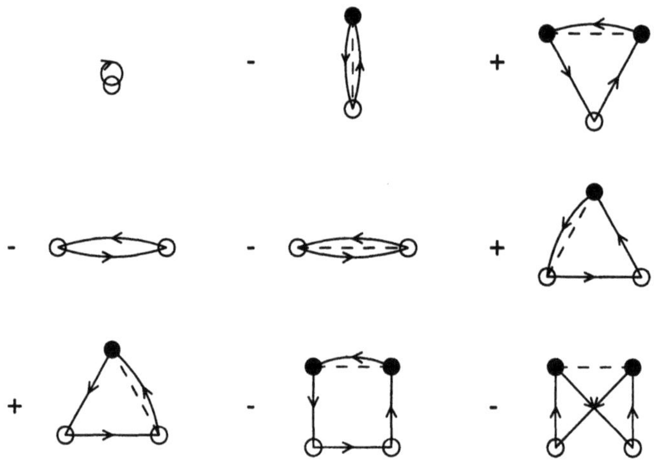

Fig. 7. The simplest insertions to the chaining of two dynamical bonds.

The requirement of consistent approximation schemes for the one- and two- body quantities is a more general one to be observed in the derivation of integral equation techniques. Consider the chaining operation of two bonds $h_2(\mathbf{r}_i, \mathbf{r}_j)$, see Fig. 7. Diagrams 1 and 4 shown in Fig. 7 represent the two simplest ways of "chaining". The remaining diagrams shown in Fig. 7 depict all corrections to the chaining operation which contain one two-body correlation function. Diagrams 2 and 3 show the first-order corrections to the density

$\rho_1(\mathbf{r})$ beyond the Hartree–Fock density (cf. Fig. 5). Using the convolution property (4.14), it is readily shown that the volume integral of the two-point function connecting the dynamical bonds $h_2(\mathbf{r}_i, \mathbf{r}_j)$ in diagrams 5–9 of Fig. 7, equals the first-order correction terms to the one-body density with opposite sign. Hence, in a systematic approximation scheme which preserves the long-wavelength properties of the expansion, the first-order corrections should either all be kept, or all be neglected.

With these preliminary remarks, we are ready to formulate the norm-conserving FHNC equations for an inhomogeneous system. We will formulate these equations at the simplest level which is consistent with optimization. To be specific, we neglect all propagator corrections (i.e. the diagrams 5–9 in Fig. 7 and their generalizations). For consistency, the density is then the optimized Hartree–Fock density $\rho_1^{\mathrm{F}}(\mathbf{r})$. In this approximation, the link between two dynamical correlation bonds is simply the Hartree–Fock static form factor.

$$S_{\mathrm{F}}(\mathbf{r}_1, \mathbf{r}_2) = \delta(\mathbf{r}_1 - \mathbf{r}_2) - \frac{1}{\nu} \frac{|\rho_1^{\mathrm{F}}(\mathbf{r}_1, \mathbf{r}_2)|^2}{\sqrt{\rho_1^{\mathrm{F}}(\mathbf{r}_1, \mathbf{r}_1)\rho_1^{\mathrm{F}}(\mathbf{r}_2, \mathbf{r}_2)}}. \tag{4.34}$$

We obtain a set of two equations for the "nodal" and "non-nodal" *dd*-diagrams

$$X_{\mathrm{dd}}(\mathbf{r}_1, \mathbf{r}_2) \equiv \exp[u_2(\mathbf{r}_1, \mathbf{r}_2) + N_{\mathrm{dd}}(\mathbf{r}_1, \mathbf{r}_2)] - 1 - N_{\mathrm{dd}}(\mathbf{r}_1, \mathbf{r}_2), \tag{4.35}$$

$$\tilde{N}_{\mathrm{dd}}(\mathbf{r}_1, \mathbf{r}_2) = [\tilde{X}_{\mathrm{dd}} * S_{\mathrm{F}} * \hat{\Gamma}_{\mathrm{dd}}](\mathbf{r}_1, \mathbf{r}_2)\ , \tag{4.36}$$

and

$$\Gamma_{\mathrm{dd}}(\mathbf{r}_1, \mathbf{r}_2) \equiv X_{\mathrm{dd}}(\mathbf{r}_1, \mathbf{r}_2) + N_{\mathrm{dd}}(\mathbf{r}_1, \mathbf{r}_2)\ . \tag{4.37}$$

In the same approximation, the static form factor of the interacting system is

$$S(\mathbf{r}_1, \mathbf{r}_2) = S_{\mathrm{F}}(\mathbf{r}_1, \mathbf{r}_2) + [S_{\mathrm{F}} * \hat{\Gamma}_{\mathrm{dd}} * S_{\mathrm{F}}](\mathbf{r}_1, \mathbf{r}_2). \tag{4.38}$$

The construction of a satisfactory pair correlation function requires some extra thought in systems with a strongly repulsive core. The representation (4.38) is sufficient for the electron gas in a metal surface.

4.5 Optimal Pair Correlations

The most complete description of the optimization algorithm and the relevant equations for the bulk Fermi system may be found in Ref. [34]; earlier applications deviate from that by technical details. But the solution of the full FHNC/Euler–Lagrange (EL) problem involves the solution of eight coupled equations instead of two in the Bose case. In addition, inclusion or at least adequate estimates of Pauli blocking effects are required to all orders. Such a calculation may well be feasible and, considering the success of the Bose optimization, worthwhile. We will stay here at the simplest implementation of the

FHNC scheme, in other words, the FHNC approximation (4.35), (4.36) in the inhomogeneous case. For consistency with the FHNC approximation (4.35), (4.36) we must also assume $\rho_1(\mathbf{r}) \equiv \rho_1^{\mathrm{F}}(\mathbf{r},\mathbf{r})$. This approximation displays the structure of the theory most clearly. We also note that with the kinetic energy expression and the one-body density as well as with the variational determination of the single-particle orbitals we have made direct contact to density functional theory.

In the chosen approximation, the one-body part of the ground-state energy is simply the expectation value of the kinetic energy and the external potential in the Hartree–Fock basis,

$$T_{\mathrm{F}} + E_{\mathrm{ext}} = \int d^3r \sum_i n(i)\phi_i^*(\mathbf{r}) \left[-\frac{\hbar^2}{2m}\nabla^2 + U_{\mathrm{ext}}(\mathbf{r})\right] \phi_i(\mathbf{r}) \ , \tag{4.39}$$

and

$$T_{\mathrm{JF}} = -\frac{\hbar^2}{8m\nu} \int d^3r_1 d^3r_2 \Gamma_{\mathrm{dd}}(\mathbf{r}_1,\mathbf{r}_2)\nabla_1 \left[\rho_1(\mathbf{r}_1)\cdot\nabla_1 \frac{|\rho_1^{\mathrm{F}}(\mathbf{r}_1,\mathbf{r}_2)|^2}{\rho_1(\mathbf{r}_1)}\right] . \tag{4.40}$$

For Fermi systems, it is better to use the "prime-equation" technique for the derivation of Euler equations since this method does not require premature approximations. Two structurally different terms appear, namely the *explicit* dependence of the energy on $u_2(\mathbf{r},\mathbf{r}')$ through the Jackson–Feenberg interaction, and the *implicit* dependence through the distribution functions. The evaluation of both terms in the Euler equation $\delta E/\delta u_2(\mathbf{r}_1,\mathbf{r}_2) = 0$ yields

$$\begin{aligned} &\frac{\hbar^2}{8m}\left[\nabla_{\mathbf{r}}\rho_1(\mathbf{r})\rho_1(\mathbf{r}')\cdot\nabla_{\mathbf{r}} + \nabla_{\mathbf{r}'}\rho_1(\mathbf{r})\rho_1(\mathbf{r}')\cdot\nabla_{\mathbf{r}'}\right] g(\mathbf{r},\mathbf{r}') = \\ &= \int d^3r_1 d^3r_2 v_{\mathrm{JF}}(\mathbf{r}_1,\mathbf{r}_2)\frac{\delta\rho_2(\mathbf{r}_1,\mathbf{r}_2)}{\delta u_2(\mathbf{r},\mathbf{r}')} + \frac{\delta T_{\mathrm{JF}}}{\delta u_2(\mathbf{r},\mathbf{r}')} \equiv \rho_2'(\mathbf{r},\mathbf{r}'). \end{aligned} \tag{4.41}$$

Each term in (4.41) has its unique graphical definition, which allows an easy derivation of integral equations consistent with the chosen level of FHNC approximation. The first term in $\rho_2'(\mathbf{r}_1,\mathbf{r}_2)$ is generated graphically from the two-body density $\rho_2(\mathbf{r}_1,\mathbf{r}_2)$ by replacing, in turn, each correlation bond $h_2(\mathbf{r}_i,\mathbf{r}_j)$ by a screened Jackson–Feenberg interaction $[1 + h_2(\mathbf{r}_i,\mathbf{r}_j)]v_{\mathrm{JF}}(\mathbf{r}_i,\mathbf{r}_j)$. It is identical to what one obtains for bosons. The term $\delta T_{\mathrm{JF}}/\delta u_2(\mathbf{r}_1,\mathbf{r}_2)$ comes from the variation of T_{JF}. It is generated by replacing, in turn, each exchange loop $|\rho_1^{\mathrm{F}}(\mathbf{r},\mathbf{r}')|^2$ contributing to $\rho_2(\mathbf{r}_1,\mathbf{r}_2)$ by the combination

$$\frac{\hbar^2}{8m}\nabla_{\mathbf{r}}\left[\rho_1(\mathbf{r})\cdot\nabla_{\mathbf{r}}\frac{|\rho_1^{\mathrm{F}}(\mathbf{r},\mathbf{r}')|^2}{\rho_1(\mathbf{r})}\right] . \tag{4.42}$$

We can now derive the FHNC equations for the generalized two-body structure function $S'(\mathbf{r}_1,\mathbf{r}_2) = \rho_2'(\mathbf{r},\mathbf{r}')/\sqrt{\rho_1(\mathbf{r})\rho_1(\mathbf{r}')}$ and formulate the Fermion Euler equation

$$-\frac{1}{4}\left[H_1 * (S-\mathbb{1}) + (S-\mathbb{1}) * H_1\right](\mathbf{r}_1,\mathbf{r}_2) = S'(\mathbf{r}_1,\mathbf{r}_2) \tag{4.43}$$
$$= \left[S * \left(\tilde{X}'_{\rm dd} - \frac{1}{4}S_{\rm F}^{-1} * H_1 - \frac{1}{4}H_1 * S_{\rm F}^{-1} + \frac{1}{2}S_{\rm F}^{-1} * H_1 * S_{\rm F}^{-1}\right) * S\right](\mathbf{r},\mathbf{r}')$$

Defining, in analogy to (2.85)

$$\tilde{V}_{\rm p-h}(\mathbf{r},\mathbf{r}') = \tilde{X}'_{\rm dd}(\mathbf{r},\mathbf{r}') - \frac{1}{4}[H_1 * \tilde{X}_{\rm dd} + \tilde{X}_{\rm dd} * H_1](\mathbf{r},\mathbf{r}') \tag{4.44}$$

one arrives readily at the final form of the fermion Euler–Lagrange equation

$$\left[S^{-1} * H_1 * S^{-1} - S_{\rm F}^{-1} * H_1 * S_{\rm F}^{-1}\right](\mathbf{r},\mathbf{r}') = 2\tilde{V}_{\rm p-h}(\mathbf{r},\mathbf{r}')\ , \tag{4.45}$$

which is the fermion version of (3.28).

It remains to derive a practical form for $V_{\rm p-h}(\mathbf{r},\mathbf{r}')$. This is done in analogy to the calculation for bosons; the FHNC equation (4.35) is used to eliminate reference to the $u_2(\mathbf{r},\mathbf{r}')$. The final form is

$$V_{\rm p-h}(\mathbf{r},\mathbf{r}') = [1 + \Gamma_{\rm dd}(\mathbf{r},\mathbf{r}')]v(|\mathbf{r}-\mathbf{r}'|) \tag{4.46}$$
$$+\frac{\hbar^2}{2m}\left[\left|\nabla_1\sqrt{1+\Gamma_{\rm dd}(\mathbf{r},\mathbf{r}')}\right|^2 + \left|\nabla_2\sqrt{1+\Gamma_{\rm dd}(\mathbf{r},\mathbf{r}')}\right|^2\right] + \Gamma_{\rm dd}(\mathbf{r},\mathbf{r}')w_{\rm I}(\mathbf{r},\mathbf{r}')\,,$$
$$\tilde{w}_{\rm I}(\mathbf{r},\mathbf{r}') = -\frac{1}{2}\left[\tilde{N}_{\rm dd} * H_1 * S_{\rm F}^{-1} + S_{\rm F}^{-1} * H_1 * \tilde{N}_{\rm dd} + \tilde{X}_{\rm dd} * H_1 * \tilde{X}_{\rm dd}\right](\mathbf{r},\mathbf{r}')\,. \tag{4.47}$$

Equations (4.45), (4.46) together with the chain equation (4.36) form the simplest fermion version of the Euler-Lagrange equations for the pair correlations. The closeness of the formulation of (4.45), (4.46) and (4.47) to the boson version (3.28), (3.30) and (3.31) is evident and intended; the only new feature are the static structure functions of the free Fermi system.

4.6 FHNC, CRPA, and Correlated Basis Functions

We have stressed in the sections on bosons the relationship between the optimization of the ground state and the time-dependent Hartree theory of excitations and it is natural to expect that a similar relationship exists for fermions. To highlight the connections, it is sufficient to study the bulk liquid. The Euler equation (4.45) takes the simple form

$$S(k) = \frac{S_{\rm F}(k)}{\sqrt{1 + \frac{2}{t(k)}S_{\rm F}^2(k)\tilde{V}_{\rm p-h}(k)}}\ . \tag{4.48}$$

Obviously, such a simple form for $S(k)$ cannot be derived from frequency integration as in the boson case. In the fermion case one would instead have an RPA equation

$$\chi(k,\omega) = \frac{\chi_0(k,\omega)}{1 - \tilde{V}_{\rm p-h}(k)\chi_0(k,\omega)} \tag{4.49}$$

where $\chi_0(k,\omega)$ is the Lindhard function and obtain the structure function by frequency integration

$$S(k) = -\int \frac{d(\hbar\omega)}{\rho\pi} \Im m\, \chi(k;\omega) \tag{4.50}$$

which does *not* lead to (4.48). Rather, one can obtain (4.48) with an *approximate* Lindhard function as follows [33]: Replace the $\chi_0(k,\omega)$ by a "collective approximation" $\chi_0^{\rm coll}(k,\omega)$ that has a single pole:

$$\chi_0^{\rm coll}(k,\omega) = \frac{2\rho t(k)}{(\hbar\omega + i\eta)^2 - (t(k)/S_F(k))^2} \tag{4.51}$$

The form of this approximation is determined by the requirement that both the ω^0 and the ω^1 sumrule are fulfilled:

$$S_{\rm F}(k) = -\int_0^\infty \frac{d\omega}{\rho\pi} \Im m\, \chi_0(k,\omega) = -\int_0^\infty \frac{d\omega}{\rho\pi} \Im m\, \chi_0^{\rm coll}(k,\omega)\,, \tag{4.52}$$

$$t(k) = -\int_0^\infty \frac{d\omega}{\rho\pi} \hbar\omega\, \Im m\, \chi_0(k,\omega) = -\int_0^\infty \frac{d\omega}{\rho\pi} \hbar\omega\, \Im m\, \chi_0^{\rm coll}(k,\omega)\,. \tag{4.53}$$

With this "collective" Lindhard function, the frequency integration (4.50) can be carried out in closed form and leads to the FHNC approximation (4.48). We conclude therefore that the FHNC/EL approximation contains the RPA ring-diagrams with the *additional* collective approximation of the Lindhard function. This is a feature of the Jastrow–Feenberg wave function and *not* of the approximations chosen here for the purpose of discussion.

In this aspect, the Jastrow–Feenberg theory for fermions differs fundamentally from that for bosons. For bosons, the task is to calculate, as well as one is prepared to do, the n-body correlation functions and the energy for a given wave function. In the fermion system this is not the end of the story. The key reason is that the wave function (2.2) is not exact for a system of fermions even if local correlations to all orders are included. Another way to phrase the problem is that the Jastrow-Feenberg wavefunction of a Fermi system has the same nodal surfaces as the non-interacting system. Several methods have been proposed to get around this problem; one that is suited for Monte Carlo calculations is to generalize the correlation operator $F(1,\ldots,N)$ to include gradient terms. These corrections have occasionally (but incorrectly) been identified with Feynman–Cohen backflow which is a useful way to think about excitations.

There is a more systematic way to go beyond Jastrow-Feenberg than inventing more complicated (and harder to handle) wave functions. The general procedure is called "Correlated Basis Functions" (CBF) theory and was introduced in the sixties by Feenberg and his collaborators [1, 35]. The idea of the theory has been reviewed in Fantoni's lectures: The correlation operator $F(1,\ldots,N)$ is used to generate a non-orthogonal, but (unless one uses pathological correlation operators) complete correlated basis of the Hilbert space through

$$|\psi_{\mathbf{m}}\rangle \equiv F(1,\ldots,N)\,|\phi_{\mathbf{m}}\rangle \quad . \tag{4.54}$$

Basic building blocks of the theory are (off-)diagonal matrix elements of the Hamiltonian and overlap matrix elements of the non-orthogonal basis states (4.54). These quantities are

(a) Normalization Integrals

$$I_{\mathbf{m},\mathbf{m}} = \langle \psi_{\mathbf{m}} \mid \psi_{\mathbf{m}} \rangle \quad , \tag{4.55}$$

(b) Ratios of normalization integrals

$$z_{\mathbf{m}} = \left(I_{\mathbf{m},\mathbf{m}}/I_{\mathbf{0},\mathbf{0}}\right)^{1/2} \quad , \tag{4.56}$$

(c) Overlap integrals between different states

$$N_{\mathbf{m},\mathbf{n}} = \frac{\langle \psi_{\mathbf{m}} \mid \psi_{\mathbf{n}} \rangle}{\left(I_{\mathbf{m},\mathbf{m}} I_{\mathbf{n},\mathbf{n}}\right)^{1/2}} \quad , \tag{4.57}$$

(d) Diagonal matrix elements elements of the Hamiltonian

$$H_{\mathbf{m},\mathbf{m}} = \frac{\langle \psi_{\mathbf{m}} \mid H \mid \psi_{\mathbf{m}} \rangle}{I_{\mathbf{m},\mathbf{m}}} \quad . \tag{4.58}$$

These matrix elements occur normally only in differences $H_{\mathbf{m},\mathbf{m}} - H_{\mathbf{0},\mathbf{0}}$, where $H_{\mathbf{0},\mathbf{0}} = E_{\text{var}}$ is the variational ground state.

(d) Off-diagonal matrix elements of the Hamiltonian

$$H'_{\mathbf{m},\mathbf{n}} = \frac{\left\langle \psi_{\mathbf{m}} \mid H - H_{\mathbf{0},\mathbf{0}} \mid \psi_{\mathbf{n}} \right\rangle}{\left(I_{\mathbf{m},\mathbf{m}} I_{\mathbf{n},\mathbf{n}}\right)^{1/2}} \quad , \tag{4.59}$$

which appear frequently in the combination

$$W_{\mathbf{m},\mathbf{n}} = H'_{\mathbf{m},\mathbf{n}} - \frac{1}{2}\left[H_{\mathbf{m},\mathbf{m}} + H_{\mathbf{n},\mathbf{n}} - 2H_{\mathbf{0},\mathbf{0}}\right] N_{\mathbf{m},\mathbf{n}} \quad . \tag{4.60}$$

The off-diagonal matrix elements are conveniently classified in terms of the number of orbitals in which the left and the right states differ. If they differ in, say, d orbitals they give rise to the definition of d-body operators. As a notation that is somewhat more intuitive than the above definition of the correlated states $|\psi_{\mathbf{m}}\rangle$ we shall write $|\psi_{i,j}\rangle = F a_j^\dagger a_j\,|\Phi_0\rangle$ where the $a_i^\dagger$ and a_j are ordinary creation and destruction operators. Similarly, $I_{i,j} \equiv \langle \psi_{i,j} \mid \psi_{i,j} \rangle$, and other quantities as needed. Ordinary Rayleigh-Schrödinger or Goldstone perturbation theory [35] as well as methods to generate whole classes of CBF diagrams [36, 37] can be derived without undue complications.

For the case $d = 2$ (we shall be concerned with this case only[7]), the overlap matrix elements can be written as an antisymmetrized matrix element of a two-body operator:

$$N_{\mathbf{m},\mathbf{n}} \equiv \langle n_1 n_2 | \mathcal{N}(1,2;\mathbf{m}) | m_1 m_2 \rangle_a \tag{4.61}$$

where $|\phi_m\rangle$ would be $|\phi_m\rangle = a^\dagger_{m_1} a^\dagger_{m_2} a_{n_2} a_{n_1} |\phi_n\rangle$, and the subscript a indicates antisymmetrization. The task of constructing the overlap matrix elements (4.61) between the non-orthogonal wavefunctions is then identical to the construction of a configuration-space representation of the (generally non-local) two-body operator $\mathcal{N}(1,2;\mathbf{m})$. The diagrammatic analysis is lengthy [38] but the results are reasonably plausible. The *local* part of the overlap operator is just the two-body function $\Gamma_{\rm dd}(r)$, that is

$$\begin{aligned} \langle n_1 n_2 | \mathcal{N}(1,2;\mathbf{m}) | m_1 m_2 \rangle_a &\approx \langle n_1 n_2 | \mathcal{N}_{\rm loc}(1,2) | m_1 m_2 \rangle_a \ , \\ \mathcal{N}_{\rm loc} &= \Gamma_{\rm dd}(r) \end{aligned} \tag{4.62}$$

Similarly, the local part of the interaction matrix element $W_{\mathbf{m},\mathbf{n}}$ can be shown to be

$$\begin{aligned} \langle n_1 n_2 | \mathcal{W}(1,2;\mathbf{m}) | m_1 m_2 \rangle_a &\approx \langle n_1 n_2 | \mathcal{W}_{\rm loc} | m_1 m_2 \rangle_a \ , \\ \mathcal{W}_{\rm loc}(r) &= \Gamma'_{\rm dd}(r) + \frac{\hbar^2}{4m} \nabla^2 \Gamma_{\rm dd}(r) \ . \end{aligned} \tag{4.63}$$

There are, of course, also more complicated, non-local, corrections to the effective interaction [38], but since our goal is to display the structure of the theory as clearly as possible using a fairly primitive approximation – the RPA – we do not need to be concerned with these issues now.

Returning to the specific problem of the "collective approximation" for the Lindhard function, it is clear that one should calculate all correlated ring-diagrams. The analysis is technically rather involved [37], but the result of the calculation is obvious from the fact that the theory should converge towards the exact ground state: The FHNC/EL theory sums (among many other diagrams) all ring-diagrams in terms of the effective interaction $\tilde{V}_{\rm p-h}(k)$ and the collective Lindhard function $\chi_0^{\rm coll}(k,\omega)$. Summing all ring diagrams in CBF theory mostly have the effect to *remove* these approximate diagrams and to *replace* them by the exact set of ring-diagrams in terms of the same interaction. Since neither the approximate nor the exact set of ring-diagrams is small, one must be aware of large-scale cancellations and extreme caution and consistency between the evaluation of the variational energy, that of CBF

---

[7] This includes, strictly speaking, also 4, 6, ... operators that factor in disconnected products of two-body operators.

matrix elements, and the calculated set of CBF diagrams is needed to obtain meaningful results.

Numerically, the discrepancy between the $S(k)$ obtained from (4.50) and from (4.48) differ only by a few percent; this is the reason why the FHNC theory for the energy is quite good. But we also can immediately spot potential problems: The Jastrow–Feenberg wave function is good as long as Fermi sea averages are calculated but it fails for effects that depend on the specifics of the correlations close to the Fermi surface. Excellent examples where the naïve use of FHNC fails are Fermi liquid parameters and pairing phenomena[39].

4.7 Time-Dependent Correlated Hartree–Fock

Instead of going through the tedious analysis of the ground-state ring diagrams [37], we turn to a related, but more generally useful project. That is the design of a time-dependent theory of excitations in strongly interacting Fermi systems analogous to the time-dependent correlated Hartree theory. The analysis is closely related to Fabrocini's lecture series, but offers, again, a "view from the top". For this project we need everything: FHNC–summations not only for the variational energy, but also for off-diagonal matrix elements. Obviously we need optimization – since it was crucial to get the correct Feynman dispersion relation for bosons, there is no reason to assume that it is less important for fermions. Finally we need correlated basis functions theory to infinite order. Our analysis will also bring us — at the very last — to a justification for why we called $V_{\rm p-h}(\mathbf{r}_1, \mathbf{r}_2)$ the "particle-hole" interaction. The intriguing aspect is that this analysis will be feasible only by looking at both, the FHNC/EL theory and the CBF perturbation theory simultaneously.

Defining the fermion generalization of the wave functions of our excited or perturbed states we must be somewhat more explicit as previously because the excitation operator does not need to be a local function any more, and hence does normally not commute with the Jastrow–Feenberg correlation operator F. In analogy to our ansatz (3.44), we write

$$
\begin{aligned}
|\Psi_{\rm X}(t)\rangle &= \exp\left[-\frac{\mathrm{i}}{\hbar}E_{\rm var}t\right]|\Psi_0(t)\rangle \ , \\
|\Psi_0(t)\rangle &= \frac{F\,|\Phi_0(t)\rangle}{\langle\Phi_0(t)\mid F^\dagger F\mid\Phi_0(t)\rangle^{1/2}} \ , \\
|\Phi_0(t)\rangle &= \exp\left[\sum_{p,h} u_{ph}(t)a_p^\dagger a_h\right]|\Phi_0\rangle \ ,
\end{aligned}
\tag{4.64}
$$

where p and h stand for "particle" and "hole" states in the model state $|\Phi_0\rangle$. The equations of motion are again derived from an action principle which is now formulated for the particle-hole amplitudes $u_{ph}(t)$:

$$\mathcal{S}\left[u_{ph}(t), u^*_{ph}(t)\right] = \int dt\mathcal{L}(t) \ ,$$

$$\begin{aligned}\mathcal{L}(t) &= \left\langle \Psi_X(t) \mid H(t) - \mathrm{i}\hbar\frac{\partial}{\partial t} \mid \Psi_X(t)\right\rangle \\ &= \left\langle \Psi_0(t) \mid H(t) - E_{\mathrm{var}} - \mathrm{i}\hbar\frac{\partial}{\partial t} \mid \Psi_0(t)\right\rangle \ . \end{aligned} \tag{4.65}$$

The connection to the Bose theory is readily drawn: If the particle-hole amplitudes are a local function of the momentum transfer, i.e. $u_{ph} = u(|\mathbf{p} - \mathbf{h}|) \equiv u(q)$, then, $\sum_{ph} u_{ph} a^\dagger_p a_h = u(q)\hat{\rho}(q)$ and the identity with (3.44) is obvious.

Since it would be the third time that we derive the equations of motion for a trial excited state (4.64) from the action principle,

$$\delta\mathcal{S}\left[u_{ph}(t), u^*_{ph}(t)\right] = 0 \ , \tag{4.66}$$

we only list the main steps and give the final results:

(i) Evaluate the time-derivative term:

$$\left\langle \Psi_0(t) \mid i\hbar\frac{\partial}{\partial t} \mid \Psi_0(t)\right\rangle = -\hbar\sum_{p,h} \Im m\, \dot{u}_{ph} \frac{\langle \Phi_0(t) \mid F^\dagger F a^\dagger_p a_h \mid \Phi_0(t)\rangle}{\langle \Phi_0(t) \mid F^\dagger F \mid \Phi_0(t)\rangle} \ . \tag{4.67}$$

(ii) Expand the time-derivative term to second order in the u_{ph}

(iii) Evaluate the expectation value of the Hamiltonian to second order,

(iv) Derive equations of motion from the action principle.

Considering now sets of states ph as row/column indices of a "supermatrix", one arrives at the general "correlated RPA" (CRPA) equations

$$\begin{pmatrix} A & B \\ B^* & A^* \end{pmatrix}\begin{pmatrix} u \\ u^* \end{pmatrix} = \mathrm{i}\hbar\frac{\partial}{\partial t}\begin{pmatrix} M & 0 \\ 0 & -M^* \end{pmatrix}\begin{pmatrix} u \\ u^* \end{pmatrix} \tag{4.68}$$

where A, B, and M are particle-hole matrices of the form

$$\begin{aligned} A_{ph,p'h'} &= z_{ph}z_{p'h'}\,\langle \psi_{ph} \mid H - H_{00} \mid \psi_{p'h'}\rangle \\ B_{ph,p'h'} &= z_{ph,p'h'}\,\langle \psi_{ph,p'h'} \mid H - H_{00} \mid \psi_0\rangle \\ M_{ph,p'h'} &= z_{ph}z_{p'h'}\,\langle \psi_{ph} \mid \psi_{p'h'}\rangle \ . \end{aligned} \tag{4.69}$$

These equations were derived by Clark and collaborators in full generality in Ref. [40]. They are superficially *similar* to the time-dependent Hartree equations derived in Sec. 2.3 but *dissimilar* to their "correlated" counterpart of Sec. 3.4 in the sense that they contain matrix elements of the interaction on the left hand side.

To see how the optimization condition comes in, and also to determine potential systematic approximation procedures, let us therefore examine the consequences of optimization on our RPA equations. Momentum conservation dictates $\mathbf{p} = \mathbf{h}+\mathbf{q}$ and $\mathbf{p}' = \mathbf{h}' \pm \mathbf{q}$. That way, one can calculate the Fermi-sea averages (projection operators on particle- or hole-states are implicit)

$$\begin{aligned}
\sum_{hh'} M_{ph,p'h'} &= \langle \rho_{-q}\rho_q \rangle = S(q) \;, \\
\sum_{hh'} B_{ph,p'h'} &= \langle \mid H - H_{00} \mid \rho_{-q}\rho_q \rangle = 0 \;, \\
\sum_{hh'} A_{ph,p'h'} &= \langle \rho_{-q} \mid H - H_{00} \mid \rho_q \rangle = \frac{1}{2} \langle [\rho_{-q}[T, \rho_q]] \rangle = \frac{\hbar^2 q^2}{2m} \;,
\end{aligned} \tag{4.70}$$

the last two lines assuming optimized correlations. In other words, if the correlations are optimized, the Fermi sea averages of the matrix elements vanish, and the equations of motion reduce to those of the boson theory (3.53). The essential physics is in the *metric matrix* $M_{ph,p'h'}$, whereas the interaction matrix is basically kinetic energy. In this sense, the formulation (4.68) is indeed the fermion-analog of (3.53).

Going back to the full equations (4.68), the normalization constants z_{ph} and $z_{ph,p'h'}$ factorize to leading order in the particle number, $z_{ph} = z(p)/z(h)$, and the diagonal portion of $A_{ph,p'h'}$, can be written in terms of single-particle energies

$$\langle \psi_{ph} | \, H - H_{00} \, | \psi_{ph} \rangle = \epsilon_p - \epsilon_h \;. \tag{4.71}$$

To make further progress, one must now look into the structure of the CBF matrix elements. To write the equations in a reasonably compact form, define

$$\begin{aligned}
\Omega &= \begin{pmatrix} \epsilon_p - \epsilon_h & 0 \\ 0 & \epsilon_p - \epsilon_h \end{pmatrix} , \qquad \mathbf{u} = \begin{pmatrix} u_{p,h} \\ u^*_{p,h} \end{pmatrix} , \\
\mathbf{N} &= \begin{pmatrix} \langle p'h \mid \mathcal{N} \mid h'p \rangle_a & \langle p'h' \mid \mathcal{N} \mid ph \rangle_a \\ \langle ph \mid \mathcal{N} \mid p'h' \rangle_a & \langle ph' \mid \mathcal{N} \mid p'h \rangle_a \end{pmatrix} , \\
\mathbf{M} &= \begin{pmatrix} \delta_{p,p'}\delta_{h,h'} + \langle p'h \mid \mathcal{N} \mid h'p \rangle_a & 0 \\ 0 & -\delta_{p,p'}\delta_{h,h'} - \langle ph' \mid \mathcal{N} \mid p'h \rangle_a \end{pmatrix}
\end{aligned}$$

as well as a corresponding matrix $\mathbf{W}$ and the new vector of particle-hole amplitudes

$$\mathbf{v} = \left[\mathbb{1} + \frac{1}{2}\mathbf{N} \right] \mathbf{u} \;. \tag{4.72}$$

Then the CRPA matrix equation can be written in the form

$$\left[\Omega + \frac{1}{2}\Omega\mathbf{N} + \frac{1}{2}\mathbf{N}\Omega + \mathbf{W} \right] \mathbf{u} = i\hbar \frac{\partial}{\partial t} \mathbf{M}\mathbf{u} \tag{4.73}$$

and can be transformed to

$$\left[\Omega + \mathbf{V}_{\mathbf{p}-\mathbf{h}} \right] \mathbf{v} = i\hbar \frac{\partial}{\partial t} \bar{\mathbf{M}}\mathbf{v} \tag{4.74}$$

with

$$\mathbf{V_{p-h}} = \left[\mathbb{1} + \frac{1}{2}\mathbf{N}\right]^{-1} \left[\mathbf{W} - \frac{1}{4}\mathbf{N}\Omega\mathbf{N}\right] \left[\mathbb{1} + \frac{1}{2}\mathbf{N}\right]^{-1} \tag{4.75}$$

$$\bar{\mathbf{M}} = \left[\mathbb{1} + \frac{1}{2}\mathbf{N}\right]^{-1} \mathbf{M} \left[\mathbb{1} + \frac{1}{2}\mathbf{N}\right]^{-1} \tag{4.76}$$

These manipulations might seem up to this point quite formal but when we look back to the boson theory we see that they in fact *simplify* everything. However, one must use at this point the diagrammatic structure of the operators $\mathcal{W}(1,2)$ and $\mathcal{N}(1,2)$ to see that, due to the apparently complicated operations, many diagrams *cancel*. We shall not go through this calculation in full detail, but rather demonstrate how an RPA approximation is obtained. Then, one must of course make the same approximations that one makes in the ordinary RPA. These are

(a) Take only the direct terms,
(b) Use the local approximations (4.62) and (4.63).

Inserting these approximations into **N** and **W** leads to

$$\mathbf{V_{p-h}} = \begin{pmatrix} \tilde{V}_{\text{p-h}}(q) & \tilde{V}_{\text{p-h}}(q) \\ \tilde{V}_{\text{p-h}}(q) & \tilde{V}_{\text{p-h}}(q) \end{pmatrix} \quad \text{and} \quad \bar{\mathbf{M}} = \begin{pmatrix} \mathbb{1} & 0 \\ 0 & -\mathbb{1} \end{pmatrix}, \tag{4.77}$$

where $q = |\mathbf{p} - \mathbf{h}|$ is the momentum transfer.

With these steps, we have managed to reformulate the CRPA equations as an ordinary RPA equation with the effective interaction $\tilde{V}_{\text{p-h}}(q)$. Of course, a large number of simplifications have been necessary to accomplish this, but we have kept the diagrammatic structures of those quantities that were retained intact. In particular we have seen the following: In the original formulation of the CRPA equations, FHNC "chain-diagrams" had in both the matrix elements of the Hamiltonian and those of the non-orthogonality corrections. These chain-diagrams have completely disappeared in the re-formulated equations. The cancellations go much farther that could be shown here [37], they also occur for large classes of exchange diagrams – those that could be legitimately identified with the exchange-diagrams of the RPA. The actual power of the result lies in the elimination of those diagrams that would normally be quite laborious to evaluate.

The results should be seen in two ways: On the one hand side, the reader might be horrified by a number of seemingly brutal approximations that we had to make to get the time-dependent correlated Hartree–Fock theory into a form that resembles an RPA. This is, however, not a statement about FHNC or CBF, but rather a statement about RPA, and FHNC estimates of classes of diagrams omitted in the "ordinary" RPA should provide quite practical tests for the validity of RPA and RPA-based theories. On the other hand, we have also seen how CBF perturbation theory can effectively improve upon the "collective" description of Fermi systems imposed by the Jastrow–Feenberg wave function. We also have seen how cautiously one should proceed when using CBF perturbations theory in order not to "split small quantities into large pieces[41]".

5 Summary

We have formulated in this review the present status of the theory of optimized variational wave functions for an inhomogeneous quantum many-body systems. Connections have been drawn to other descriptions which have mostly turned out to be facets of the very general equations of the variational theory. As far as ground state calculations are concerned, one is in a pretty satisfactory situation in the sense that the equations appear to describe all of the relevant physics — specifically short- and long-range structure — simultaneously and improvements of ground state calculations are mostly numerical challenges but do not promise any new and deep formal insights. New physical understanding and understanding of nature can, as a matter of course, be gained by applying the theory to new systems.

Our derivations were exact, within the Jastrow–Feenberg model (2.2) of the ground-state wave function, for the structure of Bose liquids and, for fermions, for the determination of the single-particle basis. In the treatment of the two-body equation we have restricted ourselves to the simplest implementation of the FHNC theory. This simplification is not mandatory; we have resorted to that approximation for the sake of a compact notation as a starting point for numerical application until some practical experience is gained.

Through the manipulations in fermion sections we have encountered a number of related problems: These were the choice of the single-particle basis, the determination of the one-body attenuation function, and the appearance of reducible diagrams due to the non conservation of linear momentum. We have succeeded in eliminating a large portion of the reducible diagrams by an optimal choice of the single-particle basis. This optimal basis minimizes the energy expectation value in the space of *all* correlated wave functions of the Feenberg form (2.2). At the same time we have decoupled the remaining problem of determining the one- and two-body correlations: The two-body equation (4.45) has an acceptable solution for any single-particle basis. This is necessary in order to permit an iterative solution.

The fermion two-body equation in the present formulation is hardly more complicated than the corresponding two-body equation for a Bose system. We anticipate therefore that the problem can be solved numerically for systems with a simple geometry, for example, films or droplets of ^{3}He atoms, metal films and surfaces, or for simple models of finite nuclei. We have in this paper deliberately refrained from presenting numerical results since details of the implementation of the theory will be somewhat different for different systems. The electron gas in a metal film or a surface is probably the system that allows for the most straightforward application of the formalism. The variational theory offers a number of significant advantages compared with conventional perturbative treatments.

Our theory for excited states and for the thermodynamics has been on a much simpler level, and it is fair, in conclusion, to mention where forefront research stands. The equations of motion method for time-dependent pair correlations in homogeneous ^{4}He has been developed very successfully [42]. It is now in accuracy competitive with simulation results for the phonon-roton spectrum [43] but offers richer physical information such as lifetime and multiparticle effects. The same theory is less developed for non-uniform systems, but has already provided valuable insight into neutron scattering from ^{4}He films, the dynamics of impurity atoms, and modern questions related to quantum sticking, transmission, and evaporation. While numerically more demanding, it is certainly worth exploring whether the most advanced version of this theory can be implemented in a non-uniform geometry.

The theory is in a less satisfactory state for ^{3}He where one should also include the particle-hole structure of the fermion system in the time-dependent pair correlations. But it looks as if the scene has been set and the main problem is to find the sometimes narrow path between the desirable and the doable.

We have described here only the simplest version of a finite temperature theory, and some rather complicated formal development that has been reported over the past few years in both the variational spirit [44, 45] and from the point of view of parquet-diagrams [46] have not been mentioned. It is too early to evaluate these methods; their success and in particular a transparent physical interpretation of the huge arrays of diagrams produced by inventing more sophisticated wave functions remains to be seen.

Acknowledgments

The research reported in this paper was, over the years, funded by a number of agencies: The Deutsche Forschungsgemeinschaft in the early stages, the National Science Foundation through regular and INT grants, the Texas Advanced Research Program, NATO, and most recently the Austrian Science Fund. This work has profited by stimulation and exchange of ideas with numerous senior colleagues, of which I can name only the most influential ones: C. E. Campbell, S. A. Chin, J. W. Clark, B. E. Clements, A. D. Jackson, A. Polls, M. Saarela, and R. A. Smith, as well as the encouragement and support of W. Kohn in the early stages of this project. Likewise, progress is also made by antagonisms and it is a pleasure to thank those colleagues with whom I had some disagreements for providing much stimulation. Finally, I would like to thank my graduate students of the past decade for the persistent challenge to keep them busy.

References

1. E. Feenberg, *Theory of Quantum Liquids* (Academic, New York, 1969)
2. E. Krotscheck and M.L. Ristig, Phys. Lett. A **48**, 17 (1974)
3. E. Krotscheck, Phys. Rev. A **15**, 397 (1977)
4. A.D. Jackson, Ann. Review Nucl. Part. Sci **33**, 105 (1983)
5. M.H. Anderson *et al.*, Science **269**, 198 (1995)
6. K.B. Davis *et al.*, Phys. Rev. Lett. **75**, 3969 (1995)
7. C.C. Bradley, C.A. Sackett, J.J. Tollett, and R.G. Hulet, Phys. Rev. Lett. **75**, 1687 (1995)
8. A.D. Jackson, A. Lande, and R.A. Smith, Phys. Rep. **86**, 55 (1982)
9. T. Wettig, Ph.D. thesis, State Univ. of New York, 1994
10. P. Hohenberg and W. Kohn, Phys. Rev. B **136**, 864 (1964)
11. W. Kohn and P. Vashishta, in *Theory of the Inhomogeneous Electron Gas*, edited by S. Lundqvist and N.H. March (Plemum, New York, 1983), p. 79
12. M. Levy, Proc. Natl. Acad. Sci. USA **76**, 6062 (1979)
13. A.K. Kerman and S.E. Koonin, Ann. Phys. (NY) **100**, 332 (1976)
14. P. Kramer and M. Saraceno, *Geometry of the Time-Dependent Variational Principle in Quantum Mechanics*, Vol. 140 of *Lecture Notes in Physics* (Springer, Berlin, Heidelberg, and New York, 1981)
15. R.P. Feynman, Phys. Rev. **94**, 262 (1954)
16. R. de Bruyn Ouboter and C.N. Yang, Physica **144B**, 127 (1986)
17. P.A. Whitlock, G.V. Chester, and M.H. Kalos, Phys. Rev. B **38**, 2418 (1988)
18. S. Giorgini, J. Boronat, and J. Casulleras, Phys. Rev. B **54**, 6099 (1996)
19. E. Krotscheck and M. Saarela, Phys. Rep. **232**, 1 (1993)
20. B.E. Clements, J.L. Epstein, E. Krotscheck, and M. Saarela, Phys. Rev. B **48**, 7450 (1993)
21. C.E. Campbell, K.E. Kürten, M.L. Ristig, and G. Senger, Phys. Rev. B **30**, 3728 (1984)
22. G. Senger, M.L. Ristig, K.E. Kürten, and C.E. Campbell, Phys. Rev. B **33**, 7562 (1986)
23. B. E. Clements, E. Krotscheck, J.A. Smith, and C.E. Campbell, Phys. Rev. B **47**, 5239 (1993)
24. C. E. Campbell and B.E. Clements, in *Elementary Excitations in Quantum Fluids*, Vol. 79 of *Springer Series in Solid-State Sciences*, edited by K. Ohbayashi and M. Watabe (Springer, New York, 1989), p. 8
25. T. Morita and K. Hiroike, Progr. Theor. Phys. **25**, 537 (1960)
26. E. Krotscheck, Q.X. Qian, and W. Kohn, Phys. Rev. B **31**, 4245 (1985)
27. E. Krotscheck and M.L. Ristig, Nucl. Phys. A **242**, 389 (1975)
28. E. Krotscheck, J. Low Temp. Phys. **27**, 199 (1977)
29. E. Krotscheck, Phys. Rev. B **31**, 4267 (1985)
30. E. Krotscheck, W. Kohn, and G.X. Qian, Phys. Rev. B **32**, 5693 (1985)
31. C.E. Campbell, E. Krotscheck, and T. Pang, Phys. Rep. **223**, 1 (1992)
32. M. Gaudin, J. Gillespie, and G. Ripka, Nucl. Phys. A **176**, 237 (1971)
33. G. Ripka, Phys. Rep. **56**, 1 (1979)
34. E. Krotscheck, Ann. Phys. (NY) **155**, 1 (1984)
35. J.W. Clark, in *Progress in Particle and Nuclear Physics*, edited by D.H. Wilkinson (Pergamon Press, Oxford, 1979), Vol. 2, pp. 89–199

36. E. Krotscheck, H. Kümmel, and J.G. Zabolitzky, Phys. Rev. A **22**, 1243 (1980)
37. E. Krotscheck, Phys. Rev. A **26**, 3536 (1982)
38. E. Krotscheck and J.W. Clark, Nucl. Phys. A **328**, 73 (1979)
39. A.D. Jackson, E. Krotscheck, D. Meltzer, and R.A. Smith, Nucl. Phys. A **386**, 125 (1982)
40. J.M.C. Chen, J.W. Clark, and D.G. Sandler, Z. Physik A **305**, 223 (1982)
41. F. Coester, in *Lectures in Theoretical Physics: Quantum Fluids and Nuclear Matter* (Gordon and Breach, New York, 1969), Vol. XI B
42. M. Saarela, Phys. Rev. B **33**, 4596 (1986)
43. J. Boronat and J. Casulleras, Europhys. Lett. **38**, 291 (1997)
44. G. Senger, M.L. Ristig, C.E. Campbell, and J.W. Clark, Ann. Phys. (NY) **218**, 160 (1992)
45. M.L. Ristig, G. Senger, M. Serhan, and J.W. Clark, Ann. Phys. (NY) **243**, 247 (1995)
46. H.W. He, Ph.D. thesis, Texas A& M University, 1993

Some Applications of Correlated Basis Function Theories in Finite and Infinite Nuclear Systems

Adelchi Fabrocini

Department of Physics, University of Pisa and INFN, Sezione di Pisa, I-56100 Pisa, Italy

Abstract. We discuss two applications of Correlated Basis Function theory to nuclear physics. First we address the problem of the evaluation of the density and electromagnetic inclusive responses in infinite nuclear matter in a nonrelativistic, correlated model, including both central and tensor nucleon-nucleon correlations. Then we study the ground state of doubly closed-shell nuclei, showing how Fermi hypernetted chain equations can be extended to these systems and presenting results for the energy and the density.

1 Introduction

This contribution will be devoted to show how and why Correlated Basis Function (CBF) theory (and CBF based perturbative approaches) can be safely considered nowadays one of the most powerful tools available to nuclear physics theorists, in the nonrelativistic realm. We will concentrate on two specific problems: the evaluation of the inclusive responses in infinite, homogeneous nuclear matter and the study of the ground state (g.s.) of doubly closed shell nuclei.

The wave function of strongly interacting nuclear systems largely differs from that provided by independent particle models (IPM). These deviations may be described in terms of nucleon-nucleon (NN) correlations, arising from the nuclear interaction.

Realistic potential theories for a system of A nucleons employ Hamiltonians such as

$$H = -\frac{\hbar^2}{2m}\sum_i \nabla_i^2 + \sum_{i<j} v_{ij} + \sum_{i<j<k} v_{ijk} \ . \tag{1.1}$$

Meson exchange processes determine rather unambiguously the behavior of the two- and three-nucleon potentials, v_{ij} and v_{ijk}, at large interparticle distances, whereas a stronger uncertainty is present at shorter distances, so generating some differences in modern potentials. However, much effort has been lately paid to produce *phase equivalent* potentials, e.g. interactions that reproduce within the same degree of accuracy the large body of available NN scattering data, despite of their different parametrizations [1].

Some of the results discussed in this paper have been obtained with v_{14} NN potentials, either the Urbana-v_{14} (U14) [2] or the Argonne-v_{14} (A14) [3] interactions.

v_{14} NN potentials are parametrized as

$$v_{14,ij} = \sum_{p=1,14} v^p(r_{ij}) O^p_{ij} \ , \tag{1.2}$$

with

$$O^{p=1,14}_{ij} = [1, \boldsymbol{\sigma}_i \cdot \boldsymbol{\sigma}_j, S_{ij}, (\mathbf{L}\cdot\mathbf{S})_{ij}] \otimes [1, \boldsymbol{\tau}_i \cdot \boldsymbol{\tau}_j] \ , \tag{1.3}$$
$$[L^2, L^2 \boldsymbol{\sigma}_i \cdot \boldsymbol{\sigma}_j, (\mathbf{L}\cdot\mathbf{S})^2_{ij}] \otimes [1, \boldsymbol{\tau}_i \cdot \boldsymbol{\tau}_j] \ ,$$

where $S_{ij} = (3\hat{\mathbf{r}}_{ij} \cdot \boldsymbol{\sigma}_i \hat{\mathbf{r}}_{ij} \cdot \boldsymbol{\sigma}_j - \boldsymbol{\sigma}_i \cdot \boldsymbol{\sigma}_j)$ is the tensor operator.

CBF theories are based on a basis set of *correlated* states (CS)

$$|\Psi_n(1,2...A)\rangle = \frac{\mathcal{G}(1,2...A)|\Phi_n(1,2...A)\rangle}{\langle\Phi_n|\mathcal{G}^\dagger\mathcal{G}|\Phi_n\rangle^{1/2}} \ , \tag{1.4}$$

where $\mathcal{G}(1,2...A)$ is an A-body correlation operator, acting on the eigenstates $|\Phi_n(1,2...A)\rangle$ of an IPM Hamiltonian. The $\mathcal{G}$ operator is expected to have a similar structure to that of the NN interaction and here is taken as

$$\mathcal{G}(1,2...A) = \mathcal{S}\left(\prod F_{ij}\right) \ , \tag{1.5}$$

e.g. a symmetrized product of two-body, state-dependent correlation operators, F_{ij}, chosen of the form:

$$F_{ij} = \sum_{p=1,8} f^p(r_{ij}) O^p_{ij} \ , \tag{1.6}$$

where the variational functions $f^p(r)$ are determined by minimizing the g.s. expectation value of the Hamiltonian, $\langle H\rangle_{00} = \langle\Psi_0|H|\Psi_0\rangle$. The many-body integrals needed for the evaluation of $\langle H\rangle_{00}$, as well as of CBF matrix elements of other operators, could be in principle sampled by Monte Carlo (MC) techniques. However, MC can be efficiently used in nuclear systems for light nuclei only , so heavier systems are studied by cluster expansion and Fermi hypernetted chain (FHNC) [4] integral equations.

FHNC allows for summing infinite classes of Mayer-like diagrams and it has been already introduced and discussed by other contributors to this volume.

The strong state dependence of F_{ij} results in the non-commutativity of the correlation operators and prevents from the development of a complete FHNC theory for this type of correlated wave function. The single operator chain (SOC) approximation [5] for the operatorial correlations ($p > 1$), in conjuction with a full FHNC treatment of the $p = 1$ (or Jastrow) part, provides an apparently accurate description of nuclear and neutron matters

(NM and nm, respectively). Its estimated accuracy in the g.s. energy has been set to less than 1 MeV at saturation density [6].

Infinite nucleonic matter is well suited for CBF applications because of the advantages coming from translational invariance. The saturation properties of NM and the nm equations of state (EOS) are in satisfactory agreement with our present knowledge [6]. The same happens for other properties as momentum distribution [7] and optical potential [8].

NM dynamical quantities have also been evaluated within CBF. A perturbative correlated expansion must be implemented, where non diagonal CBF matrix elements are computed. For these specific applications, FHNC type sums are presently the only viable techniques, even if some attempts to extend MC methods are under scrutiny. Inclusive responses and one-body Green functions were studied, describing to a good degree of accuracy the heavy nuclei experimental data or the empirical NM responses, as extracted from mass formulae (see e.g. [9] and references therein). A pedagogical description of the CBF approach to the density response will be given in the second section of this contribution and some results for physically interesting cases will be shown.

The application of CBF theory to inhomogeneous systems, and in particular to finite nuclei, has a less rich and more recent history. MC sampling has been used to study the g.s. of nuclei up to ^{16}O using the correlated wave function (1.4) and realistic Hamiltonians [10]. Low order cluster expansion were adopted by Guardiola and coworkers [11] in N=Z nuclei (up to ^{40}Ca) with central interactions and Jastrow correlations.

Clearly, finite order cluster summations become more and more unreliable as A increases. In recent years, however, the FHNC method has been succesfully extended to doubly closed-shell nuclei. So far, it has been possible to study the g.s. properties of such heavy nuclei as ^{208}Pb with semirealistic interactions, without tensor parts, and isospin dependent correlations [12, 13, 14]. FHNC/SOC has been recently applied to N=Z nuclei with non central interactions and correlations, so achieving in ^{16}O and ^{40}Ca the same accuracy of modern nuclear matter CBF studies [15].

The third section of this paper will show, in some details, how the FHNC machinery can be used to compute the g.s. energy of N=Z, doubly closed shell nuclei described by a central Hamiltonian and Jastrow correlations, providing also some results for more realistic cases.

The final section will try to briefly discuss which are the perspectives and the open problems with the CBF based approaches.

2 Responses in Nuclear Matter

The dynamical structure function, $S(q,\omega)$, represents the NM response to a density fluctuation

$$\rho_{\mathbf{q}} = \sum_{i=1,A} \mathrm{e}^{\mathrm{i}\mathbf{q}\cdot\mathbf{r}_i} \quad . \tag{2.7}$$

For the sake of brevity, we will refer to $S(q,\omega)$ as the density response (DR), bearing however in mind that the actual response, $R(q,\omega)$, is given by

$$R(q,\omega) = F^2(q,\omega)\ S(q,\omega)\ . \tag{2.8}$$

$F(q,\omega)$ is the nucleon form factor, keeping track of the finite size of the nucleon itself. Fantoni and Pandharipande [16] employed CBF theory to study the DR of NM as an approximation to the longitudinal response, giving the response of the system to a charge density fluctuation and computed, within the same approach, by Fabrocini and Fantoni [17].

2.1 CBF Theory of the Density Response

The DR is obtained as the imaginary part of the density-density correlation function:

$$S(q,\omega) = \frac{1}{\pi}\ \mathrm{i}D(q,\omega)\ , \tag{2.9}$$

$$D(q,\omega) = \langle 0|\rho^\dagger_{\mathbf{q}} \frac{1}{H - E_0 - \omega - \mathrm{i}\eta} \rho_{\mathbf{q}}|0\rangle\ , \tag{2.10}$$

where $|0\rangle$ is the exact g.s. with energy E_0. Formally, H is written as a sum of an unperturbed Hamiltonian H_0 and an interaction term H_I such that

$$\langle m|H_0|n\rangle = \langle m|H|n\rangle\delta_{mn} = H_{mn}\delta_{mn}\ , \tag{2.11}$$

$$\langle m|H_\mathrm{I}|n\rangle = \langle m|H|n\rangle(1-\delta_{mn}) = \hat{H}_{mn}\delta_{mn}\ , \tag{2.12}$$

where $|n\rangle$ are basis states. An expansion is obtained by splitting

$$H - E_0 = H_0 - E_0^\mathrm{v} + H_\mathrm{I} - \Delta E_0\ , \tag{2.13}$$

where $E_0^\mathrm{v} = \langle H\rangle_{00}$ is the zeroth order CBF g.s. energy and $\Delta E_0 = E_0 - E_0^\mathrm{v}$ its perturbative correction. The term $H_\mathrm{I} - \Delta E_0$ is then treated perturbatively. This is justified on the ground that we assume that the correlation operators give a good estimate of the exact basis states ($|\Psi_n\rangle \sim |n\rangle$), so the non-diagonal matrix elements of the Hamiltonian (defining H_I) are expected to be small. The resulting expansion is

$$\frac{1}{H - E_0 - \omega - \mathrm{i}\eta} = \frac{1}{H_0 - E_0^\mathrm{v} - \omega - \mathrm{i}\eta} \sum_{l\geq 0}(-)^l \left[(H_\mathrm{I} - \Delta E_0)\frac{1}{H_0 - E_0^\mathrm{v} - \omega - \mathrm{i}\eta}\right]^l\ . \tag{2.14}$$

Equation (2.14) is then inserted into (2.10). Finally, a perturbative expansion is obtained by inserting $\sum_n |n\rangle\langle n| = 1$ between each pair of contiguous operators and expanding the true intermediate states, $|n\rangle$, in terms of the CBF states, $|\Psi_n\rangle$, by

$$|n\rangle = \sum_{l \geq 0} (-)^l \left[\frac{1}{H_0 - E_n^{\mathrm{v}}} (H_{\mathrm{I}} - \Delta E_n) \right]^l |\Psi_n\rangle \ , \tag{2.15}$$

where $E_n^{\mathrm{v}} = \langle \Psi_n | H | \Psi_n \rangle$ and $\Delta E_n = E_n - E_n^{\mathrm{v}}$.

The CBF zeroth order DR is given by

$$S_0(q, \omega) = \sum_n |\langle \Psi_0 | \rho_{\mathbf{q}}^{\dagger} | \Psi_n \rangle|^2 \delta(\omega - E_n^{\mathrm{v}} + E_0^{\mathrm{v}}) \ . \tag{2.16}$$

We notice that $S_0(q, \omega)$, as it is constructed with CS's, may already constitute a good approximation to the full response.

In NM, the intermediate states are correlated m-particle, n-hole (mp-nh) states, obtained by applying the $\mathcal{G}$ operator (1.5) to the excited np-mh states of the Fermi gas (FG). Only the 1p-1h state contributes to the FG response. In a correlated model, all np-nh CS's contribute, as the density fluctuation operator connects to the g.s. any np-nh state, with $n \geqq 1$. [1]

Using only 1p-1h CS, we obtain

$$S_{\mathrm{ph}}(q, \omega) = \sum_{p,h} |\langle \Psi_0 | \rho_{\mathbf{q}}^{\dagger} | \Psi_{p,h} \rangle|^2 \delta(\omega - \mathrm{e}_p^{\mathrm{v}} + \mathrm{e}_h^{\mathrm{v}}) \ , \tag{2.17}$$

where $\mathrm{e}_k^{\mathrm{v}}$ is the variational single particle energy of the k-state. In the FG case, we just have $\mathrm{e}_k^{\mathrm{FG}} = \hbar^2 k^2 / 2m$ and the resulting $S_{\mathrm{FG}}(q, \omega)$ is the Lindhard function [18].

Standard cluster expansion techniques allow for computing the non-diagonal matrix element

$$\xi_{\mathrm{ph}}(\mathbf{q}) = \langle \Psi_{p,h} | \rho_{\mathbf{q}} | \Psi_0 \rangle \tag{2.18}$$

at all orders. The so called two-body factorizable (TBF) approximation is a reliable one at NM densities. It consists in summing all terms that can be expressed as products of *dressed* two-body integrals. Considering only Jastrow correlations, the TBF approximation yields:

$$\xi_{\mathrm{ph}}^{\mathrm{TBF}}(\mathbf{q}) = \delta_{\mathbf{q}-\mathbf{p}+\mathbf{h}} \frac{1}{D_{\mathrm{cc}}^{1/2}(p) D_{\mathrm{cc}}^{1/2}(h)} \left[1 + \hat{Y}_{\mathrm{d}}^{\mathrm{c}}(q) \right] \ , \tag{2.19}$$

where $D_{\mathrm{cc}}(k) = 1 - \hat{X}_{\mathrm{cc}}^{\mathrm{c}}(k)$,

$$\hat{g}(k) = \rho_{\mathrm{NM}} \int \mathrm{d}^3 r e^{\mathrm{i}\mathbf{k}\cdot\mathbf{r}} g(r) \ , \tag{2.20}$$

ρ_{NM} is the NM density ($\rho_{\mathrm{NM}} = 0.16\,\mathrm{fm}^{-3}$ at saturation) and $g = X_{\mathrm{cc}}^{\mathrm{c}}$ or $Y_{\mathrm{d}}^{\mathrm{c}}$ (FHNC quantities, defined in [16]).

The lowest order matrix element is given by

---

[1] If the CS's are orthogonalized, following the Löwdin-Schmidt procedure outlined in [16], then only 1p-1h orthogonal CS's contribute to $S_0(q, \omega)$.

$$\xi_{\rm ph}^{\rm LO}(\mathbf{q}) = \delta_{\mathbf{q}-\mathbf{p}+\mathbf{h}} \left[1 + \hat{Y}_{\rm d,LO}^{\rm c}(q) + \frac{1}{2}\hat{X}_{\rm cc,LO}^{\rm c}(p) + \frac{1}{2}\hat{X}_{\rm cc,LO}^{\rm c}(h)\right] , \quad (2.21)$$

where $Y_{\rm d,LO}^{\rm c} = [(f^1)^2 - 1]$ and $X_{\rm cc,LO}^{\rm c} = (-l/\nu)[(f^1)^2 - 1]$. $l(x = k_{\rm F}r)$ is the Slater exchange function, $l(x) = 3(\sin x - x\cos x)/x^3$, $\nu = 4$ the NM spin-isospin degeneracy and $k_{\rm F} = 6\pi^2\rho_{\rm NM}/\nu$ the Fermi momentum ($k_{\rm F} = 1.33\,{\rm fm}^{-1}$ at saturation). The net result is a negative correction to the free matrix element.

This effect is clearly shown in Figure 1, where the Jastrow and FG 1p-1h responses are compared at two momentum values.

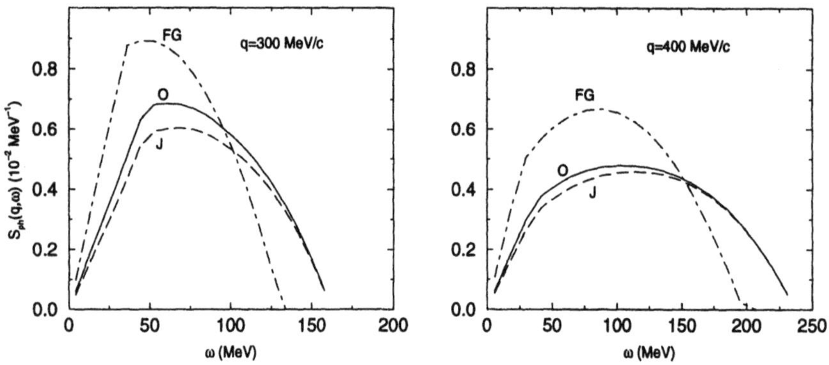

Fig. 1. 1p-1h responses per nucleon in the Jastrow (J), operatorial (O) and FG models for nuclear matter at saturation density

The most important operatorial correlations are, by far, the spin- and tensor-isospin ones, related to the long range one pion exchange potential (OPEP) and denoted as $f^{\sigma\tau}$ and $f^{t\tau}$, respectively. Their corrections to $\xi_{\rm ph}^{\rm LO}$ are:

$$\Delta Y_{\rm d,LO}^{\rm op} = 9\left[(f^{\sigma\tau})^2 + 2\left(f^{t\tau}\right)^2\right] \quad (2.22)$$

and

$$\Delta X_{\rm cc,LO}^{\rm op} = 9(-l/\nu)\left[2\left(f^1 f^{\sigma\tau}\right) + (f^{\sigma\tau})^2 + \ldots\left(f^{t\tau}\right)^2\right] . \quad (2.23)$$

The full operatorial 1p-1h response is also given in Figure 1.

The shift ($\sim 50\,{\rm MeV}$) of the correlated responses with respect to FG is due to the use of the correlated single-particle energies instead of $e_k^{\rm FG}$. This also leads to a quenching of the response, since the corresponding effective mass is ~ 0.7. An extra $\sim 10\%$ quenching is provided by the fact that the matrix element is lesser than 1. Operatorial correlations slightly enhance the Jastrow response, but their contribution decreases along q.

Other contributions come from p-h re-scattering and perturbative corrections to both $|\Psi_0\rangle$ and $|\Psi_{p,h}\rangle$. The former has been evaluated by solving a correlated Dyson equation [16] and finding little contribution, already at $q = 2.0\,\mathrm{fm}^{-1}$; the latter are obtained by proper, correlated self-energy insertions [17]. Their main effect is the appearence of a tail at large energy.

2.2 Comparison with Experimental Data

The cross section for inclusive electron scattering at intermediate three-momentum transfers ($q \lesssim 3\,\mathrm{fm}^{-1} \sim 600\,\mathrm{MeV/c}$) in the one-photon exchange approximation is

$$\frac{\mathrm{d}^2\sigma}{\mathrm{d}\Omega\mathrm{d}\omega} = \sigma_\mathrm{M}\left\{\frac{q_\mu^4}{q^4}R_\mathrm{L}(q,\omega) + \left[\tan^2\left(\frac{\theta}{2} - \frac{q_\mu^2}{2q^2}\right)R_\mathrm{T}(q,\omega)\right]\right\} , \tag{2.24}$$

where σ_M is the Mott cross section, $q_\mu^2 = \omega^2 - q^2$ the squared four-momentum transfer and θ the scattering angle. $R_{\mathrm{X=L,T}}(q,\omega)$ are the longitudinal (LR) and trasverse (TR) responses obtained by the Rosenbluth separation (for a recent discussion of the argument, see [19]).

The total cross section of medium-heavy nuclei is well described by a FG model with a scaled effective mass [20]. However, the agreement fades away when the L/T separation is carried on, since the FG overestimates the experimental LR.

The charge density operator, driving the LR, is

$$\rho_{L,\mathbf{q}} = \sum_{i=1,A}\left(\frac{1+\tau_z(i)}{2}\right) \mathrm{e}^{\mathrm{i}\mathbf{q}\cdot\mathbf{r}_i} , \tag{2.25}$$

where a small neutron contribution has been omitted.

The elecromagnetic current operator, $\mathbf{j}(\mathbf{q})$, is the one needed in the RT. It is given by the sum of several terms

$$\mathbf{j}(\mathbf{q}) = \mathbf{j}_\mathrm{M}^{(1)}(\mathbf{q}) + \mathbf{j}_\mathrm{C}^{(1)}(\mathbf{q}) + \mathbf{j}_\mathrm{MEC}^{(2)}(\mathbf{q}) , \tag{2.26}$$

$\mathbf{j}_\mathrm{M,C}^{(1)}(\mathbf{q})$ being the one-body magnetization (M) and convection (C) currents and $\mathbf{j}_\mathrm{MEC}^{(2)}(\mathbf{q})$ is the two-body meson exchange current (MEC).

Using the one-body current alone gives the impulse approximation (IA) to TR, $R_\mathrm{T}^\mathrm{IA}(q,\omega)$. As the convection current contributes by only a few percent to the RT non energy weighted sum, except that at very low q-values, we will not consider it. The magnetization current is written as:

$$\mathbf{j}_\mathrm{M}^{(1)}(\mathbf{q}) = \mathrm{i}\frac{\mu_0}{e}\sum_{i=1,A}\left(\mu_p\frac{1+\tau_z(i)}{2} + \mu_n\frac{1-\tau_z(i)}{2}\right) \mathrm{e}^{\mathrm{i}\mathbf{q}\cdot\mathbf{r}_i}\boldsymbol{\sigma}_i \times \mathbf{q} , \tag{2.27}$$

where μ_0 is the nuclear magneton and $\mu_{p,n}$ are the nucleon magnetic moments.

In the majority of the literature concernig two-body currents, MEC's associated with one-pion exchange and originating from the contact, pionic and intermediate Δ Feynman diagrams have been used. However, these currents do not satisfy the continuity equation with a realistic NN potential. To overcome this inconsistency, in [21] the authors used the continuity equation with A14 and U14 potentials to derive model-independent (MI) exchange currents. The two MEC models will be referred to as OPE and SPR, respectively. The second model includes also a model dependent (MD) current, associated to π and ρ MEC's with intermediate Δ excitations and to $\rho\pi\gamma$ and $\omega\pi\gamma$ currents.

The TR is then written as

$$R_{\mathrm{T}}(q,\omega) = R_{\mathrm{T}}^{\mathrm{IA}}(q,\omega) + R_{\mathrm{T}}^{\mathrm{IN}}(q,\omega) + R_{\mathrm{T}}^{(2)}(q,\omega) \ , \tag{2.28}$$

where $R_{\mathrm{T}}^{\mathrm{IN}}(q,\omega)$ is the interference contribution between the one- and two-body currents and $R_{\mathrm{T}}^{(2)}(q,\omega)$ is quadratic in the MEC's.

Fabrocini and Fantoni [17] used CBF to estimate LR in a realistic NM model employing U14 and the TNI many-body interaction [22] for the variational determination of the wave function. In the TNI model a small density dependence is introduced in the NN interaction and a density dependent contribution is added to the energy, whose parameters are fixed by reproducing the empirical NM saturation properties. A dipole fit of the proton form factor

$$F(q,\omega) = \left[\ 1 \ + \frac{q^2-\omega^2}{(839\ \mathrm{MeV})^2}\ \right]^{-2} \ , \tag{2.29}$$

was adopted. An equivalent CBF study of TR was later performed in [23] using the SPR two-body current model.

The 1p-1h responses are given in Figure 2, at $q = 400\,\mathrm{MeV/c}$. The left panel contains also the density response, whereas the right one shows the different contributions to TR. The dominant interference comes from the one-body/contact (OB/C) term. If only Jastrow correlations are present, then the remaining two terms, the one-body/pionic (OB/π) and the one-body/Δ (OB/Δ), are large and negative, almost cancelling the first one (this also happens in the FG model). However, the inclusion of the operatorial correlations (mostly the OPEP components) leads to almost vanishing OB/π and OB/Δ contributions [23]. The quadratic MEC term results to be very small.

Finally, we compare the CBF responses with the experimental ones, as measured in some nuclei. The density dependence of real life, finite nuclei is accounted for by using the NM results in local density approximation. Figure 3 gives LR and Figure 4 TR. In both cases, the spreading of the 1p-1h states into 2p-2h was obtained by folding the 1p-1h response with the imaginary part of the CBF optical potential [16].

Figure 3 gives also the relativistic, single-particle model response of [25]. The agreement of the CBF LT with the experiments is excellent, showing that short-range correlations are responsible for quenching the response respect to

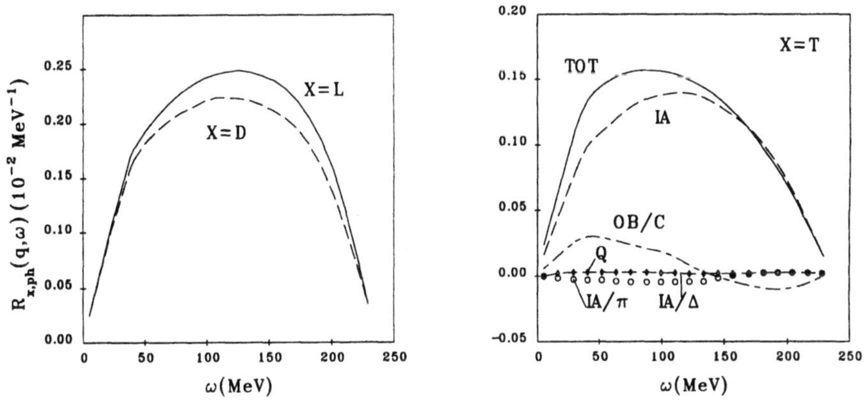

Fig. 2. CBF 1p-1h longitudinal (left panel) and transverse (right panel) NM responses per nucleon at $q = 400\,\mathrm{MeV/c}$. The right panel compares the density with the longitudinal response. The IA, interference and quadratic contributions to TR are given. Q is the quadratic MEC contribution; TOT is the total response.

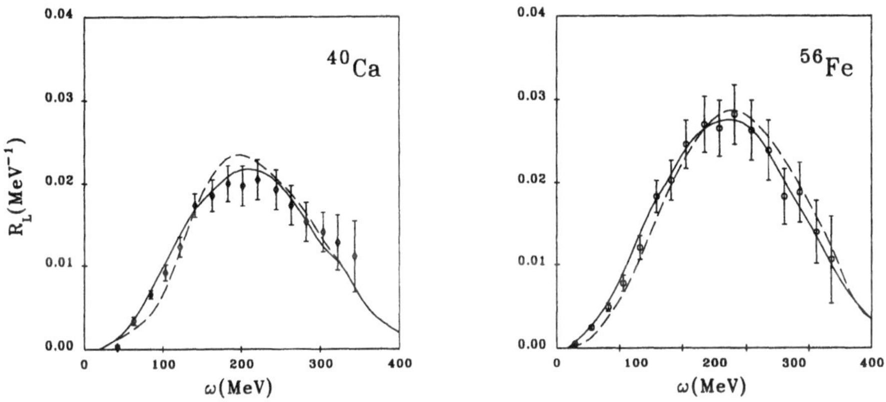

Fig. 3. Experimental and theoretical longitudinal response at $q = 570\,\mathrm{MeV/c}$ for ^{40}Ca and ^{56}Fe. Experimental data from [24]. The solid lines give the CBF response; the dashed lines are the results of [25].

the FG model. This fact is confirmed by an accurate analysis of the Coulomb sum $S_{\mathrm{L}}(q)$

$$S_{\mathrm{L}}(q) = \frac{1}{Z} \int \mathrm{d}\omega \frac{R_{\mathrm{L}}(q,\omega)}{F^2(q,\omega)} \ , \tag{2.30}$$

where Z is the atomic number of the nucleus. In a non-relativistic model, $S_{\mathrm{L}}(q \to \infty) = 1$. This limit is reached in the FG model for $q \geq 2k_{\mathrm{F}} \sim 500\,\mathrm{MeV/c}$. CBF predicts a reduction of the order of a few percent of the Coulomb sum, relative to the asymptotic limit, for $q \sim 500 - 600\,\mathrm{MeV/c}$ [26, 17]. The recent re-analysis of the world data by Jourdan [19] has found

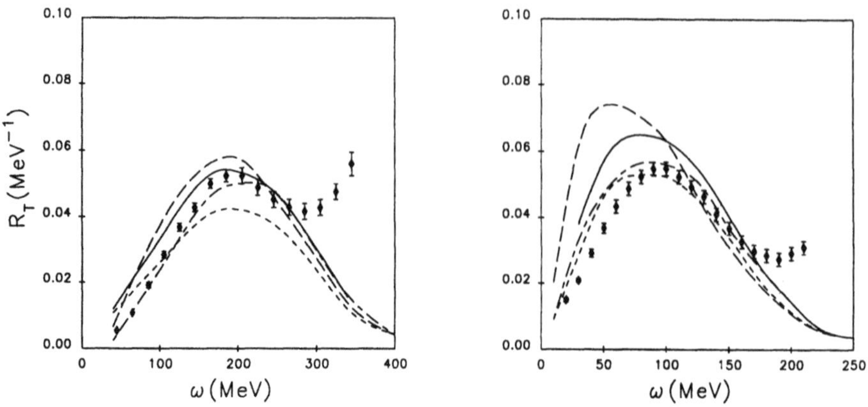

Fig. 4. ^{40}Ca experimental and CBF transverse response at $q = 380\,\text{MeV/c}$ (left) and $q = 570\,\text{MeV/c}$ (right). Experimental data from [24].

complete agreement with the CBF estimates. For instance, at $q = 570\,\text{MeV/c}$ $S_L(\text{CBF}) = 0.96$ vs. the experimental values $S_L(^{12}\text{C}) = 0.95 \pm 0.16$ and $S_L(^{56}\text{Fe}) = 0.98 \pm 0.15$.

The different curves in Figure 4 correspond to the IA (dot-dashed) and the IA+MEC (dashed) responses. Also given are the pure shell model (dotted) and shell model plus NM correlations (solid) responses. The last estimate is obtained by extracting from the NM calculations the correlation effects as the difference between the full response and the mean field one (corresponding to $\xi_{\text{ph}}(\mathbf{q}) = 1$ in (2.18)).

MEC's enhance the one-body response, worsening the agreement with the most recent data point at the lower momentum but improving it at the higher one. It has to be recalled that MEC's include a model dependent piece, corresponding to intermediate Δ-isobar excitations. This part of the current may be particularly sensible to the treatment of the Δ itself and, probably, the inclusion of explicit isobar degrees of freedom in the nuclear wave function may be needed. In addition, the transverse channel might be more heavily affected by higher-order perturbative corrections, as 1p-1h rescattering.

3 A Simple Model of Correlated Nucleus

The aim of this section is to describe how to compute the g.s. energy of a spherically symmetric, doubly closed-shell nucleus, with equal number of protons and neutrons, within a correlated model employing a Jastrow correlation operator

$$\mathcal{G}^{\text{J}}(1,2...A) = \prod f(r_{ij}) \ . \tag{3.31}$$

The mean field g.s. wave function $\Phi_0(1,2...A)$ is the Slater determinant of single-particle wave function $\phi_\alpha(i)$

$$\Phi_0(1,...A) = \mathcal{A}\left[\phi_1(1)...\phi_A(A)\right] \ , \tag{3.32}$$

where $\mathcal{A}$ is the antisymmetrization operator and $\phi_\alpha(i)$ are eigenfunctions of the single-particle Hamiltonian

$$h_{\rm sp}(r) = -\frac{\hbar^2}{2m}\nabla_r^2 + U(r) \ . \tag{3.33}$$

Fantoni and Rosati [27] originally generalized the FHNC theory with Jastrow correlation to finite Bose and Fermi systems. Later, Krotscheck et al. [28] implemented and applied the theory to study surface properties of quantum liquids. Quantitative studies of the nuclear g.s. have been then carried on by Có et al. [12, 13] and Arias de Saavedra et al. [14].

3.1 One and Two-Body Densities

Key quantities entering the calculation of $\langle H\rangle_{00}$ (and, in general, of the expectation value of any one- or two-body operator) are the one- and two-body densities (OBD and TBD)

$$\rho_1(\mathbf{r}_1) \equiv \rho(\mathbf{r}_1) \equiv \langle\Psi_0|\sum_i \delta(\mathbf{r}_1 - \mathbf{r}_i)|\Psi_0\rangle \ , \tag{3.34}$$

$$\rho_2(\mathbf{r}_1,\mathbf{r}_2) \equiv \langle\Psi_0|\sum_{i\neq j} \delta(\mathbf{r}_1 - \mathbf{r}_i)\delta(\mathbf{r}_2 - \mathbf{r}_j)|\Psi_0\rangle \ , \tag{3.35}$$

Let's outline the procedure to calculate the densities within the FHNC summation technique. Cluster diagrams contributing to the TBD are denoted as two-point renormalized FHNC (RFHNC) diagrams, to distinguish them from the standard FHNC diagrams of translationally invariant many-body systems, as nuclear matter or liquid Helium, and discussed by Fantoni elsewhere in these proceedings.

The RFHNC diagrams, besides the two external points 1 and 2, representing the arguments of $\rho_2(\mathbf{r}_1,\mathbf{r}_2)$, may have any number of internal points $(i \neq 1,2)$, irrespective of the mass number A of the nucleus. For each internal point i, the integration over $\mathbf{r}_i$ and the summation over spin-isospin states are implied.

The points are joined either by dynamical correlations (f^2-1, h-bonds), or by exchange links (e-bonds), representing the uncorrelated density matrix

$$\rho_0(\mathbf{r}_i,\mathbf{r}_j) = \sum_{\alpha\leq\alpha_{\rm F}} \phi^\dagger_\alpha(i)\phi_\alpha(j) \ , \tag{3.36}$$

where the sum on α is extended to the $A/4$ states of the Fermi sea of the A-nucleus considered.

The diagonal part of (3.36) is the uncorrelated OBD

$$\rho_0(\mathbf{r}_i) = \sum_{\alpha \leq \alpha_F} |\phi_\alpha(i)|^2 \ . \tag{3.37}$$

We stress that if the $|l^2 s^2 l_z s_z\rangle$ representation is used for the single-particle orbitals, the uncorrelated density matrix is not only diagonal in the spin variables but also independent on them. Moreover, there is no isospin dependence, as N=Z and we assume the neutrons and protons single-particle states being the same (no Coulomb interaction).

The RFHNC expansion is linked, as in the homogeneous case, since there are no isolated points. However it is not irreducible (contrary to FHNC); in fact, each point i carries a vertex correction $\xi(\mathbf{r}_i)$. There are three types of points and, correspondingly, three different vertex corrections: type d points, reached by dynamical correlations only and vertex corrected by $\xi_\mathrm{d}(\mathbf{r}_i) = \rho(\mathbf{r}_i)$, type e points, reached by two exchange links and by at least one h-bond and vertex corrected by $\xi_\mathrm{e}(\mathbf{r}_i)$, and type p points, reached by two e-bonds and by no h-bonds and whose vertex correction is $\xi_\mathrm{p}(\mathbf{r}_i) = \xi_\mathrm{e}(\mathbf{r}_i) - 1$. [2]

Apart the vertex corrections, the RFHNC equations are formally close to the FHNC ones, with the obvious spatial vector dependence coming from not being translationally invariant. In particular, the topological classification of the nodal(N_xy) and elementary(E_xy) diagrams and of the partial distribution functions, g_xy, (xy=dd, de, ed, ee and cc) is the same and it is illustrated by Fantoni in his lectures.

If the label i stands for the coordinate $\mathbf{r}_i$, the RFHNC equations in the xy=dd, de, ed and ee cases are

$$N_\mathrm{xy}(1,2) = \sum_{\mathrm{x'y'}} \int \mathrm{d}^3 r_3 \quad [g_\mathrm{xx'} - \delta_\mathrm{xd}\delta_\mathrm{x'd} - N_\mathrm{xx'}]_{13}\, \xi_\mathrm{x'y'}(3) \quad [g_\mathrm{y'y} - \delta_\mathrm{y'd}\delta_\mathrm{yd}]_{32} \ , \tag{3.38}$$

where

$$g_\mathrm{dd}(1,2) = f^2(r_{12}) \mathrm{e}^{N_\mathrm{dd}(1,2)+E_\mathrm{dd}(1,2)} \ , \tag{3.39}$$

$$g_\mathrm{de}(1,2) = g_\mathrm{dd}(1,2)\,[N_\mathrm{de}(1,2) + E_\mathrm{de}(1,2)] = g_\mathrm{ed}(2,1) \ , \tag{3.40}$$

$$g_\mathrm{ee}(1,2) = g_\mathrm{dd}(1,2)\Big[N_\mathrm{ee}(1,2) + E_\mathrm{ee}(1,2) + (N_\mathrm{de}(1,2) + E_\mathrm{de}(1,2)) (N_\mathrm{ed}(1,2) + E_\mathrm{ed}(1,2)) - 4\,(N_\mathrm{cc}(1,2) + E_\mathrm{cc}(1,2) - \rho_0(1,2))^2\Big] \tag{3.41}$$

The possible x'y' connections are dd, ed and de, $\xi_\mathrm{dd} = \xi_\mathrm{d}$ and $\xi_\mathrm{xy=de,ed} = \xi_\mathrm{e}$.

cc-diagrams are further classified according to whether or not the external points are reached by h-bonds. So, we have

$$N_\mathrm{cc}^\mathrm{hh}(1,2) = \int \mathrm{d}^3 r_3 \quad [g_\mathrm{cc} - N_\mathrm{cc} + \rho_0]_{13}\, \xi_\mathrm{e}(3) \quad [g_\mathrm{cc} - N_\mathrm{cc}^{\mathrm{h}\rho} - N_\mathrm{cc}^{\rho\rho} + \rho_0]_{32} \ , \tag{3.42}$$

[2] In ordinary FHNC diagrams containing p points exactly cancel.

$$N_{cc}^{h\rho}(1,2) = \int d^3r_3 \quad [g_{cc} - N_{cc} + \rho_0]_{13}\, \xi_e(3) \qquad (3.43)$$
$$[N_{cc}^{h\rho} + N_{cc}^{\rho\rho} - \rho_0]_{32} = N_{cc}^{\rho h}(1,2) \ ,$$

$$N_{cc}^{\rho\rho}(1,2) = -\int d^3r_3 \rho_0(13)\xi_e(3) N_{cc}^{h\rho}(32) \qquad (3.44)$$
$$- \int d^3r_3 \rho_0(13)\xi_p(3)\, [N_{cc}^{\rho\rho} - \rho_0]_{32} \ ,$$

with

$$g_{cc}(1,2) = g_{dd}(1,2)\, [N_{cc}(1,2) + E_{cc}(1,2) - \rho_0(1,2)] \ , \qquad (3.45)$$

and $N_{cc} = N_{cc}^{hh} + N_{cc}^{h\rho} + N_{cc}^{\rho h} + N_{cc}^{\rho\rho}$.

The vertex corrections have the following structure:

$$\xi_d(1) = [\rho_0(1) + U_e(1)]\, e^{U_d(1)} \ , \qquad (3.46)$$

$$\xi_e(1) = e^{U_d(1)} \ , \qquad (3.47)$$

where the functions $U_{x=d,e}(1)$ are solutions of integral equations given in Appendix A of [12].

Finally, the TBD is expressed in terms of the RFHNC functions as

$$\rho_2(1,2) = \xi_d(1)\xi_d(2)g_{dd}(1,2) + \xi_d(1)\xi_e(2)g_{de}(1,2) \qquad (3.48)$$
$$+\xi_e(1)\xi_d(2)g_{ed}(1,2) + \xi_e(1)\xi_e(2)g_{ee}(1,2) \ ,$$

Both the OBD and the TBD must satisfy the normalization conditions

$$S_1 = \frac{1}{A} \int d^3r_1 \rho(1) = 1 \ , \qquad (3.49)$$

$$S_2 = \frac{1}{A(A-1)} \int d^3r_1 d^3r_2 \rho_2(1,2) = 1 \ . \qquad (3.50)$$

If the RFHNC equations are exactly solved, then the above conditions are fulfilled. However, the elementary functions, E_{xy}, are to be viewed as an input for the solution of the equations, as no exact method has been so far devised for their evaluation. Different recipes and approximations are available in literature: the simplest one consists in setting $E_{xy} = 0$ (RFHNC/0 approximation). This seemingly crude approximation is well suited for nuclear physics because of the low densities involved, whereas it fails to give an accurate description of denser systems, as liquid Helium. The degree of violation of the normalizations (3.49) and (3.50) provides a measure of the accuracy of the solution.

3.2 Some Results

This section contains some results for the densities and the g.s. energy of ^{16}O and ^{40}Ca nuclei described by a Jastrow correlated wave function.

The best variational choices for both $f(r)$ and $\phi_\alpha(\alpha)$ would be the solutions of the Euler equations $\delta H_{00}/\delta f(r) = 0$ and $\delta H_{00}/\delta\phi_\alpha(\alpha) = 0$. However, even an approximated solution of these equations is a prohibitive task when realistic nucleon-nucleon interactions and correlations are considered. An alternative and effective procedure consists in parametrizing the wave function and minimizing H_{00} with respect to the variational parameters.

Following this approach, we have generated the single particle wave functions with an harmonic oscillator well characterized by the oscillator length $b = \sqrt{\hbar/m\omega}$. The correlation functions has the Gaussian form

$$f(r) = 1 + \alpha\,\mathrm{e}^{-\beta r^2} \ . \tag{3.51}$$

b, α and β are the variational parameters.

The energy has been minimized in the RFHNC/0 approximation and using the Jackson-Feenberg form for the kinetic energy [4]. We have employed the semi-realistic central interaction (without tensor components) S3 by Afnan and Tang [29]. S3 reproduces the s-wave two-body scattering data up to roughly 60 MeV and provides reasonable values of the binding energies and of the radii of the $A = 3,4$ nuclei. Since the original S3 potential is only defined for two-particle states of even parity, in analogy with what has been done in [11], the potential has been supplemented in the odd channels with a repulsive interaction given by the repulsive term of the even channels. Details on the actual calculation of H_{00} are given in [12].

The g.s. energies per nucleon and the variational parameters are shown in Table 1. Since we are not working in the center of mass reference frame, the true energy of the nucleus is $E_{\text{gs}} = H_{00} - T_{\text{cm}}$, where $T_{\text{cm}} = \frac{3}{4}\hbar\omega$ is the c.m. kinetic energy for a shell-model mean field. The second line for each nucleus gives the results for the potential and total energy in the FHNC-1 approximation [12]. In FHNC-1 the only elementary diagram linear in the h-bond, belonging to E_{ee}, is computed. Its contribution may be relevant for potentials having large exchange components, as S3, whereas is negligible for the other pieces of the energy.

The errors in the normalizations (3.49) and (3.50) in RFHNC/0 are well less than 1 % in both nuclei. The same degree of accuracy was found in ^{16}O [12] for the energy by comparing the RFHNC results with the available, exact Monte Carlo calculations.

Figures 5 and 6 give the OBD's and the two particle distribution functions (TPDF), $\rho_2(r_{12})$, compared to the mean field estimates. The TPDF is defined as

$$\rho_2(r_{12}) = \frac{1}{A}\int \mathrm{d}^3R_{12}\rho_2(1,2) \ , \tag{3.52}$$

where $\mathbf{R}_{12} = \frac{1}{2}(\mathbf{r}_1 + \mathbf{r}_2)$ is the center of mass coordinate.

Table 1. Energies per nucleon and variational parameters for ^{16}O and ^{40}Ca. $\langle v \rangle$ and $\langle T \rangle$ are the potential and kinetic energies. Energies in MeV and lengths in fm. See text.

	b	α	β	$\langle v \rangle$	$\langle T \rangle$	H_{00}	E_{gs}
^{16}O	1.538	-0.50	1.85	-27.38	21.07	-6.31	-7.13
				-26.59		-5.52	-6.34
^{40}O	1.654	-0.53	2.05	-36.97	25.90	-11.07	-11.36
				-36.08		-10.18	-10.47

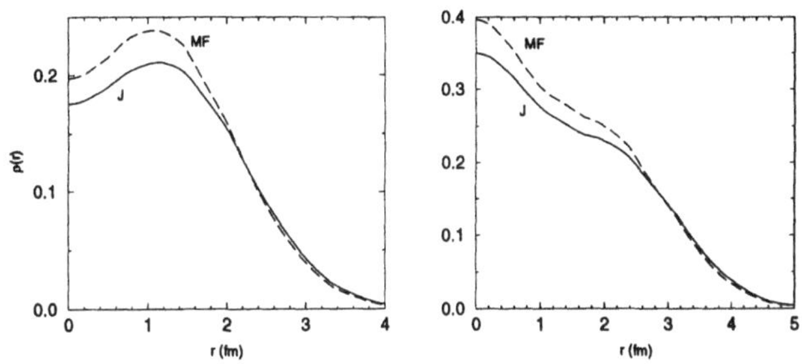

Fig. 5. One-body densities for ^{16}O (left panel) and ^{40}Ca (right panel) in Jastrow (J) and mean field (MF) models.

It is transparent from Figures 5 and 6 the effect of the correlations in reducing both the density and the distribution function because of the short-range repulsive interaction.

The semi-realistic S3 potential provides, in the Jastrow model, two bound nuclei. A comparison with the experimental binding energies is just a matter of curiosity, given the simplicity of both the potential and the correlation and the lack of a three-body potential, that was found to be important to correctly reproduce the g.s. energies of light nuclei. [3]

4 Conclusions and Perspectives

We hope to have made the readers a little more confident with the CBF theory, the technology behind it and the chain summation method. Two

[3] The experimental binding energies are -7.72 MeV in ^{16}O and -8.30 MeV in ^{40}Ca.

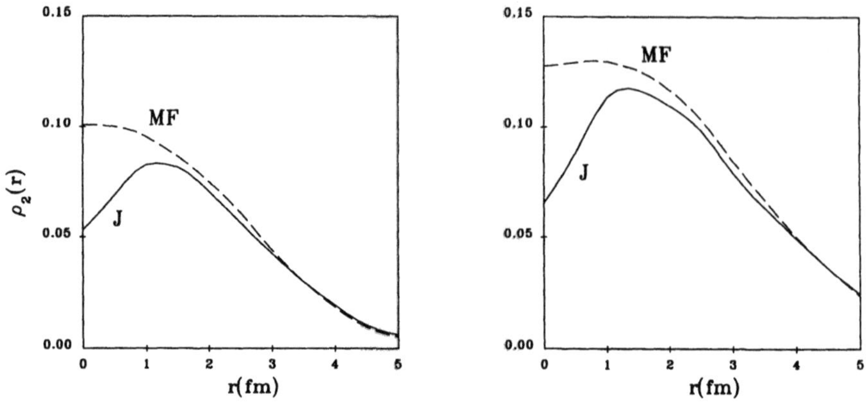

Fig. 6. Two particle distribution functions for ^{16}O (left panel) and ^{40}Ca (right panel) in Jastrow (J) and mean field (MF) models.

relatively simple models in hogeneous, infinite nuclear matter (the density response) and N=Z, doubly closed-shell nuclei (g.s. properties with Jastrow correlations and central Hamiltonian) have been illustrated, discussing some numerics and the comparison with experimental results or theoretical benchmarks.

There is a number of open theoretical and practical problems with these theories. For instance, the accuracy of the cluster expansion with non-commuting correlations must be thorougly checked and implemented. This point is of crucial relevance given the need of an accurate knowledge of the high-density equation of state of nucleonic matter requested by either modern heavy-ion physics or by neutron stars (and related physics) studies. We would like also to mention that perturbative CBF corrections to the nuclear matter g.s. energy with realistic Hamiltonians are still evaluated by low order cluster expansion [8, 30] and efforts should be made either to improve on that or to enrich the correlated wave function. A possibility presently under scrutiny is the insertion of three-body correlations, as done in light nuclei.

The g.s. of medium-heavy nuclei needs to be studied by fully realistic models. The most recent studies include tensor forces and correlations [23], but the full richness of the interaction must still be addressed. Moreover, the dynamics of these systems is an open challenge to any many-body theory and, in particular, to CBF.

References

1. R.B. Wiringa, V.G.J. Stoks and R. Schiavilla, Phys. Rev. C **51**, 38 (1995)
2. I.E. Lagaris and V.R. Pandharipande, Nucl. Phys. **A359**, 331 (1981)
3. R.B. Wiringa, R.A. Smith and T.L. Ainsworth, Phys. Rev. C **19**, 1207 (1984)

4. S. Rosati, in *From Nuclei to Particles*, Proc. Int. School E. Fermi, course LXXIX, edited by A. Molinari (North Holland, Amsterdam, 1981)
5. V.R. Pandharipande and R.B. Wiringa, Rev. Mod. Phys. **51**, 821 (1979)
6. R.B. Wiringa, V. Ficks and A. Fabrocini, Phys. Rev. C **38**, 1010 (1988)
7. O. Benhar, A. Fabrocini and S. Fantoni, Phys. Rev. C **41**, R24 (1990)
8. A. Fabrocini and S. Fantoni, Phys. Lett. **B298**, 263 (1993)
9. O. Benhar, A. Fabrocini and S. Fantoni, in *Modern Topics in Electron-Scattering*, edited by B. Frois and I. Sick (World Scientific, Singapore, 1992), p. 460
10. S.C. Pieper, R.B. Wiringa and V.R. Pandharipande, Phys. Rev. Lett. **64**, 364 (1990)
11. R. Guardiola et al., Nucl. Phys. **A371**, 79 (1981)
12. G. Có et al., Nucl. Phys. **A549**, 439 (1992)
13. G. Có et al., Nucl. Phys. **A568**, 73 (1994)
14. F. Arias de Saavedra et al., Nucl. Phys. **A605**, 359 (1996)
15. A. Fabrocini et al., Pisa Preprint IFUP-TH 40/97
16. S. Fantoni and V.R. Pandharipande, Nucl. Phys. **A473**, 234 (1987)
17. A. Fabrocini and S. Fantoni, Nucl. Phys. **A503**, 375 (1989)
18. A.L. Fetter and J.D. Walecka, *Quantum Theory of Many-Particle Systems* (McGraw-Hill, New York, 1971)
19. J. Jourdan, Phys. Lett. **B353**, 189 (1995)
20. E.J. Moniz et al, Phys. Rev. Lett. **26**, 445 (1971)
21. R. Schiavilla, V.R. Pandharipande and D.O. Riska, Phys. Rev. C **40**, 2294 (1989)
22. B. Friedman and V.R. Pandharipande, Nucl. Phys. **A361**, 501 (1981)
23. A. Fabrocini, Phys. Rev. C **55**, 338 (1997)
24. J. Jourdan, Nucl. Phys. **A603**, 117 (1996)
25. Y. Jin et al., Phys. Rev. C **45**, 1311 (1992)
26. R. Schiavilla et al., Nucl. Phys. **A473**, 267 (1987)
27. S. Fantoni and S. Rosati, Nucl. Phys. **A328**, 478 (1979)
28. E. Krotscheck et al, Phys. Rev. B **32**, 5693 (1985)
29. I.R. Afnan and Y.C. Tang, Phys. Rev. **175**, 1337 (1968)
30. S. Fantoni and V.R. Pandharipande, Nucl. Phys. **A427**, 473 (1984)

Monte Carlo Methods in Quantum Many-Body Theories

Rafael Guardiola

Departamento de Física Atómica y Nuclear, Universidad de Valencia
Avda. Dr. Moliner, 50. 46100 - BURJASSOT (Valencia) Spain
e-mail: Rafael.Guardiola@uv.es

Abstract. This is an introduction of Monte Carlo methods for beginners and their application to some quantum many-body problems. Special emphasis is done on the methodology and the general characteristics of Monte Carlo calculations. An introduction to the applications to many-body physics, specifically the Variational Monte Carlo and the Green Function Monte Carlo, is also included.

1 Basic Concepts of Statistics

1.1 Random Variables

We are going to deal extensively with *random variables*, so it is necessary a precise definition. Usually a random variable is related to the result or outcome of a given physical action of a special nature, such that we have not a precise control on all the conditions which make unique the result of the referred action.

The typical example is the result of throwing a die. If we place ourselves in a purely (like Laplace) deterministic point of view, we could imagine being able of *exactly determining* the motion of the die and, in consequence, the result of the throwing once the initial conditions are fixed (such as height, angle, velocity, and so on). It is, however, simpler to imagine our incapacity and, in consequence, it is commonly stated that the result of throwing a die is *unexpected*, or *random* or some other synonymous. This incapacity does not mean that we have absolute ignorance about the outcome, and the partial knowledge of the action is related to the so-called *probability distribution function*, which will be abbreviated from now on as *pdf*. Insisting again with our example, if it is assumed that the die has a perfect regular form, and is perfectly homogeneous, and other additional requirements, the outcome of the action *throwing a die* is a number of the set $\{1, 2, 3, 4, 5, 6\}$, each result having a *probability* of value $1/6$.

It will not escape to the reader that the above description is somehow empty and meaningless. The set of conditions of the die and its throwing are equivalent to the word *equiprobable*, and the full statement may be recast in the obviously fallacious form *if the set of results of throwing a die are equiprobable, then each result has a probability* $1/6$.

When we toss two dice simultaneously, then the possible result for the sum of values is an integer between 2 and 12. A simple combinatorial calculation shows that there is a different number of ways of obtaining a different result: to get two points, both dice must show one point, but to get four points there are three different ways ([1,3], [2,2] and [3,1]). So, to this new physical action we must assign a random variable with values between 2 and 12, but with different probabilities for each of the results.

There is a double information to be supplied in order to define a random variable: the *domain* and the *probability distribution function.* The domain will represent the set of the accessible values, and the pdf the probability or frequency with which these values appear.

Example 1.1 *The random variable* tossing two dice *is characterized by a domain* $[2,3,4,5,6,7,8,9,10,11,12]$ *and corresponding probabilities*

$$[1/36, 2/36, 3/36, 4/36, 5/36, 6/36, 5/36, 4/36, 3/36, 2/36, 1/36] \ .$$

There are some conditions to be taken in consideration with respect to the definition of a random variable. The domain must contain all possible values that the random variable may take. Moreover, the probability related to a given value must be positive or null and, finally, the sum of all probabilities must be unity, indicating certitude.

In addition to variables which take only a discrete set of values, one may define random variables in the continuum. However, it is no longer possible to talk of probability of a given value, and instead one must limit oneself to talk of probability that the result is contained in the interval $[x, x+\Delta x]$. The relevant concept when dealing with continuous variables is then the *density of probability distribution function*, which will also be abbreviated by *pdf*, and represents the limit of the quotient of the probability of having a value in the interval $[x, x+\Delta x]$ by Δx, when $\Delta x \to 0$.

The most basic concepts related with random variables are shown in Table 1, both for discrete and for continuous variables.

In Table 1 a new concept has been introduced, namely the *cumulative* distribution function, P_i or $F(X)$, which represents the probability that the random variable takes a value less than or equal to x_i or X, for the cases of discrete and continuous distributions, respectively.

The previous equations regarding pdf's have been explicitly written for a single random variable. These distributions are termed *univariate.* A logical extension is to consider *multivariate* pdf, just by replacing the random variable x by the random vector $\mathbf{x}$, as well as other related quantities. A multivariate pdf is said to define *uncorrelated* random variables if it may be factorized in the form

$$f(\mathbf{x}) \equiv f(x_1, x_2, \ldots, x_d) = \prod_{i=1}^{d} f_i(x_i) \ .$$

Table 1. Basic concepts related to random variables

	Discrete	Continuous
Domain	$\{x_1, x_2, \ldots, x_N\}$	$[a, b]$
pdf	$p_i = \text{Prob}(X = x_i)$	$f(x)dx = \text{Prob}(x \leq X \leq x + dx)$
Cumulative	$P_i = \text{Prob}(X \leq x_i)$	$F(X) = \text{Prob}(x \leq X)$
function	$P_i = \sum_{j=1}^{i} p_j$	$F(X) = \int_a^X f(x)dx$
Positivity	$0 \leq p_i \leq 1$	$f(x) \geq 0$
Bounds	$0 \leq P_i \leq 1$	$0 \leq F(X) \leq 1$
Monotonicity	$P_i \geq P_j$ if $i > j$	$F(X) \geq F(Y)$ if $X > Y$
Normalization	$P_N = \sum_{j=1}^{N} p_j = 1$	$F(b) \equiv \int_a^b f(x)dx = 1$

In the following we are going to deal almost exclusively with continuous random variables. In most of the cases, the new concepts to be introduced or properties to be discussed may be easily adapted for discrete variables from the corresponding continuous variables cases.

Just to close this section, it is worth mentioning that many of the concepts of mathematical statistics have invaded the world of physics. The assumed interpretation of quantum mechanics (the Copenhagen interpretation) assigns a statistical character to the results of any quantum experiment. The Boltzmann and Gibbs statistical mechanics is based in a statistical description of systems of many components. Even the interest in the study of chaotic phenomena has its roots in the analysis by Poincaré of the uncertainties related to random events. The reader interested in the historical evolution of the concepts of probability and statistics will find an interesting description in the book of James Newman, *The World of Mathematics* [1].

We are aware of the fact that our presentation on statistics is too short and condensed. Classical texts in mathematical statistics have been written by Feller [2] and Gnedenko [3]. The introductory part of the book by Hammersley and Handscomb [4] is both concise and precise. The book of Malvin Kalos and Paula Whitlock [5] is an excellent reference for most of the content of this paper.

1.2 Examples of Distribution Functions

The Uniform Distribution $U(a, b)$ corresponds to a domain $[a, b]$ and the related pdf is a constant, with value $1/(b - a)$. By using the step function θ it may be described by a single equation

$$f(x) = \frac{1}{b - a}\theta(x - a)\theta(b - x) \ . \tag{1.1}$$

The cumulative function is

$$F(x) = \frac{x-a}{b-a}\theta(x-a)\theta(b-x) + \theta(x-b) \ . \tag{1.2}$$

The standard form of the uniform distribution is $U(0,1)$, i.e., $f(x) = \theta(x)\theta(1-x)$. This distribution function is of primary interest in Monte Carlo methods, being the basic tool to generate random deviates from the required distributions.

The Normal or Gaussian Distribution $N(\mu,\sigma)$ characterizes a pdf of gaussian shape, defined in the interval $[-\infty,\infty]$, and given by

$$f(x) = \frac{1}{\sqrt{2\pi}\sigma}\exp\left(-\frac{(x-\mu)^2}{2\sigma^2}\right) \ . \tag{1.3}$$

The special form of writing this distribution is because the parameters μ and σ have a specific meaning, as the mean and standard deviation of the distribution, to be defined later. The standard form of the normal distribution is $N(0,1)$, i.e., with null mean and variance equal to one. The cumulative distribution function has not a simple formula, but is related to the familiar complementary error function.

This distribution is particularly interesting in relation with the central limit theorem, to be described later.

The Exponential Distribution is given by

$$f(x) = \gamma\exp[-\gamma x] \ , \tag{1.4}$$

and defined in the domain $[0,\infty]$. This distribution is of particular relevance in physics, being related with the radioactive decay and with the collisions of particles moving in material media.

1.3 Expectation Values

Let $f(x)$ be a pdf with domain $[a,b]$, and $h(x)$ a function defined in the same domain. The *expectation value* of $h(x)$ *with respect to* $f(x)$ is defined by the integral

$$\langle h\rangle = \int_a^b f(x)h(x)dx \ . \tag{1.5}$$

In quantum mechanics expectation values of operators with respect to a given quantum state are defined in a similar way. In coordinate representation one has

$$\langle\Psi|H|\Psi\rangle = \frac{\int_a^b \Psi^*(x)H(x)\Psi(x)dx}{\int_a^b \Psi^*(x)\Psi(x)dx} \ . \tag{1.6}$$

For normalized states the denominator may be removed (it is the unity), and for *local operators*, i.e., operators not involving derivatives, the quantum mechanical expectation value is converted into

$$\langle \Psi | H | \Psi \rangle = \int_a^b \left\{ \frac{|\Psi(x)|^2}{\int_a^b |\Psi(x)|^2 dx} \right\} H(x) dx \ . \tag{1.7}$$

In the case of non-local operators (e.g. operators involving derivatives) one may define an auxiliary quantity

$$\mathcal{H}(x) = \frac{1}{\Psi(x)} H \Psi(x) \tag{1.8}$$

and express the quantum mechanical expectation value in the same form as in Eq. (1.7), but putting $\mathcal{H}(x)$ instead of $H(x)$. When the operator H is the Hamiltonian, the quantity $\mathcal{H}(x)$ is known as the *local energy*. The local energy has a very important property: when the wave function corresponds to an eigenstate of the operator H, its value is a constant, and its expectation value is the same constant. If the wave function is close to an eigenstate, one may expect the local energy to have smooth fluctuations around a constant value.

Moments of a pdf. There is a special class of expectation values, known with the name of *moments* of the pdf, and defined as follows

Mean is the expectation value of $h(x) = x$, and is normally represented by the letter μ,

$$\mu = \int_a^b x f(x) dx \ . \tag{1.9}$$

Moments of a pdf are defined as the expectation value of x^n, and usually are represented by μ'_n,

$$\mu'_n = \langle x^n \rangle \ .$$

Centered Moments are usually represented by μ_n, $n = 0, 1, 2, \ldots$, and correspond to the expectation value of the function $h(x) = (x - \mu)^n$. The moment μ_0 is the unity and the moment μ_1 is null. For low values of n the centered moments have names, but we will focus on the special case μ_2.

Variance and Standard Deviation. The centered moment μ_2 is known as the *variance* of a distribution function, and its square root as the *standard deviation*, being usually represented by the letter σ,

$$\sigma = \sqrt{\langle (x - \mu)^2 \rangle} \ . \tag{1.10}$$

A practical way of determining the variance of a pdf is given by the equation

$$\sigma^2 = \langle x^2 \rangle - \mu^2 \ . \tag{1.11}$$

Example 1.2 *The moments of the normal distribution $N(\mu, \sigma)$ have the values μ (mean) and σ (standard deviation).*

For a given pdf it may well happen that the moments have infinite values. One of these cases corresponds to the so called Cauchy pdf, given by

$$f(x) = \frac{1}{1+x^2}$$

and defined in $[-\infty, \infty]$. We will always refer to gentle pdf, with a well defined (i.e., not infinite) set of moments. The Fourier transform of these regular distributions is then expressed in terms of the moments by the equation

$$f(k) \equiv \frac{1}{\sqrt{2\pi}} \int_a^b e^{ikx} f(x)dx = \frac{1}{\sqrt{2\pi}} \sum_{n=0}^{\infty} \frac{i^n}{n!} \mu'_n k^n \ .$$

For a general function, i.e., not a power of the random variable, we have defined above the analogous of the moment μ'_1, under the name of expectation value $\langle g(x) \rangle$. It is also convenient to define the analogous to the variance for a general expectation value, as

$$s^2(g) = \langle (g(x) - \langle g \rangle)^2 \rangle \equiv \int_a^b f(x) g^2(x) dx - \left(\int_a^b f(x) g(x) dx \right)^2$$

1.4 Distribution of Mean Values

In many applications of statistics pdf's are characterized by giving just the values of the lowest moments μ and σ of the distribution. The quantum Monte Carlo theory is not an exception to this rule. In practice this means that instead of studying the full distribution, one limits himself to a less complete information. Certainly, this is almost necessary when dealing with multivariate distributions or with expectation values of functions of many variables. For example, instead of giving the distribution of energies of protons emerging from an accelerator, one may just give the values of the mean energy and the variance. There is a connection between these two values and the full pdf, and to get this connection it is convenient to study the distribution of averages of a given distribution function.

From an univariate pdf $f(x)$ one may define another univariate pdf related to the mean of N random values of f,

$$z = \frac{x_1 + x_2 + \cdots + x_N}{N} \ ,$$

and the question is which is the pdf of the new random variable z.

An operative description will clarify the meaning of the variable z and the posed question. Imagine we toss a die five times in succession, and quote as the result the sum of points divided by five. This is repeated many times, and we want to know the frequency distribution of the mentioned averages. It is important to note that each of the five values of this combined event is related to the same pdf (in the example, the probability of each outcome is 1/6), and that successive tossings are *independent.*

The probability of obtaining an average value z will then be the product of probabilities of obtaining arbitrary individual values x_1, x_2, ..., x_N, but with the constraint that their average is z. For continuous pfd this statement corresponds to

$$g(z) = \int dx_1 f(x_1) \dots \int dx_N f(x_N) \delta(z - \frac{x_1 + x_2 + \cdots + x_N}{N}) \ , \qquad (1.12)$$

where $g(z)$ is the searched distribution.

The Central Limit Theorem. In general it is not possible to get a closed result for Eq. (1.12) when considering an arbitrary value of n and a general starting distribution $f(x)$. It is possible, however, to get an approximate result for $N \to \infty$, and this limit result is the so called *central limit theorem.* A formal proof may be obtained as follows. Write down the integral representation of the delta function,

$$g(z) = \frac{1}{2\pi} \int dq \int dx_1 f(x_1) \dots \int dx_N f(x_N) \exp[iq(z - \frac{x_1 + x_2 + \cdots + x_N}{N})]$$

and rearrange terms to obtain

$$g(z) = \frac{1}{2\pi} \int dq \left[\int dx f(x) \exp[\frac{iq(z-x)}{N}] \right]^N$$

Afterwards, add and subtract μ in the exponential, where μ is the mean of the pdf $f(x)$. The resulting integral with respect to x is evaluated by means of a series expansion of the exponential,

$$\int dx f(x) e^{\frac{iq(\mu - x)}{N}} \sim \int dx f(x) \left[1 + \frac{iq(\mu - x)}{N} - \frac{q^2(\mu - x)^2}{2N^2} + \cdots \right] \ .$$

The integral of the first term of the sum is 1, corresponding to the normalization of the pdf. The integral of the second term is zero, because of the presence of the mean value μ. The integral of the third term is related to the variance σ^2 of $f(x)$. Collecting these results the seeked distribution is converted into

$$\begin{aligned} g(z) &= \frac{1}{2\pi} \int dq \exp[iq(z - \mu)] \left[1 - \frac{q^2\sigma^2}{2N^2} + \cdots \right]^N \\ &\approx \frac{1}{2\pi} \int dq \exp[iq(z - \mu)] \exp[-\frac{q^2\sigma^2}{2N}] \\ &= \frac{1}{\sqrt{2\pi}[\sigma/\sqrt{N}]} e^{-\frac{(z-\mu)^2}{2[\sigma/\sqrt{N}]^2}} \ . \end{aligned}$$

The last row of above equation is the **central limit theorem**: *the pdf $g(z)$ of the average of n random values corresponding to a pdf $f(x)$ is a gaussian distribution whose mean is the mean μ of the pdf $f(x)$ and whose variance is the variance σ^2 of the distribution $f(x)$ divided by the number of values n used to compute the average.*

This theorem is satisfied by a large class of pdf, but not all of them fulfill the conditions for its validity (e.g., the Cauchy distribution is one of the exceptions). It is a theorem of asymptotic or limit validity, the larger n the better the fulfillment. In practice, even quite small values of n satisfy with good precision the central limit theorem.

Example 1.3 *The central limit theorem is exact for a gaussian pdf $f(x)$ and for any value of n. To check this statement just follow the previous derivation but computing exactly the integral which has to be raised to the power n.*

Empirical Determination of pdf. In addition to the central limit theorem, there are two other theorems which serve to connect the empirical study of pdf with their mathematical properties.

The **Chebyshev Inequality** establishes a relation between an outcome of the random variable x and the mean of the distribution. It asserts that the probability that a random variable x takes a value *outside* the interval $[\mu - k\sigma, \mu + k\sigma]$ is bounded by $1/k^2$,

$$Pr\{|x-\mu| \geq k\sigma\} \leq \frac{1}{k^2} \; .$$

The larger k, the smaller the probability that a single sample is outside the domain of width $2k\sigma$ around the mean.

A further refinement of this inequality is the **Law of Large Numbers**, very similar to Chebyshev inequality but now in reference to an estimator of the mean $\overline{x} = (x_1 + \ldots + x_N)/N$,

$$Pr\{|\overline{x}-\mu| \geq \frac{k\sigma}{\sqrt{N}}\} \leq \frac{1}{k^2} \; .$$

The last refinement is the above described *Central Limit Theorem.* All these three properties establish the link between the empirical analysis of a given pdf and the mathematical knowledge of it. In practice the central limit theorem is related to a *confidence level,* based on the values of integrals of the normal distribution function in symmetric intervals around its mean. A selection of these values is shown in the table below.

	$k=1$	$k=2$	$k=3$
$\int_{\mu-k\sigma}^{\mu+k\sigma} N[\mu,\sigma](x)dx$	0.6826	0.9545	0.9973

One standard deviation corresponds to a 68% of confidence level, which grows to 95% when talking of two standard deviations.

The central limit theorem is used in two different ways. After an empirical sampling of mean values one may look to the results as a determination of the mean and the variance of the original distribution, or as a measure of the mean of the original distribution and a confidence limit related to the new variance σ^2/N. The square root of this quantity, $\sigma/\sqrt{N}$, is normally appended to the estimator of the mean with a $\pm$ sign, i.e., $\mu \pm \sigma/\sqrt{N}$. The *error* thus written must be interpreted as a confidence limit.

The dependence on n from the central limit theorem may also inform about the amount of work to be done so as to have a predetermined error. The unpleasant fact is that in order to halve the error the empirical sampling must deal with four times more samples. Brute force is not the best way of improving the quality of the results, and the creativity in Monte Carlo work resides in the search of procedures to lower the variance σ, which is the quantity which fixes the error scale.

Generalization of the Central Limit Theorem. The previous discussion has been limited to the study of the distribution of the mean of a pdf, and we have specifically considered univariate distributions.

The theorem may be generalized to deal with other kind of expectation values as well as with multivariate distributions. For example, consider a function $h(\mathbf{x})$ which depends on a vector random variable $\mathbf{x}$ with pdf $f(\mathbf{x})$. Define the mean and variance similarly as above

$$m = \int h(\mathbf{x}) f(\mathbf{x}) d\mathbf{x}$$

and

$$s^2 = \int [h(\mathbf{x}) - m]^2 f(\mathbf{x}) d\mathbf{x} \; .$$

By following analogous steps as previously, it may be proved that the average of N samples, which defines the random variable

$$z = \frac{1}{N} \sum_{k=1}^{N} h(\mathbf{x}_k) \; ,$$

follows a normal distribution function with mean m and variance s^2/N in the limit of large N. Apparently the pdf $f(\mathbf{x})$ does not appear in this statement. In fact, it is present in the definition of the values of the mean m and the variance s^2.

Histograms. A way of getting empirically the shape of a pdf is to build up an histogram: the domain is divided into bins of a predetermined size, and the sample is placed in the appropriate bin if it is larger than the lower limit and smaller than the upper limit. *Placing* the sample means to increment by one the number of counts at the given bin.

For a bin $[x, x+\Delta]$, the probability of the trial to be there is given by

$$P = \int_x^{x+\Delta} f(z)dz$$

and for a total number of samples N_T one should expect $P \times N_T$ samples in the bin. When Δ is small and the pdf is smooth, the probability may be approximated by

$$P \simeq \Delta f(x + \Delta/2) \ .$$

This probability may we explicitly written as an expectation value of a function $h(z) = \theta(z-x)\theta(x+\Delta-z)$, and the variance may be defined in the usual manner, by computing the expectation value of $h^2(z)$. This special function is such that $h^2(z) = h(z)$ so the variance is given by

$$\sigma^2 = P - P^2$$

or, in terms of the number of counts in the bin N and the total number of counts N_T, there results

$$\sigma^2 = \frac{N}{N_T}\left(1 - \frac{N}{N_T}\right) \ .$$

This formula may generally be approximated by N/N_T. In conclusion, if the number of counts in a given bin is N one should have fluctuations within the error bar $\sigma \simeq \sqrt{N}$. It is important to have in mind the statistical meaning of the standard deviation (i.e. as a confidence limit) and expect to have sometimes results outside one standard deviation.

1.5 How Far Is the Limit?

The central limit theorem is satisfied in the $N \to \infty$ limit. In practice we may be able to deal with very large numbers of N, but of course not infinity. Nevertheless one may try to analyze if a given number of samples is large enough to satisfy quite reasonably the central limit theorem.

Empirical Analysis. The way of analyzing empirically this property is to study empirically the distribution of averages: one generates M times averages of N samples, and the individual values are binned in an histogram. It is convenient to plot also the expected normal distribution with the parameters fixed according the central limit theorem. An eye estimation or, better, a χ^2 analysis, may measure how close we are to the fulfillment of the theorem.

This kind of analysis, which may seem purely academic, is strongly recommended in practical MC calculations: if you plan to do a total of $\mathcal{N}$ samples, then classify those samples in bunches or blocks of N samples, and collect the information as mentioned above. Of course, the quotient $\mathcal{N}/N$ should be a large integer, otherwise the histogram will have too few points. After the first block of N samples one may have already a reasonable estimate of both the mean and the variance of the related normal distribution, and these values could be used to determine the range of the histogram, typically of three standard deviations around the guessed mean so as to see the shape of the resulting distribution.

Table 2. Fit to the histogram of the averages of 100 samples

pdf	μ	σ	μ (fit)	$\sigma/\sqrt{N}$ (fit)	χ^2	μ (sampling)
$U[0,1]$	1/2	$1/\sqrt{12} \approx 0.2886$	0.4998	0.0282	12.3	0.5000 ± 0.0009
$1/2\sqrt{x}$	1/3	$2/\sqrt{45} \approx 0.2981$	0.3335	0.029	6.2	0.3337 ± 0.0009

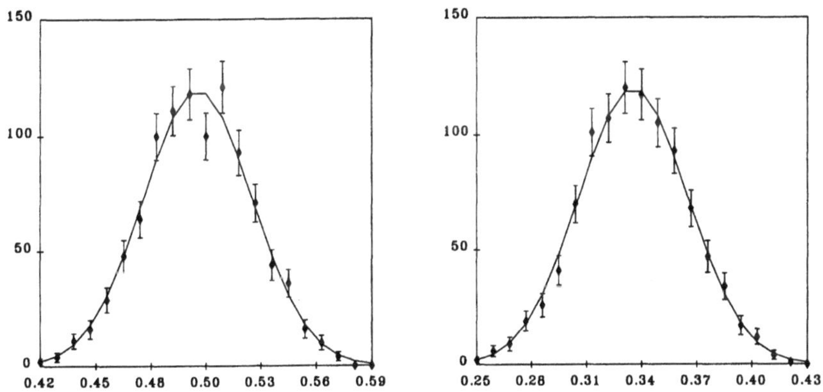

Fig. 1. Distribution of the average of 100 samples drawn from the uniform distribution $U[0,1]$ (left) and from the $1/2\sqrt{x}$ distribution (right). The gaussian in each figure is the central limit theorem prediction.

In Fig. 1 we show this kind of analysis for a total of $M = 1000$ blocks of $N = 100$ samples, for the cases of the uniform distribution $U[0,1]$ and for the inverse square root distribution $f(x) = 1/2\sqrt{x}$ defined in the domain [0,1]. From the empirical histogram one may fit a normal distribution and get the parameters. The results corresponding to these two test cases are shown in Table 2. In the table we include the mean and standard deviation of the original distribution, the mean and $\sigma/\sqrt{N}$ fitted to the gaussian and the χ^2 and, in the last column, the mean and error obtained by using the full set of samples in a single piece. There is an overall agreement between the fitted values and the expected values from the central limit theorem. Also the χ^2 value says that the empirical data are well fitted to a gaussian distribution. In conclusion, the limit of application of the central limit theorem is not very far away, even when the starting distribution is singular, like the inverse square root one.

The Central Limit Theorem for the Exponential Distribution. This is one case in which the distribution of the mean may be exactly computed. The Fourier transform of the exponential e^{-x} distribution is

$$\int_0^\infty e^{-iqx-x} dx = \frac{1}{1+iq}$$

and the distribution of the mean z of N samples is

$$g(z) = \frac{1}{2\pi} \int_{-\infty}^{\infty} dq \exp(iqz)/[1+iq/N]^N \ .$$

This integral may be computed by going to the complex $q-$ plane, obtaining

$$g(z) = \frac{N^N z^N \exp(-Nz)}{(N-1)!} \ ,$$

known as the Poisson distribution. The relation between Poisson distribution and the gaussian form related to the central limit theorem is very familiar in the statistical analysis of radioactive measurements.

1.6 Probability Distributions of Physical Interest

In section **1.3** we have already mentioned the role of the squared wave function as a probability distribution function. However, the most immediate pdf in physics is provided by Classical Statistical Mechanics, where the probability density of a system of N particles, characterized by a set of positions and momenta $\{\mathbf{r}_k, \mathbf{p}_k\}$ is proportional to $\exp(-H(\mathbf{r},\mathbf{p})/kT)$, where H is the Hamiltonian of the system. Evaluation of macroscopic properties of the system is done by computing expectation values of the required quantity with respect to this distribution function.

2 Generation of Random Numbers

2.1 Monte Carlo Simulation

Any physical problem tied to a pdf may be dealt with a *simulation* by means of random number generation. The simulation consists in statistically evolving the physical system through the generation of sequences of random numbers which are distributed with the pdf, and it is best done with the help of a computer. So, the unavoidable first step is to design algorithms to generate random numbers related to a given pdf.

Of course, if random numbers are generated by means of a deterministic algorithm, they are no longer random numbers, so the usual name is *pseudo-random* numbers. However, the fact that we can predict the sequence is independent of its randomness.

Algorithms to generate random numbers with a prescribed pdf are based on the generation of random numbers from the uniform distribution $U[0,1]$. So, the scheme of presentation will start with the uniform distribution and afterwards some methods to transform the random numbers to another pdf.

2.2 The Uniform Distribution

The main characteristics of a uniform random generator are the following:

1. First of all, it must be truly random. In other words, the generated sequence must pass all imaginable tests of randomness, including their consideration as pairs, triplets ...nth-tuples of numbers. One of the basic references regarding the analysis of randomness is the book of Knuth [6].
2. Normally, random generators are *periodic*, so that after a certain number of sequential generations the results repeat. Then one should demand a very large period, much larger than any run you need to carry out.
3. Another characteristic of interest is the speed of the generation of a random deviate. This has originated algorithms based on the hardware structure of the processor or the floating point unit tied to it. However, speed may not be so important as one may imagine, and the reason is that after generating a random number or vector one usually does a large number of calculations with it. In other words, the real overhead of the calculations is normally on subsequent processing of the random numbers. For this reason it is probably better to demand *portability*.
4. Portability means that the same sequence of random numbers may be generated by another computer or another language, just by coding appropriately the algorithm. This concept is also related to the *repeatibility* of the sequence, which obviously means the possibility of generating again a previous sequence. The main reason to demand these two properties is that a large part of the computer work is devoted to checking and tuning the code.

Usually you will find a procedure in your preferred computing language to generate uniform random deviates. Using these procedures spoils the portability of the code. One should respectfully admit that the responsibles of the compiler have taken guards with respect to the randomness, but there are known cases in which serious drawbacks have been discovered [7].

Easily coded algorithms are based on congruential methods. They consist in a simple operation on integer numbers modulus a large constant, like

$$N_i = (a \times N_{i-1} + b) \bmod M \to x_i = N_i/M \ .$$

Here a, b and M are predetermined constants, and x_i is the $U[0,1]$ random deviate. The period of the sequence is, at most, M.

Sometimes *shuffling* is added to the algorithm. This consists in maintaining several deviates and updating one of them at random.

There are many public-domain algorithms for the generation of uniform deviates [8, 9, 10].

2.3 The Method of Change of Variables

Let $x \in U[0,1]$ and apply a one-to-one transformation

$$y = h(x)$$

which maps monotonically the domain $[0,1]$ into a new domain $[a,b]$. The question is to determine the pdf corresponding to y, which will be represented by $g(y)$. The link with the starting distribution is the equality of the probabilities

$$dx = g(y)dy \ .$$

From here it is simple to obtain the form of the new pdf,

$$g(y) = \frac{dx}{dy} = \frac{dh^{-1}(y)}{dy} \ , \tag{2.13}$$

where h^{-1} is the inverse function of $h(y)$.

Unfortunately Eq. (2.13) does not solve our problem, which is to determine the change of variables $h(y)$ appropriate to our pdf $g(y)$. Solving (2.13) for h we obtain

$$h^{-1}(y) = \int_a^y g(z)dz \equiv G(y) \ , \tag{2.14}$$

where $G(y)$ is the cumulative pdf of $g(y)$. The resulting algorithm is simple: the change of variables is the inverse of the cumulative function of the pdf of interest. Of course, the algorithm will only be practical if both the cumulative function and its inverse are simple to obtain.

Some examples of the use of the method of change of variables are the following:

The Uniform Distribution $U[a,b]$, with sampling method $y = a+(b-a)x$, with $x \in [0,1]$.

The Inverse Square Root Distribution given by $1/(2\sqrt{y})$ in $[0,1]$. The cumulative distribution is $\sqrt{y}$ and its inverse is y^2, so the algorithm is to sample $x \in U[0,1]$ and square it.

The Triangular Distribution $2y$, in $[0,1]$. The sampling algorithm is $\sqrt{x}$ with $x \in [0,1]$.

The Exponential Distribution $\gamma \exp(-\gamma y)$ with $y = -\log(x)/\gamma$, for $x \in [0,1]$.

The Gaussian Distribution. The cumulative distribution function is given in terms of the complementary error function, which has not a simple form and its inversion requires the solution of a transcendental equation.

A way to bypass this problem is to consider the bivariate gaussian distribution

$$g(x,y) = \frac{1}{2\pi} e^{-(x^2+y^2)/2}$$

in polar coordinates $\{r, \phi\}$, i.e.

$$g(r,\phi) = \frac{1}{2\pi} r e^{-r^2/2} \ .$$

The angular part is the uniform distribution $U[0, 2\pi]$, and the cumulative distribution of the radial part is $\exp(-r^2/2)$, the inverse being $\sqrt{-2\log y}$. Then one may obtain pairs of gaussian distributed variates by using the pair of equations

$$\begin{aligned} y_1 &= \sqrt{-2\log x_1}\cos(2\pi x_2) \\ y_2 &= \sqrt{-2\log x_1}\sin(2\pi x_2) \ , \end{aligned} \tag{2.15}$$

with both x_1 and x_2 from the uniform distribution. This method is known as the Box–Muller method. Once y_1 and y_2 have been determined, one of the values is taken at random and the other one may be rejected or stored for a subsequent call.

If the gaussian distribution has variance σ^2 the above values must be multiplied by σ. If the mean is μ, it should be added to the generated random numbers (after the scaling).

2.4 Other Algebraic Methods

There is a large collection of methods which may not be classified within a specific scheme. Just for completeness let's mention some of them

– Given $x_1, x_2 \in U[0,1]$, $z = x_1 + x_2$ belongs to a triangular distribution

$$\begin{aligned} g(z) &= z \text{ for } z < 1 \\ g(z) &= 2 - z \text{ for } z > 1 \end{aligned}$$

defined in $[0,2]$.

- The variable $z = \max\{x_1, x_2, \ldots, x_n\}$, with $x_i \in U[0,1]$, corresponds to the pdf nz^{n-1}.
- If z is the middle value of three uniform deviates, its distribution is the parabola $6z(1-z)$
- **Sum of Random Variables:** when $z = x + y$, with $x \in f(x)$ and $y \in g(y)$, its distribution is the *convolution* product

$$h(z) = \int f(x)g(z-x)dx \ .$$

- **Product of Random Variables:** In the same conditions as above, the product $z = xy$ corresponds to the pdf

$$h(z) = \int \frac{f(x)}{x} g(z/x)dx \ .$$

In the last two methods care has to be taken with the domain boundaries of the involved distributions.

2.5 The Acceptance-Rejection Method

This method, due to von Neumann, is a general purpose method valid for *bounded distributions with finite domains.* It may be used also in multivariate distributions, but its efficiency tends to decrease with the number of variables.

Assume the pdf $f(x)$ defined in the domain $[a,b]$, and let M be an upper bound to the distribution, i.e. $f(x) \le M, \ x \in [a,b]$. To obtain a sequence of random numbers of $f(x)$ proceed as follows

1. Draw $z \in U[a,b]$
2. Draw $p \in U[0,M]$
3. If $f(z) \ge p$ then accept z as the random number, else discard both z and p and `go to 1`

The proof of this method is very simple. One should just write down mathematically the previous statements

$$\overline{f}(x) = \int_a^b \frac{1}{b-a} dz \int_0^M dp \frac{1}{M} \delta(x-z)\theta(f(z)-p) \ ,$$

where the delta function says that z will be the random number, and the theta function says that this will happen if $f(z) \ge p$.

By carrying out the integral there results the equality

$$\overline{f}(x) = \frac{1}{M(b-a)} f(x)$$

which says that we obtain a function proportional to the wanted pdf. The proportionality constant is always less than one, and simply accounts for the rejected points.

The acceptance-rejection method may be modified so as to have (on the average) efficiency one, i.e., so as to obtain as many random deviates as trials. The new scheme could be as follows:

1. Obtain $x \in U[a,b]$ and compute $f(x)$
2. Count x a total of n times, where n is a random integer with mean $(b-a)f(x)$. The way of determining this number is to compute
$$n = [p + (b-a)f(x)] \ ,$$
where $p \in U[0,1]$ and the square brackets mean the *integer part*. The result may be zero (so, x is not accepted), or 1 or larger. In the last case, the random number x must be used as many times as the value of n.
3. Go to 1

The modified von Neumann method may still be generalized to the case of a pdf $h(x)$ which may be written as the product of a pdf $f(x)$ times a *positive* rate term $g(x)$. Note that the three quantities f, g and h must be semipositive definite, and all of them defined in the same domain $[a,b]$. Then, random deviates from h are obtained by counting random deviates of $f(x)$ an integer number of times drawn from a uniform distribution with mean $g(x)$. Mathematically this statement corresponds to evaluating the integral

$$h(z) = \int_a^b f(x)dx \int_0^1 dp\delta(z-x)[g(x)+p] \equiv f(z)\int_0^1 dp[g(z)+p] \ .$$

The integer part appearing in this equation is denoted by the square brackets [ ], and it may be recast as the sum of the integer part of g plus the integer part of the sum of the fractional part of g plus p,

$$[g(z)+p] = [g(z)] + [\mathrm{frac}(g(z))+p]$$

and the second term of this sum is not null only if p is greater than $1-\mathrm{frac}(g(z))$, its value being 1. So, the integral over p is

$$\int_0^1 dp[g(z)+p] = \int_0^1 dp[g(z)] + \int_{1-\mathrm{frac}(g(z))}^1 dp \equiv [g(z)] + \mathrm{frac}(g(z)) \equiv g(z) \ ,$$

obtaining the required result.

The part assigned to $g(x)$ has been called above a *rate term*, because it acts like in a creation/annihilation process: the assumed x value may survive or reproduce or just die.

2.6 Other Methods

Just a few lines to mention two capital methods, known as Metropolis method and composition method. These two methods are closely related to the Variational Monte Carlo theory and to the Green Functions Monte Carlo method respectively, and will receive special attention in the rest of this paper.

3 Monte Carlo Quadrature

A large amount of Monte Carlo practical work is devoted to the computation of integrals. There are many numerical methods for the computation of integrals, but in some cases the Monte Carlo method is one of the most efficients. Generally speaking, Monte Carlo method is advantageous in two special circumstances: when the integration domain is complicated and/or when there are many variables involved.

In this section we are going to consider only few techniques related to Monte Carlo, with the goal of arriving to the very useful method called *importance sampling*. The review of methods is very incomplete. We will basically refer to the calculation of the integral of a bounded function of many variables in a finite many-dimensional domain.

3.1 Hit-or-Miss Method

This method is best represented by the computation of the area inside a closed line, or the volume inside a closed surface. These figures are framed with a rectangle or parallelepiped, and uniform random numbers are generated inside the rectangular figure. The ratio of the area of the figure to the area of the enclosing rectangle is the same as the (limit) ratio of the points inside the closed surface to the total number of points drawn.

This method is particularly adequate for integrals where the boundaries are defined in a quite complex form, like the phase space integrals in collision problems. The main disadvantage is that the variance is usually quite large. The method is basically the von Neumann acceptance-rejection algorithm.

3.2 Crude Monte Carlo

Crude MC is the plain application of the central limit theorem to the calculation of the integral

$$\int_a^b h(x)dx = (b-a)\int_a^b \frac{1}{b-a}h(x)dx$$

which may be interpreted as the expectation value of the function $h(x)$ with respect to the uniform distribution function $U[a,b]$.

The algorithm consists in evaluating the average of $\{h(x_i)\}$, where $x_i \in U[a,b]$. Along with the calculation of the average one may also compute the variance by evaluating the expectation value of $h(x)^2$. Practical equations are

$$\langle h\rangle = \int_a^b h(x)dx = \lim_{N\to\infty}\frac{b-a}{N}\sum_{i=1}^{N} h(x_i)\ ,\ \ x_i \in U[a,b]$$

and

$$\sigma^2 = \lim_{N\to\infty}\frac{b-a}{N}\left[\frac{1}{N}\sum_i h(x_i)^2 - \left(\frac{1}{N}\sum_i h(x_i)\right)^2\right]\ .$$

3.3 Importance Sampling

There is not an intrinsic difference between the crude MC and the importance sampling MC. Importance sampling is just another application of the central limit theorem but this time to the computation of the expectation value

$$\langle h \rangle = \int f(x) h(x) dx$$

where $f(x)$ is a pdf and $h(x)$ is an arbitrary function. The only difference, thus, with crude MC is the use of a non-uniform pdf.

Any integral may be converted into this form,

$$\int h(x) dx = \int f(x)[h(x)/f(x)] dx .$$

However, the two sides of this equation are very different, from the point of view of MC. On the lhs, one samples the uniform distribution and computes averages of $h(x)$. On the rhs, one samples the *ad hoc* introduced pdf $f(x)$ and samples the quotient of the function $h(x)$ and the pdf.

One should obtain the same value for the expectation value $\langle h \rangle$ by using both procedures. However, the main difference is in the variance of their expectation values. Crude MC will give for the variance the value

$$s^2 = \int h(x)^2 dx - \langle h \rangle^2 \ ,$$

whereas the importance sampling method will give a different value

$$s^2 = \int f(x)[h(x)/f(x)]^2 dx - \langle h \rangle^2 \ .$$

In conclusion, by cleverly choosing the importance sampling function $f(x)$ one may design a better calculation from the MC point of view.

Is there any optimal importance sampling function? The answer to this question is *yes*, the optimal function being $f(x) = C|h(x)|$, C being a proportionallity constant. Unfortunately the answer is, generally, useless. To sample from this importance function one should be able to obtain its cumulative distribution function, in other words, one should be able to compute analytically the integral.

This poses the two main problems related to the importance sampling method: first, to determine the importance sampling function so that the variance is significantly lowered and second, to find a reasonably simple sampling algorithm to draw points from the pdf.

In addition to lowering the variance, there is a side benefit of this method, namely, that it allows to deal with (null-measure) infinite integrands and/or infinite intervals, just by choosing an appropriate pdf.

To give an idea of the power of the importance sampling method the next example describes the calculation of a rather complex function of six variables.

Example 3.1 *Consider the six-dimensional integral of the function*

$$F(\mathbf{x},\mathbf{y}) = e^{-\mathbf{x}^2-\mathbf{y}^2-(\mathbf{x}-\mathbf{y})^2/2}$$

extended to the full space. The exact value is 10.9626.

Table 3 presents the calculation using the crude MC method (each variable integrated between -5 *to* $+5$*, and the calculation with the importance sampling function* $e^{-\mathbf{x}^2-\mathbf{y}^2}$*, adequately normalized. It is absolutely clear the advantage of using the importance sampling algorithm, so that the extra work required to sample from the importance sampling pdf is largely compensated by the smallness of the error. Just by comparing the errors quoted in the last row of the table we may conclude that the importance sampling method is of the order of 250 times more efficient.*

Table 3. The integral of the six-dimensional function of example 3.1. The left column shows the crude MC method, and the right column the importance sampling MC method. Each row corresponds to one hundred thousand samples, and the last row is the full average corresponding to one million samples.

Crude MC	Importance Sampling
12.80 ± 4.51	10.46 ± 0.26
8.93 ± 2.96	11.23 ± 0.26
21.90 ± 8.43	11.02 ± 0.26
12.70 ± 3.06	11.15 ± 0.26
13.84 ± 4.48	10.90 ± 0.25
4.58 ± 1.02	10.94 ± 0.26
14.84 ± 4.68	10.72 ± 0.25
8.17 ± 2.38	10.89 ± 0.25
6.79 ± 1.35	10.87 ± 0.26
5.56 ± 1.11	10.75 ± 0.25
11.01 ± 1.27	10.89 ± 0.08

Going back to the question of choosing the importance sampling method, it is already convenient to mention that in the variational MC method the importance sampling function is fixed by the problem itself, see Eq. (1.7). In addition, there is a general purpose algorithm, known as the *Metropolis* or $\mathrm{MR}^2\mathrm{T}^2$ method for sampling arbitrary pdf, which will be described below.

3.4 Some Comments

We have limited ourselves to three ways of doing MC integrals. There are many other variance reducing techniques, which however are not of primarily interest for our present discussion. The interested reader may find a variety of methods in Refs. [4, 5].

4 Markov Chains and Metropolis Algorithm

4.1 Markov Chains

A *Markov chain* is defined in terms of a discrete set of states $\{s_1, s_2, \dots s_N\}$ and a rule governing the transition from a given state s_i to any other state of the set s_j. This rule is expressed in terms of a *transition matrix* $T(j \leftarrow i) \equiv T_{ji}$ so that T_{ji} is the probability of transition from the state s_i to the state s_j. Some transition probabilities may be null, and the matrix may also contain non-null diagonal terms representing the probability of remaining at the initial state.

Given that the matrix elements are transition probabilities and that the transition probability from s_i to *any* final state must be unity, this matrix must fulfill the following two properties

$$\begin{array}{l} 0 \le T_{ji} \le 1 \\ \sum_j T_{ji} = 1 \ , \end{array}$$

so that the sum of any column is 1. Matrices fulfilling these two conditions are termed *stochastic*. If it happens that a given column has all its elements null, but the diagonal, then the state corresponding to this column is referred to as an *absorbing state*. Once it is reached, it will never move from there.

A way of visualizing a Markov chain is through its interpretation as a *random walk*. One may walk at random many physical systems. An obvious example of random walk is a cards game in which you and the contrary part bet 1 euro. Your *state* is the amount of money remaining after each game. The state with *zero euros* is an absorbing wall.

A more serious example could be the chain of events followed by a neutron traversing an absorbing medium. The neutron may scatter, or may be captured. Here there are two absorbing walls, corresponding to the events *being captured* and *having traversed the medium*.

Both examples clarify the concept of transition probability and the idea of a random walk. There is a variable, which could be called time, which signals a single step of the walk. In the case of the cards game, it is the bet. In the case of neutron absorption, it is the mean free path. Interesting examples of random walks may be found in Ref. [13]

The main difference of the second case with respect to the first case is that in the second case we may have a *continuum* of states: after a collision, the neutron may be deflected to any direction of space. In cases like this, termed *Markov processes*, instead of referring to a transition probability one must deal with a *density of probability*, $T(x', x)$ which must in turn satisfy the two properties

$$\begin{array}{l} T(x', x) \ge 0 \\ \int T(x', x) dx' = 1 \ . \end{array}$$

We will be interested in *endless* chains, i.e., with no absorbing states. We will also consider only *irreducible chains*. A chain is called *reducible* if the set of states may be divided into two disjoint sets so that there cannot be a transition from the first set to the second one, and inversely.

4.2 The Direct Problem

Which is the probability of staying at a given state s_i after a long random walk? The answer is very simple: the probability of arriving to s_i is the product of the probability of arriving to any state s_j times the transition probability from s_j to s_i,

$$P_i = \sum_j T_{ij} P_j \ ,$$

and the solution is again simple. This is an homogeneous linear system, which will have a non-trivial solution only if the set of equations is *linearly dependent.* This fact is easily checked by summing up over the free index i, and remembering the normalization condition on the transition matrix.

The solution is not uniquely defined, and it should be supplemented by a *normalization condition*

$$\sum_i P_i = 1 \ .$$

In the case of a Markov process, the equivalent to the matricial equation is the linear integral equation

$$P(x) = \int T(x, x') P(x') dx' \ ,$$

with the supplementary normalization condition $\int P(x) dx = 1$.

An alternative way of determining the probabilities P_i related to a Markov chain is to carry out a random walk. The algorithm is very simple

- Initialize `P(i)=0, i=1, ... Nstates`
- Start at predetermined state `I`
- `For k=1 to Nmoves do`
  - `J=1`
  - `Q=RAND()`
  - `While T(J,I) < Q do`
    - `Q=Q-T(J,I)`
    - `J=J+1`
    - `enddo`
  - `P(J)=P(J)+1`
- `For J=1 to Nstates do P(J)=P(J)/Nmoves`

Here `RAND()` is a procedure generating $U[0,1]$ uniform random numbers. The above algorithm is a surprising (and not very efficient) way of solving a class of linear problems. For us the best interpretation is to consider this algorithm as a method of generating random integers (the labels of the states) corresponding to a pdf P_i, without knowing the actual distribution, the generation being controlled by the Markov chain.

The case of Markov processes is more complicated. Of course, it has no sense to jump from a continuous x to another x'. Instead, the problem function $P(x)$, which is a pdf, is represented by a set of random numbers $\{\xi_i\}$ obtained sequentially, in such a form that the number ξ_{i+1} of the sequence is obtained by drawing a random number from the pfd $f(x) \equiv T(x, \xi_i)$, which is a *conditional probability* of obtaining x from a given ξ_i.

It is convenient to spend some words on the last consideration. Certainly, it is not the same to know the pfd $P(x)$ that to have a sequence of random numbers related to $P(x)$ as its own pdf. However, knowing the set of random numbers permits the determination of expectation values related to $P(x)$: just generate the sequence of random numbers as described above and evaluate the expectation value by using the central limit theorem. Of course, the method will only be practical when it would be possible to sample from the pdf $T(x, \xi_i)$.

This so simple idea is the core of the so called Green's Functions Monte Carlo method, for solving many-body Schrödinger equations.

4.3 The Inverse Problem and Metropolis Algorithm

The inverse problem is stated in very simple words: given a pdf P_i or $P(x)$ may we find a stochastic matrix T_{ij} or $T(x, x')$ defining a random walk which gives rise to the mentioned pdf's?

This question, ant its relevance with physics, was answered by N. Metropolis, A.W. Rosenbluth, M.N. Rosenbluth, A.H. Teller and E. W. Teller in 1953 [12], giving a specific rule to construct it. In addition, the random walk was so designed to be easily carried out. The method is known as *Metropolis* or MR^2T^2 method. It is a general purpose algorithm to generate random numbers from an arbitrary pdf.

For discrete distributions the solution is as follows. Let S_{ji} be an auxillary *stochastic* and *symmetric* matrix, otherwise arbitrary. The seeked stochastic matrix is given by

$$\begin{array}{lll} T_{ji} = S_{ji} & \text{if } P_j \geq P_i, & i \neq j \\ T_{ji} = S_{ji} P_j / P_i & \text{if } P_j < P_i, & i \neq j \\ T_{ii} = S_{ii} + \sum_k' S_{ki}(1 - P_k/P_i) & \text{for } k \in \{P_k < P_i\} & \end{array}$$

The prime on the sum indicates that only states k such that $P_k < P_i$ are considered. The values of the diagonal elements have been fixed so as to fulfill the property $\sum_j T_{ji} = 1$.

In order to prove that the matrix so defined satisfies the wanted relation $\sum_i T_{ji} P_i = P_j$, an intermediate step is to prove the *detailed balance* property

$$T_{ji} P_i = T_{ij} P_j \ .$$

In this relation there is no sum over repeated indices. To check this relation it is convenient to distinguish the two cases

$$
\begin{array}{lll}
P_j \geq P_i, & T_{ji}P_i = S_{ji}P_i & T_{ij}P_j = S_{ij}\frac{P_i}{P_j}P_j = S_{ij}P_i \\
P_j < P_i, & T_{ji}P_i = S_{ji}\frac{P_j}{P_i}P_i = S_{ji}P_j & T_{ij}P_j = S_{ij}P_j
\end{array}
$$

and to remember the symmetry of the auxiliary transition matrix S_{ij}.

Summing up on i in the detailed balance equation there results the desired condition $P_j = \sum_i T_{ji}P_i$.

In the case of continuous processes one may write down a similar form. Start from a stochastic and symmetric auxiliary transition function $S(y,x)$ and define

$$
\begin{aligned}
T(y,x) &= S(y,x)\theta(P(y) - P(x)) \\
&+ S(y,x)\frac{P(y)}{P(x)}\theta(P(x) - P(y)) \\
&+ \delta(y-x)\int dz S(z,y)\left(1 - \frac{P(z)}{P(y)}\right)\theta(P(x) - P(z))
\end{aligned}
$$

which again satisfies the detailed balance property

$$
T(x,y)P(y) = T(y,x)P(x) \ .
$$

Metropolis algorithm is not the only solution to the problem, but it is a procedure which may be easily used. It is interesting to present the algorithm in words. Let's assume that we are at the state s_i. By means of the auxiliary transition matrix S_{ji} we select a candidate s_j for the next state. If the probability of the candidate P_j is greater than the probability of the starting state, the transition is accepted, i.e., the next state of the chain is s_j. If the probability P_j is smaller than P_i, the new state is accepted with probability P_j/P_i. In case of rejection this means that we remain in s_i, i.e., the next state in the chain is the same as the old state, and it should be counted. Here is the algorithm:

- Let `I`, starting state
- Draw `J` from `S(J,I)`
- `if P(J)/P(I) > RAND() then I=J`
- `I` is the next state

The algorithm will be practical if we know the way of generating random numbers from the distribution S_{ji}, where i acts as a parameter and j as the random variable. The freedom in choosing the auxiliary transition function S should be used to choose the most convenient one.

In the case of continuous variables the procedure is quite similar. Usually, but not necessarily, the transition operator $S(y,x)$ is taken as a uniform distribution $U[x - \Delta/2, x + \Delta/2]$ centered at x, so the first step is to draw $y \in U[x - \Delta/2, x + \Delta/2]$, and then accept it if $P(y)/P(x) > \xi$, where $\xi \in$

$U[0,1]$. Let me insist again that if the attempted move is rejected this should be interpreted as *take x as the next random number*, so x is counted twice (or more times).

As in any other random generating algorithm, Metropolis algorithm is of asymptotic nature, i.e., it has meaning after many samples have been drawn. Moreover, specially in the continuous case, it is prone to *strong correlations* between successive moves. The main reason for this is that usually the pdf $P(x)$ is peaked around some maximum value, and also the auxiliary distribution $S(y,x)$ is of a rather short range around x. In consequence, successive random numbers are quite close, and to sample the full domain where $P(x)$ is defined it is necessary to carry out many moves.

Here we should note that there are two ways of *not moving appreciably*: when the step Δ is very small, in which case the rate of acceptances is large, but the real displacement is small, or when Δ is large, in which case the rate of rejections is large, and the effective displacement is again small. There are not specific rules to say when a certain rate of acceptances is reasonable. It is normally assumed that a rate between 50% and 80% is acceptable.

4.4 Thermalization in Markov Chains

An interesting property of stochastic matrices is that the product of two stochastic matrices is again a stochastic matrix. The proof is straightforward.

The interesting consequence of this property is that any integer power of a given stochastic matrix, which of course is again stochastic, gives rise to a pdf which is the same as the pdf related to the original transition matrix. Again the proof is very simple, and we show it here for the square of the matrix,

$$\sum_i [T^2]_{ji} P_i = \sum_{ki} T_{jk}[T_{ki}P_i] = \sum_k T_{jk}P_k = P_j \quad .$$

This property permits to *thermalize* the random walk, in the following sense. Starting at a given state s_i we proceed carrying out a given number n of steps, $\{s_{i+1}, s_{i+2} \dots s_{i+n}\}$, but we only count or collect the state s_{i+n} after n steps, and so on. In other words, $n-1$ intermediate steps are discarded. Nevertheless, the resulting collection of states is a set of random deviates corresponding to the pdf under study.

The objective of this thermalization is to minimize the effect of correlations among successive states when executing the random walk.

5 The Variational Monte Carlo Method

5.1 General Description

Given a Hamiltonian H and a trial wave function Ψ, the variational principle states that the quantity

$$\langle H \rangle = \frac{\int \Psi^* H \Psi}{\int |\Psi|^2}$$

is an upper bound to the energy of the ground state of the Hamiltonian. One may be still more precise, and consider the case in which the Hamiltonian has certain symmetries, so that the eigenstates may be classified as families with precise quantum numbers related to these symmetries. Then, if the trial function has a given symmetry, the expectation value of the Hamiltonian is an upper bound to the energy of the lowest state with that symmetry.

For example, for a harmonic oscillator Hamiltonian p^2+x^2 the eigenstates have a good parity. Then, trial functions like $\Psi(x) = C\exp(-ax^2)$ provide upper bounds to the even parity ground state, and trial functions like $\Psi(x) = Cx\exp(-ax^2)$ will provide upper bounds to the first excited state, which has negative parity.

One of the properties of a many-body Hamiltonian involving identical particles is that it will commute with any permutation of the labels of the particles. Among the operators related to permutations there are two of them of specific importance: the full antisymmetrizer and the full symmetrizer. Note that the symmetry of the trial state is not dictated by the Hamiltonian, and we should impose it externally by selecting the adequate trial function (bosonic or fermionic).

Let $\mathbf{R}$ represent the set of all coordinates of the many-body system. If the system has N particles, $\mathbf{R}$ will have $3N$ components. The expectation value of H is usually written in terms of a multivariate pdf

$$f(\mathbf{R}) = \frac{|\Psi(\mathbf{R})|^2}{\int d\mathbf{R}|\Psi(\mathbf{R})|^2}$$

and the quantity called *local energy*, already mentioned in section **1.3**, given by

$$E_L(\mathbf{R}) = \frac{1}{\Psi(\mathbf{R})} H\Psi(\mathbf{R}) \ ,$$

i.e., the action of the Hamiltonian on the trial wave function divided by the trial wave function. For the typical Hamiltonians which are the sum of a kinetic energy term, with second derivatives, and a momentum-independent interaction potential, the contribution to the local energy from the potential part is just the numerical value of the potential.

Then the expectation value of the Hamiltonian is given by

$$\langle H \rangle = \int d\mathbf{R} f(\mathbf{R}) E_L(\mathbf{R})$$

and the Monte Carlo way of evaluating this integral is to consider $f(\mathbf{R})$ as the importance sampling distribution function. This is the Variational Monte Carlo (VMC) method.

Here we should mention two *dangerous* shortcuts related to the computation of the local energy, particularly the kinetic energy part. One of them consists in taking advantage of the spherical symmetry of the wave function, so that instead of computing the laplacian $\sum_i \nabla_i^2 \Psi$, one may be tempted to compute just the second derivatives with respect to a coordinate axis (say, X axis) $\sum_i \partial^2 \Psi / \partial x_i^2$ and multiply by 3. On the average both procedures will produce the same value, but the variance of the second method will probably be larger.

The other typical shortcut is to transform the expectation value $\langle \Psi | \nabla^2 | \Psi \rangle$ into $-\langle \nabla \Psi | \nabla \Psi \rangle$, by integration by parts, and compute the second form which is simpler and quicker. The transformed contribution to the local kinetic energy will then be proportional to $|\nabla \Psi / \Psi|^2$. Again this second procedure will probably increase the variance of the calculation.

Consider the second case. The basic relation to be used is

$$\int \partial(\Psi \partial \Psi) \equiv 0 = \int \partial \Psi \partial \Psi + \int \Psi \partial^2 \Psi$$

and the integral is null because of the boundary conditions corresponding to a localized system. The integrands of the rhs, however, are not the same, but the following relation is still valid

$$-\Psi \partial^2 \Psi = -\partial(\Psi \partial \Psi) + \partial \Psi \partial \Psi \ .$$

If the trial state were the exact g.s. wave function, when adding the potential times $|\Psi|^2$ to the above relation, the lhs will be converted into a constant times the squared wave function, $E_0 \Psi^2$, and the variance of the integral will be zero. This will not be the case for the rhs, so the variance using the transformation will be larger. One may expect the same to be true for wave functions close to the exact g.s. wave function. For this reason we added above the word *probably*. The argument does not apply to the computation of expectation values of other operators.

The previous discussion may serve to put a general statement about the variance of the calculation of expectation values other than the Hamiltonian. One should expect to have, in general, a variance larger than the variance of the energy, for sufficiently good trial functions. Consequently, if one samples separately the kinetic and the potential part of the Hamiltonian, again one should expect for these observables a variance larger than the variance of the complete Hamiltonian.

In order to run a VMC code one should prepare in advance the required formulae for the calculation. There are two quantities of interest: the calculation of the wave function and the calculation of the local energy. These two procedures are called almost as frequently as the random generator, so one should be very careful in preparing a very efficient code.

The algorithm for the VMC evaluation may go along the following steps

Starting

- Initialize random routine `RAND()`
- Get number of moves `NM`, number of thermalization moves `Ntherm` and Metropolis `STEP`
- Determine an initial point (3N coordinates) `R` and the value of the squared wave function `PSI(R)**2` at this point
- Initialize accumulation variables for energy `SE=0` and energy squared `SE2=0`

Loop of Moves: `for i=1 to NM do`

Thermalization Loop: `for j=1 to NTHERM do`

  - Trial step `RP = R + (RAND()-0.5) * STEP`
  - Compute new wave function `PSI(RP)**2`
  - Ask Metropolis question. If move is accepted, copy `RP` to `R`, and also the squared wave function
  - `enddo` of thermalization loop

Sampling:

  - Compute local energy `EL(R)`
  - Update accumulation `SE=SE+EL` and `SE2=SE2+EL**2`
  - Eventually, other expectation values

End Loop of Moves

Final Calculations

- Compute averages and errors
- Print results

5.2 A Simple Example: the Harmonic Oscillator

Consider the one-dimensional harmonic oscillator defined by the Hamiltonian $H = -d^2/dx^2 + x^2$. The exact g.s. wave function is

$$\Psi_0(x) = \frac{1}{\pi^{1/4}} e^{-x^2/2} ,$$

the exact energy being 1. To practice with VMC method let use a similar wave function

$$\Psi(x) = \frac{\sqrt{\alpha}}{\pi^{1/4}} e^{-\alpha^2 x^2/2} . \tag{5.16}$$

The local energy operator has the value $E_L(x) = \frac{H\Psi}{\Psi} = \alpha^2 + x^2(1 - \alpha^4)$. The upper-bound requires the computation of the integral

$$E_V = \int_{-\infty}^{\infty} |\Psi(x)|^2 E_L(x) dx ,$$

its value being

$$E_V = \frac{\alpha^2}{2} + \frac{1}{2\alpha^2} .$$

For the purpose of evaluating the MC method, it is also of interest to determine thc variance at the variational upper bound, whose value is

$$\sigma^2 = \frac{(\alpha^4 - 1)^2}{2\alpha^4} \ .$$

If we had used the integration-by-parts way of computing the energy, then the equivalent to the local energy would have been

$$\tilde{E}_L(x) = x^2(1+\alpha^4)$$

and the new variance

$$\tilde{\sigma}^2 = \frac{(\alpha^4 + 1)^2}{2\alpha^4} \ ,$$

i.e., larger than the previous one.

The MC calculation is very simple: just generate a large number N of random numbers $\{x_i\}$ corresponding to the pdf Ψ^2 (it is a gaussian) and for each value compute the local energy. The MC estimators of interest will be

$$\langle E \rangle = \frac{1}{N}\sum_i E_L(x_i) \text{ and } \langle E^2 \rangle = \frac{1}{N}\sum_i E_L^2(x_i) \ .$$

Here is a quick and dirty FORTRAN code for the evaluation of the upper bound:

```
      implicit real*8 (a-h,o-z)
      parameter (seed=1902231.d0, alpha=0.5d0 , N=10000)
      data e,e2/0.d0,0.d0/
      z=randini(seed)                    ! Initialize RANDOM
      do 10 i=1,N                        ! Loop of samples
      x=rangauss()/sqrt(2.d0)/alpha      ! Sample point (scaled!)
      el=alpha**2+x**2*(1-alpha**4)      ! Compute LOCAL energy
      e=e+el                             ! Accumulate for MEAN
10    e2=e2+el**2                        ! and VARIANCE
      write(*,*) e/N, '+/-',sqrt((e2/N-(e/N)**2)/N)
      end
```

This code corresponds to $\alpha = 0.5$, where we know that the upper bound is 2.125 and the variance is $\sigma = 2.6516$. For a total of 10000 samples (as the code above) we should expect a result like 2.13 ± 0.03. However, there is only a 63 % of probability of getting a number between one standard deviation around the mean value. Just running this code 100 times (with different random sequences) I have obtained 56 values within one standard deviation, 98 between two standard deviations and 100 within three standard deviations. These numbers are compatible with the known confidence limits.

Note that this minimal statistical analysis has just shown us the content of the central limit theorem: the set of averages of the local energy is distributed as a normal distribution, with known mean and variance.

It may have some interest to analyze the probability distribution of the local energy. The problem at hand is simple enough to permit the analytic construction of this density, and in other problems one may always study it directly by means of random sampling. The calculation shown above proceeds in two steps. First, a random number x is generated from the pdf $|\Psi(x)|^2$, and then we compute the value of the local energy $E_L(x)$. This quantity is what is ultimately recorded and averaged. In correspondence with the distribution function of x we indirectly generate a distribution function for the energy, and the computed average is just the mean of the later pdf.

The mathematical form of the energy pdf is obtained from the equation

$$g(E) = \int |\Psi(x)|^2 \delta(E - E_L(x)) dx$$

which simply states that the probability of obtaining the value E for the local energy is the integral over all possible x such that $E_L(x)$ equals E. The integral is evaluated by using the known rule of Dirac delta function

$$g(E) = \sum_{i \in \{E_L(x_i)=E\}} \frac{1}{|dE_L(x)/dx|_{x=x_i}} |\Psi(x_i)|^2$$

the sum extending to all roots of the equation $E_L(x) - E = 0$. For $\alpha < 1$ the domain of the new pdf $g(E)$ is $[\alpha^2, \infty]$, and the actual value of the pdf is

$$g(E) = e^{-\alpha^2 \frac{E - \alpha^2}{1 - \alpha^4}} \frac{\alpha}{\sqrt{\pi (E - \alpha^2)(1 - \alpha^4)}} \,.$$

This is a quite surprising form: it has a (null measure) singularity at $E = \alpha^2$, and decreases exponentially up to infinity.

The way of obtaining the energy pdf and the resulting form is shown in Fig. 2. In the left part we show (arbitrary scale) both the x–space pdf and the local energy. The lines limiting an energy interval establish the correspondence with the coordinate values (the x_i). At the rhs we show the actual pdf for the energy. The VMC method has simply evaluated $g(E)$ in an indirect way, and the quoted upper bound is just the expectation value $\langle E \rangle = \int E\, g(E) dE$. The distribution is far from being a gaussian, but the central limit theorem ensures that the distribution of the mean values is a normal distribution.

5.3 Finding the Minimum

The class of trial functions used in the precedent exercise is very special, because it contains the exact ground state when $\alpha = 1$. In consequence, the variance for this special case is strictly null, and if you run the code for $\alpha = 1$ you will obtain exactly 1 with error 0.

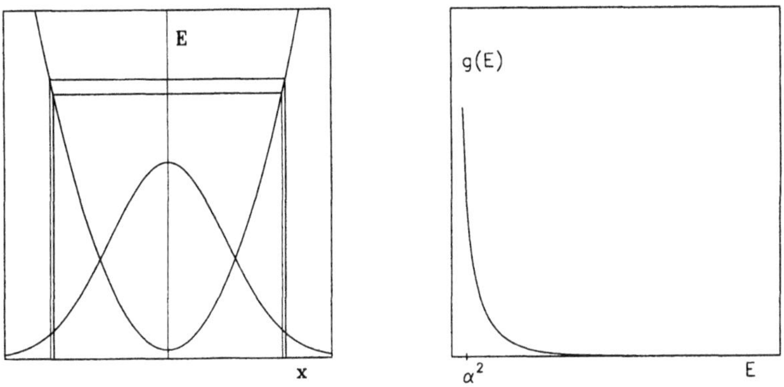

Fig. 2. The pdf related to the squared trial function and the local energy versus the coordinate x (left), and the resulting pdf $g(E)$ of the energy (right).

If we were dealing with other Hamiltonian, but the same family of functions, this will not be the case. In a region around the minimum we should expect something similar to a parabola, within the expected statistical fluctuations. It may very well happen that these fluctuations hide the position of the minimum, giving the aspect of a flat curve or even an inverted parabola.

In order to present the common situation we may consider the problem of determining the optimal upper bound for the Hamiltonian $H = p^2 - 5\exp(-x^2)$ by using gaussian trial functions like the above Eq.(5.16). The minimum may be determined algebraically (solving a transcendental equation), resulting in an energy of -3.123266 at $\alpha = 1.2389$. In the left part of Fig. 3 we show the actual running of a VMC code for several values of the variational parameter α. As mentioned above, the fluctuations around the minimum make it difficult to estimate the lowest point.

A way to overcome this problem is to carry out *correlated samplings* in the region near the minimum. The idea is the following: assume a fixed parameter β close to the minimum, and compute several values of the energy for several values of α by using the following equality

$$E_\alpha \equiv \frac{\int |\Psi(\alpha,x)|^2 E_L(\alpha,x)dx}{\int |\Psi(\alpha,x)|^2 dx} = \frac{\int |\Psi(\beta,x)|^2[|\Psi(\alpha,x)|^2/|\Psi(\beta,x)|^2 E_L(\alpha,x)]dx}{\int |\Psi(\beta,x)|^2[|\Psi(\alpha,x)|^2/|\Psi(\beta,x)|^2]dx}$$

so that one evaluates independently numerator and denominator of the rhs equation by taking as importance sampling function the fixed-parameter quantity $|\Psi(\beta,x)|^2$. The crucial point is that random numbers *must be* generated only once, corresponding to the importance sampling function, and used for all values of α. The various computations are not statistically independent, but the hope is that the fluctuations will go always in the same direction. This may be checked by observing the figure at the right of Fig. 3, which is very smooth, compared with the figure at the left which corresponds to independent samplings. The last consideration is that once the position of the minimum is found, one should carry out a longer VMC calculation, with the variational parameter found, to check the result and to have a smaller variance.

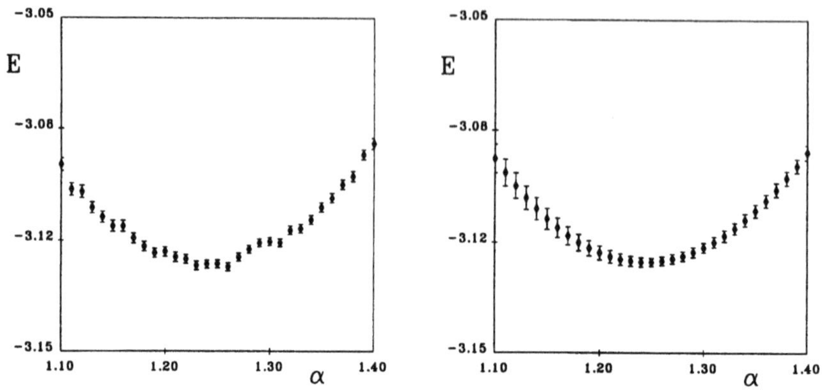

Fig. 3. Search for the minimum by using independent sampling (left) or correlated sampling (right).

5.4 The Structure of Variational Trial Functions

In this section we are going to describe briefly the class of trial functions currently used in the study of many-body systems of strongly interacting particles.

There is a large collection of systems of this kind. Without attempting to be exhaustive, let us mention

1. *Quantum fluids* or *quantum drops*, whose constituents interact through a Lennard-Jones like interaction, with a strongly repulsive core at very short distances and a quite smooth attraction at medium and large distances.
2. *Nuclear systems* and *nuclear matter*, in which there is a soft-core repulsion at very short distances, and attraction at short distances.
3. *Electron systems* like atoms, molecules or the so called *jellium* which describes a cloud of electrons in a substrate of positive charge. Here the main complication comes from the long range of the Coulomb interaction, as well as its singularity at the origin.
4. *Quantum crystals*, *atomic clusters*, and so on.

Some of these systems allow for a zeroth-order description in terms of a mean field or independent particle model. For strongly repulsive interactions this is not possible, because the matrix elements of the potential will be infinite or very large. A simple but very efficient way of describing strongly interacting systems is known with the name of Jastrow [14], which assumes for the trial function the product of an *uncorrelated state*, which takes care of the symmetry or antisymmetry as well as other global properties of the system, and a *correlation factor* which, in its original form, is the product of two-body correlations

$$\Psi(\mathbf{R}) = F_J(\mathbf{R})\Phi(\mathbf{R}) \ ,$$

with

$$F_J(\mathbf{R}) = \prod_{i<j} f(r_{ij}) \ .$$

In this form the two-body correlation depends only on the distance of the pair of particles. For interactions having a core one should have correlations which go to zero at short distances and to one at long distances. One may even assume *state-dependent correlations*, i.e., depending on the spin or isospin of the particles involved. For this kind of correlations the Jastrow factor must be properly symmetrized with respect to all orderings of the operators.

5.5 The ^{4}He Nucleus

A simple application of the VMC method is the study of the g.s. of the ^{4}He nucleus for a spin/isospin independent nuclear interaction. We will consider the Afnan-Tang [11] S3 interaction, which was adjusted so as to reproduce the low-energy properties of the nucleon-nucleon system and defined in the two $l = 0$ channels, corresponding to spin triplet and singlet. For the description of a saturated nuclear system, it may be simplified to a purely central potential by taking the average of these two channels, having

$$V(r) = 1000e^{-3r^2} - 163.35e^{-1.05r^2} - 21.5e^{-0.6r^2} - 83e^{-0.8r^2} - 11.5e^{-0.4r^2} \ ,$$

where the potential is given in MeV and the internucleon distances should be measured in fm.

A simple yet efficient trial wave function is a Jastrow factor acting on the $1s^4$ shell-model configuration with radial functions from the harmonic oscillator,

$$\Psi(\mathbf{r}_1,\ldots,\mathbf{r}_4) = \prod_{i<j}^{4} f(r_{ij}) e^{-\alpha^2(r_1^2+\ldots+r_4^2)/2}$$

or the explicit translationally invariant form

$$\Psi(\mathbf{r}_1,\ldots,\mathbf{r}_4) = \prod_{i<j}^{4} f(r_{ij}) e^{-\gamma\sum_{k<l}(\mathbf{r}_k-\mathbf{r}_l)^2} \ ,$$

where $\gamma = \alpha^2/8$. In this example we will consider a simple form for the correlation $f(r) = 1 + a\exp[-\beta r^2]$ so that the variational function has three parameters, α, β and a. Looking to the translationally invariant form, an equivalent description of the variational wave function is just the product of pair correlations $g(r_{ij})$ where the new correlation is

$$g(r) = e^{-\gamma r^2} + ae^{-(\beta+\gamma)r^2} .$$

Here we see the double role of the Jastrow correlation: it may describe both the short range behavior and the long-range confinement property of the finite system. This form is easily generalized to a system of N bosons.

Note that we have not considered the spin-isospin dependent part, which will be a 4×4 Slater determinant built up with the four possible spin-isospin orientations.

Calculations with this kind of correlations and several Wigner nuclear forces may be found in [15]. In that paper the calculations were carried out by a FORTRAN coding of an algebraic algorithm. They may serve as a good check of the MC code.

A simple MC code is presented below. In order to simplify the code the potential and the correlation have been defined in a single sentence. Moreover, the kinetic term of the local energy has been calculated by approximating second derivatives by second differences.

Table 4. Several calculations of ^{4}He with the Wigner part of the S3 interaction and with the Malfliet-Tjon MT-V potential. The g.s. energies are in MeV. ATMS stands for *amalgamation of two-body matrix elements*, CRC is the *coupled rearrangement channel method*, HH is the *hyperspherical harmonics Jastrow correlated method*, CCM is the *coupled cluster method*, FY the *Fadeev-Yakubowsky* method and DMC the *diffusion Monte Carlo method.*

Method	GS energy (S3)[11]	GS energy (MT-V) [17]
HO shell model	−5.89	−6.40
Jastrow: One gaussian [15]	−24.35	−29.01
Jastrow: Two gaussians[15]	−27.15	−30.83
ATMS [18]		−31.36
CRC [19]		−31.357
Jastrow+HH [20]	−27.369	−31.344
CCM [21]		−31.36
FY [22]		−31.36
DMC [15]	-27.35 ± 0.02	-31.32 ± 0.02

Some of the results corresponding to this code are shown in Fig. 4. We have plotted there the first 20 moves (omitting the 10 intermediate thermalization moves of each step), showing how effectively the Metropolis algorithm makes the nucleons to move within the nucleus, and also thousand positions of one particle, with and without thermalization. The latter case shows that the particle is sampling mainly the central part of the wave function, and it is the

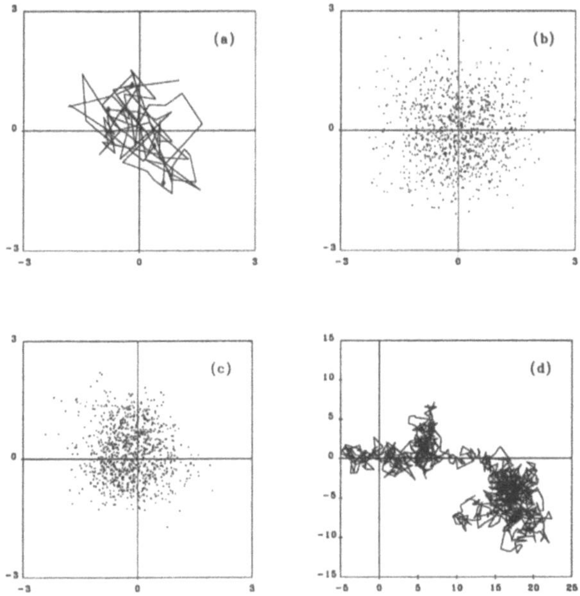

Fig. 4. Some results corresponding to the calculation of ^{4}He. **a**: The first 20 moves of the particles, without showing the intermediate thermalization moves. The diamonds are the starting positions with respect to the cm. **b**: Plot of the position of one particle after 1000 moves, with thermalization, referred to the cm. **c**: Same as **b** but without thermalization. **d**: Same as **b** but referred to an external origin. All positions are in fm.

thermalization which extends the sampling also to the tail part. In these three plots, corresponding to parts **a**, **b** and **c** of the figure, the actual positions of the particles have been referred to the center-of-mass of the system. Finally, part **d** shows the position of the particle, but this time referred with respect to an external system. It shows clearly that the center-of-mass is wandering around, with very large displacements. This is harmless, because the wave function is translationally invariant and there is no binding related to the center-of-mass.

The code may easily be modified to deal with other interactions and correlations. In Ref. [15] you may find parameters adequate for other nucleon-nucleon potentials. Also, the code may be generalized to other finite boson systems, but it will be quite slow. In Ref. [16] one can find results corresponding to invented bose systems of up to 40 particles with the nucleon mass interacting by means of the S3 potential.

How good is this simple Jastrow correlation? Table 4 presents some selected results for the ^{4}He nucleus corresponding to several nucleon-nucleon interactions and to several computational techniques. For a Jastrow wave function of four parameters (sum of two gaussians) the result differs very little from the exact Diffusion Monte Carlo calculation of the energy. In conclusion, two-body Jastrow correlations contain almost hundred per cent of the structure of the exact wave function. The table also contains the results of other few-body techniques applied to the same problem. It is very rewarding to see the overall agreement between all of them and the high level of precision reached in this (and related) problems.

```
c       ==========================================================
c       Variational Monte Carlo for BOSONS
c       Parameters adjusted for He-4 with S3 potential
        include 'varhe.inc'
        step=1                  ! Metropolis step (to be adjusted)
        z=RANDINI(12231.d0)      ! Initialize RAND routine
        do 10 i=1, NA
        do 10  k=1, 3
10      R(i,k)=10*step*(RAND()-0.5d0)    ! Starting  walker
        wfold=WF(R)                      ! Wave function
        Nmoves=1000                      ! Start random walk
        Ntherm=10
        sume=0
        sume2=0
        Nacceptances=0
        do 100 imoves=1, Nmoves
           do 200 itherm=1, Ntherm
                do 300 i=1, NA
                do 300 k=1, 3
300             RP(i,k)=R(i,k)+step*(RAND()-0.5d0)     ! Attempt Move
           wfnew= WF(RP)
c          Metropolis question
           if((wfnew/wfold)**2 .lt. RAND() ) GOTO 200  ! Move rejected
           Nacceptances=Nacceptances+1                 ! Move accepted
                do 400 i=1, NA
                do 400 k=1,3
400             R(i,k)=RP(i,k)                     ! Update walker
                wfold=wfnew                        ! and wave function
200     continue
        energy=ELOCAL(WFOLD,R,DRIFT)               ! Compute energy
        sume=sume+energy                           ! Accumulate for
100     sume2=sume2+energy**2                      ! averages
c       Print results
        e=sume/Nmoves
        e2=sume2/Nmoves
        write(*,'(A10, F10.3)') 'Energy=   ', e
        write(*,'(A10, F10.3)') 'Error =   ', sqrt((e2-e**2)/Nmoves)
        write(*,'(A10, F10.3)') '% Accept= ',
     1       (100.d0*Nacceptances)/(Nmoves*Ntherm)
        end
c       ==========================================================
c       Computes the wave function
```

```
      double precision function WF (R)
      include 'varhe.inc'
c     Correlation function -> R2 is the DISTANCE SQUARED
      f(r2)=exp(-0.08597d0*r2)-0.7191d0*exp(-2.13796d0*r2)
      wf=1
      do 10 i=1, NA-1
      do 10 j=i+1, NA
      r2=(R(i,1)-R(j,1))**2+(R(i,2)-R(j,2))**2+(R(i,3)-R(j,3))**2
10    wf=wf*f(r2)
      return
      end
c     ==========================================================
c     Computes the local energy and gradient (to be used in DMC)
      double precision function ELOCAL(wfold , R, DRIFT)
      include 'varhe.inc'
c     Two-nucleon potential-> WIGNER part of S3
      V(r2)=1000.d0*exp(-3.d0*r2)-163.35d0*exp(-1.05d0*r2)-83.d0*
     1 exp(-0.8d0*r2)-21.5d0*exp(-0.6d0*r2)-11.5d0*exp(-0.4d0*r2)
      do 40 i=1, NA              ! To compute derivatives
      do 40 k=1,3                ! define RPlus and Rminus
      RP(i,k)=R(i,k)             ! as R
40    RM(i,k)=R(i,k)
      ekin=0
      do 20 i=1, NA              ! For each particle and each
      do 20 k=1, 3               ! coordinate, compute
      RM(i,k)=R(i,k)-h           ! Second derivatives
      RP(i,k)=R(i,k)+h           ! as centered SECOND differences
      wfplus=wf(RP)              ! and 2*dif(log(WF),R(I,K))
      wfminus=wf(RM)             ! as centered FIRST differences
      ekin=ekin+(WFplus+WFminus-2*WFold)/h**2
      drift(i,k)=(WFplus-WFminus)/(WFold*h)
      RM(i,k)=R(i,k)             ! Restore positions
20    RP(i,k)=R(i,k)
      epot=0                     ! Potential energy
      do 30 i=1, NA-1
      do 30 j=i+1, NA
      r2=(R(i,1)-R(j,1))**2+(R(i,2)-R(j,2))**2+(R(i,3)-R(j,3))**2
30    epot=epot+V(r2)
      ELOCAL=-h2div2m/WFOLD*ekin+epot
      return
c     ==========================================================
c     INCLUDE file, to be named VARHE.INC
      implicit real*8 (a-h,o-z)
      parameter (NA=4, h2div2m=20.74d0, h=0.001d0)
      dimension R(NA,3), RP(NA,3), RM(NA,3), DRIFT(NA,3)
```

5.6 Extended Systems

When dealing with extended systems we cannot attempt the simulation with an infinite number of particles. The usual way of dealing with these systems, inherited from Statistical Mechanics [12, 23], consists in representing the system in a simulation cell with periodic boundary conditions. If the side of the cells is L, the density of the system is determined by putting a given number of particles in the L^3 cell, so that $\rho = N/L^3$.

Metropolis moves are always recast into the simulation cell, but in the actual calculations one should also consider particles in the periodic images. To carry out consistently the Metropolis moves it has to be assumed that the correlation function has a range shorter than $L/2$. Then, to decide if a move of a single particle is accepted or not, only the set of particles contained in a sphere of radius $L/2$, centered in the referred particle, have to be considered, as shown in figure 5. These particles are not necessarily in the simulation cell. If the move is accepted, one imagines that all images have undergone the same move. The evaluation of the potential energy does not require this limitation, but one should carefully check that there is no double counting.

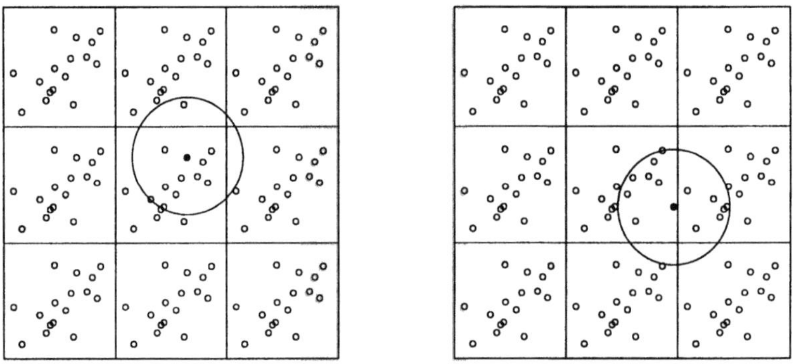

Fig. 5. The simulation cell and its replicas. On the rhs the black dot is the point to be moved, and the circle centered in it determines the selected nearest neighbors which should be counted for the interaction of the particle represented by a black circle. The lhs figure shows the equivalent situation once the selected particle has moved.

There are sets of *magic* numbers in the practical application of the VMC method. In the case of bosonic systems, the idea of choosing a specific number of particles is tied to the possibility of allowing the system *crystallize*, i.e., when particles prefer to place themselves around an ordered and periodic structure. One expects the highest density crystalline order, which corresponds to the face centered cubic structure. The fcc unit cell has 8 particles at the vertices (each particle shared by eight adjacent cells) and six particles at the faces (each one shared by two cells) which give a total of four *full* particles per unit cell. So, the number of particles is taken as $4n^3$, in such a form that the simulation cell contains an integer number of unit cells. Typical numbers are 108, 256, 500, 864, and so on. For bosonic systems the uncorrelated wave function is simply a constant $(1/L^{3N/2})$, and the integrals are computed, with the prescription above mentioned, in the simulation cell. Once the number of particles has been chosen, to vary the density one simply changes the value of L.

The situation is quite different in the case of fermionic systems. Here the number of particles in the simulation cell is determined by the quantization rules. The single-particle states are eigenvectors of the kinetic energy operator p^2 with periodicity L in each of the X, Y and Z directions. The free solutions are of the form $\psi_k = \exp(i\mathbf{kr})$, and the possible values of $\mathbf{k}$ must follow the rule

$$\mathbf{k} = \frac{2\pi}{L}(n_x, n_y, n_z)$$

where the n are positive or negative integer numbers. These single-particle levels are filled up in order of increasing energy, up to a maximum value, as indicated in Table 5. The cumulative numbers are 1, 7, 19, 27, 33, 57, ..., to be multiplied by the degeneracy of the system.

Table 5. The values of momenta ordered by increasing energy values. The number in the third column considers the $\pm$ and the reorders of the assignation.

$k^2_{\rm max}$	(n_x, n_y, n_z)	Number	Cumulative
0	$(0,0,0)$	1	1
1	$(\pm1,0,0)$	2×3	7
2	$(\pm1,\pm1,0)$	4×3	19
3	$(\pm1,\pm1,\pm1)$	8	27
4	$(\pm2,0,0)$	2×3	33
5	$(\pm2,\pm1,0)$	4×6	57
6	$(\pm2,\pm1,\pm1)$	8×3	81

A detailed example of the use of periodic boundary conditions may be found in [24], for the specific case of liquid ^{4}He interacting in terms of a Lennard-Jones potential. There a simple FORTRAN code is also included. This example uses the simple McMillan correlation factor [25],

$$f(r) = \exp(-(b/r)^5/2),$$

where b is the only variational parameter and which gives a quite good upper bound to the g.s. energy. This quantum liquid was extensively studied by Schiff and Verlet [26] and by Kalos and coworkers (see, e.g., [27]).

5.7 Fermionic Systems

There are two main complications in the VMC procedures when referring to fermionic systems. The first one is the need of dealing with antisymmetric states (usually Slater determinants) and the second is the role played by the spin and isospin operators which may appear both in the interaction and in the correlation terms.

We may distinguish three levels of complexity, which are briefly described here. For simplicity we will only deal with spin degrees of freedom, the extension to cope with isospin degrees of freedom being straightforward.

Spin Independent Forces and Correlations. This is the simplest case. There are two extreme situations quite easily dealt with: the case of *saturated* systems, in which half of the particles are assumed to have spin up, and the other half spin down, and the case of *fully polarized* systems, in which all particles have the spin in the same direction. This second case is the simplest one, because the wave function is the product of a Slater determinant depending only on the coordinates and the spin part, which is a fully symmetric term.

The first job is to compute the pdf for the Monte Carlo evaluation. Here it is worth mentioning the inadequacy of one of the important simplifications very common in the evaluation of matrix elements with Slater determinants. The usual simplification consists in considering only the antisymmetrization of the, say, ket state, and selecting for the bra state a simple assignment of particles to states. To be more precise, let $\{\phi_1, \phi_2, \ldots \phi_N\}$ be the set of occupied orbitals. The antisymmetric wave function is given by

$$\Psi(1,2,\ldots N) = \sum_P \epsilon_P \phi_1(P1)\phi_2(P2)\ldots\phi_N(PN) \ ,$$

the sum being extended to the $N!$ permutations. The norm requires the computation of

$$\langle\Psi|\Psi\rangle = \langle\sum_P \epsilon_P \phi_1^*(P1)\phi_2^*(P2)\ldots\phi_N^*(PN)|\sum_Q \epsilon_Q \phi_1(Q1)\phi_2(Q2)\ldots\phi_N(QN)\rangle$$

and one may just remove the sum over one set of permutations to obtain the equivalent value

$$\langle\Psi|\Psi\rangle = N!\langle\phi_1^*(1)\phi_2^*(2)\ldots\phi_N^*(N)|\sum_Q \epsilon_Q \phi_1(Q1)\phi_2(Q2)\ldots\phi_N(QN)\rangle \ .$$

The elimination of the sum over P may be done under the integral sign, but the integrands are not the same. In fact, the integrand in the second case is not necessarily positive.

In conclusion, to compute the pdf one must evaluate the Slater determinant and square it. In the case of polarized systems, the Slater determinant is just a function of the space coordinates. In the case of saturated systems one has two alternatives. The most economical corresponds to consider the spin up and spin down particles as distinguishable, representing the wave function as the product of two Slater determinants of dimension $N/2$. This method has, unfortunately, a quite large variance, because the pdf is not symmetrical under the exchange of any pair of particles. It has, however, partial symmetry under the exchange of particles with the same spin. Full symmetry may be restored by applying a subset of permutations to the pdf (not to the wave functions, and without including the permutation parity) which are the ways of selecting $N/2$ particles from the N particles. The number of these permutations is large even for moderate values of N (for example, for 16 particles there are 12870 permutations) and each of them requires a costly work, namely the evaluation of two determinants. In consequence, unless the number of particles is very small, it may be only possible to deal with the partially symmetrized form of the pdf.

Spin Dependent Forces. The next case corresponds to the situation in which the trial wave function is assumed to be spin *independent*. Here Metropolis moves are carried out as above, but to compute the local energy one should allow for jumps of particles from one spin status to another spin status. For the simple $(\boldsymbol{\sigma}_i \cdot \boldsymbol{\sigma}_j)$ force, its action will be adequately represented with swaps of spins.

Systems with Coupled Channels. The spin dependent problem may be imagined as a *coupled channel problem*. The most familiar coupled channel problem in Nuclear Physics appears in the description of the deuteron with tensor components in the nuclear force [28]. It corresponds to a two-channel differential equation of the form

$$\begin{bmatrix} -\frac{\hbar^2}{2\mu}\nabla^2 & 0 \\ 0 & -\frac{\hbar^2}{2\mu}\nabla^2 \end{bmatrix} \begin{bmatrix} \Psi_s \\ \Psi_d \end{bmatrix} + \begin{bmatrix} V_{ss} & V_{sd} \\ V_{sd} & V_{dd} \end{bmatrix} \begin{bmatrix} \Psi_s \\ \Psi_d \end{bmatrix} = E \begin{bmatrix} \Psi_s \\ \Psi_d \end{bmatrix} ,$$

where there is a diagonal kinetic term and a full matrix potential term coupling the s and d channels. The normalization condition of the wave function reads now

$$\int [|\Psi_s|^2 + |\Psi_d|^2] = 1 .$$

From the variational point of view one may deal with the coupled-channel problem exactly in the same way as in the ordinary case, considering as pdf the quantity $f(r) = |\Psi_s(r)|^2 + |\Psi_d(r)|^2$ properly normalized. The Metropolis algorithm is simply adapted to this problem: starting at point r one tries a new point r' and is accepted accordingly with the probability quotient $f(r')/f(r)$. The equivalent to the local energy is now the quantity

$$E_L(r) = \frac{1}{|\Psi_s(r)|^2 + |\Psi_d(r)|^2} [\Psi_s^*(r) \;\; \Psi_d^*(r)] \begin{bmatrix} H_{ss} & H_{sd} \\ H_{sd} & H_{dd} \end{bmatrix} \begin{bmatrix} \Psi_s(r) \\ \Psi_d(r) \end{bmatrix} ,$$

and the obvious generalization to more channels.

As an illustration of the VMC procedure for coupled channels let's consider the simple problem of two energy-displaced and coupled harmonic oscillators, with a Hamiltonian matrix

$$H = \begin{bmatrix} p^2 + x^2 & x \\ x & p^2 + x^2 + 1 \end{bmatrix}$$

This simple problem is a schematic representation of the nuclear-electronic coupling of diatomic molecules, which is used to explain fluorescence phenomena in diatomic molecules (see e.g. [29, 30]). The ground state has an energy of 0.82867, the upper component being of even parity and the lower component of odd parity. A simple variational function is

$$\Psi_T = \begin{bmatrix} e^{-x^2/2} \\ axe^{-x^2/2} \end{bmatrix} ,$$

where a should be determined variationally. This may be done analytically, with the result $a = 3 - \sqrt{11} \approx -0.31662$ and $E_{\text{var}} \approx 0.84168$, not far from the exact result.

Here is the simple FORTRAN code for this evaluation

```
c         Variational coupled oscillators problem
          implicit real*8 (a-h,o-z)
          parameter (Nm=1000, Nt=10, step=2.d0)
          EL(x,a)=4+2/a-(3*a+2)/(a*(1+a**2*x**2))
          f(x,a)=(1+a**2*x**2)*exp(-x**2)
          data a, e, e2/ -0.3166d0, 0.d0, 0.d0/
          x=randini(19923.d0)
          do 10 imoves=1,nm
          do 20 itherm=1, nt
          y=x+step*(rand()-0.5d0)
          if(f(y,a)/f(x,a) .gt. rand()) x=y
20        continue
          e=e+EL(x,a)
          e2=e2+EL(x,a)**2
10        continue
          write(*,*) e/Nm,'+', sqrt((e2/nm-(e/nm)**2)/Nm)
          end
```

Running this code I have obtained the value $E_V = 0.838 \pm 0.006$ for the parameter a corresponding to the minimum.

The main conclusions regarding this discussion is that, even if the trial function has two (in general, more) components, i.e., two coordinate dependent functions, there is not an independent sampling for each one of the components. The other important point is the role of the mixing parameter a of our example. It measures essentially the relative weight of the two components of the wave function, and has a prominent variational role. Even if the coupled Schrödinger equation is homogeneous, this only allows for an arbitrary global constant and does not permit to fix arbitrarily the value (and even the phase) of the various components of the wave function.

The general spin/isospin variational method may be carried out in the coupled channel formalism described above. However, the enumeration and construction of the spin/isospin channels may be quite involved, and it turns out more convenient to generate the channels by applying a spin/isospin dependent correlation operator to a model state, frequently a shell-model wave function. If Jastrow correlations are used, the correlation operator has the general structure

$$f_{ij} = \sum_k f_k(r_{ij}) \mathcal{O}^k_{ij} \ ,$$

where the pair operators are of the type spin-spin, or tensor, or any reasonable structure related with the two-body interaction. The $N-$particle correlation operator is the *symmetrized* product of the pair correlations, the role of the symmetrization being to maintain the fermionic statistics.

I will not enter in more details regarding this question. A quite detailed description of the algebraic treatment of this kind of correlation operators, applied to few-nucleon problems, may be found in [48].

5.8 Additional Comments

The brief presentation of the VMC methods may give a flavor of the power of Monte Carlo methods applied to the variational problem. The required Monte Carlo techniques are conceptually simple, but the practical application may turn out to be both complex and tedious. The main problem is to construct the variational functions and an adequate parameterization, so as to include as much physics of the problem as possible. The tedious part is the search for the variational minimum. A good knowledge of the system is required to carry out reasonable VMC calculations.

6 The Diffusion Monte Carlo Method

6.1 Introduction

The diffusion Monte Carlo (DMC) method is a way of solving exactly (in a statistical sense) the many-body Schrödinger equation by means of a stochastic procedure.

In general, the statement *Quantum Monte Carlo*, abbreviated as QMC, refers to a wide set of methods which attempt to integrate the many-body Schrödinger equation. There was the habitude of referring as *Green Function Monte Carlo* (GFMC) or *Domain Green Function Monte Carlo* the approach in which the time-independent Green Function is approximated by means of a Born series in a certain domain, whereas the name *Diffusion Monte Carlo* was associated with algorithms which approximate the time-dependent Green Function. However, these two methods do not exhaust the possibilities of stochastic integration.

The basic ideas of QMC were established already around 1950 [31, 32]. The computers at that time were not fast enough to apply the methods to many-body systems. The reborn of the stochastic methods took place in 1962, in a paper by Malvin H. Kalos [33], which actually was involved with the inverse problem, i.e., to determine the depth of a gaussian potential so as to reproduce the binding energy of three and four-body nuclei.

Later, in 1975, Anderson [34] introduced the short-time approximation to the time-dependent Green function, initiating the DMC algorithm. The work of Anderson was concentrated in the application to atomic and molecular systems [35, 36], i.e., to systems with a finite number of particles. The first application to extended systems was carried out by Ceperley and Alder [37] in a famous paper dealing with the electron gas (jellium).

From the didactical point of view we suggest the reader to consult the article of Reynolds et al [38] as well as the book of Hammond, Lester Jr. and Reynolds [39] *Monte Carlo Methods in Ab Initio Quantum Chemistry.*

These lectures will be limited to DMC methods. However, one should cite, for reasons of historical completeness, the work developed in parallel in the GFMC method. Actually, many of the ideas (importance sampling, fixed-node approximation and so on) are shared by the two methods. After the above mentioned article of Kalos [33] in 1962, there were a series of papers also involving Malvin Kalos [40]-[42] which culminated in 1974 with a precise study of the liquid ^{4}He properties [44]. A good review of GFMC method may be found in [27, 45].

The list of references is not complete, and since the 80's there have been many papers dealing with extensions and applications of QMC methods. Some of the applications will be described by J. Boronat and S. Pieper in this volume, so we refer to their lectures for more recent developments. Nevertheless, there have been several methodological advances since that time, some of them being

- Improvement of the short-time approximation in DMC, by S.A. Chin [47]
- Dealing with coupled-channel equations and discrete degrees of freedom, by J. Carlson [48, 49]
- Description of the excitation spectrum, by Ceperley and Bernu [50] and Chin and Krotscheck [51].

Books are not common in QMC methods. I have already referred to the applications to Quantum Chemistry [39] as well as my own didactical paper [24]. Another review-like book is [46].

6.2 Basic Ideas of DMC

The basic ingredients of DMC are the following

1. It considers the Schrödinger equation in imaginary time,
$$-\frac{\partial \Psi(\mathbf{R},t)}{\partial t} = [H(\mathbf{R}) - E]\Psi(\mathbf{R},t) \ , \tag{6.17}$$
where $\mathbf{R}$ represents the set of all coordinates. The formal solution of (6.17) is
$$\Psi(\mathbf{R},t) = e^{-[H - E]t}\Psi(\mathbf{R},0) \tag{6.18}$$
where $\exp[-(H-E)t]$ is called the *Green function*, and E is a convenient energy shift.
2. The wave function is positive definite everywhere, as it happens with the ground state of a bosonic system, so it may be considered as a probability distribution function.
3. The wave function is represented by a set of *random vectors* or *walkers* $\{\mathbf{R}_1, \mathbf{R}_2 \ldots \mathbf{R}_{N_c}\}$, in such a form that the time evolution of the wave function is actually represented by the evolution of the set of walkers.
4. The actual computation of the time evolution is done in small time steps τ, and the Green function is accordingly approximated,
$$e^{-[H - E]t} = \prod_{i=1}^{n} e^{-[H - E]\tau} \ ,$$
where $\tau = t/n$.
5. The imaginary time evolution of an arbitrary starting state $\Psi(\mathbf{R},0)$, once expanded in the basis of stationary states of the Hamiltonian
$$\Psi(\mathbf{R},0) = \sum_{\nu} C_\nu \Phi_\nu(\mathbf{R})$$
is given by
$$\Psi(\mathbf{R},t) = \sum_{\nu} e^{-[E_\nu - E]t} C_\nu \Phi_\nu(\mathbf{R}) \ ,$$
in such a way that the lowest energy components will have the largest amplitudes after a long elapsed time, and in the $t \to \infty$ limit the most important amplitude will correspond to the ground state (if $C_0 \neq 0$).
6. An important improvement of this scheme is the introduction of the *importance sampling*, which will be described below.

In conclusion, the scheme is quite simple: once one has found an appropriate approximation for the short time Green function, and a starting state has been determined, the job consists in representing the starting state by a collection of walkers and letting them evolve in time, i.e., obtaining a new collection of walkers from the old collection of walkers, up to a time large enough so that only the g.s. amplitude is of importance.

There is an important question related to the imaginary time evolution. The evolution operator *is not unitary*, so that the norm of the state is not conserved in time. Actually, it may grow without limit or it may go to zero, depending on the value of E. If E equals the g.s. energy, $E = E_0$, then all amplitudes will go to zero at large time, with the exception of the amplitude of Φ_0. On the other hand, if $E > E_0$, all amplitudes will go to infinity and, conversely, for $E < E_0$, all amplitudes will go to zero. In any case, always the C_0 amplitude will dominate. A way of determining the g.s. energy is then to adjust E so that the norm of the state tends to a constant. This way is called, in MC language, the *growth estimator* of the energy.

6.3 The Green Function and the Small-Time Approximation

In coordinate representation the Green function is given by the matrix element

$$G(\mathbf{R}', \mathbf{R}, t) = \langle \mathbf{R}' | e^{-(H-E)t} | \mathbf{R} \rangle \ ,$$

and the time evolution equation is

$$\Psi(\mathbf{R}', t) = \int d\mathbf{R} G(\mathbf{R}', \mathbf{R}, t) \Psi(\mathbf{R}, 0) \ . \tag{6.19}$$

At $t = 0$ the value of the Green function is

$$G(\mathbf{R}', \mathbf{R}, 0) = \delta(\mathbf{R}' - \mathbf{R}) \ . \tag{6.20}$$

From the operatorial representation of the Green function one may easily obtain the formal differential equation

$$-\frac{\partial G}{\partial t} = [H - E] G \ ,$$

which is transformed in coordinate representation into the differential equation

$$-\frac{\partial G(\mathbf{R}', \mathbf{R}, t)}{\partial t} = \left[-\frac{\hbar^2}{2m} \sum_i \nabla_i^2 + \sum_{i<j} V_{ij} - E \right] G(\mathbf{R}', \mathbf{R}, t)$$

with the boundary condition (6.20). This equation is very similar to (6.17), so we have not progressed too much. The important point, however, is that one may obtain approximations to the value of the Green function for short time intervals.

The main problem in obtaining the Green function is the non commutativity of the kinetic energy operator and the potential energy operator. However, the individual pieces, i.e., the Green functions related exclusively to the kinetic or potential operators, may be readily determined.

For the kinetic operator we have the coordinate representation

$$G_K(\mathbf{R}',\mathbf{R},t) = \langle \mathbf{R}'|e^{-D\nabla^2 t}|\mathbf{R}\rangle \equiv \frac{1}{(2\pi)^{3N}} \int d\mathbf{k} e^{-i\mathbf{k}\mathbf{R}'} e^{-Dtk^2} e^{i\mathbf{k}\mathbf{R}} \;,$$

where last equation corresponds to the momentum representation of the position bra and ket states. Here $D = \hbar^2/2m$ is called the *diffusion constant* and all vectors have $3N$ components.

The integral is easily carried out, obtaining

$$G_K(\mathbf{R}',\mathbf{R},t) = \frac{1}{[4\pi Dt]^{3N/2}} e^{-(\mathbf{R}-\mathbf{R}')^2/4Dt} \;.$$

It is easily checked that this Green function is the Dirac delta function $\delta(\mathbf{R}-\mathbf{R}')$ in the $t \to 0$ limit.

The case of the Green function related to the potential is even simpler, for momentum-independent interactions. The potential is then a *local operator* and there results

$$G_V(\mathbf{R}',\mathbf{R},t) = e^{-V(\mathbf{R})t}\delta(\mathbf{R}-\mathbf{R}') \;.$$

Approximate forms for the full Green function are obtained by using the following approximations

$$\begin{aligned}
\langle \mathbf{R}'|e^{-(K+V)\tau}|\mathbf{R}\rangle &= \langle \mathbf{R}'|e^{-K\tau}|\mathbf{R}''\rangle\langle \mathbf{R}''|e^{-V\tau}|\mathbf{R}\rangle + O(\tau^2) \\
&= \langle \mathbf{R}'|e^{-V\tau}|\mathbf{R}''\rangle\langle \mathbf{R}''|e^{-K\tau}|\mathbf{R}\rangle + O(\tau^2) \\
&= \langle \mathbf{R}'|e^{-V\tau/2}|\mathbf{R}''\rangle\langle \mathbf{R}''|e^{-K\tau}|\mathbf{R}'''\rangle\langle \mathbf{R}'''|e^{-V\tau/2}|\mathbf{R}\rangle + O(\tau^3)
\end{aligned}$$

In all cases an integration over the internal coordinates $\mathbf{R}''$ or $\mathbf{R}'''$ must be understood, being easily carried out by using the delta functions appearing in G_V. The practical equations are

$$\begin{aligned}
G(\mathbf{R}',\mathbf{R},\tau) &= G_K(\mathbf{R}',\mathbf{R},\tau)e^{[E-V(\mathbf{R})]\tau} + O(\tau^2) \\
&= e^{[E-V(\mathbf{R}')]\tau}G_K(\mathbf{R}',\mathbf{R},\tau) + O(\tau^2) \\
&= e^{[E-\{V(\mathbf{R}')+V(\mathbf{R})\}/2]\tau}G_K(\mathbf{R}',\mathbf{R},\tau) + O(\tau^3)
\end{aligned}$$

In these equations we have also added the energy shift appearing in the original time-dependent equation. It should also be noted that the average of the first two forms also generates an $O(\tau^3)$ approximate Green function,

$$G(\mathbf{R}',\mathbf{R},\tau) = G_K(\mathbf{R}',\mathbf{R},\tau)\frac{e^{[E-V(\mathbf{R})]\tau} + e^{[E-V(\mathbf{R}')]\tau}}{2} + O(\tau^3) \;.$$

The Small-Time Limit. It is quite simple to check formally the validity of the previous forms of the short time Green function in the $\tau \to 0$ limit.

For this purpose we require the expansion of the kinetic part of the Green function in the small-time limit. This expansion, in the simpler one-dimensional problem, is given by

$$\frac{1}{\sqrt{4\pi D\tau}} e^{-x^2/4D\tau} = \delta(x) + D\tau\delta''(x) + \frac{1}{2}D^2\tau^2\delta^{iv}(x) + O(\tau^4) .$$

To obtain this strange expansion one should consider the integral of the gaussian times a smooth function $f(x)$, and Taylor expand this function in powers of x. The resulting integrals are of the form $\int_{-\infty}^{\infty} x^n exp(-x^2)$ multiplied by the n-th derivatives of $f(x)$. Only the $n =$ even terms survive.

By using the expansion of the kinetic Green function, and when stability has been reached, i.e., $\Psi(x,\tau) = \Psi(x,0) \equiv \Psi(x)$, the time evolution equation becomes

$$\Psi(x') = e^{[E-V(x')]\tau}\Psi(x') + D\tau\frac{d^2}{dx^2}\left[e^{[E-V(x')]\tau}\Psi(x)\right]_{x=x'} + O(\tau^2)$$

Expanding the term with second derivatives up to $O(\tau^2)$, there results

$$\Psi(x') = \Psi(x') + [E - V(x')]\tau\,\Psi(x') + D\tau\,\Psi''(x') + O(\tau^2) .$$

The coefficient in τ^0 of this equation is null, so that effectively the equality becomes an $O(\tau)$ expansion. The leading term is exactly the Schrödinger equation

$$-D\Psi''(x) + V(x)\Psi(x) = E\Psi(x) + O(t) .$$

In conclusion, an $O(\tau^2)$ Green function has generated energies which are only $O(\tau)$ accurate.

A similar procedure, but more involved, may be carried out for the $O(\tau^3)$ approximations, showing that the energies obtained are accurate up to $O(\tau^2)$.

6.4 The One-Dimensional Harmonic Oscillator

The purpose of this section is to understand the meaning and use of the short-time Green functions defined above. We will not use here MC methods. Instead, all calculations will be done analytically.

Consider the simple Hamiltonian

$$H = p^2 + x^2 ,$$

which corresponds to $D = 1$, and a starting wave function, normalized as a probability distribution function, of gaussian shape

$$\Psi(x,0) = \sqrt{\frac{a}{\pi}}e^{-ax^2} .$$

Here a is an arbitrary parameter, but not of variational nature: it just fixes the starting state.

By applying the $O(\tau^3)$ Green function there results a new wave function also of gaussian form, i.e.,

$$\Psi(x,\tau) = \sqrt{\frac{a'}{\pi}} e^{-a'x^2} \left[e^{E\tau} \sqrt{\frac{a}{a+\tau+2a\tau^2+\tau^3}} \right] , \tag{6.21}$$

where

$$a' = \frac{a+\tau+2a\tau^2+\tau^3}{1+2\tau^2+4a\tau} ,$$

so that, apart from the change in the normalization, a time step is equivalent to the replacement

$$a \leftarrow \frac{a+\tau+2a\tau^2+\tau^3}{1+2\tau^2+4a\tau} . \tag{6.22}$$

Repeated time steps correspond to repeated application of this replacement rule. If the iteration tends to a limiting value, the limiting value is obtained by solving the equation obtained from (6.22) by putting an equal sign instead of the arrow. The solution, i.e., the stabilized gaussian parameter $\tilde{a}$, is

$$\tilde{a} = \frac{\sqrt{1+\tau^2}}{2}$$

and if the time-step is close to 0, it turns out to be $\tilde{a} = 1/2$, the well known exact result. The first correction has the value $\tau^2/4$, so even if the approximate Green function has an $O(\tau^3)$ error, the resulting wave function, after stability, has an error $O(t^2)$, confirming the results of the calculation of $\tau \to 0$ limit of the last section. For example, when $\tau = 0.01$, the stabilized value is $\tilde{a} = 0.500025$, and for $\tau = 0.1$ its value is $\tilde{a} = 0.5025$.

There are some interesting questions easily answered with the help of this simple example, which are briefly commented.

The Stability Time. The question is how many iteration steps have to be done in order to arrive to stability. Of course, the answer will depend on the starting value of a but *does not depend on the value of the time step* τ.

This is shown in Fig. 6.4 for two starting values of the gaussian parameter a and for two values of the time step. It is worth noting that the case corresponding to $a = 1$ at the beginning (lhs plot) reaches the stability noticeably before the case corresponding to start at $a = 0$ (rhs plot). This simply indicates that the starting wave function with $a = 1$ is closer to the exact g.s. wave function than the case of $a = 0$.

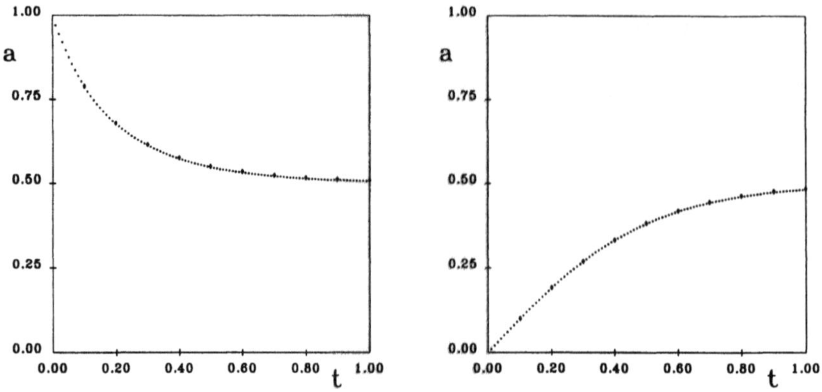

Fig. 6. Time evolution of the parameter a. The figure at the left starts at $a = 1$, and the figure of the right starts at $a = 0$. In Monte Carlo language, the second case corresponds to a starting wave function represented by random points uniformly distributed in $[-\infty, \infty]$. The small dots correspond to the calculation with a time step $\tau = 0.01$, and the big dots to $\tau = 0.1$. The x axis is the total evolved time. The lines end up at different values of a, but this cannot be appreciated in the figure.

The Mixed Estimator for the Energy. The value of the energy is readily obtained from the stabilized solution by computing the integral

$$E = \frac{\int H\Psi \mathrm{d}x}{\int \Psi \mathrm{d}x}$$

and applying the Hamiltonian on the bra state, whose value is 1. Note that this equation is exact when Ψ is the *exact ground state* wave function. If it is not, the equation is not even an upper bound, in the variational sense.

Given that the second derivative of 1 is null, one may replace H by $V(x) = x^2$ and compute the integral. The result is

$$E = \frac{2}{2 + \tau^2} ,$$

and the Taylor expansion is $1 - \tau^2/2 \cdots$, showing the $O(\tau^2)$ error term. Here we check that the energy estimator has given a value *smaller* than the exact energy (which is 1), showing the non-variational character of the mixed estimator.

Of relevance for the Monte Carlo way of dealing with this method is the computation of the variance. This requires to compute the matrix element of x^4, and the resulting value for the variance is

$$\sigma = \frac{1}{\sqrt{2}a} ,$$

and the stabilized limit is $\sigma = \sqrt{2/(1 + \tau^2)}$. Knowing a priori the value of the variance will help in deciding the number of samples to compute, in Monte

Carlo method, so as to have a prescribed value for the error. In our case, the variance is quite large, and for sufficiently small values of the time step τ its value is close to $\sqrt{2}$. If we would like to obtain the energy with an error of 10^{-3} then the required number of samples will be of two millions. This drawback shown in the simple example may be corrected by using an appropriate importance sampling algorithm.

This way of computing the expectation value of the energy has been called a *mixed estimator*, because one could have introduced as bra state any other trial wave function instead of just putting 1. In general, the mixed estimator is given by

$$E = \frac{\int \Psi_T^* H \Psi \mathrm{d}x}{\int \Psi_T^* \Psi \mathrm{d}x} ,$$

or, in terms of the same local energy introduced in the description of the variational method, the mixed estimator is

$$E = \frac{\int \Psi_T^* E_L(x) \Psi \mathrm{d}x}{\int \Psi_T^* \Psi \mathrm{d}x} .$$

If the trial function is very close to the exact g.s. wave function, $E_L(x)$ will be a rather flat function and the variance of the integral will be smaller than in the case of putting $\Psi_T = 1$. This will not affect the analytic calculations, however, but will be of primarily importance in the MC evaluation.

We suggest the reader to carry out a simple calculation using for the trial function $\Psi_T = \exp(-x^2)$, and checking that the energy expectation value is the same, but the variance has changed.

There is another alternative for this algebraic calculation which, however, cannot be used in Monte Carlo methods, namely to act with the Hamiltonian on the ket state. In the $\tau = 0$ limit the ket state is an eigenstate of the Hamiltonian, with eigenvalue 1 and the variance is null. This has no interest in DMC calculations because one does not know how to compute the second derivative of a function represented by a set of walkers.

The Normalization of the Solution. After each step the resulting gaussian is not, in general, normalized. There is a correcting factor which appears in Eq. (6.21) in square braces. This factor may be used to determine the energy E by requiring it to be 1. This way of evaluating the eigenvalue is known as the *growth estimator* of the energy. Once the stability has been reached, the equation to be solved is

$$e^{E\tau} = \tau + \sqrt{1+\tau^2}$$

and the Taylor expansion gives $E = 1 - \frac{\tau^2}{6} + \cdots$, so that again an $O(\tau^2)$ error for the energy with this new method.

6.5 Importance Sampling

Importance sampling has a relevant role in QMC calculations. As we have shown in the previous section, its use may help in lowering the variance of the energy estimators, reducing the required number of samples for a given value of the desired error.

However, this is not the only role of importance sampling. Actually, the potential part of the Green function depends on the potential $V(\mathbf{R})$, and when the potential is large and negative, the exponential may take unbounded values.

One of the ways of dealing with importance sampling in QMC is to consider the (imaginary) time evolution of the new quantity

$$f(\mathbf{R},t) = \Psi_T(\mathbf{R})\Psi(\mathbf{R},t) \ , \tag{6.23}$$

where Ψ is the wave function satisfying the Schrödinger equation and Ψ_T is a trial wave function, time independent, which should preferably be close to the exact ground state wave function. It is not necessarily the starting wave function, even if normally it is taken also as the starting wave function.

The time evolution of $f(\mathbf{R},t)$ is given by the differential equation

$$-\frac{\partial f(\mathbf{R},t)}{\partial t} = -\frac{\hbar^2}{2m}\boldsymbol{\nabla}^2 f(\mathbf{R},t) + \frac{\hbar^2}{2m}\boldsymbol{\nabla}\cdot[\mathbf{F}(\mathbf{R})\, f(\mathbf{R},t)] + [E_L(\mathbf{R}) - E] f(\mathbf{R},t) \ , \tag{6.24}$$

where we have used again the *local energy*, which was already defined in the section devoted to variational MC methods.

A new quantity appearing in the evolution equation is the so called *drift force* (it should better be called drift velocity), given by

$$\mathbf{F} = \frac{2}{\Psi_T(\mathbf{R})}\boldsymbol{\nabla}\Psi_T(\mathbf{R}) \ . \tag{6.25}$$

The transformed Hamiltonian acting on f in the rhs of Eq. (6.24) may be written as the sum of three terms

$$\begin{aligned} H &= K + F + L \\ K &= -\frac{\hbar^2}{2m}\boldsymbol{\nabla}^2 \\ F &= \frac{\hbar^2}{2m}[(\boldsymbol{\nabla}\cdot\mathbf{F}(\mathbf{R})) + \mathbf{F}(\mathbf{R})\cdot\boldsymbol{\nabla} \\ L &= E_L(\mathbf{R}) \end{aligned}$$

corresponding respectively to the kinetic part, the drift part and the local energy part. To obtain the previous non importance sampling formula, just put $\Psi_T = 1$, $F = 0$ and $E_L(\mathbf{R}) = V(\mathbf{R})$.

One may define as above a Green function corresponding to this transformed operator, obeying a similar, but more complex, differential equation. Formally, the new Green function is

$$G = e^{-[K+F+L-E]t} \ ,$$

and an $O(\tau^3)$ approximation may be obtained from

$$G = e^{E\tau} e^{-L\tau/2} e^{-F\tau/2} e^{-K\tau} e^{-F\tau/2} e^{-L\tau/2} + O(\tau^3) \ .$$

It is important to maintain the order of the operators.

In order to get the coordinate representation one must first obtain the representation of the three pieces, referred as G_L, G_F and G_K. The first has the same form as G_V above, simply putting E_L instead of V. The G_K term is the same as above.

The most complicated term corresponds to the drift force. Its explicit form is (see [52])

$$\langle \mathbf{R}' | G_F | \mathbf{R} \rangle = \delta(\mathbf{R}' - \boldsymbol{\mathcal{R}}(t)) \ , \tag{6.26}$$

where $\boldsymbol{\mathcal{R}}(t)$ is the *trajectory* obtained by integrating the differential equation

$$\frac{d\boldsymbol{\mathcal{R}}(t)}{dt} = D\mathbf{F}(\boldsymbol{\mathcal{R}}(t)) \tag{6.27}$$

with the boundary condition $\boldsymbol{\mathcal{R}}(0) = \mathbf{R}$. In equation (6.27) the quantity $\mathbf{F}(\boldsymbol{\mathcal{R}}(t))$ of the rhs must be understood as a (vector) function $\mathbf{F}$ of the function to be integrated $\boldsymbol{\mathcal{R}}(t)$. On the other hand, $\mathbf{R}$ is the initial position. Eq. (6.27) shows that the right name for $\mathbf{F}$ is *velocity*, instead of *force*.

Understanding the Drift Green Function. The form just discussed for the Green function G_F related to the drift force is quite involved, and an explicit example will clarify its structure.

Assume a one-dimensional problem with a drift force $F(x) = -x$. This kind of force will appear when the trial wave function has a gaussian shape, specifically for $\Psi_T = \exp(-x^2/4)$, accordingly with Eq. (6.25). Note that the drift always points in the direction in which the trial function grows, and its actual role will be to displace the walkers towards regions where Ψ_T is large.

To obtain the Green function we must solve the differential equation (6.27). Taking $D = 1$ for simplicity and given that the value of the drift force is just its argument with a minus sign, there results the simple differential equation

$$\frac{d\mathcal{R}(t)}{dt} = -\mathcal{R}(t) \ ,$$

whose solution is

$$\mathcal{R}(t) = Re^{-t} \ ,$$

satisfying the boundary condition that at $t = 0$ its value is R. So, the drift Green function Eq. (6.26) is

$$\langle R' | G_F | R \rangle = \delta(R' - Re^{-t}) \ .$$

To check this result we must go back to the integral form of the time evolution equation, $f(R',t) = \int \langle R'|G_F|R\rangle f(R,0)dR$. To carry out the integration on R we should follow the usual steps of solving the argument of the delta function for R and introducing the jacobian of the argument. In this way there results

$$f(R',t) = e^{-t} f(R'e^{-t},0) \ ,$$

where $f(R',0)$ is the boundary condition, i.e., the value of $f(R',t)$ at $t=0$. This is an *arbitrary function*, which may be explicitly defined by saying, e.g., that at $t=0$ our function is some $h(x)$. Then the solution at t is

$$f(R',t) = e^{-t} h(R'e^{-t}) \ .$$

The last step required to be convinced that the obtained result is correct is to go back to the importance-sampling differential equation (6.24), removing the first and the last terms of the rhs, and check that our $f(R,t)$ satisfies it. For our force $F=-R$ the equation reads

$$-\frac{\partial f(R,t)}{\partial t} = \frac{\partial[-Rf(R,t)]}{\partial R}$$

and it is satisfied for any value of the starting function $h(R)$.

6.6 The Explicit Importance-Sampling Green Function

The form of the Green function involving the computation of the classical trajectory is not very much practical. One may obtain a simpler form by solving the classical equation up to some given order in τ, resulting in the explicit Green function [52]

$$\langle \mathbf{R}'|G|\mathbf{R}\rangle = \frac{1}{(4\pi D\tau)^{3N/2}} e^{-\{\mathbf{R}'-\mathbf{R}-D\tau[\mathbf{F}(\mathbf{R}')+\mathbf{F}(\mathbf{R})]/2\}^2/4D\tau} \qquad (6.28)$$
$$e^{\tau E - \tau[E_L(\mathbf{R}')+E_L(\mathbf{R})]/2} \det\left[\mathbf{I} - D\tau\mathbf{H}(\mathbf{R}')/2\right] + O(\tau^3) \ ,$$

where $\mathbf{I}$ is the $3N\times 3N$ unit matrix, $\mathbf{H}$ contains the second derivatives of the logarithm of the importance sampling function

$$H_{ij} = \frac{\partial[2\log \Psi_T(\mathbf{R}')]}{\partial R_i \partial R_j}$$

and the determinant of the matrix with $\mathbf{I}$ and $\mathbf{R}'$ is the $O(\tau^3)$ normalization condition of the gaussian part of the Green function (when integrated with respect to $\mathbf{R}$).

A simpler $O(\tau^2)$ form of the importance-sampling Green function is [38]

$$\langle \mathbf{R}'|G|\mathbf{R}\rangle = \frac{1}{(4\pi D\tau)^{3N/2}} e^{-\{\mathbf{R}'-\mathbf{R}-D\tau\mathbf{F}(\mathbf{R})\}^2/4D\tau} \qquad (6.29)$$
$$e^{\tau E - \tau[E_L(\mathbf{R}')+E_L(\mathbf{R})]/2} + O(\tau^2) \ .$$

The use of any of these forms of the Green function was already discussed in section **2.5** as an improvement of von Neumann acceptance-rejection method. In both cases there is a gaussian-like part, properly normalized, and a rate term represented by the exponential involving the energy and the local energies. The Monte Carlo interpretation of the integral equation with these Green functions is an algorithm to obtain new points $\mathbf{R}'$ (none, one or more) from each member in the old set $\{\mathbf{R}\}$ by applying the exponential rate term to $\mathbf{R}'$ obtained after a gaussian isotropic diffusion centered at $\mathbf{R} + D\tau\mathbf{F}(\mathbf{R})$ (case of eq (6.29)). This last step requires first to displace the starting point $\mathbf{R}$ by $D\tau\mathbf{F}(\mathbf{R})$ and afterwards to select another displacement governed by the normal distribution characterized from the parameter $\sigma = \sqrt{2D\tau}$.

The first displacement is known with the name of *drift*, and is related to our trial function Ψ_T: it points always in the growing direction of the trial function, so as to bias the distribution of final points by increasing their number in the regions where the trial function is larger. This drift is modulated by the rate term, which will favor the areas where the local energy is smaller, i.e., favoring the g.s. components.

In the case of the more complex $O(\tau^3)$ Green function (6.28) there is a double drift, before and after the gaussian diffusion, with a half step.

These forms of the Green function are the most frequently used. However, they have not been generalized for the use in the coupled-channel problems. An alternative way of introducing importance sampling is by multiplying the time evolution integral equation (6.19) by the time-independent trial function,

$$\Psi_T(\mathbf{R}')\Psi(\mathbf{R}',\tau) = \int \left[\Psi_T(\mathbf{R}')\langle\mathbf{R}'|G|\mathbf{R}\rangle\Psi_T^{-1}(\mathbf{R})\right]\Psi_T(\mathbf{R})\Psi(\mathbf{R},0)d\mathbf{R} \ ,$$

where $\langle\mathbf{R}'|G|\mathbf{R}\rangle$ is one of the forms shown in section **6.3**. As in the previous formalism, the quantity with evolves in time is again $f(\mathbf{R},\tau) = \Psi_T(\mathbf{R})\Psi(\mathbf{R},\tau)$, but the new Green function is $\Psi_T(\mathbf{R}')\langle\mathbf{R}'|G|\mathbf{R}\rangle\Psi_T^{-1}(\mathbf{R})$.

We have already mentioned the role of the importance sampling trial function in regard with the bias of the random walk which selects the areas where the trial function is larger and the trial energy is smaller. Another important result is that the stabilized population of walkers refers to the quantity $\Psi_T(\mathbf{R})\Psi(\mathbf{R},t)$, which is required to compute the mixed estimator of the energy introduced in section **6.4**. To compute the energy one simply averages the values of the local energy $E_L(\mathbf{R})$ over the set of walkers. At the same time, one may also obtain an estimate of the variance.

A final role of the importance sampling method is the ability of dealing with states whose wave function is *not* positive definite. The way of proceeding is to assume that the new probability distribution function $f(\mathbf{R},t)$ will be positive or null, i.e., tying the nodes of the seeked wave function to the nodes of the trial function. This means to put impenetrable walls along the nodal surfaces defined by the zeros of the trial function, and this means that any time a walker tries to cross one of these walls, it is killed. This approach, known as the *fixed node* approximation, is of fundamental importance in the case or fermionic systems.

6.7 QMC and Numerical Methods

The MC evaluation of the time evolution equation by means of a combination of drift and diffusion is a very clever way of computing a quite complicated integral, and here we would like to mention the relation of the MC method with the Gauss-Hermite quadrature rule.

The integral to be carried out has a piece consisting in a very narrow gaussian, characterized by a width $\sigma = \sqrt{2D\tau}$. The validity of the step-by-step time evolution requires σ to be a small quantity, much smaller than the typical dimension characterizing the system, so that it is a very peaked function. If one attempts to compute numerically the integral in some one-dimensional problem by means of the usual trapezoidal or Simpson rules, the required number of points must be very large, making the calculation quite time consuming.

The proper way of computing numerically the evolution integral is by means of an algorithm like the one shown in the section **6.5**, i.e., using the delta-function expansion of the gaussian. This method is very similar to the Gauss-Hermite quadrature rule, where a given integral involving a gaussian is approximated by

$$\int_{-\infty}^{\infty} e^{-x^2} f(x)dx = \sum_k w_k f(x_k) ,$$

the weights w_k and the points x_k being chosen such as to have an exact rule for the highest possible polynomial. This method is conceptually equivalent to the MC sampling algorithm, where the new random points are selected accordingly with the (conditional) gaussian distribution.

The small-time approximation to the Green function is not very much adequate for ordinary quadrature rules, but is well tailored for MC methods.

6.8 The DMC Algorithm

Here we collect all results presented in the previous section and present the basic algorithm of DMC method.

Previous to the preparation of a computer code an importance sampling trial function must be determined. Adequate procedures must be devised to compute the local energy and the logarithmic gradient of the trial function.

The remaining tasks are written orderly:

Initialize:

- Get a value for the time step. This time step should be varied and from the set of pairs (time step, energy) carry out a least-squares extrapolation to zero.

- Guess a value for the energy shift E
- Create a starting set of walkers, representing the function $f(\mathbf{R}, t = 0)$. Normally, the importance trial function is used as a variational trial function, and both E and the starting set of walkers are determined by a previous VMC Metropolis–like algorithm.
- For each walker $\mathbf{R}$ compute and store the local energy and the drift function
- Initialize accumulators for energy and its square, as well as for other observables

Time Evolution Loop: Repeat up to large enough time the next steps

Thermalization Loop: Successive positions are strongly correlated, because the time step has to be very small. So, go on doing moves but only count a subset of them to get the averages

Loop over Old Walkers:

- Generate a trial move

$$\mathbf{R}' = \mathbf{R} + Dt\mathbf{F}(\mathbf{R}) + \boldsymbol{\xi}$$

  where $\boldsymbol{\xi}$ is drawn from the multivariate gaussian distribution with null mean and $\sigma = \sqrt{2Dt}$
- Compute the *replication factor*

$$n = [\exp\{t(E_L(\mathbf{R})/2 + E_L(\mathbf{R}')/2 - E)\}] \quad ,$$

  where the square brackets mean *integer part.*
- Unless $n = 0$ make n copies of $\mathbf{R}'$, $E_L(\mathbf{R}')$ and $\mathbf{F}(\mathbf{R}')$ to the new set of walkers

End of Loop over Walkers

New Generation:

- Copy the new set to the old set
- Test the growth of the population, and adjust the energy E

End of Thermalization Loop

Update Accumulators: Once the stabilization time has elapsed, update the accumulators for averages and errors

End of Time Iterations Loop

Final Work: Compute averages, errors ... and print results

6.9 Example: the ^{4}He Nucleus

Below there is a simplified code for a practical calculation of the g.s. energy of the ^{4}He nucleus described by a Wigner nucleon interaction. It is the same problem described in section **5.5** and uses the same routines described there for the calculation of the local energy as well as the gradient required for the determination of the drift term.

The code has been simplified by defining many parameters in a FORTRAN `PARAMETER` statement, just to reduce the number of lines. Among the required improvements, there is one of particular concern, namely the control on the population growth. There are two variables, `NWstart` and `NWend`, representing the number of walkers before and after a time step respectively, which may grow or decline in correspondence with the value of the energy `E0`, and it should be convenient to add a procedure to vary `E0` so as to maintain a stabilized population.

Two sets of time steps have been distinguished, labeled by a positive or negative value of the counter `itime`, so as to discard the energy averages obtained during the stabilization time. There is also a call to a `RANGAUSS` routine which generates random numbers from the $N[0,1]$ normal distribution.

The code has two phases, the initialization and the time evolution. The starting set of walkers has been obtained by generating nucleon coordinates at random. In a serious application the Metropolis algorithm based on the trial function should have been used to get both a good starting set of walkers and an estimate of the parameter `E0`.

The main MC operation is carried out by the statement labeled with the number 40, which selects a possible destination point obtained from the previous point by adding the drift (`Grad`) and the isotropic diffusion (`rangauss`). Branching is decided by computing the variable `Nsons` which is an integer number from a pdf taking two values, $n_1 = [\exp(tE - t[E_L(\mathbf{R}) + E_L(\mathbf{R}')]/2)]$ and $n_2 = n_1+1$ with mean value $\exp(tE - t[E_L(\mathbf{R}) + E_L(\mathbf{R}')]/2)$ (see section **2.5**).

```
c       ================================================
c       Diffusion Monte Carlo for He-4 with S3 force
c       Must be linked with ELOCAL and WF subroutines
        include 'varhe.inc'
        parameter(NWmax=3000, Nstab=20, Ntime=100, Nth=10)
        dimension Walker (NA,3,NWmax,2), EL(NWmax,2),
     1            Grad(Na,3,NWmax,2)
        parameter(D=h2div2m, t=0.001d0)
c       =============Starting configuration ===================
        sig=sqrt(2*D*t)              ! Parameter for gauss distrib.
        z=randini(1234.5d0)          ! Initialize gaussian generator
        NWstart=1000                 ! Starting population
        E0=-27                       ! and guess energy
        IN=1                         ! To swap OLD and
        IO=2                         ! new generations
        EAV=0
        do 10 kw=1, Nwstart          ! Raw initial set of
        do 20 i=1, NA                ! walkers
        do 20 j=1, 3
20      Walker(i,j,kw,IN)=rangauss()
        wfold=WF (Walker(1,1,kw,IN))
10      EL(kw,IN)=ELOCAL(wfold, Walker(1,1,kw,IN), Grad(1,1,kw,IN))
c       ============= START ITERATIONS ==========================
        EF=0                           ! To compute averages and
```

```
      EF2=0                          ! error
      do itime=-Nstab, Ntime         ! Nstab to reach stability
      do ith=1, Nth                  ! Thermalization loop
      eav=0
      NWend=0                        ! New set of walkers
      do 30 kw=1, NWstart
      do 40 ip=1, Na                 ! Diffusion and drift
      do 40 j= 1, 3                  ! O(t**2) Green function
40    R(ip,j)=Walker(ip,j,kw,IN)+D*t*Grad(ip,j,kw,IN)+sig*rangauss()
      wfold=WF(R)
      ELnew=ELOCAL(wfold, R , DRIFT)
      Nsons=exp((E0-(ELnew+EL(kw,IN))/2)*t)+RAND()    ! Branching
              do 60 isons=1, nsons
              NWend=NWend+1
              do 50 ip=1, Na
              do 50 j=1, 3
              walker(ip,j,NWend,IO)=R(ip,j)
50            GRAD  (ip,j,NWend,IO)=DRIFT(ip,j)
              EL(NWend,IO)=ELnew
60            eav=eav+ELnew
30    continue
      IN=3-IN                          ! Swap OLD/NEW generations
      IO=3-IO
      NWstart=NWend
      write(*,*) NWend, eav/NWend      ! Monitor evolution
      enddo                            ! End of thermalization loop
      if(itime.gt.0) then
      EF=EF+eav/NWend                  ! Compute averages only
      EF2=EF2+eav**2/NWend**2          ! after stability
      endif
      enddo                            ! End of TIME iterations
c     ==============   Compute and print results ============
      EF=EF/Ntime
      EF2=EF2/Ntime
      write(*,*) ' Energy=', EF        ! Write results
      write(*,*) ' Error =', SQRT((ef2-ef**2)/Ntime)
      end
```

Results obtained with the above code are shown in figure 7. The wavy-like aspect of the time evolution is typical of this specific Markov process with a small time step, and shows how strong are the correlations between two successive moves. Thermalization is required in order to have a good estimate of the variance. The left part of the figure shows the $t = 0$ limit extrapolation, which results in a value of $E = -27.18 \pm 0.12$ MeV. To obtain the value mentioned in Table 4 of section **5.5** and in Ref. [15] a very long run will be required with this simplified code. It would be better to use a more precise importance sampling wave function [15].

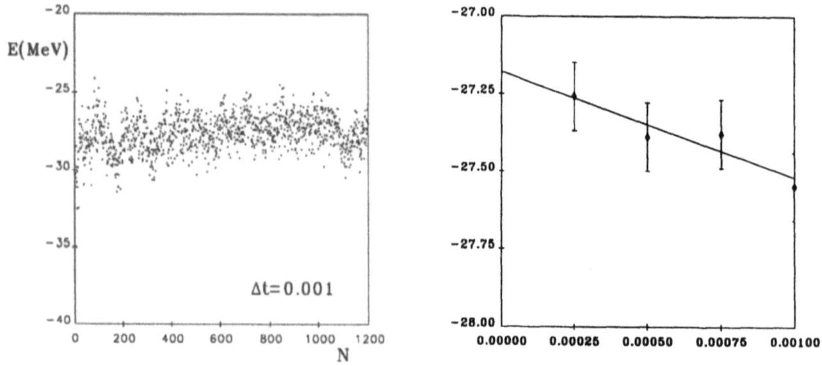

Fig. 7. The time evolution of the average energy (left) as a function of the number of time steps N and the evaluation of the average for different time steps (right). The continuous line represents a linear fit for $t = 0$ extrapolation.

6.10 Some Physical Systems

Here I briefly present some calculations which have mainly an historical interest, because they show the effort of many research groups to attain the maximum precision by using various many-body techniques. The possibility of obtaining essentially exact results by QMC methods was a strong stimulus to improve other theories, as we shall see.

More complete descriptions of nowadays studies in the QMC framework are discussed in this volume by S. Pieper, referred to the Argonne–Los Alamos–Urbana collaboration on the nuclear few-body problem, and by J. Boronat, applied to the study of zero temperature quantum liquids.

Jellium. One of the most impressive applications of the DMC method was carried out by Ceperley and Alder [37] who studied the system known as *jellium*, consisting in an electron gas on a uniform background of positive charge. This is a simple model of conductivity of metals, and is characterized by a single length parameter r_s, conveniently measured in units of Bohr radius a_0, which defines the electron density as

$$\rho = \frac{3}{4\pi a_0^3 r_s^3} \ .$$

Many years ago Wigner [53] proved that this system will have a crystal structure, body centered cubic, at very low densities (large r_s).

The table below presents a set of results for this system, computed with several many-body theories. This is a nice example of several many-body theories properly working in the domain in which they are adequate: RPA and theories, like CCM, containing RPA, are well suited in the high density regime. FHNC is appropriate at densities not very large.

Table 6. Correlation energy per particle in the jellium model, computed with several theories. RPA is the *random phase approximation*, CCM is the *coupled cluster method*, FHNC is the *fermionic hypernetted chain* approximation and HNC/Lado is a bosonization of the FHNC equation. Energies are quoted in millirydbergs.

Method	$r_s \to 0$	$r_s = 1$	$r_s = 3$	$r_s = 5$	$r_s = 20$
QMC [37]		-120		-56.3	-23.0
RPA [54]	$62.2 \log(r_s)$	-157	-105	-84.9	-42.8
CCM(2) [55]	$62.2 \log(r_s)$	-123	-75.1	-56.8	
CCM(4) [56]	$62.2 \log(r_s)$	-122	-73.8	-56.0	-23.6
FHNC [57]		-138	-79	-58	
HNC/Lado [58]	$57 \log(r_s)$	-114	-71.0	-54.1	-21.8
VMC [59]		-122	-72.2	-55.0	-22.5

It should be mentioned that the QMC calculations are not a single calculation: one of the unpleasant properties of the importance sampling trial function is that forces the walkers to adapt to its shape. So, the row labeled QMC is a subset of a series of calculations which assume for the trial function a paramagnetic (unpolarized) Fermi fluid, a ferromagnetic (polarized) Fermi fluid, and a bcc Wigner crystal, mentioned in order of decreasing density. It would be nice to have a theory able to select the appropriate shape for the wave function, but this goal has not yet been reached.

The Diatomic Molecule LiH. Atomic and molecular physicists are used to do calculation with a very high degree of precision. The energy scale in atomic physics is the Rydberg (13.6 eV) or the Hartree (2 Rydberg), and the theoretical studies attain precisions of the order of one thousandth of Rydbergs.

The LiH is a rather simple molecule, with two centers and four electrons, one from H and the other three from Li. In the Born–Oppenheimer approximation the centers are assumed to be fixed and the calculations are carried out for several values of the intercenter distance, quoting the minimum as the energy of the molecule. The quantity which is experimentally measured is the *dissociation energy*, which is the energy which must be provided so as to put the two atoms far away. The comparison with experiment requires two corrections: relativistic effects and zero-point vibrational motion of the atomic centers.

The results of several theoretical calculations are presented in Table 7, which shows the value of the *correlation energy*, the difference between the total energy of the molecule and the energy computed in the self-consistent field calculation. As in the previous case, the agreement between the various theoretical methods is the most important result of this comparison.

Table 7. The correlation energy, in millihartree, for the LiH molecule, as computed with several methods. CI is configuration interaction, CCM is coupled-cluster method. The VMC calculation quoted refers to a quite poor trial function, which was used as importance sampling for the DMC calculation.

Method	Comment	Ref.	Corr. energy
CI	Slater Orbitals	[60]	-76.5
CI	Elliptic Orbitals	[61]	-78.2
CCM	SUB(2) approx	[62]	-81.5
VMC		[38]	-54±8
DMC		[38]	-80±2
Exact	Mixed Theor. Exper.	[63]	-83.2 ±1

6.11 Fermionic Systems

There are two complications related to the study of fermionic systems. On one side, the g.s. fermionic wave function is not in general positive definite, the exception being systems of distinguishable particles, like the example of ^{4}He nucleus above presented. Then, it cannot be interpreted as a probability distribution function. The other complication arises from the spin/isospin degrees of freedom, which convert the Schrödinger equation into a set of coupled equations: relative phases between the various components may be negative or even complex, and again putting troubles to make a correspondence between the wave function and a probability distribution function.

The first case is usually solved in the so called *fixed node approximation*, followed by a *nodal relaxation*. The idea is to use the importance-sampling method and to assume that the zeros of the trial function, called *nodal surfaces*, define impenetrable barriers which cannot be crossed by the walkers: every time a walker attemps to cross one of these boundaries, is is killed, i.e., its *replication number* is made zero.

The fixed node approximation then solves the many-body Schrödinger equation in a series of non-intersecting domains subjected to Dirichlet boundary conditions. The problem is mathematically well posed, but any imaginable analytic solution will result, in general, in a different value for the energy. The stochastic MC method will generate some weighted average of these energies.

Nodal Surfaces. There are several questions difficult to answer related to the nodal surfaces. Considering the case of a polarized system, the space part of the trial wave function will be fully antisymmetric, and the equation $\Psi_T(\mathbf{R}) = 0$ may correspond to a single hypersurface or more than one. The set hypersurfaces must contain all the lower dimension *surfaces* or *hyperpoints* [64] defined by the equations $\mathbf{r}_i = \mathbf{r}_j$, $i < j$, required by Pauli principle. So, questions like how many nodal surfaces, how many non-intersecting regions, and so on are normally very difficult to answer.

Intuition as well as simpler problems may be a bad allied. For example, one-dimensional systems have the peculiarity that the minimal Pauli conditions (now $x_i = x_j$) define actually full nodal surfaces. Also, symmetries of systems of few particles may define uniquely the nodal surfaces.

Example: The HeI Atom. Triplet states of helium atom have a fully symmetric spin wave function, and the space part must be fully antisymmetric.

The lowest energy triplet state has $L = 0$, and its most general wave function depends only on r_1, r_2 and $\cos(\theta_{12})$, as requested by the rotational invariance. Pauli principle requires the equality $\Psi(r_1, r_2, \cos(\theta_{12})) = -\Psi(r_2, r_1, \cos(\theta_{12}))$ which in turn implies that $r_1 = r_2$ must be a nodal surface. This does not exclude trial functions having other nodal surfaces, but it is reasonable to expect that the g.s. will have the smaller number of nodes compatible with the symmetries of the system.

In conclusion, the triplet $L = 0$ g.s. will have a single nodal surface in a three-dimensional space $\{r_1 \in [0, \infty], r_2 \in [0, \infty], \cos(\theta_{12}) \in [-1, 1]\}$ which is a plane along the $\cos(\theta_{12})$ axis with $r_1 = r_2$. The space will be divided in two regions, $r_1 < r_2$ and $r_1 > r_2$ and for a given point in one region there is a corresponding point in the other region obtained by applying the permutation (12), i.e., by exchanging r_1 and r_2. Any reasonable trial function will posess only this nodal surface.

A different situation happens with the triplet $L = 1$ states. Considering the $M = 1$ state, a trial wave function could be the one provided by the central field model

$$\Psi_T(L = 1, M = 1) = \det \begin{bmatrix} \psi_s(r_1) & \psi_s(r_2) \\ \psi_p(r_1)z_1 & \psi_p(r_2)z_2 \end{bmatrix} .$$

The description of the system involves now four coordinates, $\{r_1 \in [0, \infty], z_1 \in [-r_1, r_1], r_2 \in [0, \infty], z_2 \in [-r2, r2]\}$. If $\psi_s = \psi_p$ the nodal surface will be determined by the equation $z_1 = z_2$. However, a different nodal surface will result for ψ functions not satisfying the above equality.

These two examples have shown the two kinds of situations to face up: the nodal surfaces are fully determined by the symmetries of the system, or there may be families of nodal surfaces compatible with the symmetries of the system. Using the fixed node approximation in systems like the former will ensure to obtain exactly its energy. On the contrary, in systems where the nodal surfaces are not uniquely defined the fixed node approximation will not provide the exact value of the energy.

Ceperley Theorem. Ceperley [65, 38] and almost simultaneously Moskowitz and collaborators [66] proved that the energy obtained with the mixed estimator in the fixed node approximation is an upper bound to the fermion g.s. energy. If the nodal surfaces of the trial function were the exact surfaces, then this upper bound will coincide with the g.s. energy.

The proof of the theorem is as follows. Assume that the nodal surfaces have divided the space in domains $\{D_k, \ k = 1, M\}$. The fixed node approximation solves the Schrödinger equation in each of these domains, with Dirichlet boundary conditions at the edges, providing values for the energy E_k and wave function Φ_k in each domain (actually, the QMC algorithm will only give a weighted average of the domain energies). From any of the domain solutions one may construct a fully antisymmetric wave function by applying the antisymmetrizer

$$\bar{\Phi}_k(\mathbf{R}) = \sum_P \epsilon_P \Phi_k(P\mathbf{R}) \ ,$$

having in this form an adequate variational trial function. The expectation value of the Hamiltonian with respect to $\bar{\Phi}_k$, which has the value E_k, will be an upper bound to the fermion g.s. energy.

A Counter Example. Just searching for an example to explain the meaning of Ceperley's theorem I found a counter example. Imagine a system of two particles in a one-dimensional space, with a translationally invariant interaction which, in addition, commutes with the parity operator. Concentrating on the relative wave function the exchange of the two particles is equivalent to the parity operation. Assume that we look for solutions *with a wrong nodal surface*, which in our case will just be a wrong node. Figure 8 shows, in the lhs part, the two solutions which are obtained, one of them (labelled R) with energy *higher* than the energy of the first excited state (which will be the *fermionic* gs), but the other (labelled L) will have an energy *lower* than the fermionic g.s. energy. This means that Ceperley's theorem cannot be satisfied, or, in other words, that the expectation value of the Hamiltonian with respect to the antisymmetrized wave functions cannot be the same than the domain energies.

The reason lies in the presence of discontinuities of the derivative of function R at the imposed nodal point and its image. The presence of these discontinuities, and the corresponding effect on the kinetic energy, was already realized in Ref. [38], but there it was assumed that the wave function will be null at those points. This is not the case for the trial function L of our example, but it is the case of the other wave function R.

The reason of having non-null values of the antisymmetrized wave function at the discontinuities is that the permutations may overlap a given domain with itself. It should be said that normally importance sampling functions with such wrong nodal surfaces will not be used, if one is able to recognize how unappropriate they are, as in our simple example.

In conclusion, either the theorem is wrong or some extra conditions must be added for its fulfillment. This is a point that should be studied more in depth.

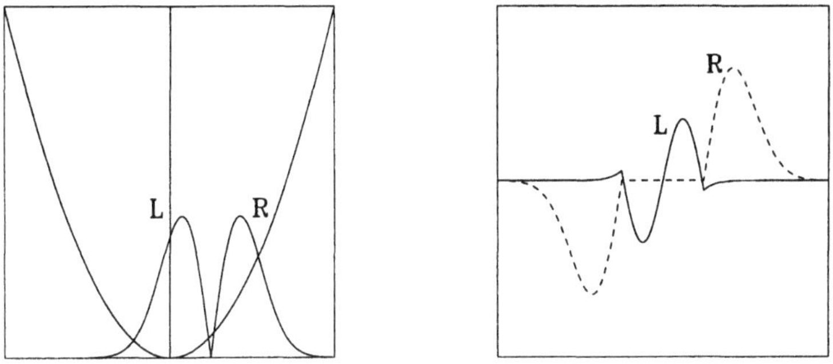

Fig. 8. Solution of a one-dimensional problem with a wrong nodal surface. The lhs shows the potential and the solutions confined in the left (L) and right (R) domain. The rhs shows the fermionic trial functions constructed from the domain solution by antisymmetrization.

Nodal Relaxation. The fixed-node QMC may be followed by a *nodal relaxation* process. Once the fixed-node MC walk has stabilized, one allows walkers to cross the nodal surfaces, adding a *signature* which indicates if their contribution to the energy should be added or substracted, accordingly with the number of crosses (even or odd) undergone.

The danger with crossing nodes is that the boson g.s. may start to grow up, basically because of the statistical noise. Then it may well happen that the variance of the result (nodes relaxed) is too high to ascertain if there has been a real gain in energy, so the extra work will be useless. Sooner or later, the boson g.s. will start to be the most important piece of the set of walkers, so the relaxation may only be continued for a limited number of time steps. Such fact is the reason for adding the adjective *transient* to the energy estimator. A qualitative discussion of the releasing of nodes may be fund in [67].

7 Closing Remarks

This is the end of these quite incomplete notes on Monte Carlo methods applied to many-body systems. The objective was to show the basic ideas, and adding examples was considered as a complementary didactic tool which, hopefully, will be appreciated by the reader.

Specific applications require some extra work. Even if MC methods are very powerful, they require our help (in other words, a good importance sampling driving function) to become really effective. This means that one must know in depth the physical system to be described, and this knowledge may be more complex than the mere MC techniques.

A last comment regarding to QMC methods, of special interest for newcommers to this technique: most of the effort in MC is actually devoted to have a variance as low as possible. Brute force is not the best way of diminishing the variance.

Acknowledgements

This work is suported by DGICyT under contract N. PB92–0820. The author wishes to thank J. Navarro and A. Polls for their kind invitation to lecture in this excellent many-body school. The collaboration of Mariela Portesi in the search of misprints and other kind of errors is warmly acknowledged.

References

1. J. R. Newman, *The World of Mathematics* (Simon and Schuster, New York, 1956)
2. W. Feller, *An Introduction to Probability Theory and its Applications* (J. Wiley, New York, 1957)
3. B. Gnedenko, *The theory of Probability* (Mir Publ., Moscow, 1975)
4. J.M. Hammersley and D.C. Handscomb, *Monte Carlo Methods* (Methuen, Norwich, 1965)
5. M.H. Kalos and P.A. Whitlock, *Monte Carlo Methods* (J. Wiley, New York, 1986)
6. D.E. Knuth, *The Art of Computer Programmer*, Vol. II *Seminumerical Algorithms*, Chap. 3 (Addison Wesley, New York, 1971)
7. K. Binder and Stauffer, in *Applications of the Monte Carlo Method in Statistical Physics*, edited by K. Binder, (Springer–Verlag, Berlin 1984)
8. J.A. Carlson and R.B. Wiringa, in *Computational Nuclear Physics 1*, edited by K. Langanke, J.A. Maruhn and S.E. Koonin (Springer–Verlag, Berlin, 1991). The algorithm appears in the accompanying diskette, and is coded in FORTRAN.
9. G. Marsaglia and A. Zaman, Computers in Phys. **8**, 117 (1994)
10. W.H. Press, S.A. Teukolsky, W.T. Vetterling and B.P. Flannery, *Numerical Recipes: The Art of Scientific Computing* (Cambridge U. Press, Cambridge Mass., 1992), Chap. 7
11. I.R. Afnan and Y.C. Tang, Phys. Rev. **175**, 1337 (1968)
12. N. Metropolis, A.W. Rosenbluth, M.N. Rosenbluth, A.H. Teller and E. W. Teller , J. Chem. Phys. **21**, 1087 (1953)
13. Martin Gardner, *Mathematical Circus*, spanish translation *Circo Matemático* (Alianza, Madrid, 1985)
14. R. Jastrow, Phys. Rev. **98**, 1479 (1955)
15. R.F. Bishop, E. Buendía, M.F. Flynn and R. Guardiola, J. Phys. **G18**, L21 (1992)
16. R. Guardiola, P.I. Moliner and J. Navarro, Phys. Lett. **B383**, 243 (1996)
17. R. Malfliet and J.A. Tjon, Nucl. Phys **A127**, 161 (1969)
18. Y. Akaishi, Nucl. Phys. **A416**, 409 (1984)

19. H. Kameyama, M. Kamimura and Y. Fukushima, Phys. Rev. C **40**, 974 (1989)
20. A. Kievsky, M. Viviani and S. Rosati, *Condensed Matter Theories* **6**, edited by S. Fantoni and S. Rosati (New York, Plenum, 1991) p. 391
21. J.G. Zabolitzky, Phys. Lett. **B100**, 5 (1981)
22. H. Kamada and W. Glöckle, Nucl. Phys. **A584**, 205 (1992)
23. W.W. Wood and F.R. Parker, J. Chem. Phys. **27**, 720 (1957)
24. R. Guardiola, in *First International Course on Condensed Matter*, edited by D. Prosperi, S. Rosati and G. Violini (World Sci. Singapore, 1988) p. 155
25. W.L. McMillan, Phys. Rev A **138**, 442 (1965)
26. S. Schiff and L. Verlet, Phys. Rev. A **160**, 208 (1967)
27. D.M. Ceperley and M.H. Kalos, in *Monte Carlo Methods in Statistical Physics*, edited by K. Binder (Springer Verlag, Berlin, 1979) p. 145
28. M.A. Preston, *Physics of the Nucleus*, chap. 5, Addison Wesley (Reading, Mass 1965)
29. B.H. Bransden and C.J. Joachain, *Physics of Atoms and Molecules* (Longman Sci. Tech., Essex, 1983)
30. F.M. Fernández and R. Guardiola, J. Phys **A30**, 5825 (1997)
31. N. Metropolis and S. Ulam, J.Am. Stat. Assoc. **44**, 335 (1949)
32. M.D. Donskar and M. Kac, J. Res. Nat. Bur. Standards **44**, 551 (1950)
33. M.H. Kalos, Phys. Rev. **128**, 1791 (1962)
34. J.B. Anderson, J. Chem. Phys. **63**, 1499 (1975)
35. J.B. Anderson, J. Chem. Phys. **65**, 4121 (1976)
36. J.B. Anderson, J. Chem. Phys. **73**, 3897 (1980)
37. D.M. Ceperley and B.J. Alder, Phys. Rev. Lett. **45**, 566 (1980)
38. P.J. Reynolds, D.M. Ceperley, B.J. Alder and W.A. Lester Jr., J. Chem. Phys. **77**, 5593 (1982)
39. B.J. Hammond, W.A. Lester Jr. and P.J. Reynolds *Monte Carlo Methods in Ab Initio Quantum Chemistry* (World Scientific, Singapore, 1994)
40. M.H. Kalos, J. Comput. Phys. **2**, 257 (1967)
41. M.H. Kalos, Phys. Rev. A **2**, 250 (1970)
42. M.H. Kalos, D. Levesque and L. Verlet, Phys. Rev. A **9**, 2178 (1974)
43. D.M. Ceperley, G.V. Chester and M.H. Kalos, Phys. Rev. B **16**, 3081 (1977)
44. M.H. Kalos, M.A. Lee, P.A. Whitlock and G.V. Chester, Phys. Rev. B **24**, 115 (1981)
45. K. Schmidt and M.H. Kalos, in *Applications of the Monte Carlo Method in Statistical Physics*, edited by K. Binder (Springer–Verlag, Berlin, 1984), p. 125
46. S. Caracciolo and A. Fabrocini Eds., *Monte Carlo Methods in Theoretical Physics*, ETS Editrice (Pisa, 1991)
47. S.A. Chin, Phys. Rev. A **42**, 6991 (1990)
48. J. Carlson, Phys. Rev. C **36**, 2026 (1987)
49. J. Carlson, Phys. Rev. C **38**, 1879 (1988)
50. D.M. Ceperley and R. Bernu, J. Chem. Phys **89**, 6316 (1988)
51. S.A. Chin and E. Krotscheck, Phys. Rev. B **45**, 852 (1992)
52. J. Vrbik and S.M. Rothstein, J. Comput. Phys. **63**, 130 (1986)
53. E.P. Wigner, Phys. Rev. **46**, 1002 (1934)
54. R.F. Bishop and K.H. Lührmann, Phys. Rev. B **17**, 3757 (1978)
55. R.F. Bishop and K.H. Lührmann, Phys. Rev. B **26**, 5523 (1982)
56. K. Emrich and J.G. Zabolitzky, Phys. Rev. B **30**, 2049 (1984)

57. L. Lantto and P. J. Siemens, Nucl. Phys. **A317**, 55 (1979)
58. J.G. Zabolitzky, Phys. Rev. B **22**, 2353 (1980)
59. D.M. Ceperley, Phys. Rev. B **18**, 3126 (1978)
60. H. Partridge and S.R. Laughoff, J. Chem. Phys **74**, 2361 (1981)
61. D.M. Bishop and L.M. Cheung, J. Chem. Phys **78**, 1396 (1983)
62. B. Jeziorski, H.J. Monkhorst, K. Szalevicz and J.G. Zabolitzky, J. Chem. Phys **81**, 368 (1984)
63. K. Szalevicz, B. Jeziorski, H.J. Monkhorst, and J.G. Zabolitzky, J. Chem. Phys **79**, 5543 (1983)
64. D.J. Klein and H.M. Pickett, J. Chem. Phys **64**, 4811 (1976)
65. D.M. Ceperley, in *Recent Progress in Many-Body Theories*, edited by J.G. Zabolitzky, M. de Llano, M. Fortes and J.W. Clark, Lecture Notes in Physics **142** 262 (Springer Verlag, Berlin, 1981)
66. J. Moskowitz, K.E. Schmidt, M.A. Lee and M.H. Kalos, J. Chem. Phys. **77**, 349 (1982)
67. D. Ceperley and B.J. Alder, Science **231**, 555 (1986)

Monte Carlo Calculations of Nuclei

Steven C. Pieper

Physics Division, Argonne National Laboratory,
Argonne, IL 60439, USA

Abstract. Nuclear many-body calculations have the complication of strong spin- and isospin-dependent potentials. In these lectures I discuss the variational and Green's function Monte Carlo techniques that have been developed to address this complication, and present a few results.

1 Introduction

A major goal in nuclear physics is to understand how nuclear binding, stability, and structure arise from the underlying interactions between individual nucleons. To achieve this goal, we must both determine the Hamiltonian to be used, and devise reliable methods for many-body calculations with it. In principle quantum chromodynamics can prescribe the nuclear Hamiltonian, but it will be a long time before this will be done with useful accuracy. Thus the nuclear Hamiltonian is determined phenomenologically, and our knowledge of it is refined, in part, by the many-body calculations we make with it. A large amount of empirical information about the nucleon-nucleon scattering problem has been accumulated over time, resulting in ever more sophisticated *NN* potential models. These models have strong spin and isospin dependence, and spin and orbital angular momentum are mixed by a strong tensor interaction. In addition the three-nucleon interaction must be considered in realistic calculations, however there is very little experimental knowledge of it.

Thus the nuclear many-body Hamiltonian is significantly more complicated than those encountered in most atomic and condensed-matter problems, and progress in accurate nuclear ground state calculations has been quite slow. As Table 1 shows, it took 30 years to progress from the two-body ground state to the three-body one; the four-body problem was solved in a few more years. It then took some time until ongoing advances in computational resources, particularly the advent of massively parallel computers, allowed us to apply sophisticated quantum Monte Carlo methods to the study of light *p*-shell nuclei, up to 8-body nuclei.

Quantum Monte Carlo methods for central interactions are discussed in R. Guardiola's contribution to this volume; here I will concentrate on the complications due to the state-dependence of the nuclear forces. The nuclear Hamiltonian is presented in the next section, variational Monte Carlo (VMC) for nuclei is presented in Sect. 3 and Green's function Monte Carlo (GFMC) in Sect. 4. Finally a very few recent results are given in the last section. The

Table 1. Exact (1%) realistic ground-state solutions.

Nucleus	Method	First Done	Floating Ops.	Computer	Time
^{2}H	Variational,	1940's –			
	Coupled Dif Eq	1953	$\sim 50 \times 10^3$	Illiac-I	15 min
					(+ 5 for printing)
^{3}H	34 Chn Faddeev	1984	100×10^9	Cray XMP	30 min
^{4}He	GFMC	1987	15×10^{12}	Cray 2	40 hr
^{5}He	GFMC	1993	100×10^{12}	Cray C90	100 hr
^{6}Li	GFMC	1995	1×10^{13}	IBM SP1	40 nodes, 150 hr
^{7}Li	GFMC	end 95	1.5×10^{15}	IBM SP2	100 nodes, 50 hr
^{8}Be	GFMC	end 96	7×10^{15}	IBM SP2	25 nodes, 120 hr

specific methods and results presented here are from the work of the Argonne, Los Alamos, and Urbana groups; a complete description of our VMC and GFMC calculations may be found in [1]. A good pedagogical discussion of the nuclear VMC method, including a complete, simplified, program may be found in [2].

2 The Nuclear Hamiltonian

In this section I discuss a slightly simplified version of the Hamiltonian used in our calculations. Specifically the complications of charge dependence and of the full electromagnetic potential will be only briefly mentioned. The complexity of the nuclear Hamiltonian is dictated by a number of experimental observations:

Nuclear forces are of short but finite range. The binding energy for few-body nuclei increases as the number of pairs, from 2.2 MeV in ^{2}H to 2.8 MeV/pair in ^{3}H to 4.7 MeV/pair in ^{4}He. For larger nuclei the binding energy per nucleon is approximately constant at 8 MeV/A. This indicates that nuclear forces are short-range, roughly equal to the radius of ^{4}He, or ~ 1.7 fm.

As is indicated by the fact that the low-energy S-wave phase shifts are positive, the nuclear force is attractive at intermediate and long ranges. However, the nuclear force has a short-ranged repulsive core. The $^1S_0$ phase shifts turn negative for $E_{lab} \sim 250$ MeV, while the $^1D_2$ phase shifts remain positive up to 800 MeV. This can be understood as a repulsive core of ~ 0.6 fm, which is masked in the D-wave by the centrifugal barrier.

The force depends on the spins and isospins of the nucleons. Only the S = 1, T = 0 channel has a bound state, while the phase shifts are very different in different S, T channels. In addition, the structure of the deuteron indicates that there is a tensor force. The deuteron has a quadrupole moment, a magnetic moment that requires a D-state, and an asymptotic D/S state ratio. In scattering states there are finite mixing parameters ϵ_J (transitions between L = J - 1 and L = J + 1 states). Finally, there is also a spin-orbit force; the energy-dependence of the $^3P_{0,1,2}$ partial wave phase shifts requires both tensor and spin-orbit forces.

Our Hamiltonian includes a nonrelativistic one-body kinetic energy, the Argonne v_{18} two-nucleon potential [3] and the Urbana IX three-nucleon potential [4],

$$H = \sum_i K_i + \sum_{i<j} v_{ij} + \sum_{i<j<k} V_{ijk} \ . \tag{2.1}$$

Ignoring the difference in proton and neutron masses, the kinetic energy operator is

$$K_i = -\frac{\hbar^2}{2m}\nabla_i^2 \ . \tag{2.2}$$

The Argonne v_{18} potential is one of a number of new, highly accurate NN potential models developed since 1990. It can be written as a sum of electromagnetic and one-pion-exchange terms and a shorter-range phenomenological part,

$$v_{ij} = v_{ij}^\gamma + v_{ij}^\pi + v_{ij}^R \ . \tag{2.3}$$

The electromagnetic terms include one- and two-photon-exchange Coulomb interactions, vacuum polarization, Darwin-Foldy, and magnetic moment terms, with appropriate proton and neutron form factors.

Ignoring the difference between charged and neutral pion masses, the one-pion-exchange part of the potential can be written as

$$v_{ij}^\pi = f_\pi^2 \tfrac{1}{3} m_\pi c^2 X_{ij}^\pi \ \boldsymbol{\tau}_i \cdot \boldsymbol{\tau}_j \ , \tag{2.4}$$

$$X_{ij}^\pi = Y(r) \ \boldsymbol{\sigma}_i \cdot \boldsymbol{\sigma}_j + T(r) S_{ij} \ . \tag{2.5}$$

Here $Y(r)$ and $T(r)$ are the normal Yukawa and tensor functions with a cutoff specified by a parameter c:

$$Y(r) = \frac{e^{-m_\pi r}}{m_\pi r} C(r) \ , \tag{2.6}$$

$$T(r) = [1 + \frac{3}{m_\pi r} + \frac{3}{(m_\pi r)^2}] \frac{e^{-m_\pi r}}{m_\pi r} C^2(r) \ , \tag{2.7}$$

$$C(r) = 1 - e^{-cr^2} \ , \tag{2.8}$$

The $\boldsymbol{\sigma}_i$ are the 2×2 Pauli spin matrices; the nucleon spin operator is $\mathbf{S}_i = \frac{1}{2}\boldsymbol{\sigma}_i$. The $\boldsymbol{\tau}_i$ are the same Pauli matrices acting in isospin space; our convention is that $\tau(z) = +1$ for protons and -1 for neutrons. The tensor operator, S_{ij}, is introduced below.

The one-pion-exchange and the remaining phenomenological part of the potential can be written as a sum of eighteen operators, which is where the name v_{18} comes from:

$$v_{ij}^{\pi} + v_{ij}^{R} = \sum_{p=1,18} v_p(r_{ij}) O_{ij}^{p} \ . \tag{2.9}$$

The first fourteen are charge-independent,

$$O_{ij}^{p=1,14} = [\ 1, (\boldsymbol{\sigma}_i \cdot \boldsymbol{\sigma}_j), S_{ij}, (\mathbf{L} \cdot \mathbf{S}), \mathbf{L}^2, \mathbf{L}^2(\boldsymbol{\sigma}_i \cdot \boldsymbol{\sigma}_j), (\mathbf{L} \cdot \mathbf{S})^2] \otimes [1, (\boldsymbol{\tau}_i \cdot \boldsymbol{\tau}_j)] \ , \tag{2.10}$$

and the last four break charge independence. In addition to the one-pion $Y(r)$ and $T(r)$, the $v_p(r)$ contain intermediate- and short-range functions which are determined by fitting *NN* scattering data.

The potential was fit directly to the Nijmegen *NN* scattering data base [5, 6], which contains 1787 *pp* and 2514 *np* data in the range 0 – 350 MeV, with a χ^2 per datum of 1.09. It was also fit to the *nn* scattering length measured in $d(\pi^-, \gamma)nn$ experiments and the deuteron binding energy.

A brief discussion of the effects of some of the operators in the potential follows. Nuclear isospin is (approximately) conserved by the strong interaction. This means that it is useful to project a pair of nucleons into states of good total isospin:

$$T = 0 \text{ pair}: \quad \tfrac{1}{\sqrt{2}} [\, |pn\rangle - |np\rangle \,] \ ; \tag{2.11}$$

$$T = 1 \text{ pairs}: \quad |pp\rangle \,, \ \tfrac{1}{\sqrt{2}} [\, |pn\rangle + |np\rangle \,] , \ \ |nn\rangle \ . \tag{2.12}$$

These are eigenstates of $\boldsymbol{\tau}_i \cdot \boldsymbol{\tau}_j$:

$$\boldsymbol{\tau}_i \cdot \boldsymbol{\tau}_j \, |T = 0\rangle = -3 |T = 0\rangle \ ; \tag{2.13}$$

$$\boldsymbol{\tau}_i \cdot \boldsymbol{\tau}_j \, |T = 1\rangle = 1 |T = 1\rangle \ . \tag{2.14}$$

Thus operators with or without $\boldsymbol{\tau}_i \cdot \boldsymbol{\tau}_j$ allow for different potentials in $T = 0$, $T = 1$ states. Similarly $\boldsymbol{\sigma}_i \cdot \boldsymbol{\sigma}_j$ results in different forces in spin-singlet and spin-triplet states.

The tensor operator is

$$S_{12} = 3 \boldsymbol{\sigma}_1 \cdot \hat{\mathbf{r}}_{12} \boldsymbol{\sigma}_2 \cdot \hat{\mathbf{r}}_{12} - \boldsymbol{\sigma}_1 \cdot \boldsymbol{\sigma}_2 \ . \tag{2.15}$$

This produces a D-state in the $S = 1$, $J = 1$ deuteron:

$$|J = 1, M_J\rangle = \sqrt{1 - \beta^2} \ |^3S_1\rangle + \beta \, |^3D_1\rangle \ ; \ \ \beta^2 \sim 6\% \ . \tag{2.16}$$

The spin-orbit operator is

$$\mathbf{L} \cdot \mathbf{S} = \left[\frac{1}{2i} (\mathbf{r}_i - \mathbf{r}_j) \times (\boldsymbol{\nabla}_i - \boldsymbol{\nabla}_j) \right] \cdot \left[\frac{1}{2} (\boldsymbol{\sigma}_i + \boldsymbol{\sigma}_j) \right] , \tag{2.17}$$

and gives different forces in $^3P_0$, $^3P_1$, and $^3P_2$ waves. $\mathbf{L}^2$ and $(\mathbf{L}\cdot\mathbf{S})^2$ operators are one (non-unique) way of fine-tuning the interaction in $L = 2,3$ (D,F) waves. Other potential models use $\mathbf{p}^2$ terms instead of $\mathbf{L}^2$, but this can result in much more difficult many-body calculations.

The tensor-isospin potential is very important in nuclear physics both because it provides most of the binding and because it greatly complicates calculations. It arises from pion exchange which provides the longest-range strong interaction ($v^\pi \sim e^{-m_\pi r}/r$, $m_\pi \sim 0.7$ fm^{-1}), and dominates the nucleon-nucleon interaction. As an example, consider the binding of the deuteron:

$$\begin{aligned} E_{2_H} &= 2.224 \text{ MeV} , \\ &= +(19.81 \text{ kinetic}) \; - (21.28\ v^\pi) \; - (0.75 \text{ other}\, v_{ij}) . \end{aligned} \tag{2.18}$$

Typically v^π provides 70% of v_{ij} in other nuclei. As shown above, v^π has $\boldsymbol{\sigma}_i\cdot\boldsymbol{\sigma}_j\ \boldsymbol{\tau}_i\cdot\boldsymbol{\tau}_j$ and $S_{ij}\ \boldsymbol{\tau}_i\cdot\boldsymbol{\tau}_j$ terms and it turns out that S_{ij} is typically 70% of v^π; without the $S_{ij}\ \boldsymbol{\tau}_i\cdot\boldsymbol{\tau}_j$ term, no nuclei would be bound! But,

$$\int d^2\Omega_{r_{12}}\, S_{12} = \int d^2\Omega_{r_{12}}\, S_{12}\,\boldsymbol{\sigma}_1\cdot\boldsymbol{\sigma}_2 = 0 , \tag{2.19}$$

while

$$\int d^2\Omega_{r_{12}} (S_{12})^2 = 24\pi . \tag{2.20}$$

So in $E = \langle\Psi_{G.S.}|H|\Psi_{G.S.}\rangle$ we need a S_{ij} operator in $\Psi_{G.S.}$ to get the v^π contribution. This results in the 6% D-state in the deuteron.

The Urbana series of three-nucleon potentials is written as a sum of two-pion-exchange and shorter-range phenomenological terms,

$$V_{ijk} = V_{ijk}^{2\pi} + V_{ijk}^{R} . \tag{2.21}$$

The two-pion-exchange term can be expressed simply as

$$V_{ijk}^{2\pi} = \sum_{cyclic} A_{2\pi}\{X_{ij}^\pi, X_{jk}^\pi\}\{\boldsymbol{\tau}_i\cdot\boldsymbol{\tau}_j, \boldsymbol{\tau}_j\cdot\boldsymbol{\tau}_k\} + C_{2\pi}[X_{ij}^\pi, X_{jk}^\pi][\boldsymbol{\tau}_i\cdot\boldsymbol{\tau}_j, \boldsymbol{\tau}_j\cdot\boldsymbol{\tau}_k] , \tag{2.22}$$

where X_{ij}^π is defined in Eq. (2.5), For the Urbana models $C_{2\pi} = \frac{1}{4}A_{2\pi}$, as in the original Fujita-Miyazawa model [7]. The shorter-range phenomenological term is given by

$$V_{ijk}^R = \sum_{cyclic} U_0 T^2(r_{ij}) T^2(r_{jk}) . \tag{2.23}$$

The parameters for model IX are $A_{2\pi} = -0.0293$ MeV and $U_0 = 0.0048$ MeV. They have been determined by fitting the density of nuclear matter and the binding energy of ^{3}H in conjunction with the Argonne v_{18} interaction. The V_{ijk}^R certainly should have other terms [8], however we need additional data to obtain their strengths; presumably a part of it is also really due to relativistic effects [9, 10, 11].

In light nuclei we find

$$\langle V^{NNN}\rangle \sim (.02 - .05)\,\langle v^{NN}\rangle \sim (.15 - .3)\,\langle H\rangle\,. \tag{2.24}$$

where the large fraction of $\langle H\rangle$ is due to the large cancelation of K and v. We expect a similar ratio for the four-body potential:

$$\langle v^{4N}\rangle \sim (.04 - .05)\,\langle V^{NNN}\rangle \sim .01\,\langle H\rangle\,. \tag{2.25}$$

Calculations are just approaching this accuracy.

3 Variational Monte Carlo

The variational method can be used to obtain approximate solutions to the many-body Schrödinger equation, $H\Psi = E\Psi$, for a wide range of nuclear systems, including few-body nuclei, light closed shell nuclei, and nuclear and neutron matter [12]. A suitably parameterized trial wave function, Ψ_T, is used to calculate an upper bound to the exact ground-state energy,

$$E_T = \frac{\langle\Psi_T|H|\Psi_T\rangle}{\langle\Psi_T|\Psi_T\rangle} \geq E_0\ . \tag{3.26}$$

The parameters in Ψ_T are varied to minimize E_T, and the lowest value is taken as the approximate ground-state energy.

3.1 The Variational Wave Function

Here I discuss a simplified form of Ψ_T; see [1] for the complete form used in current calculations,

$$|\Psi_T\rangle = \mathcal{S}\prod_{i<j}^{A} F_{ij}|\Phi\rangle\ ; \tag{3.27}$$

The complete Ψ_T also includes three-body correlations induced by V_{ijk}.

The Φ is the one-body part; it determines the J, M_J, T, T_z of the nucleus, and is completely antisymmetric. The F_{ij} are the two-body correlations induced by v_{ij}. Therefore, F_{ij} is written in terms of a subset of the operators in v_{ij}:

$$\begin{aligned} F_{ij} &= \sum_p f_p(r_{ij}) O_{ij}^{(p)} \\ &= f_c(r_{ij}) + f_\tau(r_{ij})\,\boldsymbol{\tau}_i\cdot\boldsymbol{\tau}_j + \cdots f_{t\tau}(r_{ij})\,S_{ij}\boldsymbol{\tau}_i\cdot\boldsymbol{\tau}_j + \cdots\ . \end{aligned} \tag{3.28}$$

The $f_p(r)$ are the solutions of Euler-Lagrange equations; see [2] for an introduction to these and [13] for details on the complete $f_p(r)$ used in modern calculations.

These operators do not commute, for example

$$[\boldsymbol{\sigma}_i \cdot \boldsymbol{\sigma}_j , \boldsymbol{\sigma}_i \cdot \boldsymbol{\sigma}_k] = 2i\, \boldsymbol{\sigma}_i \cdot (\boldsymbol{\sigma}_j \times \boldsymbol{\sigma}_k) . \tag{3.29}$$

Hence, $[F_{ij} , F_{ik}] \neq 0$, and we must use a symmetrized product. As an example, for ^{3}H,

$$\mathcal{S} \prod_{i<j} F_{ij} = \frac{1}{6} (F_{12}F_{13}F_{23} + F_{12}F_{23}F_{13} + F_{13}F_{12}F_{23} + F_{23}F_{13}F_{12} + F_{13}F_{23}F_{12} + F_{23}F_{12}F_{13}) . \tag{3.30}$$

Note that it has $[A(A-1)/2]!$ terms!

The Ψ_T must be translationally invariant. The F_{ij} are functions of $|\mathbf{r}_i - \mathbf{r}_j|$ and hence satisfy this. If Φ has dependence on $\mathbf{r}_i$, then we use $\tilde{\mathbf{r}}_i$ in Φ, where

$$\tilde{\mathbf{r}}_i = \mathbf{r}_i - \mathbf{R}_{cm} ; \quad \mathbf{R}_{cm} = \frac{1}{A} \sum_i \mathbf{r}_i , \tag{3.31}$$

or a similar translationally invariant coordinate. The Ψ_T must also describe a localized (bound state) system. This is achieved by having

$$F_{ij}(r_{ij}) \to 0 , \quad r_{ij} \to \infty , \tag{3.32}$$

or

$$\Phi(\mathbf{r}_1, \mathbf{r}_2, \cdots, \mathbf{r}_A) \to 0 , \quad r_i \to \infty , \tag{3.33}$$

or some combination of these.

Spin–Isospin Structure of Ψ_T. Each nucleon can be a proton or neutron but total charge is conserved. Thus for ^{3}H we can have *pnn*, *npn*, or *nnp*. Each nucleon can be spin up or down and tensor correlations do not conserve the total spin ($J = L + S$ is of course conserved); for ^{3}H we can have $\uparrow\uparrow\uparrow$, $\uparrow\uparrow\downarrow$, $\cdots$, $\downarrow\downarrow\downarrow$. Thus ^{3}H has $3 \times 8 = 24$ components in its spin-isospin structure.

The number of components can be reduced by noting that ^3H has almost pure total isospin $T = \frac{1}{2}$, so one really needs only two states:

$$|T = \tfrac{1}{2}, a\rangle = \tfrac{1}{\sqrt{2}}(|pnn\rangle - |npn\rangle) , \tag{3.34}$$

and

$$|T = \tfrac{1}{2}, b\rangle = \tfrac{1}{\sqrt{3}}[\tfrac{1}{\sqrt{2}} (|pnn\rangle + |npn\rangle) - \sqrt{2}\, |nnp\rangle] . \tag{3.35}$$

The 3rd linearly independent state does not contribute:

$$|T = \tfrac{3}{2}\rangle = \tfrac{1}{\sqrt{3}}(|pnn\rangle + |npn\rangle + |nnp\rangle) . \tag{3.36}$$

The number of components in Ψ_T grows rapidly with the number of nucleons, as is shown in Table 2. Calculations of the sort being described here are currently feasible up to only $A = 8$. Cluster methods have been used for VMC calculations of larger nuclei [14], and these are being developed for the GFMC.

Table 2. Number of spin-isospin components

A	Z	T	$N_{\text{spin}}{}^{a}$	$N_{\text{charge}}{}^{b}$	$N_{\text{isospin}}{}^{c}$	$N_{\text{total}}{}^{d}$
3	1	1/2	8	3	2	16
4	2	0	16	6	2	32
6	2	1	64	15	9	576
6	3	0	64	20	5	320
7	2	3/2	128	21	14	1792
7	3	1/2	128	35	14	1792
8	2	2	256	28	20	5120
8	3	1	256	56	28	7168
8	4	0	256	70	14	3584
12	6	0	4096	924	132	540,672
16	8	0	65,536	12,870	1430	93,716,480
40	20	0	1.0995E12	1.3785E11	6.5641E9	7.2173E21

$^{a}\ 2^{A}\ ;\ ^{b}\binom{A}{Z}\ ;\ ^{c}\ \frac{2T+1}{A/2+T+1}\binom{A}{A/2+T}\ ;\ ^{d}\ N_{\text{spin}} \times N_{\text{isospin}}$.

For ^{3}H and ^{4}He, Φ contains just the spin and isospin structure and has no spatial dependence:

$$|\Phi(^3H, M_J = \tfrac{1}{2})\rangle = \tfrac{1}{\sqrt{6}}\ (\ |p\!\uparrow n\!\uparrow n\!\downarrow\rangle - |p\!\uparrow n\!\downarrow n\!\uparrow\rangle + |n\!\downarrow p\!\uparrow n\!\uparrow\rangle - |n\!\uparrow p\!\uparrow n\!\downarrow\rangle + |n\!\uparrow n\!\downarrow p\!\uparrow\rangle - |n\!\downarrow n\!\uparrow p\!\uparrow\rangle\,)\ . \tag{3.37}$$

Thus the 24-component Φ has 6 elements $= \pm\frac{1}{\sqrt{6}}$; the rest are 0. It is totally antisymmetric and has total isospin and angular momentum $T = 1/2$, $J = 1/2$. For ^{4}He we have

$$|\Phi(^4He)\rangle = \tfrac{1}{\sqrt{24}}\ (\ |p\!\uparrow p\!\downarrow n\!\uparrow n\!\downarrow\rangle - |p\!\downarrow p\!\uparrow n\!\uparrow n\!\downarrow\rangle + |p\!\downarrow p\!\uparrow n\!\downarrow n\!\uparrow\rangle - |p\!\uparrow p\!\downarrow n\!\downarrow n\!\uparrow\rangle - |p\!\uparrow n\!\uparrow p\!\downarrow n\!\downarrow\rangle + \cdots)\ . \tag{3.38}$$

It has 24 non-zero terms in a 96-component vector. An antisymmetric state of more than four nucleons cannot be made in just spin and isospin; one needs spatial degrees of freedom (the p-shell).

Operators in Ψ_T. The full $|\Psi_T\rangle$ is a sum of terms like

$$F_{nm} \cdots F_{23} F_{13} F_{12} |\Phi\rangle \tag{3.39}$$

where all orderings of the F_{ij} are used. Each F_{ij} is a (very) sparse matrix operating on a column vector in spin-isospin space to produce a resultant vector.

For example, consider the $\boldsymbol{\sigma}_i \cdot \boldsymbol{\sigma}_j$ operator

$$\boldsymbol{\sigma}_i \cdot \boldsymbol{\sigma}_j = 2(\sigma_i^+ \sigma_j^- + \sigma_i^- \sigma_j^+) + \sigma_i^z \sigma_j^z = 2P_{ij}^{\sigma} - 1 \tag{3.40}$$

where P_{ij}^{σ} exchanges the spin of i and j. For ^3H and considering just the spin of the nucleons:

$$\Psi = \begin{pmatrix} a_{\uparrow\uparrow\uparrow} \\ a_{\uparrow\uparrow\downarrow} \\ a_{\uparrow\downarrow\uparrow} \\ a_{\uparrow\downarrow\downarrow} \\ a_{\downarrow\uparrow\uparrow} \\ a_{\downarrow\uparrow\downarrow} \\ a_{\downarrow\uparrow\downarrow} \\ a_{\downarrow\downarrow\downarrow} \end{pmatrix} ; \quad \boldsymbol{\sigma}_2 \cdot \boldsymbol{\sigma}_3 \Psi = \begin{pmatrix} a_{\uparrow\uparrow\uparrow} \\ 2a_{\uparrow\downarrow\uparrow} - a_{\uparrow\uparrow\downarrow} \\ 2a_{\uparrow\uparrow\downarrow} - a_{\uparrow\downarrow\uparrow} \\ a_{\uparrow\downarrow\downarrow} \\ a_{\downarrow\uparrow\uparrow} \\ 2a_{\downarrow\downarrow\uparrow} - a_{\downarrow\uparrow\downarrow} \\ 2a_{\downarrow\uparrow\downarrow} - a_{\downarrow\downarrow\uparrow} \\ a_{\downarrow\downarrow\downarrow} \end{pmatrix} . \tag{3.41}$$

Note that $\boldsymbol{\sigma}_1$ is a "spectator" and that $\boldsymbol{\sigma}_i \cdot \boldsymbol{\sigma}_j$ will not mix different isospin components. In 2-nucleon spin space this is the 4×4 matrix

$$\boldsymbol{\sigma}_i \cdot \boldsymbol{\sigma}_j = \begin{pmatrix} 1 & 0 & 0 & 0 \\ 0 & -1 & 2 & 0 \\ 0 & 2 & -1 & 0 \\ 0 & 0 & 0 & 1 \end{pmatrix} \text{acting on} \begin{pmatrix} a_{\uparrow\uparrow} \\ a_{\uparrow\downarrow} \\ a_{\downarrow\uparrow} \\ a_{\downarrow\downarrow} \end{pmatrix} . \tag{3.42}$$

Similarly, the tensor operator (Eq. 2.15) is

$$S_{ij} = 3 \begin{pmatrix} z^2 - 1/3 & z(x - iy) & z(x - iy) & (x - iy)^2 \\ z(x + iy) & -z^2 - 1/3 & x^2 + y^2 - 2/3 & -z(x - iy) \\ z(x_i y) & x^2 + y^2 - 2/3 & -z^2 + 1/3 & -z(x - iy) \\ (x + iy)^2 & -z(x + iy) & -z(x + iy) & z^2 - 1/3 \end{pmatrix} . \tag{3.43}$$

This is dense in the space of two nucleons but in an A-body nucleus, there will be $A - 2$ spectators whose spins remain unchanged. Thus the A-body operator will be sparse with many 4×4 blocks. Efficient programs for these calculations contain explicit code for the operators and do not use general sparse-matrix routines.

Considerations for $A > 4$. For nuclei with $A > 4$, the one-body part must have spatial dependence in order to be completely antisymmetric. As an example, for ^{6}Li, the one-body part has the general form

$$|\Phi\rangle = \mathcal{A}\sum_{LS} \beta_{LS}|\Phi_6(LSJMTT_3)_{1234:56}\rangle \ . \tag{3.44}$$

The LS components of the single-particle wave function are given by

$$|\Phi_6(LSJMTT_3)_{1234:56}\rangle = |\Phi_4(0000)_{1234}\phi_p^{LS}(R_{\alpha 5})\phi_p^{LS}(R_{\alpha 6})$$
$$\left\{[Y_{1m_l}(\Omega_{\alpha 5})Y_{1m_l'}(\Omega_{\alpha 6})]_{LM_L} \times [\chi_5(\frac{1}{2}m_s)\chi_6(\frac{1}{2}m_s')]_{SM_S}\right\}_{JM}$$
$$\times[\nu_5(\frac{1}{2}t_3)\nu_6(\frac{1}{2}t_3')]_{TT_3}\rangle \ . \tag{3.45}$$

The $\phi_p^{LS}(R_{\alpha k})$ are p-wave solutions of a particle of reduced mass $\frac{4}{5}m_N$ in an effective α-N potential. They are functions of the distance between the center of mass of the α core (which contains particles 1-4 in this partition) and nucleon k, and may be different for different LS components. The parameters of the effective α-N potential are variational parameters and do not necessarily represent a phenomenological α-N interaction. The antisymmetry operator $\mathcal{A}$ sums, with appropriate minus signs, over all partitions of 6 nucleons into $4+2$ subdivisions.

After other parameters in the trial function have been optimized, we make a series of calculations in which the β_{LS} may be different in the left- and right-hand-side wave functions to obtain the diagonal and off-diagonal matrix elements of the Hamiltonian and the corresponding normalizations and overlaps. We diagonalize the resulting energy matrices to find the β_{LS} eigenvectors. The one-body wave functions, Φ are orthonormal, but the correlated Ψ_T are not. Hence the diagonalizations use generalized eigenvalue routines including overlap matrices.

3.2 The VMC Method

The energy expectation value of Eq. (3.26) is evaluated using Monte Carlo integration. A detailed technical description of the methods used here can be found in Refs. [13, 2, 15]. Monte Carlo sampling is done both in configuration space and in the order of operators in the symmetrized product of Eq. (3.27) by following a Metropolis random walk. The expectation value for an operator O is given by

$$\langle O\rangle = \frac{\sum_{p,q}\int d\mathbf{R}\Psi_p^\dagger(\mathbf{R})O\Psi_q(\mathbf{R})}{\sum_{p,q}\int d\mathbf{R}\Psi_p^\dagger(\mathbf{R})\Psi_q(\mathbf{R})} \ . \tag{3.46}$$

The subscripts p and q specify the order of operators in the left and right hand side wave functions, while the integration runs over the particle coordinates $\mathbf{R} = (\mathbf{r}_1, \mathbf{r}_2, \ldots, \mathbf{r}_A)$. This multidimensional integration is facilitated by introducing a probability distribution, $W_{pq}(\mathbf{R})$, such that

$$\langle O\rangle = \frac{\sum_{p,q}\int d\mathbf{R}\left[\Psi_p^\dagger(\mathbf{R})O\Psi_q(\mathbf{R})/W_{pq}(\mathbf{R})\right]W_{pq}(\mathbf{R})}{\sum_{p,q}\int d\mathbf{R}\left[\Psi_p^\dagger(\mathbf{R})\Psi_q(\mathbf{R})/W_{pq}(\mathbf{R})\right]W_{pq}(\mathbf{R})} . \tag{3.47}$$

This probability distribution is taken to be

$$W_{pq}(\mathbf{R}) = |\mathrm{Re}(\langle\Psi_{T,p}^\dagger(\mathbf{R})\Psi_{T,q}(\mathbf{R})\rangle)| , \tag{3.48}$$

which is constructed from the wave function, Ψ_T, but with only one operator order of the symmetrized product.

Expectation values of the kinetic energy and spin-orbit potential require the computation of first derivatives and diagonal second derivatives of the wave function. These are obtained by evaluating the wave function at $6A$ slightly shifted positions of the coordinates $\mathbf{R}$ and taking finite differences, as discussed in Ref. [13]. Potential terms quadratic in $\mathbf{L}$ require mixed second derivatives, which can be obtained by additional wave function evaluations and finite differences. A rotation trick can be used to reduce the number of additional locations at which the wave function must be evaluated [16].

The heart of the VMC calculation is the Metropolis algorithm, an inherently serial algorithm which must be adapted to modern parallel computers. Since the bulk of the work in our variational calculations lies in the energy expectation value, the straightforward division of labor is to have one master processor perform the Metropolis walk while several slave processors calculate the energy and other expectation values for the configurations that the master generates. The number of slave processors that can be efficiently used is the ratio of the CPU time needed for expectation values to that needed to walk from one configuration to the next. We find that typically 50 processors can be used efficiently in a ^{7}Li VMC calculation.

4 Green's Function (Diffusion) Monte Carlo

GFMC projects out the lowest energy ground state using

$$\Psi_0 = \lim_{\tau\to\infty} \exp[-(H-E_0)\tau]\Psi_T. \tag{4.49}$$

The eigenvalue E_0 is calculated exactly while other expectation values are generally calculated neglecting terms of order $|\Psi_0 - \Psi_T|^2$ and higher. In contrast, the error in the variational energy, E_T, is of order $|\Psi_0 - \Psi_T|^2$, and other expectation values calculated with Ψ_T have errors of order $|\Psi_0 - \Psi_T|$. GFMC for scalar potentials is presented in R. Guardiola's and J. Boronat's contributions to this volume; here I will discuss some of the complications due to the state-dependence of the nuclear Hamiltonian. It is not possible to present all of the details necessary for a successful nuclear GFMC calculation; a rather complete discussion may be found in [1].

We use the Ψ_T of Eq. (3.27) as our initial trial function and define the propagated wave function $\Psi(\tau)$ as

$$\Psi(\tau) = e^{-(H-E_0)\tau}\Psi_T; \tag{4.50}$$

obviously $\Psi(\tau = 0) = \Psi_T$ and $\Psi(\tau \to \infty) = \Psi_0$. Introducing a small time step, $\Delta\tau$, $\tau = n\Delta\tau$, gives

$$\Psi(\tau) = \left[e^{-(H-E_0)\Delta\tau}\right]^n \Psi_T = G^n\Psi_T . \tag{4.51}$$

where G is the short-time Green's function. The $\Psi(\tau)$ is represented by a vector function of $\mathbf{R}$, and the Green's function, $G_{\alpha\beta}(\mathbf{R}',\mathbf{R})$ is a matrix function of $\mathbf{R}'$ and $\mathbf{R}$ in spin-isospin space, defined as

$$G_{\alpha\beta}(\mathbf{R}',\mathbf{R}) = \langle \mathbf{R}',\alpha|e^{-(H-E_0)\Delta\tau}|\mathbf{R},\beta\rangle. \tag{4.52}$$

It is calculated with leading errors of order $(\Delta\tau)^3$ as discussed below. Omitting spin-isospin indices for brevity, $\Psi(\mathbf{R}_n,\tau)$ is given by

$$\Psi(\mathbf{R}_n,\tau) = \int G(\mathbf{R}_n,\mathbf{R}_{n-1})\cdots G(\mathbf{R}_1,\mathbf{R}_0)\Psi_T(\mathbf{R}_0)d\mathbf{R}_{n-1}\cdots d\mathbf{R}_1 d\mathbf{R}_0 \tag{4.53}$$

4.1 The Short-Time Propagator

The short-time propagator should allow as large a time step $\Delta\tau$ as possible, since the total computational time for propagation is proportional to $1/\Delta\tau$. The first nuclear GFMC calculations [4, 17, 18] used the propagator obtained from the Feynman formulae. Ignoring three-nucleon interaction terms in H, it is given by

$$e^{-H\Delta\tau} = \left[\mathcal{S}\prod_{i<j} e^{-v_{ij}\Delta\tau/2}\right] e^{-K\Delta\tau} \left[\mathcal{S}\prod_{i<j} e^{-v_{ij}\Delta\tau/2}\right] + \mathcal{O}(\Delta\tau^3) . \tag{4.54}$$

Note that it is useful to symmetrize the product of $e^{-v_{ij}\Delta\tau/2}$ when $[v_{ij}, v_{jk}] \neq 0$, in order to reduce the error per iteration. The nuclear v_{ij} has a repulsive core of order GeV. The main error in the above propagator comes from terms in $e^{-H\Delta\tau}$ having multiple v_{ij}, like $v_{ij}Kv_{ij}(\Delta\tau)^3$ for example, which can become large when particles i and j are very close. In order to make them negligible, a rather small $\Delta\tau \sim 0.1$ GeV^{-1} is used with the above propagator.

The matrix elements of the propagator are given by:

$$G_{\alpha\beta}(\mathbf{R}',\mathbf{R}) = \langle\alpha| \left[\mathcal{S}\prod_{i<j} e^{-v_{ij}(\mathbf{r}'_{ij})\Delta\tau/2}\right] \times G_0(\mathbf{R}',\mathbf{R}) \left[\mathcal{S}\prod_{i<j} e^{-v_{ij}(\mathbf{r}_{ij})\Delta\tau/2}\right] |\beta\rangle , \tag{4.55}$$

$$G_o(\mathbf{R}',\mathbf{R}) = \langle \mathbf{R}'|e^{-K\Delta\tau}|\mathbf{R}\rangle = \left[\sqrt{\frac{m}{2\pi\hbar^2\Delta\tau}}\right]^{3A} \exp\left[\frac{-(\mathbf{R}'-\mathbf{R})^2}{2\hbar^2\Delta\tau/m}\right] \quad (4.56)$$

Remember that v_{ij} is an operator in spin-isospin space, so evaluating $e^{-\Delta\tau v_{ij}/2}$ is nontrivial. Consider the first 6 operators:

$$O_{1-6} = 1\,,\boldsymbol{\tau}\cdot\boldsymbol{\tau}\,,\boldsymbol{\sigma}\cdot\boldsymbol{\sigma}\,,\boldsymbol{\sigma}\cdot\boldsymbol{\sigma}\boldsymbol{\tau}\cdot\boldsymbol{\tau}\,,S_{12}\,,S_{12}\boldsymbol{\tau}\cdot\boldsymbol{\tau}\,. \quad (4.57)$$

These form a closed algebra:

$$O_iO_j = \sum_{k=1}^{6} K^{ijk}O_k\ , \quad (4.58)$$

where the K^{ijk} are just numbers. For example

$$\begin{aligned}(\boldsymbol{\sigma}_1\cdot\boldsymbol{\sigma}_2)^2 &= (2P_\sigma-1)^2 = 4P_\sigma^2 - 4P_\sigma + 1 = 5 - 4P_\sigma \\ &= 5 - 4\tfrac{1}{2}(\boldsymbol{\sigma}_1\cdot\boldsymbol{\sigma}_2 + 1) = 3 - 2\boldsymbol{\sigma}_1\cdot\boldsymbol{\sigma}_2\,. \end{aligned} \quad (4.59)$$

Tables of K^{ijk} may be found in [19]. Thus

$$\begin{aligned} e^{-a\boldsymbol{\sigma}_i\cdot\boldsymbol{\sigma}_j} &= 1 - a\boldsymbol{\sigma}_i\cdot\boldsymbol{\sigma}_j + \tfrac{1}{2}a^2(\boldsymbol{\sigma}_i\cdot\boldsymbol{\sigma}_j)^2 - \cdots\ , \\ &= 1 + \tfrac{3}{2}a^2 - \cdots - (a - a^2 + \cdots)\boldsymbol{\sigma}_i\cdot\boldsymbol{\sigma}_j\ , \\ &= \tfrac{1}{4}(3e^{-a} + e^{3a}) + \tfrac{1}{4}(e^{-a} - e^{3a})\boldsymbol{\sigma}_i\cdot\boldsymbol{\sigma}_j\ , \end{aligned} \quad (4.60)$$

where the last line may be verified by expanding the e^{-a} and e^{3a}. More complete formulas, including those for exponentiating S_{ij}, are in [15].

Evaluating $L\cdot S$ involves gradients, so a first-order expansion is used when using the propagator of Eq. (4.54):

$$e^{-\frac{\Delta\tau}{2}v_{LS}(L\cdot S)_{ij}} \simeq \left[1 - \frac{\Delta\tau}{2}v_{LS}(L\cdot S)_{ij}\right]\,. \quad (4.61)$$

The dominant term comes from $\mathbf{L}\cdot\mathbf{S}$ acting on G_o:

$$(\mathbf{L}\cdot\mathbf{S})_{ij}G_o(\mathbf{R}',\mathbf{R}) = \frac{1}{4i}\,\frac{m}{\hbar^2\Delta\tau}\,(\mathbf{r}_{ij}\times\mathbf{r}'_{ij})\cdot(\boldsymbol{\sigma}_i + \boldsymbol{\sigma}_j)G_o\,. \quad (4.62)$$

Note that

$$\mathbf{r}\times\mathbf{r}' = \mathbf{r}\times(\mathbf{r}+\Delta\mathbf{r}) = \mathbf{r}\times\Delta\mathbf{r} \sim \sqrt{\Delta\tau}\ , \quad (4.63)$$

since $\Delta\mathbf{r}$ will be sampled from the free Green's function. Thus even though the explicit $\Delta\tau$ cancel in Eqs. (4.61- 4.62), the contribution of v_{LS} still, on the average, decreases as $\Delta\tau$ decreases.

It is well known from studies of liquid helium [20] that including the exact two-body propagator allows much larger time steps. This short-time propagator is

$$G_{\alpha\beta}(\mathbf{R}',\mathbf{R}) = G_0(\mathbf{R}',\mathbf{R})\langle\alpha|\left[\mathcal{S}\prod_{i<j}\frac{g_{ij}(\mathbf{r}'_{ij},\mathbf{r}_{ij})}{g_{0,ij}(\mathbf{r}'_{ij},\mathbf{r}_{ij})}\right]|\beta\rangle\ , \tag{4.64}$$

where g_{ij} is the exact two-body propagator,

$$g_{ij}(\mathbf{r}'_{ij},\mathbf{r}_{ij}) = \langle\mathbf{r}'_{ij}|e^{-H_{ij}\Delta\tau}|\mathbf{r}_{ij}\rangle\ , \tag{4.65}$$

$$H_{ij} = -\frac{\hbar^2}{m}\nabla^2_{ij} + v_{ij}\ , \tag{4.66}$$

and $g_{0,ij}$ is the free two-body propagator,

$$g_{0,ij}(\mathbf{r}'_{ij},\mathbf{r}_{ij}) = \left[\sqrt{\frac{\mu}{2\pi\hbar^2\Delta\tau}}\right]^3 \exp\left[-\frac{(\mathbf{r}'_{ij}-\mathbf{r}_{ij})^2}{2\hbar^2\Delta\tau/\mu}\right]\ , \tag{4.67}$$

where $\mu = m/2$ is the reduced mass. All terms containing any number of the same v_{ij} and K are treated exactly in this propagator, as we have included the imaginary-time equivalent of the full two-body scattering amplitude. Eq. (4.64) still has errors of order $(\Delta\tau)^3$, however they are from commutators of terms like $v_{ij}Tv_{ik}(\Delta\tau)^3$ which become large only when both pairs ij and ik are close. Since this is a rare occurrence, the $\Delta\tau$ of Eq. (4.65) can be five or more times larger than that used with Eq. (4.54).

To calculate g_{ij}, we use the techniques developed by Schmidt and Lee[21] for scalar interactions. The basic idea is to use the convolution equation to write g_{ij} as a product over N steps:

$$g_{ij}(\mathbf{r}_N,\mathbf{r}_0;\Delta\tau) = \prod_{i=1}^{N} g_{ij}(\mathbf{r}_i,\mathbf{r}_{i-1};\epsilon), \tag{4.68}$$

with $\epsilon = \Delta\tau/N$ and an implied integration over intermediate points. The very-short-time propagator, $g_{ij}(\mathbf{r}_i,\mathbf{r}_{i-1};\epsilon)$, is evaluated using the methods of Eqn. (4.55–4.62) applied to one pair. The errors for $g_{ij}(\Delta\tau)$ contain only even powers of $1/N$, starting with $1/N^2$ [21]. By evaluating $g_{ij}(\mathbf{r},\mathbf{r}';\Delta\tau)$ for several values of N (and consequently ϵ) and extrapolating $\epsilon \to 0$ the g_{ij} can be calculated with high ($\sim$ 10 digit) accuracy.

Calculating this propagator is computationally intensive. Therefore, prior to the GFMC calculation, we compute and store the the propagator on a grid. For a spin-independent interaction, the propagator g_{ij} would depend only upon the two magnitudes r' and r and the angle $cos(\theta) = \hat{\mathbf{r}}'\cdot\hat{\mathbf{r}}$ between them. Here, though, there is also a dependence upon the spin quantization axis. Rotational symmetry allows one to calculate the spin-isospin components of $g_{ij}(\mathbf{r}',\mathbf{r})$ for any $\mathbf{r}'$ and $\mathbf{r}$ by simple SU3 spin rotations and values of g_{ij} on a grid of initial points $\mathbf{r} = (0,0,z)$ and final points $\mathbf{r}' = (x',0,z')$. In addition, the fact that the propagator is Hermitian allows us to store only the values for $z > z'$.

Finally, including the three-body forces and the E_0 in Eq. (4.52), the complete propagator is given by

$$G_{\alpha\beta}(\mathbf{R}',\mathbf{R}) = e^{E_o\Delta\tau}G_0(\mathbf{R}',\mathbf{R})\exp[-\sum(V^R_{ijk}(\mathbf{R}')+V^R_{ijk}(\mathbf{R}))\frac{\Delta\tau}{2}]$$
$$\langle\alpha|I_3(\mathbf{R}')|\gamma\rangle\langle\gamma|\left[\mathcal{S}\prod_{i<j}\frac{g_{ij}(\mathbf{r}'_{ij},\mathbf{r}_{ij})}{g_{0,ij}(\mathbf{r}'_{ij},\mathbf{r}_{ij})}\right]|\delta\rangle\langle\delta|I_3(\mathbf{R})|\beta\rangle \quad (4.69)$$

$$I_3(\mathbf{R}) = \left[1-\frac{\Delta\tau}{2}\sum V^{2\pi}_{ijk}(\mathbf{R})\right] . \quad (4.70)$$

The exponential of $V^{2\pi}_{ijk}$ is expanded to first order in $\Delta\tau$ thus, there are additional error terms of the form $V^{2\pi}_{ijk}V^{2\pi}_{i'j'k'}(\Delta\tau)^2$. However, they have negligible effect since $V^{2\pi}_{ijk}$ has a magnitude of only a few MeV. It was verified that the results for ^{4}He do not show any change, outside of statistical errors, when $\Delta\tau$ is decreased from 0.5 GeV^{-1}.

The exact propagator of Eq. (4.65) can be computed for the full v_{18} potential, however the $\mathbf{L}^2$ and $(\mathbf{L}\cdot\mathbf{S})^2$ terms in the potential correspond to state-dependent changes of the mass appearing in the free Green's function. Since we do not know how to sample such a free Green's function, we cannot use the exact g_{ij} for the full potential, but rather must use one constructed for an approximately equivalent potential that does not contain quadratic $\mathbf{L}$ terms. The difference between the desired and approximate potentials is computed perturbatively. For the rest of this paper, I ignore this complication and present results and equations as if the full potential is used in the GFMC propagator.

4.2 Mixed Estimates

The mixed expectation value of an operator O is defined as:

$$\langle O\rangle_{Mixed} = \frac{\langle\Psi_T|O|\Psi(\tau)\rangle}{\langle\Psi_T|\Psi(\tau)\rangle}$$
$$= \frac{\int d\mathbf{P}_n\Psi^\dagger_T(\mathbf{R}_n)OG(\mathbf{R}_n,\mathbf{R}_{n-1})\cdots G(\mathbf{R}_1,\mathbf{R}_0)\Psi_T(\mathbf{R}_0)}{\int d\mathbf{P}_n\Psi^\dagger_T(\mathbf{R}_n)G(\mathbf{R}_n,\mathbf{R}_{n-1})\cdots G(\mathbf{R}_1,\mathbf{R}_0)\Psi_T(\mathbf{R}_0)} , \quad (4.71)$$

where $\mathbf{P}_n = \mathbf{R}_0,\mathbf{R}_1,\cdots,\mathbf{R}_n$ denotes the 'path', and $d\mathbf{P}_n = d\mathbf{R}_0 d\mathbf{R}_1\cdots d\mathbf{R}_n$. In GFMC, the integral over the paths is carried out stochastically. Generally, the required expectation values are calculated approximately from the variational Ψ_T and mixed expectation values. Let

$$\Psi(\tau) = \Psi_T + \delta\Psi(\tau). \quad (4.72)$$

Retaining only the terms of order $\delta\Psi(\tau)$, we obtain

$$\begin{aligned}\langle O(\tau)\rangle &= \frac{\langle\Psi(\tau)|O|\Psi(\tau)\rangle}{\langle\Psi(\tau)|\Psi(\tau)\ \rangle}\,, \\ &\approx \langle O(\tau)\rangle_{Mixed} + [\langle O(\tau)\rangle_{Mixed} - \langle O\rangle_T]\,,\end{aligned} \tag{4.73}$$

where

$$\langle O\rangle_T = \frac{\langle\Psi_T|O|\Psi_T\rangle}{\langle\Psi_T|\Psi_T\rangle}\,. \tag{4.74}$$

More accurate evaluations of $\langle O(\tau)\rangle$ are possible [22],
essentially by measuring the observable at the mid-point of the path. However, such estimates require a propagation twice as long as the mixed estimate.

An important exception to the above is the energy, E_0, given by $\langle H(\tau\to\infty)\rangle$. The $\langle H(\tau)\rangle_{Mixed}$ can be re-expressed as (Ref. [23])

$$\langle H(\tau)\rangle_{Mixed} = \frac{\langle\Psi_T|e^{-(H-E_0)\tau/2}He^{-(H-E_0)\tau/2}|\Psi_T\rangle}{\langle\Psi_T|e^{-(H-E_0)\tau/2}e^{-(H-E_0)\tau/2}|\Psi_T\rangle}\,, \tag{4.75}$$

since the propagator $\exp[-(H-E_0)\tau]$ commutes with the Hamiltonian. Thus $\langle H(\tau)\rangle_{Mixed}$ approaches E_0 in the limit $\tau\to\infty$, and furthermore, being an expectation value of H, it obeys the variational principle

$$\langle H(\tau)\rangle_{Mixed} \geq E_0\,. \tag{4.76}$$

Because the expectation values of the individual energy components, such as K, v_{ij}, and V_{ijk}, are extrapolated by Eq. (4.73), while the total E_0 is not, they may not sum to the correct total energy. Indeed, there must be a collective error in these individual terms equal to the total difference between the GFMC $\langle H\rangle_{Mixed}$ and the VMC $\langle H\rangle_T$. This is illustrated in Table 3 for the case of ^{6}Li, where the difference $\langle H\rangle_{Mixed} - \langle H\rangle_T$ is -4.3 MeV, and the sum of the individual $\langle O\rangle$ is an additional -4.3 MeV lower than $\langle H\rangle$. In this case, the individual corrections are comparable in magnitude to the collective error, but small compared to the total expectation values.

4.3 Dependence of Results on τ

For more than four nucleons, GFMC calculations suffer significantly from the well-known Fermion sign problem which is discussed elsewhere in this volume. The resulting exponential growth of the statistical errors as one propagates to larger τ limits our calculations to $\tau \leq 0.1$ MeV^{-1}, and for most cases we do not go beyond $\tau = 0.06$ MeV^{-1}. This means that any errors in Ψ_T corresponding to excitations of less than ~ 10 MeV will be damped out by less than $1/e$. If we write

$$\Psi_T = \Psi_0 + \sum_i \alpha_i\Psi_i\,, \tag{4.77}$$

Table 3. Contributions to the GFMC $\langle O(\tau)\rangle$ of Eq. (4.73) for ^{6}Li. All quantities are in MeV.

	$\langle O\rangle_T$	$\langle O\rangle_{Mixed}$	$\langle O\rangle_{Mixed} - \langle O\rangle_T$	$\langle O\rangle$
K	143.8(4)	147.3(5)	3.5(7)	150.8(10)
v_{ij}	-165.6(4)	-172.4(5)	-6.8(7)	-179.3(10)
V_{ijk}	-5.2(1)	-6.2(1)	-1.0(1)	-7.2(1)
Sum	-27.0(1)	-31.3(1)	-4.3(1)	-35.6(1)
H	-27.0(1)	-31.3(1)	N.A.	-31.3(1)

where Ψ_i are states with energies $E_0 + E_i^\star$, then the GFMC $E(\tau)$ will be

$$E(\tau) = E_0 + \frac{\sum_i \alpha_i^2 E_i^\star exp(-E_i^\star \tau)}{1 + \sum_i \alpha_i^2 exp(-E_i^\star \tau)}\ , \tag{4.78}$$

and by fitting such a form to the computed $E(\tau)$ one can hope to extract E_0.

Figure 1 shows such fits made for ^{4}He, for which we could afford calculations with very small statistical errors. The solid curve is a fit with $E_0 = -28.335$ MeV, excitation energies $E_i^\star = 20.2$, 341, and 1477 MeV, and corresponding $\alpha_i^2 = 0.0062$, 0.0018, and 0.00046. The lowest 0^+ excitation of ^{4}He is at 20.2 MeV and this energy was not varied in the fit. The χ^2 of the fit is 19 for 31 $E(\tau)$ (25 degrees of freedom), so the $E(\tau)$ are not statistically independent. The dashed curve shows a fit without the 20.2 MeV excitation; it gives $\chi^2 = 23$ and $E_0 = -28.28$ MeV. Finally, the heavy solid line with short dashed error bars shows the average of the $E(\tau)$ for $0.04 \leq \tau \leq 0.1$: $-28.300(15)$. It appears that in this most favorable case, with high statistics, high first excited state, and large maximum τ, we can see that including the first excited state improves the extrapolation marginally. However, the extrapolated E_0 is not significantly lower than a simple average of the $E(\tau)$ for $0.04 \leq \tau \leq 0.1$. Therefore we simply use such an average for all other nuclei.

Because of the difficulties in making useful extrapolations in τ, it is important to understand contaminations in Ψ_T, particularly from low-excitation-energy states which will not be fully filtered out in limited-τ calculations. We have made several calculations of the ground-state of ^{6}Li to study the effects of changes in Ψ_T on the GFMC $E(\tau)$. Figure 2 shows the effects of removing some of the noncentral correlations in Ψ_T; the solid circles are from a calculation with the full Ψ_T which includes the NNN correlation based on V_{ijk}. The open diamonds were computed by omitting this NNN correlation. This makes the energy at $\tau = 0$ worse by ~ 1.7 MeV. However by $\tau = 0.01$, the GFMC has fully corrected for this defect and thereafter the differences are just statistical fluctuations. Hence removing this NNN correlation enhances the admixtures of excitations > 250 MeV.

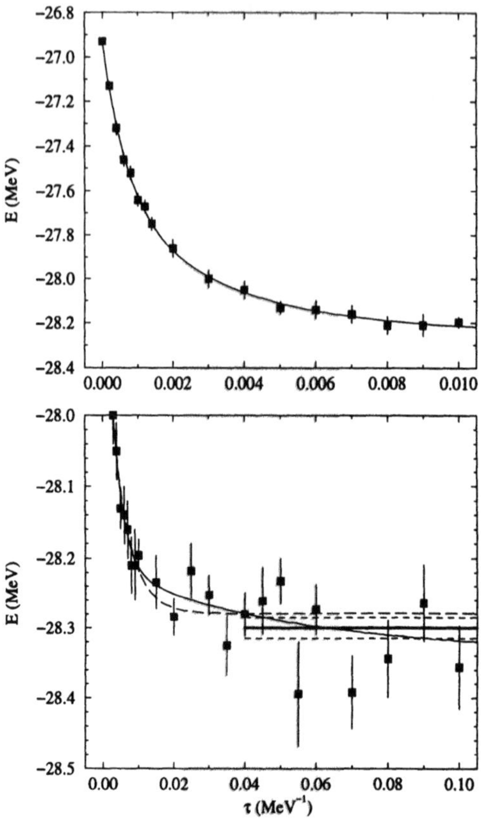

Fig. 1. ^{4}He GFMC energy as a function of imaginary time. The fits are described in the text.

The open squares in Fig. 2 come from a much more drastic approximation of Ψ_T. Here the tensor components of F_{ij} have also been omitted, resulting in a four-operator wave function. In such a wave function, the dominant tensor components of the two-body potential have zero expectation value and the energy at $\tau = 0$ is +41 MeV. It is completely corrected by $\tau = 0.03$; again the rate of correction indicates excitation energies ~ 250 MeV. However, the statistical errors from such a bad Ψ_T are much larger.

These two tests indicate that defects in the non-central parts of the correlation, which have been the subject of much optimization in VMC studies, are easily corrected by the GFMC. Deficiencies in the one-body part of Ψ_T present more of a problem. The excitation energies corresponding to different LS components in Eq. (3.44) are typically 5 to 10 MeV. Thus the β_{LS} must be reliably determined in Ψ_T.

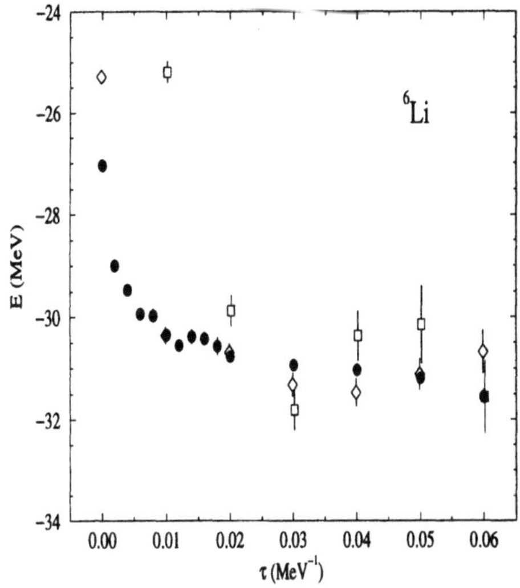

Fig. 2. ^{6}Li GFMC energy as a function of imaginary time for various truncations of the noncentral parts of Ψ_T.

5 Some Results

These lectures are primarily about Monte Carlo methods for nuclear physics, so I will present only results for the computed energies of nuclei here. Many other results may be found in [1]. Table 4 shows the ground-state energies and excitation energies of several excited states for nuclei up to $A = 8$. Only statistical errors are shown for the Monte Carlo calculations; we believe that the GFMC calculations are converged to ~ 0.3 MeV for $A = 6$ and ~ 0.6 MeV for $A = 8$. The GFMC ground state results show that the Hamiltonian being used underbinds the p-shell nuclei more and more as A increases. Also the underbinding becomes worse as one moves away from $Z = N$, indicating an isospin problem. A comparison of the VMC and GFMC results shows that our variational wave functions are much poorer for the p-shell than for the 3- and 4-body nuclei. However, wherever we have done GFMC calculations for excited states we find that the VMC excitation energies are reliable. Thus the failure of the variational wave function appears to be a bulk property with little state dependence. The computed excitation energies are in quite good agreement with experiment, indicating that failure of the Hamiltonian is also principally a bulk feature.

As is shown in Table 1, there has been much recent progress in nuclear QMC calculations. The present results show that we must make improve-

Table 4. VMC, GFMC, and experimental ground-state (negative numbers) and excitation (positive) energies of $A = 2 - 8$ nuclei in MeV

$^{A}Z(J^{\pi};T)$	VMC	GFMC	Expt
^{2}H$(1^{+};0)$	-2.2248(5)		-2.2246
^{3}H$(\frac{1}{2}^{+};\frac{1}{2})$	-8.32(1)	-8.47(1)	-8.48
^{4}He$(0^{+};0)$	-27.78(3)	-28.30(2)	-28.30
^{6}He$(0^{+};1)$	-24.87(7)	-27.64(14)	-29.27
^{6}He$(2^{+};1)$	1.86(10)	1.80(18)	1.80
^{6}Li$(1^{+};0)$	-28.09(7)	-31.25(11)	-31.99
^{6}Li$(3^{+};0)$	2.93(10)	2.72(36)	2.19
^{6}Li$(0^{+};1)$	3.84(10)	3.94(23)	3.56
^{6}Li$(2^{+};0)$	4.23(11)	4.43(39)	4.31
^{6}Li$(2^{+};1)$	5.64(10)		5.37
^{6}Li$(1^{+};0)$	5.68(11)		5.65
^{7}He$(\frac{3}{2}^{-};\frac{3}{2})$	-20.43(12)	-25.16(16)	-28.82
^{7}Li$(\frac{3}{2}^{-};\frac{1}{2})$	-32.78(11)	-37.44(28)	-39.24
^{7}Li$(\frac{1}{2}^{-};\frac{1}{2})$	0.33(16)	0.76(41)	0.48
^{7}Li$(\frac{7}{2}^{-};\frac{1}{2})$	5.48(16)	5.72(41)	4.63
^{7}Li$(\frac{5}{2}^{-};\frac{1}{2})$	6.64(16)	6.56(45)	6.68
^{8}He$(0^{+};2)$	-19.71(18)	-25.77(61)	-31.41
^{8}He$(2^{+};2)$	2.34(25)		2.6
^{8}Li$(2^{+};1)$	-29.70(19)	-38.26(19)	-41.28
^{8}Li$(1^{+};1)$	1.19(26)		0.98
^{8}Li$(3^{+};1)$	2.81(25)		2.26
^{8}Li$(4^{+};1)$	5.63(26)		6.53
^{8}Be$(0^{+};0)$	-48.06(27)	-54.66(64)	-56.50
^{8}Be$(2^{+};0)$	3.90(37)		3.04
^{8}Be$(4^{+};0)$	11.96(35)		11.4

ments on all fronts: the Hamiltonian, the variational wave function, and GFMC methods for calculations to larger τ.

Acknowledgments

I am indebted to R. B. Wiringa for much help in preparing this paper. This work was supported by the U. S. Department of Energy, Nuclear Physics Division, under contract No. W-31-109-ENG-38. This paper was completed while I was a guest at the Institute for Nuclear Theory at the University of Washington.

References

1. B. S. Pudliner, V. R. Pandharipande, J. Carlson, S. C. Pieper, and R. B. Wiringa, Phys. Rev. C **56**, to be published (Oct. 1997)
2. J. A. Carlson and R. B. Wiringa, in *Computational Nuclear Physics 1*, edited by K. Langanke, J. A. Maruhn, and S. E. Koonin (Springer Verlag, Berlin, 1991)
3. R. B. Wiringa, V. G. J. Stoks, and R. Schiavilla, Phys. Rev. C **51**, 38 (1995)
4. B. S. Pudliner, V. R. Pandharipande, J. Carlson, and R. B. Wiringa, Phys. Rev. Lett. **74**, 4396 (1995)
5. J. R. Bergervoet, P. C. van Campen, R. A. M. Klomp, J.-L. de Kok, T. A. Rijken, V. G. J. Stoks, and J. J. de Swart, Phys. Rev. C **41**, 1435 (1990)
6. V. G. J. Stoks, R. A. M. Klomp, M. C. M. Rentmeester, and J. J. de Swart, Phys. Rev. C **48**, 792 (1993)
7. J. Fujita and H. Miyazawa, Prog. Theor. Phys. **17**, 360 (1957)
8. S. A. Coon, M. T. Peña, and D. O. Riska, Phys. Rev. C **52**, 2925 (1995)
9. G. E. Brown, W. Weise, G. Baym, and J. Speth, Comments Nucl. Part. Phys. **17**, 39 (1987)
10. J. L. Forest, V. R. Pandharipande, and J. L. Friar, Phys. Rev. C **52**, 568 (1995)
11. J. L. Forest, V. R. Pandharipande, J. Carlson, and R. Schiavilla, Phys. Rev. C **52**, 576 (1995)
12. R. B. Wiringa, Rev. Mod. Phys. **65**, 231 (1993)
13. R. B. Wiringa, Phys. Rev. C **43**, 1585 (1991)
14. S. C. Pieper, R. B. Wiringa, and V. R. Pandharipande, Phys. Rev. C **46**, 1741 (1992)
15. B. S. Pudliner, Ph.D. Thesis, University of Illinois (1996)
16. R. Schiavilla, V. R. Pandharipande, and A. Fabrocini, Phys. Rev. C. **40**, 1484 (1989)
17. J. Carlson, Phys. Rev. C **36**, 2026 (1987)
18. J. Carlson, Phys. Rev. C **38**, 1879 (1988)
19. V. R. Pandharipande and R. B. Wiringa, Rev. Mod. Phys. **51**, 821 (1979)
20. D. M. Ceperley, Rev. Mod. Phys. **67**, 279 (1995)
21. K. E. Schmidt and M. A. Lee, Phys. Rev. E **51**, 5495 (1995)
22. M. H. Kalos, J. Comp. Phys. **2**, 257 (1967)
23. D. M. Ceperley and M. H. Kalos, in *Monte Carlo Methods in Statistical Physics*, edited by K. Binder (Springer-Verlag, Heidelberg, 1979)

Diffusion Monte Carlo for Excited States: Application to Liquid Helium

Jordi Boronat

Departament de Física i Enginyeria Nuclear, Universitat Politècnica de Catalunya, Campus Nord B4-B5, E-08034 Barcelona, Spain

Abstract. The extension of ab initio Monte Carlo methods to the study of excited states is a challenging problem at the forefront of present research in the field of computational techniques applied to the quantum many-body problem. The aim is to calculate excited states with similar accuracy to the one achieved in the study of the ground-state properties. We present a state of the art of such applications devoting special attention to the formalism behind the algorithms recently used: fixed-node diffusion Monte Carlo, released-node diffusion Monte Carlo, and methods for a concurrent evaluation of excited states based on a generalization of the transient estimates. As examples of these calculations, we present results on the structure of a vortex excitation and the well-known phonon-roton branch in superfluid ^{4}He at zero temperature.

1 Introduction

Ab initio quantum Monte Carlo (QMC) methods have become in the last two decades a standard tool for the study of ground-state properties of interacting many-body systems [1, 2, 3, 4, 5]. During this time the range of applicability of the QMC techniques has been continuously enlarging and now there are current studies of small molecules and nuclei, helium clusters and bulk systems as liquid and solid helium or the electron gas. This wide-spread growth has been possible for the rapid evolution of the computer architecture and the appearance of faster processors, on one hand, and the development of new and more efficient algorithms, on the other. These two features have made that the run times become accessible even for medium computer centers, and therefore the number of research groups in QMC has appreciably increased.

The accuracy of the QMC methods is not the same for all systems and states. Thus, for boson systems as liquid ^{4}He it is possible to obtain exact information, apart from statistical uncertainties, both at zero [1] and finite temperature [6, 7, 3]. In contrast, robust algorithms to deal with the calculation of the ground state of fermion systems as liquid ^{3}He and the calculation of excited-state properties for both boson and fermion systems are presently lacking [2]. In fact, both problems are related and correspond to the well-known sign problem. In the absence of a final algorithm to tackle this hard problem, several methods exist that allow for the study of these fundamental problems with an accuracy that is nowadays continuously improving. In the present work, we present the QMC methods that have been developed and successfully applied to the study of excited states in interacting quantum systems. Two of them, the fixed-node [8] and released-node methods [9, 10], were first introduced to study the ground state of fermion systems whereas a third one, developed by Ceperley and Bernu [11], was specifically designed to evaluate concurrently many excited states of a quantum system. The latter is a generalization of the transient-estimate method [10] also used previously in the study of fermions. The number of QMC calculations of excited states is presently increasing but to date there are still few applications. Among them we could mention excited states of some molecules [12], the two-dimensional electron gas [13], and both the phonon-roton branch [14] and the vortex excitation [15] in superfluid ^{4}He.

The more difficult problem of extracting dynamical information from imaginary-time quantum Monte Carlo data is far from being satisfactorily solved. This topic, which we do not discuss in this work, is a challenging problem that limits the usefulness of quantum simulations. At present, only the combination of Bayesian inference and the maximum entropy criterion has reached a relative success in the description of relevant quantities as the dynamic structure factor [16]. However, the choice of an a priori model, which is needed in this method, can significantly condition the final predictions.

The diffusion Monte Carlo method is briefly introduced in Sect. 2 together with a scheme of the basic algorithm. In Sect. 3, the specific methods to deal with the calculation of excited states are presented. The formalism ends in Sect. 4 with a procedure to analytically improve the excited-state wave function relying directly in the Schrödinger equation. In Sect. 5, we present results on the vortex excitation and the phonon-roton spectrum in superfluid ^{4}He using the methodology of the precedent sections. Finally, a brief summary and future prospects are contained in Sect. 6.

2 The Diffusion Monte Carlo Method

The starting point in diffusion Monte Carlo (DMC) methods [8, 17] is the Schrödinger equation for N particles written in imaginary time

$$-\frac{\partial \Psi(\mathbf{R},t)}{\partial t} = (H-E)\,\Psi(\mathbf{R},t) \ , \tag{2.1}$$

where $\mathbf{R} \equiv (\mathbf{r}_1,\ldots,\mathbf{r}_N)$ and t is measured in units of $\hbar$. $\Psi(\mathbf{R},t)$ can be expanded in terms of a complete set of eigenfunctions $\phi_i(\mathbf{R})$ of the Hamiltonian,

$$\Psi(\mathbf{R},t) = \sum_i c_i \exp\left[-(E_i-E)t\right]\,\phi_i(\mathbf{R}) \ , \tag{2.2}$$

where E_i is the eigenvalue associated to $\phi_i(\mathbf{R})$. The asymptotic solution of Eq. (2.1) for any value E close to the energy of the ground state, and for long times ($t \to \infty$), gives $\phi_0(\mathbf{R})$ provided that there is a non-zero overlap between $\Psi(\mathbf{R}, t=0)$ and the ground-state wave function $\phi_0(\mathbf{R})$.

In a computer simulation of Eq. (2.1) it is crucial to use the importance sampling technique in order to reduce the statistical fluctuations to a manageable level. Following this method, one rewrites the Schrödinger equation for the function

$$f(\mathbf{R}, t) \equiv \psi(\mathbf{R}) \Psi(\mathbf{R}, t) \ , \tag{2.3}$$

where $\psi(\mathbf{R})$ is a time-independent trial function. Considering a Hamiltonian of the form

$$H = -\frac{\hbar^2}{2m} \nabla^2_{\mathbf{R}} + V(\mathbf{R}) \ , \tag{2.4}$$

Equation (2.1) turns out to be

$$\begin{aligned} -\frac{\partial f(\mathbf{R}, t)}{\partial t} &= -D \nabla^2_{\mathbf{R}} f(\mathbf{R}, t) + D \nabla_{\mathbf{R}} \left(\mathbf{F}(\mathbf{R}) f(\mathbf{R}, t)\right) + (E_{\mathrm{L}}(\mathbf{R}) - E) f(\mathbf{R}, t) \\ &\equiv (A_1 + A_2 + A_3) f(\mathbf{R}, t) \equiv A f(\mathbf{R}, t) \ , \end{aligned} \tag{2.5}$$

where $D = \hbar^2/(2m)$, $E_{\mathrm{L}}(\mathbf{R}) = \psi(\mathbf{R})^{-1} H \psi(\mathbf{R})$ is the local energy, and

$$\mathbf{F}(\mathbf{R}) = 2\, \psi(\mathbf{R})^{-1} \nabla_{\mathbf{R}} \psi(\mathbf{R}) \tag{2.6}$$

is called the drift force. $\mathbf{F}(\mathbf{R})$ acts as an external force which guides the diffusion process, implied by the first term in Eq. (2.5), to regions where $\psi(\mathbf{R})$ is large.

The formal solution of Eq. (2.5) is

$$f(\mathbf{R}', t + \Delta t) = \int G(\mathbf{R}', \mathbf{R}; \Delta t)\, f(\mathbf{R}, t)\, d\mathbf{R} \tag{2.7}$$

with

$$G(\mathbf{R}', \mathbf{R}; \Delta t) = \langle \mathbf{R}' \,|\, \exp(-A \Delta t) \,|\, \mathbf{R} \rangle \ . \tag{2.8}$$

While the Green's function Monte Carlo (GFMC) method works with the whole Green's function, DMC algorithms rely on reasonable approximations of $G(\mathbf{R}', \mathbf{R}; \Delta t)$ for small values of the time-step Δt. Then, Eq. (2.8) is not directly solved but iterated repeatedly to obtain the asymptotic solution $f(\mathbf{R}, t \to \infty)$.

In the quadratic diffusion Monte Carlo algorithm we have used [18, 19], the Green's function $G(\mathbf{R}', \mathbf{R}; \Delta t)$ is approximated by

$$\begin{aligned} &\exp(-A\Delta t) = \\ &\exp\left(-A_3 \frac{\Delta t}{2}\right) \exp\left(-A_2 \frac{\Delta t}{2}\right) \exp(-A_1 \Delta t) \exp\left(-A_2 \frac{\Delta t}{2}\right) \exp\left(-A_3 \frac{\Delta t}{2}\right) . \end{aligned} \tag{2.9}$$

This decomposition, which is not unique, is exact up to order $(\Delta t)^2$. Assuming this approximation, Eq. (2.7) becomes

$$f(\mathbf{R}',t+\Delta t) = \int \left[G_3\left(\mathbf{R}',\mathbf{R}_1;\frac{\Delta t}{2}\right) G_2\left(\mathbf{R}_1,\mathbf{R}_2;\frac{\Delta t}{2}\right) G_1\left(\mathbf{R}_2,\mathbf{R}_3;\Delta t\right) \times G_2\left(\mathbf{R}_3,\mathbf{R}_4;\frac{\Delta t}{2}\right) G_3\left(\mathbf{R}_4,\mathbf{R};\frac{\Delta t}{2}\right)\right] f(\mathbf{R},t)\, d\mathbf{R}_1 \ldots d\mathbf{R}_4 d\mathbf{R}\,, \tag{2.10}$$

with

$$G_1(\mathbf{R}',\mathbf{R};t) = (4\pi D t)^{-\frac{3N}{2}} \exp\left[-\frac{(\mathbf{R}'-\mathbf{R})^2}{4Dt}\right], \tag{2.11}$$

$$G_2(\mathbf{R}',\mathbf{R};t) = \delta\left(\mathbf{R}'-\mathbf{R}(t)\right), \qquad \text{where} \begin{cases} \mathbf{R}(0) = \mathbf{R} \\ \dfrac{\mathrm{d}\mathbf{R}(t)}{\mathrm{d}t} = D\,\mathbf{F}(\mathbf{R}(t)), \end{cases} \tag{2.12}$$

and

$$G_3(\mathbf{R}',\mathbf{R};t) = \exp\left[-(E_{\mathrm{L}}(\mathbf{R}) - E)\,t\right]\,\delta(\mathbf{R}'-\mathbf{R}). \tag{2.13}$$

In the Monte Carlo computations, $f(\mathbf{R},t)$ is represented by n_w walkers $\mathbf{R}_i$, each one representing a set of the $3N$ coordinates of the N particles. The algorithm used for the implementation of Eq. (10) goes through the following steps:

1. Move the walkers, under the drift force $\mathbf{F}(\mathbf{R})$ for an interval $\Delta t/2$ with accuracy $(\Delta t)^2$.
2. Apply to each walker a displacement χ randomly drawn from the $3N$ Gaussian distribution $\exp(-\chi^2/(4D\Delta t))$.
3. Repeat step 1.
4. Randomly replicate each walker n_r times, in such a way that

$$\langle n_r \rangle = \exp\left[-\Delta t\left(\frac{E_{\mathrm{L}}(\mathbf{R}') + E_{\mathrm{L}}(\mathbf{R})}{2} - E\right)\right].$$

5. Go to step 1 for the next walker $\mathbf{R}_i$, until the set of walkers is exhausted. The new set obtained corresponds to $f(\mathbf{R},t+\Delta t)$.

The whole procedure is repeated as many times as it is needed to reach the asymptotic limit ($t \to \infty$). From then on, the walkers $\mathbf{R}_i$ are used to obtain the expectation values of the magnitudes to be determined.

3 Excited Properties with the DMC Method

As commented in Sect. 2, when the imaginary time goes to infinity only the contribution of the ground-state survives in Eq. (2.2) provided there is a non-zero overlap between $\Psi(\mathbf{R},t=0)$ and $\phi_0(\mathbf{R})$. If, on the contrary, the overlap between $\Psi(\mathbf{R},t=0)$ and $\phi_0(\mathbf{R})$ is zero ($c_0 = 0$), the wave function $\Psi(\mathbf{R},t)$

will converge to the next-lowest energy state $\phi_1(\mathbf{R})$. This condition can be implemented through the trial wave function used for importance sampling $\psi(\mathbf{R})$. If $\psi(\mathbf{R})$ is by construction orthogonal to the ground-state wave function $\phi_0(\mathbf{R})$, then the function $f(\mathbf{R},0) = \psi(\mathbf{R})\Psi(\mathbf{R},t=0)$, which is iterated in the DMC method, has no overlap with the ground state. Such an approach is used in the calculation of the ground-state energy of a fermion system which may be considered as the first fully antisymmetric excited state of the Hamiltonian. The main difficulty of the Monte Carlo calculation of excited-state energies is then common to the one of the solution of the ground state for fermions, i.e., the interpretation of the wave function as a density (positive) of random walks. The fact that now the wave function in not positively defined introduces the well-known sign problem: the Monte Carlo simulation of these systems needs to work with a positive density of walkers and simultaneously enforce the existence of nodal surfaces delimiting the positive and negative regions of the real wave function.

At present, there is not a method so robust as the boson algorithm introduced in Sect. 2 but several alternatives exist that have proved their reliability in a number of systems. In the rest of this section, we will describe three Monte Carlo methods that have been used to calculate excited-state properties: fixed node, released node, and a Monte Carlo implementation of the configuration interaction method.

3.1 Fixed Node

The Monte Carlo interpretation of the imaginary-time Schrödinger equation requires that $f(\mathbf{R},t) = \psi(\mathbf{R})\Psi(\mathbf{R},t)$ be a density, i.e., $\psi\Psi \geqq 0$ in all the domain. This boundary condition can be satisfied if ψ and Ψ change sign together and thus share the same nodes. This approximation, known as fixed-node (FN) method [8], has been extensively used in the ground-state calculations of fermions [20] and also in the calculations of excited-state properties of small molecules [21] and quantum liquids [15, 14].

In the asymptotic regime

$$f(\mathbf{R},t \to \infty) = \psi(\mathbf{R})\phi^{\mathrm{FN}}(\mathbf{R}) \ , \tag{3.14}$$

where $\phi^{\mathrm{FN}}(\mathbf{R})$ is an approximation to the exact eigenfunction $\phi(\mathbf{R})$ of the Schrödinger equation. As in the FN solution the nodes are imposed by the trial wave function, and not modified in the calculation, it can be proved that the fixed-node energies are variational upper bounds to the exact eigenvalues for a given symmetry [8]. In particular, if the nodes of ψ were exact then ϕ^{FN} would also be exact. Therefore, the FN results depend significantly on the "quality" of the trial wave function that can be analytically improved following the procedure proposed in Sect. 4.

The requirement of common nodes between ψ and Ψ is naturally fulfilled due to the infinite drift force in the nodal surface of $\psi(\mathbf{R})$. The use of a finite time step Δt originates, however, some spurious crossings of the nodal surface which are deleted in the configuration list to ensure the positiveness of $f(\mathbf{R},t)$. The probability of a crossing increases with Δt and this fact introduces an additional time-step bias that has to be analyzed in each application.

3.2 Released Node

In order to go beyond the variational bound provided by the FN approximation it is necessary to generalize the problem considering the excited (or fermionic) wave function as

$$\Psi(\mathbf{R}) = \Phi^+(\mathbf{R}) - \Phi^-(\mathbf{R}) \ , \tag{3.15}$$

where $\Phi^+(\mathbf{R})$ and $\Phi^-(\mathbf{R})$ are positive and then can be considered as densities [2]. With this wave function $\Psi(\mathbf{R})$ the Schrödinger equation can be split in two separated equations, one for Φ^+ and another for Φ^-, which may be translated into the Monte Carlo language due to its monotonic sign. From their definition it is clear that Φ^+ and Φ^- have a finite overlap with both the exact ground state and the excited state. A simple realization of $\Phi^\pm$ at the beginning of the walk could be

$$\Phi^\pm(t=0) = \frac{1}{2}\left(|\Psi| \pm \Psi\right) \tag{3.16}$$

which are ever positive and trivially satisfy Eq. (3.15). Once the higher energy excited states have decayed only the lowest excited state and the ground state components contribute to $\Phi^\pm$

$$\Phi^\pm(t \to \infty) = \pm c_1\phi_1 + c_0\phi_0 e^{(E_1-E_0)t} \ . \tag{3.17}$$

The difference $\Phi^+ - \Phi^-$ is then proportional to the excited wave function ϕ_1 but the ground-state component in Φ^+ and Φ^- is not canceled and increases exponentially with time. The attainment of a regime in which the “signal” of ϕ_1 is not masked by the ground-state component depends crucially on the difference $E_1 - E_0$ which is always positive. As a general trend, the existence of this regime is related to the degree of fulfillment of the inequality $E_2 - E_1 > E_1 - E_0$, E_2 being the next excited state above E_1.

In the MC calculation of the energy, the ground-state component does not contribute due to the orthogonality between ϕ_0 and ψ:

$$E_{\mathrm{MC}} = \frac{\int d\Omega\, \Psi(t\to\infty)\, H\,\psi}{\int d\Omega\, \Psi(t\to\infty)\,\psi} \tag{3.18}$$

$$= \frac{\int d\Omega\, \Phi^+\, H\,\psi}{\int d\Omega\, (\Phi^+ - \Phi^-)\,\psi} - \frac{\int d\Omega\, \Phi^-\, H\,\psi}{\int d\Omega\, (\Phi^+ - \Phi^-)\,\psi} = E_1 \ . \tag{3.19}$$

However, in the estimation of the variance of the energy, where the calculation of $(\Psi H)^2$ is required, the ground-state component is not canceled and introduces a noise proportional to $\exp[(E_1 - E_0)t]$. The key point is thus the achievement of a quick decay to the lowest excited state before the ground-state noise overwhelms the excited signal. Obviously, the success of the procedure also depends on the proximity between the nodes of the trial excited-state wave function ψ and the exact ones.

A Monte Carlo realization of the above formalism is the released-node (RN) method [9]. In this approach, the walkers are allowed to cross the nodal surface and survive for a finite lifetime $t_{\rm r}$. A positive or negative sign is attached at each walker if it has crossed an even or an odd number of times, respectively. The unequal flux of walkers through the model nodal surface is the origin of the small displacement of the nodes towards their exact localization.

In order to make feasible a flux of walkers through the nodal surface they are guided by a trial wave function $\psi_{\rm g}(\mathbf{R})$ positively defined. For a given trial wave function $\psi(\mathbf{R})$, the guiding function $\psi_{\rm g}(\mathbf{R})$ has to approach $|\psi(\mathbf{R})|$ away from the nodal surface and must be non-zero in the nodes to make possible the flux of walkers through it. A simple choice for $\psi_{\rm g}$ is [22, 14]

$$\psi_{\rm g}(\mathbf{R}) = \left(\psi(\mathbf{R})^2 + a^2\right)^{1/2} \; , \tag{3.20}$$

which satisfies both requirements for a proper choice of the value of the parameter a. In fact, the value given to the parameter a, which has to be of the same order of magnitude than the mean value of $|\psi(\mathbf{R})|$, governs the flux of walkers through the nodes. The relaxation time of the release process is a function of a (when the value of a is increased the relaxation time is reduced and vice versa) but the RN excitation energies are independent of the explicit value of a, and also of the explicit analytic form of $\psi_{\rm g}(\mathbf{R})$, the influence of both being only on the variance of the Monte Carlo estimation.

The released-node energy is obtained projecting out the excited state modeled by $\psi(\mathbf{R})$. This projection is carried out assigning to each walker a weight $W(\mathbf{R})$ given by

$$W(\mathbf{R}) = \sigma(\mathbf{R}) \, \frac{|\psi(\mathbf{R})|}{\psi_{\rm g}(\mathbf{R})} \; , \tag{3.21}$$

$\sigma(\mathbf{R})$ being $+1$ (-1) for an even (odd) number of crossings. The released-node energy is thus determined through

$$E_{\rm RN}(t_{\rm r}) = \frac{\sum_{t \leq t_r} W(\mathbf{R}) \, E_{\rm L}(\mathbf{R})}{\sum_{t \leq t_{\rm r}} W(\mathbf{R})} \; , \tag{3.22}$$

where the sums are extended to all the surviving walkers with a lifetime less than $t_{\rm r}$, and $E_{\rm L}(\mathbf{R}) = \psi(\mathbf{R})^{-1} H \psi(\mathbf{R})$.

The evolution towards an asymptotic regime is usually analyzed plotting a sequence of energies $E_{\rm RN}(t_{\rm r}^1)$, $E_{\rm RN}(t_{\rm r}^2)$,...,$E_{\rm RN}(t_{\rm r}^{\rm max})$ with $t_{\rm r}^1 < t_{\rm r}^2 < \ldots < t_{\rm r}^{\rm max}$. In this form, one can determine the size of the systematic error of the released-node energy and the evolution in time of the stochastic noise due to the growth of the population of negative walkers.

3.3 Matrix Quantum Monte Carlo

The matrix quantum Monte Carlo method, originally proposed by Ceperley and Bernu [11], is a Monte Carlo implementation of the configuration interaction method in which the excited-state energies are obtained from the diagonalization of the Hamiltonian matrix $\langle\psi_i|H|\psi_j\rangle$ for a given set of variational wave functions $\{\psi\}$. The starting point of this QMC approach is the variational method applied to the calculation of excited states. In this way, let us first expand the eigenfunctions of H in a basis of known wave functions

$$\phi_i = \sum_{j=1}^{m} d_{ij}\, \psi_j \ . \tag{3.23}$$

The best approximation to ϕ_i using the variational basis $\{\psi\}$ are obtained by finding the stationary points of the quotient

$$\Lambda_i = \frac{\int d\mathbf{R}\, \phi_i^*(\mathbf{R})\, H\, \phi_i(\mathbf{R})}{\int d\mathbf{R}\, \phi_i^*(\mathbf{R})\, \phi_i(\mathbf{R})} \tag{3.24}$$

with respect to the parameters d_{ij}. This results in a conventional eigenvalue problem that, in the iteration method, is solved using a function of the Hamiltonian $C(H)$ to project out some states. Having in mind a Monte Carlo interpretation, one can choose $C(H) = \mathrm{e}^{-tH/2}$ and thus

$$\phi_i(\mathbf{R},t) = \sum_{j=1}^{m} d_{ij} \mathrm{e}^{-tH/2}\, \psi_j(\mathbf{R}) \ . \tag{3.25}$$

The result is a generalized eigenvalue equation with imaginary-time dependent eigenfunctions and eigenvalues:

$$\sum_{j=1}^{m} H_{ij}(t) d_{kj}(t) = \Lambda_k(t) \sum_{j=1}^{m} N_{ij}(t) d_{kj}(t) \ , \tag{3.26}$$

$d_k(t)$ (vector of m components) and $\Lambda_k(t)$ being the k-eigenfunction and the k-eigenvalue, respectively. The matrix elements $H_{ij}(t)$ and $N_{ij}(t)$ are defined as:

$$H_{ij}(t) = \int d\mathbf{R}_1 d\mathbf{R}_2\, \psi_i^*(\mathbf{R}_2)\, \langle \mathbf{R}_2 \,|\, H \mathrm{e}^{-tH} \,|\, \mathbf{R}_1 \rangle\, \psi_j(\mathbf{R}_1) \tag{3.27}$$

$$N_{ij}(t) = \int d\mathbf{R}_1 d\mathbf{R}_2\, \psi_i^*(\mathbf{R}_2)\, \langle \mathbf{R}_2 \,|\, \mathrm{e}^{-tH} \,|\, \mathbf{R}_1 \rangle \psi_j(\mathbf{R}_1) \ , \tag{3.28}$$

which are mutually related by

$$H_{ij}(t) = -\frac{\mathrm{d}N_{ij}(t)}{\mathrm{d}t} \ . \tag{3.29}$$

The terms $H_{ij}(t)$ and $N_{ij}(t)$ are multidimensional integrals in which the Green's function appears explicitly, and one knows how to calculate it by a random walk. Introducing a guiding wave function $\psi_{\rm g}(\mathbf{R})$, positively defined everywhere, one can write

$$H_{ij}(t) = \int d\mathbf{R}_1 d\mathbf{R}_2\, F_i^*(\mathbf{R}_2) E_{\mathrm{L}j}(\mathbf{R}_2)\, G(\mathbf{R}_2, \mathbf{R}_1; t)\, F_i(\mathbf{R}_1) \psi_{\rm g}^2(\mathbf{R}_1) \quad (3.30)$$

$$N_{ij}(t) = \int d\mathbf{R}_1 d\mathbf{R}_2\, F_i^*(\mathbf{R}_2)\, G(\mathbf{R}_2, \mathbf{R}_1; t)\, F_i(\mathbf{R}_1) \psi_{\rm g}^2(\mathbf{R}_1) \ , \quad (3.31)$$

where

$$G(\mathbf{R}_2, \mathbf{R}_1; t) = \frac{\psi_{\rm g}(\mathbf{R}_2)}{\psi_{\rm g}(\mathbf{R}_1)} \langle \mathbf{R}_2 \,|\, \mathrm{e}^{-tH} \,|\, \mathbf{R}_1 \rangle \quad (3.32)$$

is the importance-sampled Green's function, $F_i(\mathbf{R}) = \psi_{\rm g}(\mathbf{R})^{-1}\, \psi_i(\mathbf{R})$, and $E_{\mathrm{L}j} = \psi_j(\mathbf{R})^{-1}\, H\, \psi_j(\mathbf{R})$.

As commented in Sect. 2, the Green's function may be well approximated in the limit $\Delta t \to 0$ by

$$G(\mathbf{R}_2, \mathbf{R}_1; \Delta t) \simeq G_{\rm b}(\mathbf{R}_2, \mathbf{R}_1; \Delta t)\, G_{\rm d}(\mathbf{R}_2, \mathbf{R}_1; \Delta t) \ , \quad (3.33)$$

$G_{\rm b}$ being the branching term and $G_{\rm d}$ the diffusion plus drift term. In the strategy developed by Ceperley and Bernu [11] the configurations evolve according to G_d, and consequently follow the probability distribution $\psi_{\rm g}^2$. The branching term is taken into account only as a weight and therefore the population of walkers is not changed. In fact, the present approach is a generalization of the transient-estimate method used in fermion calculations [10] and it can be proved that, in the limit of $t \to \infty$, the excitation energies coming from the eigenvalue of $H - \Lambda N = 0$ are exact [2]. However, the noise also increases with time and, in practice, the method seems to work better for the lower states than for the higher. To date, this method has been applied to the calculation of the vibrational states of simple molecules [12] and the two-dimensional electron gas [13]. It is clear that the main advantage of the matrix quantum Monte Carlo method is the concurrent evaluation of many states but it could be difficult, in some cases, to choose a common guiding wave function $\psi_{\rm g}$ from the requirement of a maximum overlap with all the ψ_i of the basis set.

4 Analytic Improvement of the Excited-state Wave Functions

The Monte Carlo results obtained with the different schemes developed in Sect. 3 depend significantly on the trial wave function used as importance sampling. As commented, the quality of the variational upper bound provided by the FN method, and the success of the RN and transient estimates in discriminating the excited signal above the statistical noise are only a function of ψ for a given system. It is therefore useful, and complementary to the numerical optimization of these methods, to have available a mechanism for analytically improving the trial wave function.

Let us write the imaginary-time dependent wave function as the product of a phase and a modulus

$$\Psi(\mathbf{R},t) = e^{i\Omega(\mathbf{R},t)}\,\Phi(\mathbf{R},t) \ . \tag{4.34}$$

With this decomposition, the Schrödinger equation for $\Psi(\mathbf{R},t)$ splits in two coupled equations, one for the modulus and one for the phase:

$$-\frac{\partial\Phi}{\partial t} = D(\nabla_{\mathbf{R}}\Omega)^2\,\Phi - D(\nabla^2_{\mathbf{R}}\Phi) + (V(\mathbf{R}) - E)\Phi \tag{4.35}$$

$$\frac{\partial\Omega}{\partial t} = D\left[(\nabla^2_{\mathbf{R}}\Omega) + 2(\nabla_{\mathbf{R}}\Omega)\cdot\frac{(\nabla_{\mathbf{R}}\Phi)}{\Phi}\right] \ . \tag{4.36}$$

If an initial guess for Φ and Ω is given (Φ_0, Ω_0), the above equations provide the first order correction for both functions. In the case of the modulus, and writing $\Phi(\mathbf{R},t) = \exp(U(\mathbf{R},t))$, the first correction to an initial U_0 is

$$U = U_0 - t\left[D\left((\nabla_{\mathbf{R}}\Omega_0)^2 - (\nabla^2_{\mathbf{R}}U_0) - (\nabla_{\mathbf{R}}U_0)^2\right) + (V(\mathbf{R}) - E)\right] \ . \tag{4.37}$$

Let us consider, for example, a boson system in which $\Omega = 0$. If a pair potential is assumed $V(\mathbf{R}) = \sum_{i<j} v(r_{ij})$, and we start with U_0 =const., the Jastrow form for the boson wave function is recovered, $U = \sum_{i<j} u(r_{ij})$ with $u(r) \sim -v(r)$. In a next iteration, considering U_0 as the Jastrow wave function one may identify in Eq. (4.37) a three-body term with an analytic form extensively used in variational calculations of quantum liquids [23],

$$(\nabla_{\mathbf{R}}U_0)^2 \sim \sum_k \mathbf{G}_k\cdot\mathbf{G}_k \tag{4.38}$$

with

$$\mathbf{G}_k = \sum_{l\neq k}\frac{u'(r_{kl})}{r_{kl}}\,\mathbf{r}_{kl} \ . \tag{4.39}$$

More important in the excited-state problem is the improvement of the phase. The first-order correction to an initial choice Ω_0 is given by [24, 14]

$$\Omega = \Omega_0 + Dt\left[(\nabla^2_{\mathbf{R}}\Omega_0) + 2(\nabla_{\mathbf{R}}\Omega_0)\cdot(\nabla_{\mathbf{R}}U_0)\right] \ . \tag{4.40}$$

A straightforward application of this approximation is the case of the spectrum of an impurity dissolved in a boson fluid as liquid ^{4}He. Considering a Jastrow approximation for the modulus, and the Feynman model for the excitation $\Omega_0 = \mathbf{q}\cdot\mathbf{r}_I$, Eq. (4.40) generates the first correction which contains the backflow correlations introduced by Feynman and Cohen [25]

$$\Omega = \mathbf{q}\cdot\left(\mathbf{r}_I + 2Dt\sum_j\frac{u'(r_{Ij})}{r_{Ij}}\,\mathbf{r}_{Ij}\right) \ . \tag{4.41}$$

It is worth noticing that the splitting of the Schrödinger equation has also been used to study, from a Monte Carlo viewpoint, states which require a complex wave function. The method which at present has been implemented is the fixed-phase (FP) approximation which consists in taking a model for the phase and solve by DMC the Schrödinger equation for the modulus [24]. The result is variational for the phase, that enters in the Schrödinger equation for the modulus as an external potential $(D(\nabla_{\mathbf{R}}\Omega)^2)$.

5 Excited States in Superfluid ^{4}He

We have applied the Monte Carlo methods presented in Sect. 3 to study two different classes of collective excitations in superfluid ^{4}He: a vortex excitation and the phonon-roton spectrum. Both problems share a quite long history [26, 27] in which physicists as Landau, Bogoliubov, Feynman, and Onsager wrote the more fundamental contributions. The results presented below are an example of the capabilities of the ab initio Monte Carlo methods to microscopically study the energetics of the excited states with an accuracy that is progressively approaching the one obtained in the calculation of the ground state of boson systems.

5.1 A Vortex Excitation in Two-Dimensional Liquid ^{4}He

A vortex excitation is an eigenstate of the N-particle Hamiltonian H and of the z-component of the angular momentum L_z with eigenvalue $\hbar N\ell$, corresponding to an integer number ℓ of quanta of circulation [26]. The simplest microscopic wave function to describe a vortex state was introduced by Feynman [28]:

$$\psi^{\mathrm{F}}(\mathbf{R}) = \mathrm{e}^{\mathrm{i}\ell\varphi_{\mathrm{F}}} \prod_{i=1}^{N} f(r_i)\Phi_0(\mathbf{R}) \ , \tag{5.42}$$

where $\varphi_{\mathrm{F}} = \sum_{i=1,N} \theta_i$ is the Feynman phase with θ_i the azimuthal angle of the i-th particle, $\Phi_0(\mathbf{R})$ describes the ground state of the system and $f(r_i)$ is a function of the radial distance of each particle from the vortex axis, which models the density near the core. In what follows we only consider vortices with one quantum of circulation, i.e., $\ell = \pm 1$. Starting from the Feynman phase φ_F as zeroth order ansatz, the first correction to the phase can be obtained following the method described in Sect. 4. The new phase, which includes backflow correlations, is of the form [24]

$$\varphi_{\mathrm{BF}} = \varphi_{\mathrm{F}} - 2Dt \sum_{j \neq i} \frac{r_j}{r_i} \frac{u'(r_{ij})}{r_{ij}} \sin(\theta_i - \theta_j) \ . \tag{5.43}$$

The wave function constructed with the phase φ_{BF} is the vortex analogue of the Feynman-Cohen [25] backflow wave function for the phonon-roton excitation branch.

In order to deal with a real walker probability distribution function $f(\mathbf{R},t)$, we choose as trial function the superposition of two vortex states, one with positive and one with negative circulation, which are degenerate in energy. The importance sampling function has therefore the form

$$\psi(\mathbf{R}) = \cos(\varphi(\mathbf{R})) \prod_{i=1}^{N} f(r_i)\psi^0(\mathbf{R}) \ , \tag{5.44}$$

where ψ^0 is the importance sampling function for the ground state and for the phase φ we have used both the Feynman phase φ_{F} and the backflow phase φ_{BF}. The sign problem associated with the use of the above trial wave function in the DMC algorithm has been dealt within the framework of the fixed-node approximation.

Vortex states are characteristic of systems with rotational invariance, but, at the same time, simulations must deal with a finite number of particles. A simple choice is to restrict the N particles to be inside a cylindric box with the vortex at the center and rigid boundary conditions on the walls. This is the geometry chosen for example in Ref. [24]. By confining the particles some problems arise, such as the choice of the confining potential and surface effects which can be relevant if the box size is not large enough. We have removed the confinement in order to keep surface effects as small as possible. An important point one needs to solve when attempting to simulate a vortex excitation in an infinite system, is the calculation of the collective phase $\varphi(\mathbf{R})$ that explicitly enters in the calculation of the drift force and the local energy. For a very large system one expects that the features of the vortex state near the core are weakly influenced by the behaviour of particles which are far from the vortex axis. It is thus reasonable to assume that, if the collective phase is decomposed in the sum

$$\varphi(\mathbf{R}) = \sum_{r_i \leq \bar{r}} \varphi_i + \sum_{r_i > \bar{r}} \varphi_i \ , \tag{5.45}$$

where the first term sums the contributions to the phase of all the particles with distance from the vortex axis within the cutoff length $\bar{r}$ and the second term gives the contribution coming from all the other particles, the phase fluctuations in the second term are irrelevant for the core structure of the vortex and it can be safely approximated by its mean value $\sum_{r_i > \bar{r}} \varphi_i = 0$. If this prescription is employed in the calculation of the collective phase one expects that for a large enough cutoff length $\bar{r}$ the properties of the vortex state near the axis are properly simulated [15].

In the calculation we consider 64 particles in a two-dimensional box of length $L = 15.0\ \sigma$ ($\sigma = 2.556$ Å), which corresponds approximately to the equilibrium density of the two-dimensional homogeneous liquid. The atoms interact through the two-body HFD-B(HE) potential [29]. The ground-state trial wave function we have chosen is the McMillan two body function

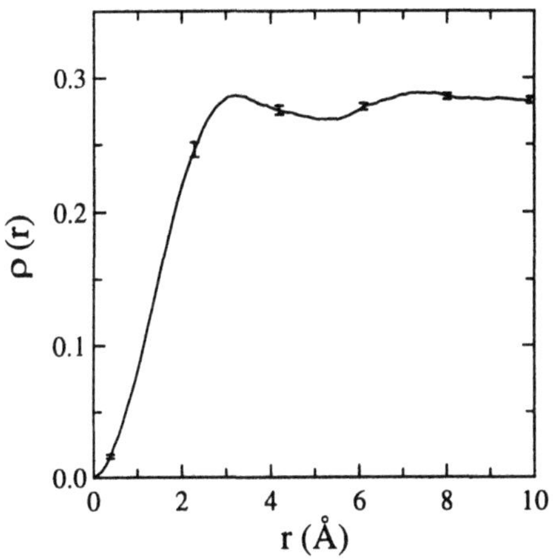

Fig. 1. Pure density profile of the vortex at the equilibrium density.

$\psi^0(\mathbf{R}) = \prod_{i<j} \exp(-b^5/2r_{ij}^5)$ with $b = 1.205\ \sigma$ as in the ground-state calculation of Ref. [30]. For the radial function $f(r)$, which models the structure of the vortex core, we consider two different options: $f_1(r) = 1 - e^{(-r/a)}$, and $f_2(r) = 1$. The first function gives a density in the trial function which decreases to zero at the vortex axis over a distance of order a, for which we take the value $a = 1$ Å, whereas the second one does not contain any parameter associated with the vortex core. For the backflow function entering Eq. (5.43) we have used the same functional form and the same values for the parameters as in Ref. [24]. We have performed the calculation using three different trial wave functions: ψ^{F1} and ψ^{F2} which correspond to the Feynman phase φ_{F} with the radial terms f_1 and f_2 respectively, and ψ^{BF} corresponding to the backflow phase φ_{BF} with f_2.

The common pure-estimated [31] profile for the particle density $\rho(r)$ is presented in Fig. 1, where it clearly appears that a zero particle density is reached on the vortex axis. This result is in contrast with the prediction of a significant non-zero particle density on the axis obtained with the backflow phase in Ref. [24]. In our opinion the result of Ref. [24] can be influenced by the extrapolation technique used by the authors to improve the mixed estimator result. In fact, the usual linear extrapolation technique, which is accurate to the same order as the one employed in Ref. [24], would change their result for the particle density on the vortex axis in an amount comparable to their prediction.

The excitation energy $E_{\mathrm{v}}(r)$ of a vortex inside a disk of radius r is obtained as the difference between the total energy of the disk with and without the vortex $E_{\mathrm{v}}(r) = E(r) - E_0(r)$. For large distances from the vortex axis $E_{\mathrm{v}}(r)$ is usually decomposed in an hydrodynamic tail, which depends logarithmically on r, and a core energy E_{c}: $E_{\mathrm{v}}(r) = \pi\hbar^2\rho_0/m\ \log(r/\xi) + E_{\mathrm{c}}$, where ρ_0 is

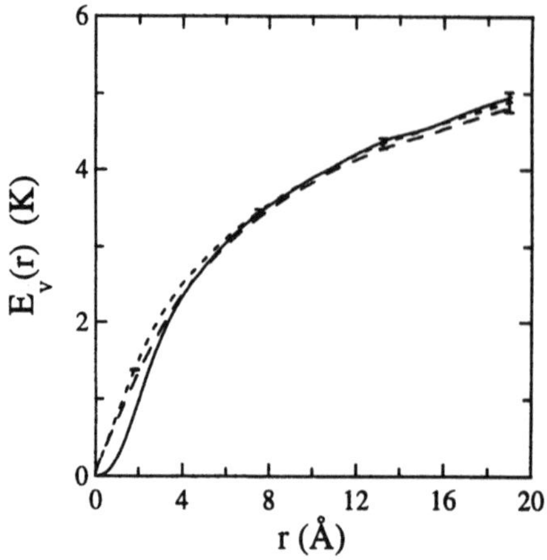

Fig. 2. Radial dependence of the excitation energy. Solid line, with ψ^{F1}; short-dashed line, with ψ^{F2}, and long-dashed line with ψ^{BF}.

the homogeneous density far from the axis and ξ is the vortex core radius. In Fig. 2 we show the vortex excitation energy $E_v(r)$ for the different choices of the trial wave function. For distances $r \gtrsim 6$ Å $E_v(r)$ shows the expected hydrodynamic behaviour with a small negative shift of the backflow energy with respect to the Feynman one. For small r's, the estimation of $E_v(r)$ is not exact and exhibits the influence of the trial wave function used. It is worth noticing the absence of spurious oscillations at large r, present in the results of Ref. [32], which are induced by the confining rigid walls.

The core radius ξ can be estimated as the position of the maximum in the azimuthal circulating current $J_\theta(r)$. The radial dependence of $J_\theta(r)$ has been estimated from the pure density profile using the expression for the current at the Feynman level, $J_\theta(r) = \rho_p(r)/r$. The value obtained is $\xi = 2.10 \pm 0.20$ Å which is in agreement with the result reported in Ref. [24]. By calculating the hydrodynamic contribution to the total energy for our square simulation box, we get the core energy $E_c^{F1} = 1.23 \pm 0.25$ K, $E_c^{F2} = 1.18 \pm 0.26$ K and $E_c^{BF} = 1.00 \pm 0.26$ K for the Feynman and backflow phases, respectively. These values coincide with the results obtained by a fit to $E_v(r)$ for $r > 6$ Å. Our results for E_c are significantly smaller from the ones obtained in Ref. [24] and the values of E_c^F are close to the variational results of Ref. [33] based on the Feynman phase. On the other hand, the results on the core energies point to a very small influence of the backflow correlations in the excited-state wave function and therefore the use of the RN method in this problem would not significantly change the FN estimations.

5.2 Phonon–Roton Branch in Superfluid ^{4}He

The wave function $\psi(\mathbf{R})$ corresponding to a phonon-roton excitation is an eigenstate of the momentum operator. As pointed out by Feynman [34], this requirement is achieved with the simple model wave function

$$\psi^{\mathrm{F}}(\mathbf{R}) = \sum_{i=1}^{N} \mathrm{e}^{\mathrm{i}\mathbf{q}\cdot\mathbf{r}_i}\, \psi^0(\mathbf{R}) \ , \tag{5.46}$$

$\psi^0(\mathbf{R})$ being the ground-state wave function. The first correction to $\psi^{\mathrm{F}}(\mathbf{R})$, originally proposed by Feynman and Cohen [25], includes backflow correlations. In this case, the excited-wave function is given by

$$\psi^{\mathrm{BF}}(\mathbf{R}) = \sum_{i=1}^{N} \mathrm{e}^{\mathrm{i}\mathbf{q}\cdot\tilde{\mathbf{r}}_i}\, \psi^0(\mathbf{R}) \ , \tag{5.47}$$

where

$$\tilde{\mathbf{r}}_i = \mathbf{r}_i + \sum_{j\neq i} \eta(r_{ij})\, \mathbf{r}_{ij} \ . \tag{5.48}$$

The inclusion of backflow correlations improves appreciably the variational results of the excitation spectrum $\varepsilon(q)$ with respect to the Feynman's choice but the quantitative agreement with experiment is still poor, specially near the roton minimum.

We choose as importance sampling wave function $\psi(\mathbf{R})$ the superposition of two excitations of momenta $\mathbf{q}$ and $-\mathbf{q}$ which are degenerate in energy:

$$\psi^{\mathrm{F}}(\mathbf{R}) = \sum_{i=1}^{N} \cos(\mathbf{q}\cdot\mathbf{r}_i)\, \psi^0(\mathbf{R}) \tag{5.49}$$

at Feynman's level, and

$$\psi^{\mathrm{BF}}(\mathbf{R}) = \sum_{i=1}^{N} \cos(\mathbf{q}\cdot\tilde{\mathbf{r}}_i)\, \psi^0(\mathbf{R}) \tag{5.50}$$

when backflow correlations are included.

In a first step, the sign problem associated to the excited wave function has been dealt within the framework of the fixed-node approximation. Beyond this approximation, we have then used the released-node technique to remove the nodal constraint of the FN approach.

In our computation [14, 35] we have considered a simulation box with 108 ^{4}He atoms interacting through the HFD-B(HE) Aziz potential [29], and at three densities, $\rho_{\mathrm{N}} = 0.285\ \sigma^{-3}$, $\rho_0 = 0.365\ \sigma^{-3}$ and $\rho_{\mathrm{P}} = 0.438\ \sigma^{-3}$. The density ρ_{N} is close to the spinodal density, ρ_0 corresponds to the equilibrium density, and ρ_{P} is close to the freezing density. The ground-state correlations have been modeled by a two-body wave function originally proposed by Reatto [36],

$$\psi^0(\mathbf{R}) = \prod_{i<j} \exp\left\{ -\frac{1}{2}\left(\frac{b}{r_{ij}}\right)^5 - \frac{L}{2}\exp\left[-\left(\frac{r_{ij}-\lambda}{\Lambda}\right)^2\right]\right\} , \tag{5.51}$$

which we have previously employed in ground-state calculations [19]. The function $\eta(r)$, entering in the backflow wave function Eq. (5.48), has been chosen to be a gaussian

$$\eta(r) = A \exp\left[-\left(\frac{r-r_{\mathrm{b}}}{\omega_{\mathrm{b}}}\right)^2\right] . \tag{5.52}$$

The values of the parameters in Eqs. (5.51,5.52) are $b = 1.20\ \sigma$, $L = 0.2$, $\Lambda = 0.6\ \sigma$, $\lambda = 2.0\ \sigma$, $A = 0.15$, $r_{\mathrm{b}} = 0.8\ \sigma$, and $\omega_{\mathrm{b}} = 0.44\ \sigma$, which are roughly the optimal ones at the equilibrium density ρ_0. The parameter a appearing in the guiding wave function $\psi_{\mathrm{g}}(\mathbf{R})$ (3.20) has been taken as $a = 3.0$ for all the q values, and the largest lifetime t_{r} used corresponds to 190 DMC sweeps. The same set of parameters have been used at the other densities ρ_{N} and ρ_{P}.

As mentioned in Sect. 3, the released-node energy estimation would be exact in the limit of large lifetimes. The empirical way in which we have studied the asymptotic regime of the RN energy is by fitting the function

$$E(t_{\mathrm{r}}) = E_\infty + C\, e^{-t_{\mathrm{r}}/\tau} \tag{5.53}$$

to the largest t_{r} values. In the interpretation of the MC results, we have followed the guide-line of accepting only the RN values that do not present discrepancies between the largest t_{r} data and the asymptotic limit E_∞. The fit (5.53) has been used to decide whether to trust or not the MC values but not to provide the asymptotic limit.

We have verified that at ρ_0 and ρ_{N}, and for values $q < 2.5$ Å^{-1}, the RN energies using the Feynman wave function (5.49) do reach the expected constant regime, the difference between the largest t_{r} calculated energy and the value of E_∞ predicted by the χ^2-fit (5.53) being less than the statistical error. This is not the case for ρ_{P}. At this high density, that agreement only subsists for the lowest q value and for the value of q nearest to the roton. For the other values of q we have had to include backflow correlations (5.50) to reach the asymptotic limit. This fact is illustrated in Fig. 3, where the excitation energy

$$\varepsilon(q) = \frac{\langle \psi(q)|H|\Phi(q)\rangle}{\langle \psi(q)|\Phi(q)\rangle} - \frac{\langle \psi^0|H|\Phi^0\rangle}{\langle \psi^0|\Phi^0\rangle} \tag{5.54}$$

per particle is plotted as a function of t_{r} for $q = 1.11$ Å^{-1} and $q = 1.84$ Å^{-1}. Near the roton, $q = 1.84$ Å^{-1}, both the Feynman and backflow results show a coincident asymptotic value without a significant slope. At $q = 1.11$ Å^{-1}, near the maxon energy, the situation is clearly different: the backflow results have reached a constant behaviour whereas the Feynman ones show a

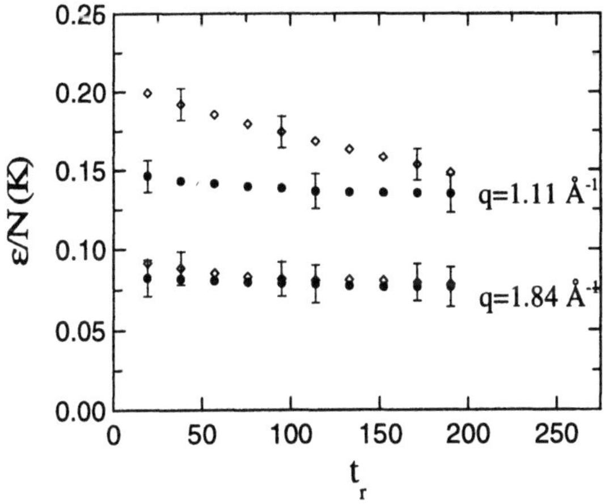

Fig. 3. Excitation energies per particle as a function of the lifetime t_r at ρ_P. The full circles are obtained using ψ^{BF} and the diamonds using ψ^{F}.

slow approach to the correct value, which has not been achieved yet for the maximum value of t_r. The latter behaviour is also observed for the highest value of q ($q \simeq 2.6$ Å^{-1}) both at ρ_0 and ρ_P. In this case, the inclusion of backflow correlations in the wave function is not enough to eliminate the bias and a significant difference exists between the largest t_r energy and the asymptotic value predicted by the numerical fit (5.53).

The released-node mechanism suppresses the fixed-node constraints and drives the calculation to the exact excitation energy, as shown in Table 1. In the table, fixed-node values using ψ^F and ψ^{BF}, and the released-node estimation are compared with experimental data [37] at the equilibrium density ρ_0. The FN results with backflow correlations improve the Feynman ones for the three values of q in a magnitude which depends on q. Thus, the inclusion of backflow correlations seems slightly more relevant in the roton than in

Table 1. Excitation energies at ρ_0 in comparison with experimental data. The FN-ψ^F and FN-ψ^{BF} columns are the fixed-node energies using ψ^F and ψ^{BF}, respectively. The RN column corresponds to the released-node estimation. Experimental data is taken from Ref. [37].

q (Å^{-1})	FN-ψ^F (K)	FN-ψ^{BF} (K)	RN (K)	Expt. (K)
0.369	7.56 ± 0.49	7.24 ± 0.38	7.02 ± 0.49	7.0
1.106	18.47± 0.49	16.52 ± 0.43	13.82 ± 0.43	13.8
1.844	13.82± 0.54	10.37 ± 0.59	9.18 ± 0.59	8.9

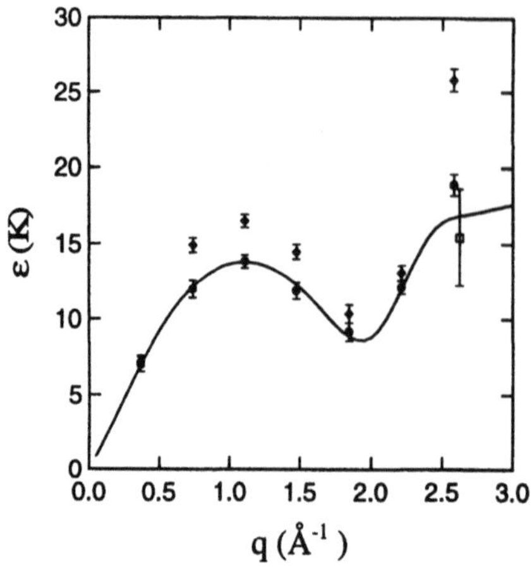

Fig. 4. Phonon-roton spectrum at the equilibrium density ρ_0. The full circles are the RN results and the diamonds correspond to a FN calculation with ψ^{BF}. The open square, which has been slightly shifted to the right for clarity, is the result of the extrapolation with the fit (5.53). The solid line is the experimental data from Ref. [37].

the maxon. On the other hand, the RN excitation energies agree with the experimental data for the three values of q within the statistical errors.

In Fig. 4 the RN excitation energies are compared with the experimental spectrum at ρ_0. The RN results correspond, for each q, to the last point in the release process, the error bars being only the statistical errors. As commented before, the systematic errors are less than the statistical ones except for the highest q result ($q = 2.58$ Å^{-1}). For this latter value of q we also report an estimation coming from the extrapolation supplied by the fit (5.53). Apart from this point, where the RN method shows the shortcomings of the backflow wave function at a so high value of q, the agreement between the RN results and the experiment is excellent. As a matter of comparison, the FN results using ψ^{BF} are also plotted. It is worth noticing the difference between the FN energies in the maxon and in the roton regimes; the roton is reproduced quite accurately whereas in the maxon the backflow correlations overestimate appreciably the excitation energy. At the highest q, where the spectrum bends down, the FN energy is quite far from the experimental data.

As is well-known from neutron scattering data, the location and depth of the roton minimum depends on the density. Thus, when the density increases the roton appears shifted to higher momenta and its energy decreases. The energies in the maxon region increase with the density but in an amount not so well experimentally known as in the roton. Our results for the behaviour of the phonon (P) (the smallest q value), the maxon (M) and the roton (R) with the density are plotted in Fig. 5. Notice the almost linear increase of

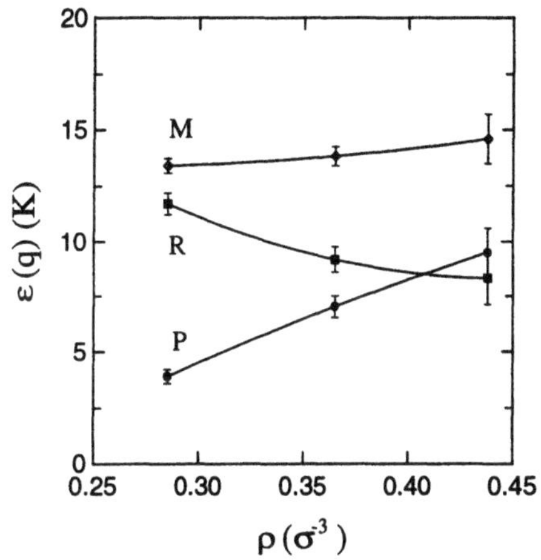

Fig. 5. Phonon (P), roton (R) and maxon (M) energies as a function of the density; the lines are "guides to the eye".

P in accordance with the dominant linear dependence with ρ of the sound velocity.

The results presented above show that the RN-DMC method improves the trial wave function for the excited states upon the exact answer. Although this improvement has been carried out numerically, it is also possible to obtain some information on the leading contributions beyond the model wave function $\psi(\mathbf{R})$ emerging from the RN-DMC method using the method presented in Sect. 4. In this way, entering with $\psi^{\mathrm{BF}}(\mathbf{R})$ the next-order correction to the excited wave function can be obtained [14]. The result is

$$\psi(\mathbf{R}) = \sum_{i=1}^{N} e^{iq_a \cdot \tilde{r}_i^a} \, \psi^0(\mathbf{R}) \ , \tag{5.55}$$

with

$$\tilde{r}_i^a = r_i^a + AZ_i^a + BT_i^a + C((W_i)_b^a Z_i^b - Y_i^a) \ , \tag{5.56}$$

A, B, C being constants, and

$$Z_i^a = \sum_{j \neq i} \eta(r_{ij})(r_{ij})^a \tag{5.57}$$

$$T_i^a = \sum_{j \neq i} (5\beta(r_{ij})(r_{ij})^a + \beta'(r_{ij}) r_{ij} (r_{ij})^a) \tag{5.58}$$

$$(W_i)_b^a = \sum_{j \neq i} (\beta(r_{ij})(r_{ij})^a (r_{ij})_b + \eta(r_{ij}) \delta_b^a) \tag{5.59}$$

$$Y_i^a = \sum_{\substack{j \neq i \\ k \neq j}} (\beta(r_{ij})\eta(r_{jk})(r_{ij})^a (r_{ij})_b (r_{jk})^b + \eta(r_{ij})\eta(r_{jk})(r_{jk})^a) \ . \tag{5.60}$$

The function $\eta(r)$ is the one used in $\psi^{\mathrm{BF}}(\mathbf{R})$ (5.48), and $\beta(r) \equiv \eta'(r)/r$. In Eq. (5.56), the first two terms on the r.h.s. correspond to $\psi^{\mathrm{BF}}(\mathbf{R})$ and the other three are new contributions. It is worth noticing the appearance of explicit three-body correlations $((W_i)^a_b Z^b_i - Y^a_i)$ in the excitation operator.

6 Summary and Outlook

We have presented several Monte Carlo methods which have been used in the calculation of excited-state properties in different systems. In spite of the difficulties associated to the sign problem, these methods have proved to be efficient in the study of excited states. The accuracy of the calculations is nowadays increasing and approaching the exactness of the boson applications.

A more complete information on the excited states would also require the calculation of dynamical properties. The study of the quantum dynamics, as the dynamic structure function, from imaginary-time dependent data is an ill-posed problem and only the combination of Bayesian inference and the maximum entropy criterion has obtained sound results in some systems. Progress in this hard problem, together with the search of more robust algorithms to deal with fermion systems and excited states, are the most challenging quantum Monte Carlo problems for the future.

Acknowledgments

Much of this work has been carried out during the last five years in close collaboration with Joaquim Casulleras who introduced me in the Monte Carlo world. I also acknowledge the contribution of Stefano Giorgini in the calculation of the properties of the vortex excitation.

References

1. D. M. Ceperley and M. H. Kalos, in *Monte Carlo Methods in Statistical Physics*, edited by K. Binder (Springer, Berlin, 1979)
2. B. L. Hammond, W. A. Lester Jr., and P. J. Reynolds, *Monte Carlo Methods in Ab Initio Quantum Chemistry* (World Scientific, Singapore, 1994)
3. D. M. Ceperley, Rev. Mod. Phys. **67**, 279 (1995)
4. K. E. Schmidt and D. M. Ceperley, in *The Monte Carlo Method in Condensed Matter Physics*, edited by K. Binder (Springer, Berlin, 1995)
5. D. M. Ceperley and L. Mitas, in *Advances in Chemical Physics*, Vol. XCIII, edited by I. Prigogine and A. Rice (John Wiley & Sons, NY, 1996)
6. D. Chandler and P. G. Wolynes, J. Chem. Phys. **74**, 7 (1981)
7. M. J. Gillan, in *Computer Modelling of Fluids Polymers and Solids*, edited by C. R. A. Catlow et al. (Kluwer, Dordrecht, 1990)
8. P. J. Reynolds, D. M. Ceperley, B. J. Alder, and W. A. Lester, J. Chem. Phys. **77**, 5593 (1982)

9. D. M. Ceperley and B. J. Alder, Phys. Rev. Lett. **45**, 566 (1980)
10. D. M. Ceperley and B. J. Alder, J. Chem. Phys. **81**, 5833 (1984)
11. D. M. Ceperley and B. Bernu, J. Chem. Phys. **89**, 6316 (1988)
12. B. Bernu, D. M. Ceperley, and J. W. A. Lester, J. Chem. Phys. **93**, 552 (1990)
13. Y. Kwon, D. M. Ceperley, and R. M. Martin, Phys. Rev. B **53**, 7376 (1996)
14. J. Boronat and J. Casulleras, Europhys. Lett. **38**, 291 (1997)
15. S. Giorgini, J. Boronat, and J. Casulleras, Phys. Rev. Lett. **77**, 2754 (1996)
16. M. Jarrell and J. E. Gubernatis, Phys. Rep. **269**, 133 (1996)
17. R. Guardiola, in *First International Course on Condensed Matter*, edited by D. Prosperi et al. (World Scientific, Singapore, 1988), and contribution to this volume
18. S. A. Chin, Phys. Rev. B **42**, 6991 (1990)
19. J. Boronat and J. Casulleras, Phys. Rev. B **49**, 8920 (1994)
20. R. M. Panoff and J. Carlson, Phys. Rev. Lett. **62**, 1130 (1989)
21. P. J. Reynolds, R. N. Barnett, B. L. Hammond, and W. A. Lester Jr., J. Stat. Phys. **43**, 1017 (1986)
22. J. B. Anderson, C. A. Traynor, and B. M. Boghosian, J. Chem. Phys. **95**, 7418 (1991)
23. K. E. Schmidt, M. H. Kalos, M. A. Lee, and G. V. Chester, Phys. Rev. Lett. **45**, 573 (1980)
24. G. Ortiz and D. M. Ceperley, Phys. Rev. Lett. **75**, 4642 (1995)
25. R. P. Feynman and M. Cohen, Phys. Rev. **102**, 1189 (1956)
26. R. J. Donnelly, *Quantized Vortices in Helium II* (University Press, Cambridge, 1991)
27. H. R. Glyde, *Excitations in Liquid and Solid Helium* (Clarendon Press, Oxford, 1994)
28. R. P. Feynman, in *Progress in Low Temperature Physics I*, edited by C. J. Gorter (North-Holland, Amsterdam, 1955)
29. R. A. Aziz, F. R. W. McCourt, and C. C. K. Wong, Mol. Phys. **61**, 1487 (1987)
30. S. Giorgini, J. Boronat, and J. Casulleras, Phys. Rev. B **54**, 6099 (1996)
31. J. Casulleras and J. Boronat, Phys. Rev. B **52**, 3654 (1995)
32. S. A. Vitiello, L. Reatto, and M. H. Kalos, Phys. Rev. B **54**, 1205 (1996)
33. M. Saarela and F. V. Kusmartsev, Phys. Lett. A **202**, 317 (1995)
34. R. P. Feynman, Phys. Rev. **94**, 262 (1954)
35. J. Boronat and J. Casulleras, J. Low Temp. Phys. (1997), in press
36. L. Reatto, Nucl. Phys. A **328**, 253 (1979)
37. R. J. Donnelly, J. A. Donnelly, and R. N. Hills, J. Low Temp. Phys. **44**, 471 (1981)

Lecture Notes in Physics

For information about Vols. 1–469
please contact your bookseller or Springer-Verlag

Vol. 475: J. Klamut, B. W. Veal, B. M. Dabrowski, P. W. Klamut, M. Kazimierski (Eds.), Recent Developments in High Temperature Superconductivity. Proceedings, 1995. XIII, 362 pages. 1996.

Vol. 476: J. Parisi, S. C. Müller, W. Zimmermann (Eds.), Nonlinear Physics of Complex Systems. Current Status and Future Trends. XIII, 388 pages. 1996.

Vol. 477: Z. Petru, J. Przystawa, K. Rapcewicz (Eds.), From Quantum Mechanics to Technology. Proceedings, 1996. IX, 379 pages. 1996.

Vol. 478: G. Sierra, M. A. Martín-Delgado (Eds.), Strongly Correlated Magnetic and Superconducting Systems. Proceedings, 1996. VIII, 323 pages. 1997.

Vol. 479: H. Latal, W. Schweiger (Eds.), Perturbative and Nonperturbative Aspects of Quantum Field Theory. Proceedings, 1996. X, 430 pages. 1997.

Vol. 480: H. Flyvbjerg, J. Hertz, M. H. Jensen, O. G. Mouritsen, K. Sneppen (Eds.), Physics of Biological Systems. From Molecules to Species. X, 364 pages. 1997.

Vol. 481: F. Lenz, H. Grießhammer, D. Stoll (Eds.), Lectures on QCD. VII, 276 pages. 1997.

Vol. 482: X.-W. Pan, D. H. Feng, M. Vallières (Eds.), Contemporary Nuclear Shell Models. Proceedings, 1996. XII, 309 pages. 1997.

Vol. 483: G. Trottet (Ed.), Coronal Physics from Radio and Space Observations. Proceedings, 1996. XVII, 226 pages. 1997.

Vol. 484: L. Schimansky-Geier, T. Pöschel (Eds.), Stochastic Dynamics. XVIII, 386 pages. 1997.

Vol. 485: H. Friedrich, B. Eckhardt (Eds.), Classical, Semiclassical and Quantum Dynamics in Atoms. VIII, 341 pages. 1997

Vol. 486: G. Chavent, P. C. Sabatier (Eds.), Inverse Problems of Wave Propagation and Diffraction. Proceedings, 1996. XV, 379 pages. 1997.

Vol. 487: E. Meyer-Hofmeister, H. Spruit (Eds.), Accretion Disks – New Aspects. Proceedings, 1996. XIII, 356 pages. 1997.

Vol. 488: B. Apagyi, G. Endrédi, P. Lévay (Eds.), Inverse and Algebraic Quantum Scattering Theory. Proceedings, 1996. XV, 385 pages. 1997.

Vol. 489: G. M. Simnett, C. E. Alissandrakis, L. Vlahos (Eds.), Solar and Heliospheric Plasma Physics. Proceedings, 1996. VIII, 278 pages. 1997.

Vol. 490: P. Kutler, J. Flores, J.-J. Chattot (Eds.), Fifteenth International Conference on Numerical Methods in Fluid Dynamics. Proceedings, 1996. XIV, 653 pages. 1997.

Vol. 491: O. Boratav, A. Eden, A. Erzan (Eds.), Turbulence Modeling and Vortex Dynamics. Proceedings, 1996. XII, 245 pages. 1997.

Vol. 492: M. Rubí, C. Pérez-Vicente (Eds.), Complex Behaviour of Glassy Systems. Proceedings, 1996. IX, 467 pages. 1997.

Vol. 493: P. L. Garrido, J. Marro (Eds.), Fourth Granada Lectures in Computational Physics. XIV, 316 pages. 1997.

Vol. 494: J. W. Clark, M. L. Ristig (Eds.), Theory of Spin Lattices and Lattice Gauge Models. Proceedings, 1996. XI, 194 pages. 1997.

Vol. 495: Y. Kosmann-Schwarzbach, B. Grammaticos, K. M. Tamizhmani (Eds.), Integrability of Nonlinear Systems. XX, 404 pages. 1997.

Vol. 496: F. Lenz, H. Grießhammer, D. Stoll (Eds.), Lectures on QCD. VII, 483 pages. 1997.

Vol. 497: J.P. De Greve, R. Blomme, H. Hensberge (Eds.), Stellar Atmospheres: Theory and Observations. Proceedings, 1996. VIII, 352 pages. 1997.

Vol. 498: Z. Horváth, L. Palla (Eds.), Conformal Field Theories and Integrable Models. Proceedings, 1996. X, 251 pages. 1997.

Vol. 499: K. Jungmann, J. Kowalski, I. Reinhard, F. Träger (Eds.), Atomic Physics Methods in Modern Research. IX, 448 pages. 1997.

Vol. 500: D. Joubert (Ed.), Density Functionals: Theory and Applications. XVI, 194 pages. 1998.

Vol. 501: J. Kertész, I. Kondor (Eds.), Advances in Computer Simulation. VIII, 166 pages. 1998.

Vol. 502: H. Aratyn, T. D. Imbo, W.-Y. Keung, U. Sukhatme (Eds.), Supersymmetry and Integrable Models. XI, 379 pages. 1998.

Vol. 503: J. Parisi, S. C. Müller, W. Zimmermann (Eds.), A Perspective Look at Nonlinear Media. From Physics to Biology and Social Sciences. VIII, 372 pages. 1998.

Vol. 504: A. Bohm, H.-D. Doebner, P. Kielanowski (Eds.), Irreversibility and Causality. Semigroups and Rigged Hilbert Space. XIX, 385 pages. 1998.

Vol. 505: D. Benest, C. Froeschlé (Eds.), Impacts on Earth. XVII, 223 pages. 1998.

Vol. 506: D. Breitschwerdt, M. J. Freyberg, J. Trümper (Eds.), The Local Bubble and Beyond. Proceedings, 1997. XXVIII, 603 pages. 1998.

Vol. 507: J. C. Vial, K. Bocchialini, P. Boumier (Eds.), Space Solar Physics. Proceedings, 1997. XIII, 296 pages. 1998.

Vol. 508: H. Meyer-Ortmanns, A. Klümper (Eds.), Field Theoretical Tools for Polymer and Particle Physics. XVI, 258 pages. 1998.

Vol. 510: J. Navarro, A. Polls (Eds), Microscopic Quantum Many-Body Theories and Their Applications. Proceedings, 1997. XIII, 379 pages. 1998.

New Series m: Monographs

Vol. m 2: P. Busch, P. J. Lahti, P. Mittelstaedt, The Quantum Theory of Measurement. XIII, 165 pages. 1991. Second Revised Edition: XIII, 181 pages. 1996.

Vol. m 3: A. Heck, J. M. Perdang (Eds.), Applying Fractals in Astronomy. IX, 210 pages. 1991.

Vol. m 4: R. K. Zeytounian, Mécanique des fluides fondamentale. XV, 615 pages, 1991.

Vol. m 5: R. K. Zeytounian, Meteorological Fluid Dynamics. XI, 346 pages. 1991.

Vol. m 6: N. M. J. Woodhouse, Special Relativity. VIII, 86 pages. 1992.

Vol. m 7: G. Morandi, The Role of Topology in Classical and Quantum Physics. XIII, 239 pages. 1992.

Vol. m 8: D. Funaro, Polynomial Approximation of Differential Equations. X, 305 pages. 1992.

Vol. m 9: M. Namiki, Stochastic Quantization. X, 217 pages. 1992.

Vol. m 10: J. Hoppe, Lectures on Integrable Systems. VII, 111 pages. 1992.

Vol. m 11: A. D. Yaghjian, Relativistic Dynamics of a Charged Sphere. XII, 115 pages. 1992.

Vol. m 12: G. Esposito, Quantum Gravity, Quantum Cosmology and Lorentzian Geometries. Second Corrected and Enlarged Edition. XVIII, 349 pages. 1994.

Vol. m 13: M. Klein, A. Knauf, Classical Planar Scattering by Coulombic Potentials. V, 142 pages. 1992.

Vol. m 14: A. Lerda, Anyons. XI, 138 pages. 1992.

Vol. m 15: N. Peters, B. Rogg (Eds.), Reduced Kinetic Mechanisms for Applications in Combustion Systems. X, 360 pages. 1993.

Vol. m 16: P. Christe, M. Henkel, Introduction to Conformal Invariance and Its Applications to Critical Phenomena. XV, 260 pages. 1993.

Vol. m 17: M. Schoen, Computer Simulation of Condensed Phases in Complex Geometries. X, 136 pages. 1993.

Vol. m 18: H. Carmichael, An Open Systems Approach to Quantum Optics. X, 179 pages. 1993.

Vol. m 19: S. D. Bogan, M. K. Hinders, Interface Effects in Elastic Wave Scattering. XII, 182 pages. 1994.

Vol. m 20: E. Abdalla, M. C. B. Abdalla, D. Dalmazi, A. Zadra, 2D-Gravity in Non-Critical Strings. IX, 319 pages. 1994.

Vol. m 21: G. P. Berman, E. N. Bulgakov, D. D. Holm, Crossover-Time in Quantum Boson and Spin Systems. XI, 268 pages. 1994.

Vol. m 22: M.-O. Hongler, Chaotic and Stochastic Behaviour in Automatic Production Lines. V, 85 pages. 1994.

Vol. m 23: V. S. Viswanath, G. Müller, The Recursion Method. X, 259 pages. 1994.

Vol. m 24: A. Ern, V. Giovangigli, Multicomponent Transport Algorithms. XIV, 427 pages. 1994.

Vol. m 25: A. V. Bogdanov, G. V. Dubrovskiy, M. P. Krutikov, D. V. Kulginov, V. M. Strelchenya, Interaction of Gases with Surfaces. XIV, 132 pages. 1995.

Vol. m 26: M. Dineykhan, G. V. Efimov, G. Ganbold, S. N. Nedelko, Oscillator Representation in Quantum Physics. IX, 279 pages. 1995.

Vol. m 27: J. T. Ottesen, Infinite Dimensional Groups and Algebras in Quantum Physics. IX, 218 pages. 1995.

Vol. m 28: O. Piguet, S. P. Sorella, Algebraic Renormalization. IX, 134 pages. 1995.

Vol. m 29: C. Bendjaballah, Introduction to Photon Communication. VII, 193 pages. 1995.

Vol. m 30: A. J. Greer, W. J. Kossler, Low Magnetic Fields in Anisotropic Superconductors. VII, 161 pages. 1995.

Vol. m 31: P. Busch, M. Grabowski, P. J. Lahti, Operational Quantum Physics. XI, 230 pages. 1995.

Vol. m 32: L. de Broglie, Diverses questions de mécanique et de thermodynamique classiques et relativistes. XII, 198 pages. 1995.

Vol. m 33: R. Alkofer, H. Reinhardt, Chiral Quark Dynamics. VIII, 115 pages. 1995.

Vol. m 34: R. Jost, Das Märchen vom Elfenbeinernen Turm. VIII, 286 pages. 1995.

Vol. m 35: E. Elizalde, Ten Physical Applications of Spectral Zeta Functions. XIV, 228 pages. 1995.

Vol. m 36: G. Dunne, Self-Dual Chern-Simons Theories. X, 217 pages. 1995.

Vol. m 37: S. Childress, A.D. Gilbert, Stretch, Twist, Fold: The Fast Dynamo. XI, 410 pages. 1995.

Vol. m 38: J. González, M. A. Martín-Delgado, G. Sierra, A. H. Vozmediano, Quantum Electron Liquids and High-T_c Superconductivity. X, 299 pages. 1995.

Vol. m 39: L. Pittner, Algebraic Foundations of Non-Commutative Differential Geometry and Quantum Groups. XII, 469 pages. 1996.

Vol. m 40: H.-J. Borchers, Translation Group and Particle Representations in Quantum Field Theory. VII, 131 pages. 1996.

Vol. m 41: B. K. Chakrabarti, A. Dutta, P. Sen, Quantum Ising Phases and Transitions in Transverse Ising Models. X, 204 pages. 1996.

Vol. m 42: P. Bouwknegt, J. McCarthy, K. Pilch, The W_3 Algebra. Modules, Semi-infinite Cohomology and BV Algebras. XI, 204 pages. 1996.

Vol. m 43: M. Schottenloher, A Mathematical Introduction to Conformal Field Theory. VIII, 142 pages. 1997.

Vol. m 44: A. Bach, Indistinguishable Classical Particles. VIII, 157 pages. 1997.

Vol. m 45: M. Ferrari, V. T. Granik, A. Imam, J. C. Nadeau (Eds.), Advances in Doublet Mechanics. XVI, 214 pages. 1997.

Vol. m 46: M. Camenzind, Les noyaux actifs de galaxies. XVIII, 218 pages. 1997.

Vol. m 47: L. M. Zubov, Nonlinear Theory of Dislocations and Disclinations in Elastic Body. VI, 205 pages. 1997.

Vol. m 48: P. Kopietz, Bosonization of Interacting Fermions in Arbitrary Dimensions. XII, 259 pages. 1997.

Vol. m 49: M. Zak, J. B. Zbilut, R. E. Meyers, From Instability to Intelligence. Complexity and Predictability in Nonlinear Dynamics. XIV, 552 pages. 1997.

Vol. m 50: J. Ambjørn, M. Carfora, A. Marzuoli, The Geometry of Dynamical Triangulations. VI, 197 pages. 1997.

Vol. m 51: G. Landi, An Introduction to Noncommutative Spaces and Their Geometries. XI, 200 pages. 1997.

Vol. m 52: M. Hénon, Generating Families in the Restricted Three-Body Problem. XI, 278 pages. 1997.

Vol. m53: M. Gad-el-Hak, A. Pollard, J.-P. Bonnet (Eds.), Flow Control. Fundamentals and Practices. XII, 527 pages. 1998.

MIX
Papier aus verantwortungsvollen Quellen
Paper from responsible sources
FSC® C105338

If you have any concerns about our products,
you can contact us on
ProductSafety@springernature.com

In case Publisher is established outside the EU,
the EU authorized representative is:
Springer Nature Customer Service Center GmbH
Europaplatz 3, 69115 Heidelberg, Germany

Printed by Libri Plureos GmbH
in Hamburg, Germany